THIN FILMS ON SILICON

Electronic and Photonic Applications

World Scientific Series in Materials and Energy*

ISSN: 2335-6596

Series Editor: Leonard C. Feldman *(Rutgers University)*

Published

Vols. 1 & 2 Handbook of Instrumentation and Techniques for Semiconductor Nanostructure Characterization
edited by Richard Haight (IBM TJ Watson Research Center, USA), Frances M. Ross (IBM TJ Watson Research Center, USA), James B. Hannon (IBM TJ Watson Research Center, USA) and Leonard C. Feldman (Rutgers University, USA)

Vol. 3 The World Scientific Handbook of Energy
edited by Gerard M. Crawley (University of South Carolina, USA)

Vol. 4 Materials for Biofuels
edited by Arthur J. Ragauskas (Georgia Institute of Technology, USA)

Vol. 5 Handbook of Green Materials: Processing Technologies, Properties and Applications (In 4 Volumes)
edited by Kristiina Oksman (Luleå University of Technology, Sweden), Aji P. Mathew (Luleå University of Technology, Sweden), Alexander Bismarck (Vienna University of Technology, Austria), Orlando Rojas (North Carolina State University, USA), and Mohini Sain (University of Toronto, Canada)

Vol. 6 Handbook of Solid State Batteries (2nd Edition)
edited by Nancy J. Dudney (Oak Ridge National Laboratory, USA), William C. West (Nagoya University, Japan) and Jagjit Nanda (Oak Ridge National Laboratory, USA)

Vol. 7 The WSPC Reference on Organic Electronics: Organic Semiconductors (In 2 Volumes)
edited by Jean-Luc Brédas (King Abdullah University of Science & Technology, Saudi Arabia & Georgia Institute of Technology, USA) and Seth R. Marder (Georgia Institute of Technology, USA)

Vol. 8 Thin Films on Silicon: Electronic and Photonic Applications
edited by Vijay Narayanan (IBM Thomas J Watson Research Center, USA), Martin M. Frank (IBM Thomas J Watson Research Center, USA) and Alexander A. Demkov (The University of Texas at Austin, USA)

*The complete list of titles in the series can be found at
http://www.worldscientific.com/series/mae

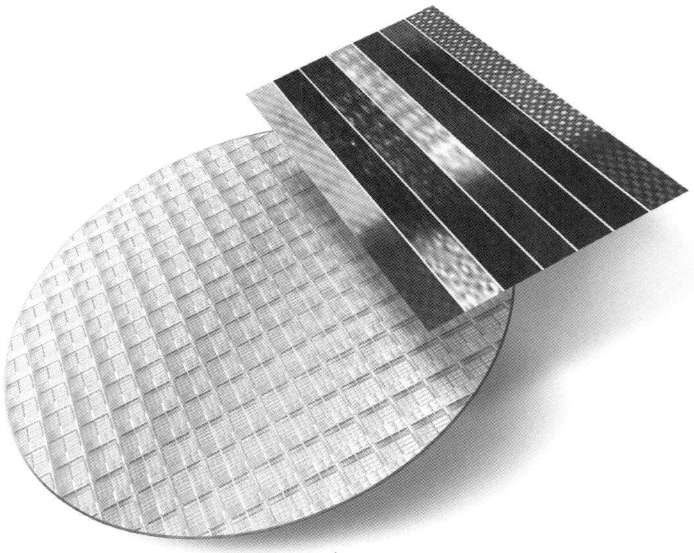

MATERIALS AND ENERGY – Vol. 8

THIN FILMS ON SILICON

Electronic and Photonic Applications

Editors

Vijay Narayanan
IBM Thomas J Watson Research Center, USA

Martin M. Frank
IBM Thomas J Watson Research Center, USA

Alexander A. Demkov
The University of Texas at Austin, USA

World Scientific

NEW JERSEY • LONDON • SINGAPORE • BEIJING • SHANGHAI • HONG KONG • TAIPEI • CHENNAI • TOKYO

Published by

World Scientific Publishing Co. Pte. Ltd.
5 Toh Tuck Link, Singapore 596224
USA office: 27 Warren Street, Suite 401-402, Hackensack, NJ 07601
UK office: 57 Shelton Street, Covent Garden, London WC2H 9HE

Library of Congress Cataloging-in-Publication Data
Names: Narayanan, V. (Vijay), editor. | Frank, Martin M., editor. | Demkov, Alexander A., editor
Title: Thin films on silicon : electronic and photonic applications / editors Vijay Narayanan, Martin M. Frank (IBM Thomas J. Watson Research Center) and
Alexander A. Demkov (The University of Texas at Austin, USA)
Description: Singapore ; Hackensack, NJ : World Scientific Publishing Co. Pte. Ltd, [2016] | 2016
Identifiers: LCCN 2015047283| ISBN 9789814740470 (hardcover ; alk. paper) |
ISBN 9814740470 (hardcover ; alk. paper)
Subjects: LCSH: Thin films. | Silicon.
Classification: LCC TK7871.15.F5 T44 2016 | DDC 621.3815/2--dc23
LC record available at http://lccn.loc.gov/2015047283

British Library Cataloguing-in-Publication Data
A catalogue record for this book is available from the British Library.

Desk Editor: Rhaimie Wahap

Typeset by Stallion Press
Email: enquiries@stallionpress.com

Printed in Singapore by Mainland Press Pte Ltd.

Foreword

"By 2018 computers will touch, see, hear, taste and smell"- a provocative headline in the press that captures the challenges and the future of the computer age. Accomplishing this fascinating goal is dependent on the clever and innovative manipulation of materials. And, because the link to computers is through silicon technology, each innovation will depend on coupling new materials to silicon.

Advances in basic science and fundamental understanding of nature leads to transformations in technology. The identification of the electron in 1896 marked the beginning of the atomic description of matter. This discovery was followed by the rapid development of quantum mechanics, first applied to the atom and then to the understanding of a collection of atoms which make up molecules and solids. We learned how to describe and control the motion of electrons in materials.

The transistor was first demonstrated in 1947 — perhaps the most far-reaching "product" of quantum science. (The first transistor was actually germanium, not silicon.) And in 1959, the silicon MOSFET, metal-oxide semiconductor field effect transistor, was first demonstrated. It has since evolved into the key component of the silicon revolution.

Silicon is the dominant material of our time. We are in the Silicon Age. The silicon transistor and development of the integrated circuit have produced a massive change in technology and economy, as well as culture and thinking. The advances commensurate with the great materials revolutions of time.

The history of the silicon MOSFET and the evolution to the billions of MOSFETs which make up the silicon chip is the story of the clever and innovative manipulation of materials. The watchwords "smaller", "faster" and "lower power" have guided the technology advance. "Smaller" — for more memory (an encyclopedia on a chip), "faster" — to access reams of data such as on the internet and "lower power" — for battery driven portability as in our smartphones.

In the last 50 years, silicon-based materials science has received an enormous amount of attention. These issues have become even more critical as the device has scaled down and obeyed the famous "Moore's Law". Major materials innovations have taken place — changes to the basic oxide material, subtle changes in the semiconductor itself such as strain, new methods of materials processing and, with astounding success, new ways to carve silicon to create an electrical device of dimensions comparable to dimensions of ~10 nm, about 40 atoms. This extraordinary development has impact beyond silicon technology. The silicon community was the forerunner of the larger nanotechnology revolution. Silicon was the proof that you could manipulate materials at the atomic scale, showed that you can manufacture such devices in massive scale, and be profitable in doing so.

We are close to some fundamental limits. Nevertheless progress in the short term demands still more innovation and more materials manipulation of the basic structure. Progress in the long term demands innovation and creativity to connect to our basic senses, to connect efficiently to the "cloud", to handle the information from billions of sensors and to achieve the shared vision of the future.

We demand much more in the way of functionality and performance. Unfortunately silicon, by itself, cannot do all we seek. It does not emit light. It does not "see" well. By itself it cannot detect a toxin — not nearly as well as a physician sampling a patients' breath. These are but a few examples. The future surely contains robotics, driverless cars, non-polluting energy, medical self-diagnosis, security, and a myriad of functions to improve the human condition. Almost all of these goals will be achieved by inventing and coupling materials to the silicon-based computer.

The next 20 years will demand even greater materials innovation as we couple silicon to the human experience. Internet eyeglasses and micro cameras will couple optical devices to silicon for information selection and processing. Complex algorithms programmed into silicon for pattern recognition and people recognition (perhaps with a short biography) are the future. Driverless cars are just on the horizon. Enormous wall screens coupling optical materials able to detect and emit light to silicon processors. And, perhaps most significantly, revolutionized medical care with data from home sensors wirelessly sent to the physician for diagnosis. All of these advances will depend on the on-going silicon revolution AND the coupling of silicon to new materials which are human compatible — able to "see", capable of smell, sensitive to touch through a sensitized glove etc. Research is underway on all these aspects as we go to press. Success depends on materials based developments — the continuing silicon evolution and the ability to couple new materials with new functionality to silicon.

A materials change in the established silicon technology is daunting. The task is not easy! Each materials innovation to this point has been the contribution of a scientific and engineering community of industry, government and academic scientists numbering in the thousands. Introduction of a small change is a daunting task. Is the new structure reliable, will it handle temperature changes, how does it react to humidity, is it possibly toxic and, importantly, is it manufacturable and at reasonable cost. And each innovation starts with the strong scientific underpinning created through detailed atomic scale examination of the materials structure. First question: does it work in the lab?

This is the starting point of this book. The scientific articles represent the state of the art in applying new physics techniques, new chemical insights, and new theoretical modeling to the description of silicon combined with new materials. The goal is to provide the scientific underpinnings to this complex system, which has proved to be more enduring and reliable than ever thought possible. It is tempting to ask "Why now?" — it has worked for 50 years. The answer is at least two-fold. To keep the technology moving forward, to continue to follow Moore's Law, we require an absolute understanding at a level which has not been tested. This book provides the latest of such insights and understanding, enriching our knowledge of materials science and providing the basis of the continuing silicon revolution.

Leonard C. Feldman
Institute for Advanced Materials, Devices and Nanotechnology
Rutgers University
New Brunswick, NJ

Preface

Historically, the bulk of our knowledge of the physics and chemistry of thin films on silicon comes from the tremendous worldwide effort to optimize complementary-metal-oxide-semiconductor (CMOS) technology. For example, oxides and nitrides are used as gate insulators, sidewall spacers, and interlayer dielectrics; silicide alloys serve as gate and source/drain contacts; and elemental metals are used for interconnects. Indeed, since the 1990s, silicon manufacturing has been opened up to a large variety of novel materials with complex crystal structure and chemistry.

The best studied thin film on silicon is silicon dioxide (SiO_2), used as a dielectric in the MOS field-effect-transistor (FET) gate stack. The 'high-k' revolution that swept through the semiconductor industry almost a decade ago introduced a new class of insulating materials — transition metal oxides — into silicon manufacturing. Solving the problem of gate leakage through the use of hafnium dioxide (HfO_2), a material with a relatively high dielectric constant of ca. 20 that allows the gate capacitor to be sufficiently thick to reduce the tunneling contribution to the gate leakage, spelled the end to the SiO_2 monopoly. Ironically, this also may have very well spelled the end of Si as the single channel material of choice, as the best property of this semiconductor is, undoubtedly, its stable wide band gap amorphous dioxide. Indeed, silicon germanium channel transistors are already in production. Germanium, III-V semiconductors, and more exotic materials are under intense investigation to further improve MOSFET performance and reduce power consumption.

It may thus seem that silicon will soon be a material of the past. However, it is unlikely that new channel materials will prove superior to silicon for all applications in logic and memory. In addition, the global CMOS manufacturing infrastructure that has allowed shrinking the MOSFET in accordance with Moore's Law for more than five decades now requires substrates ('wafers') with a diameter of 300 mm, and even larger wafer sizes are under consideration. Silicon wafers in such sizes are readily available, unlike for other semiconductors. Therefore, silicon is here to stay! And the field of thin films on silicon has never been richer than today: A wide variety of novel materials, some with complex crystal structure and chemistry, are now under investigation, revealing a plethora of intriguing physical properties and promising enhanced and perhaps entirely new device functionalities.

This volume is meant to be a snapshot of this exciting research across a broad selection of materials and possible applications in electronics and photonics. In many cases, the complexity of the systems involved allows for the emergence of phenomena relying on long range order such as ferroelecticity or magnetism. In addition, scaling to nanometer size features generates important quantum effects. Thus the division between fundamental and applied research has been significantly eroded. These problems require a multidisciplinary approach that closely couples processing, film growth, advanced materials characterization and modeling. Multiple journals and conferences bring together electrical engineers, physicists, materials scientists and chemists. We hope you will enjoy the book and perhaps also get a chance to consider research areas that, while quite different from your own, nevertheless can offer a fresh look at familiar problems.

Vijay Narayanan and Martin M. Frank
IBM Thomas J. Watson Research Center, USA

Alexander A. Demkov
The University of Texas at Austin, USA

Contents

Semiconductors on Silicon

Chapter 1

Si, SiGe, and $Si_{1-y}C_y$ on Si: Epitaxy of Group-IV Semiconductors for Nanoelectronics

J. M. Hartmann
Université Grenoble Alpes and
CEA, LETI, Minatec Campus,
17, Rue des Martyrs
38054 Grenoble Cedex 9, France

1. Current Trends in Group-IV Semiconductor Epitaxy for Nanoelectronics

1.1. *Introduction*

Group-IV epitaxy is nowadays considered as a key enabler in Complementary Metal Oxyde Semiconductor (CMOS) devices, Bipolar and BiCMOS transistors, near infra-red photo-detectors or mid infra-red waveguides and so on [BAU 11], [WAD 15]. It is used in nanoelectronics for the fabrication of stressors (e.g. *in-situ* boron-doped SiGe and *in-situ* phosphorous-doped SiC recessed or raised sources and drains in short gate length devices), SiGe channels and so on, as shown in Figure 1 (see Refs. [HAS 14], [Liu 14] and references therein).

In the majority of these applications, epitaxy has to be selective on patterned wafers, e.g. without any deposition on dielectrics that surround Si (or SiGe) planar areas or fin-like structures. It is integrated in very complex

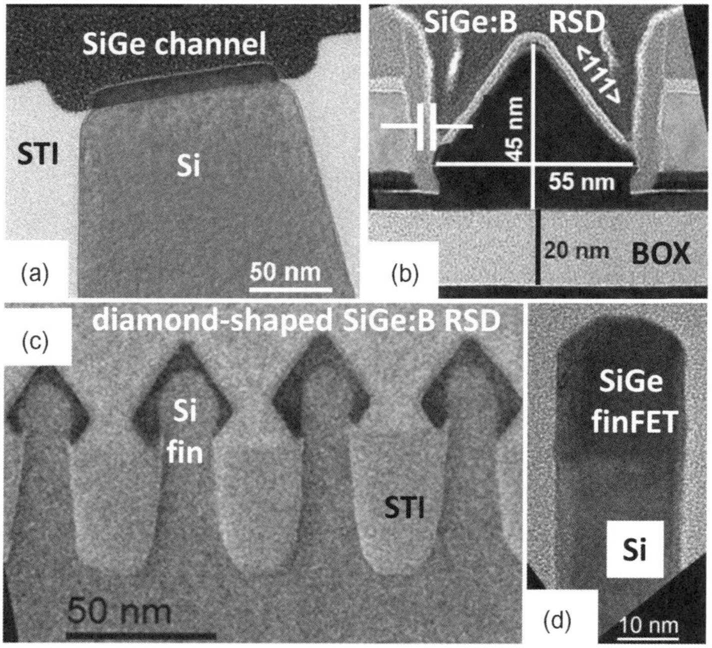

Fig. 1. Cross-sectional transmission electron microscopy images of (a) a $Si_{0.73}$ $Ge_{0.27}$ layer used as a high mobility channel for holes on bulk Si (b) facetted SiGe: B raised sources and drains on each side of Extra-Thin Silicon-On-Insulator FET devices, (c) diamond-shaped $Si_{0.3}$ $Ge_{0.7}$: B raised sources and drains around Si fins and (d) a $Si_{0.58}$ $Ge_{0.42}$ fin (adapted from [DUT 14] and [HYK 14]).

process flows, sometimes with severe thermal budget constraints, notably within the CoolCube^{TM} integration scheme.[7] In addition, growth occurs in very small areas (10–40 nm), often with significant thicknesses (8–30 nm). The vertical to horizontal aspect ratio then becomes an important process parameter. As current devices make full use of the third dimension, with fins or trigate structures instead of planar, bulk channels, growth will occur simultaneously on various surface orientations, leading to the appearance of facets and of variable thickness and composition films.

In the coming sections, we will first of all define what epitaxy is and introduce some important concepts such as critical thickness for plastic relaxation and so on. We will then move over to surface preparation prior to epitaxy, which is crucial when performing Selective Epitaxial Growth (SEG) on patterned wafers. The *in-situ* H_2 bake temperature (that follows a "HF-Last" wet cleaning) indeed needs to be lowered in order to avoid thin film islanding and/or shape change; innovative surface preparation schemes have then to be used in order to obtain high crystalline quality epitaxial layers. We will subsequently detail some of the challenges one faces when growing on patterned wafers Si and SiGe layers with a chlorinated chemistry. We will notably cover the impact precursor mass-flows, temperature, substrate orientation, buried oxide thickness, mask opening and so on have on the Si and SiGe growth kinetics. We will then show how Si and SiGe(:B) SEG enabled us to obtain high performance p-type Field Effect Transistors on Extra-Thin Silicon-On-Insulator (ET-SOI) substrates and Si trigate structures. Finally, we will present the salient features of SiC:P and SiGe:B Cyclic Deposition/Etch (CDE) processes, which are very promising for low temperature epitaxy.

1.2. *A few ideas about epitaxy*

We will first of all explain what epitaxy stands for. To perform an *epitaxy* is to deposit a *single-crystalline* layer that will adopt the same crystalline structure as the *single-crystalline* substrate on which it is deposited, that is, a structure dictated by the underlying substrate. It is a *homo-epitaxy* when the substrate and the layer are of the same type (for example: Si on Si) and a *hetero-epitaxy* when they are not (for instance: SiGe on Si).

A layer is in *tension* (*compression*) when its lattice parameter a_L is lower (higher) than that of the substrate a_S. For a *pseudomorphic* layer, the in-plane lattice parameter of the layer is equal to that of the substrate (i.e. $a_L^{//} = a_S$). For a layer in tension, we then have $a_L^{\perp} < a_L < a_L^{//} = a_S$; for one in compression, $a_L^{\perp} > a_L > a_L^{//} = a_S$. The perpendicular lattice parameter is given (in cubic layers grown on (001) surfaces) by $a_L^{\perp} = a_L + 2c_{12}/c_{11}(a_L - a_S)$ (with $2c_{12}/c_{11}(Si) = 0.77$, $2c_{12}/c_{11}(Ge) = 0.74$ and $2c_{12}/c_{11}(C) = 0.61$).

Intrinsic or *in-situ* doped Si/SiGeC stacks (known as *hetero-structures*) are most of the time grown on Si substrates with a (001) surface orientation. Those substrates can be bulk-type or not (presence then of a buried oxide layer), blanket or patterned and so on. Let us summarize the main characteristics of the elementary compounds of a SiGeC alloy: Si, Ge and C crystallize in the diamond phase, i.e. two overlapping sphalerite structures shifted one from the other by a quarter of the large diagonal (cf. Figure 2).

The lattice parameter of C (i.e. the length of one of the sides of the cubes in Figure 1) is much smaller than that of Si, which in itself is smaller than that of Ge ($a_C = 3.567$ Å $\Leftrightarrow$ $a_{Si} = 5.43105$ Å $\Leftrightarrow$ $a_{Ge} = 5.65785$ Å). The lattice parameter increase is accompanied by a sharp energy bandgap decrease: E_g (C) = 5.48 eV $\Leftrightarrow$ E_g (Si) = 1.11 eV $\Leftrightarrow$ E_g (Ge) = 0.66 eV. Mixing Si, Ge and C in a SiGeC alloy will induce strong bandgap modifications; built-in stress will also play a role. Incorporating C by substitution into a Si or Ge matrix

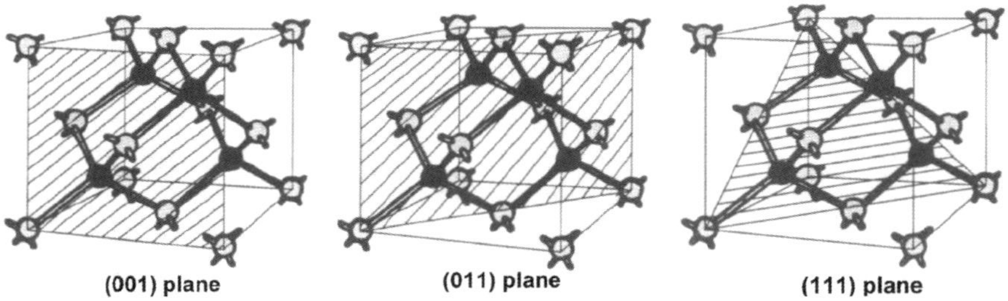

Fig. 2. Schematics of the diamond structure of group-IV semiconductors and some of the surfaces on which growth will occur when processing 3D objects such as fins or tri-gates. The tetra-coordinated atoms are symbolized by light or dark grey spheres, while bonds appear as elongated cylinders.

is otherwise quite difficult. The miscibility of C in Si is only approximately 10^{-4} % at thermodynamic equilibrium (i.e. at 1400°C), with an unfortunate tendency towards SiC precipitate formation for high concentrations. Si and Ge, however, are miscible in all proportions. During hetero-epitaxy, there will be discrepancies between the lattice parameters of the deposited layer and the substrate, which will lead to an accumulation of elastic energy in the layer. For thin layers, the substrate's atomic columns will be extended into the *pseudomorphic* epitaxial layer.

However, when the thickness of the layer increases, it becomes energetically favorable to introduce misfit dislocations to minimize the elastic energy that has accumulated and accommodate the lattice parameter mismatch between the layer and the substrate. The layer is then said to be *plastically relaxed*. The thickness at which the transition from pseudomorphic to plastically relaxed layer occurs is called *critical thickness for plastic relaxation*. Schematic diagrams in Figure 3 show the different configurations of the {epitaxial SiGe layer on Si substrate} system.

We have recently revisited the critical thickness for plastic relaxation of SiGe on Si(100). For this, 200 mm Si substrates (of a higher crystalline quality than those used in studies referenced in [HOU 91], i.e. with a lower pre-existing dislocation density) were used as templates for the Reduced Pressure — Chemical Vapor Deposition (RP-CVD) of variable composition and thickness SiGe layers.

The growth temperature was reduced from 700°C to 650°C, then to 600°C and finally to 550°C as the Ge concentration increased from 12% to 52% in order to reduce surface roughness. Due to differences between Molecular Beam Epitaxy (MBE) and RP-CVD in terms of Si and Ge precursors (solid ingots of Si and Ge which are sublimated by MBE ⇔ ultra-pure gaseous precursors for CVD), the way very low partial pressures of

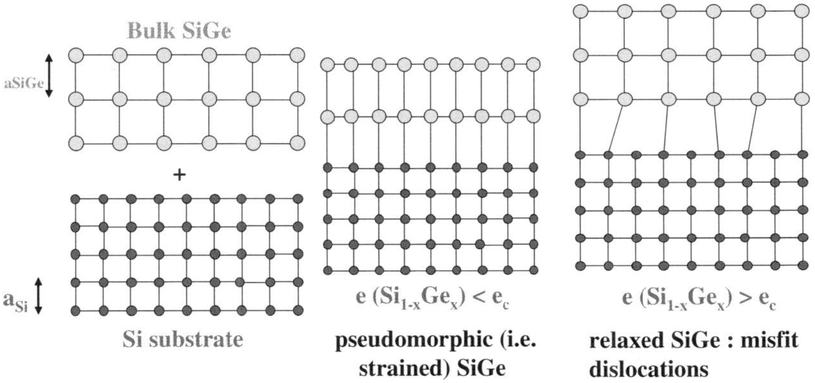

Fig. 3. Schematic diagrams of the atomic arrangement when growing a SiGe layer epitaxially on a Si substrate.

impurities such as oxygen or water are obtained (ultrahigh vacuum in MBE ⇔ large fluxes of ultra-pure hydrogen in CVD) and so on, we expect differences between [BEA 84, HOU 91, HAR 11a] findings.

Figure 4 shows the summary of the results [HAR 11a]. The dotted line shows those of Bean [BEA 84], the full squares correspond to SiGe layers seen as fully compressively-strained in X-ray Diffraction or XRD and the squares with a cross in their center represent partially relaxed SiGe layers. h_c is in fact two to three times higher than predicted by People and Bean. However, for high Ge concentrations, some layers seen as pseudomorphic in XRD are actually characterized by the presence of a limited amount of misfit dislocations in their midst. The surface signature of their propagation ("plough" lines along the ⟨110⟩ crystallographic directions) was detected thanks to Atomic Force Microscopy (hatched zone in Figure 4).

1.3. *Surface preparation prior to epitaxy*

The substrate surface preparation before any epitaxy is of prime importance. The goal is indeed to duplicate the atomic columns of the substrate into the layer. The presence on the surface of amorphous silicon oxide layers (even very thin ones), polymers or etching residues (on masked substrates) has indeed a crippling effect on the epitaxy quality. Depending on the toolset available for surface preparation and the nature of the substrate itself, three strategies may be used to obtain contamination-free Si (or SiGe) surfaces.

The starting substrate may be bulk Si or, to a lesser degree, a Silicon-On-Insulator — type substrate (i.e. a {thin layer of Si/ buried layer

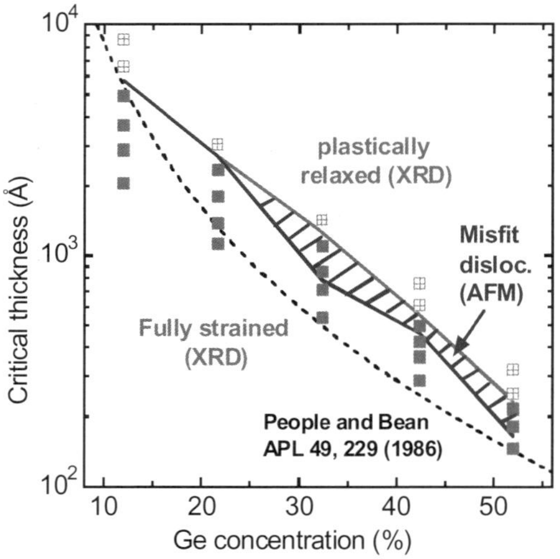

Fig. 4. Critical thickness for plastic relaxation in $Si_{1-x}Ge_x$ epitaxial layers on Si (001) versus the Ge concentration x [HAR 11a]. 200 mm data.

of SiO_2/Si substrate} stack, usually fabricated using the Smart Cut™ technique [BRU 95]). It is then possible to carry out high temperature bakes, during which dozens of liters of ultra-pure H_2 (typically only a few parts per billion of impurities thanks to dedicated gas purifiers) are injected into the epitaxy chamber, in order to remove the 0.8-1 nm thick layer of native or chemical SiO_2 which is initially present on the surface and smooth it. The simplified chemical reaction for the removal of the oxide layer is:

$$SiO_2(s) + 2H_2(g) \rightarrow Si(s) + 2H_2O(g)$$

The need for high thermal budgets (typically higher than 1050°C for 1 minute; 1100°C for 2 minutes as the CEA-LETI standard) prohibits the use of this surface preparation on patterned wafers (i.e. with active Si zones surrounded by dielectrics) or ion implanted substrates. The active zones could facet due to strain generated by masking dielectrics, especially if they are in a Shallow Trench Isolation (or STI) configuration. An exo-diffusion of pre-implanted atoms and a consequent auto-doping of the epitaxial layer may also take place [CHA 85, JER 99]. This type of bake is prohibited for thin Si films (e.g. less than 20 nm) on top of buried oxides. If these criteria are not respected, moat recess and islanding of the films will take place [ISH 99]. It is also wise to optimize the power delivered to the lamps used to heat up the

SOI substrates so that the temperature at the surface is uniform; otherwise slip lines will form.

The surface preparation described above (where a native or chemical oxide present at the surface is removed by an *in-situ* H_2 bake in suitable conditions) is far from being universal given the high thermal budgets needed. In order to minimize the latter, the following sequence is then typically used:

(i) The native SiO_2 oxide is removed by a "HF-Last" wet cleaning [ABB 04]. During the next to last step of such a cleaning, the wafer is dipped in hydrofluoric acid diluted in deionized and de-oxygenated water (dilution typically between 0.2% and 1%), to etch surface SiO_2. The dissolution reaction is as follows:

$$SiO_2 + 4HF \rightarrow 2H^+ + SiF_6^{2-} + 2H_2O \quad [BUH97]$$

The etch rate of thermal SiO_2 (the most resistant of all) is of the order of 1.2 to 1.4 nm min^{-1} for 0.2% HF. During the last step, the wafer is rinsed in deionized and deoxygenated water to remove all traces of HF. The wafer is then dried by for instance isopropyl alcohol vapors, before being loaded as quickly as possible in the load-lock chambers of the epitaxy tool, in an inert atmosphere (purified N_2). After such a wet cleaning, approximately 85% of the Si dangling bonds are occupied by hydrogen atoms. The remaining 15% are mainly occupied by fluorine atoms as well as oxygen and carbon contaminants [MEY 90, TRU 90].

(ii) An H_2 *in-situ* bake then takes place at temperatures above 800°C to remove all O, F or C surface contaminants (if compatible with the technology it will be used in). A surface perfectly passivated by hydrogen atoms is then obtained which is ideal for epitaxy.

An "HF-last" surface is only stable for a few dozens of minutes up to a couple of hours (before the re-growth of a native oxide a few Å thick can be detected by spectroscopic ellipsometry). Minimizing the time between a "HF-Last" wet cleaning and the loading of wafers inside chambers purged with high purity inert gases (such as N_2) is thus mandatory [WOS 14a]. A perfectly hydrogen-passivated Si surface obtained by a high temperature H_2 bake is by contrast stable for several days [HER 01].

We will now focus on the impact of the H_2 bake temperature on the properties of the epitaxial layers and the interfacial contamination, this for Si, SiGe and pure Ge surfaces of (100) and (110) crystallographic orientations.

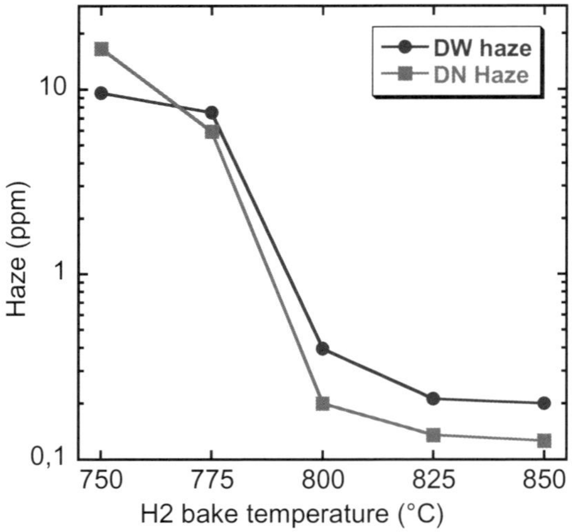

Fig. 5. Dark Wide (DW) and Dark Narrow (DN) haze measurements associated with 100 nm thick Si layers grown at 700°C on Si(001) after a "HF-Last" wet cleaning and 225s, 80 Torr H_2 bakes at various temperatures. 300 mm data. Y. Bogumilowicz *et al.*, unpublished results.

We have plotted in Figure 5 the surface haze values associated with 100 nm thick Si layers grown with SiH_4 at 700°C, 80 Torr in LETI's 300 mm ASM Epsilon 3200 Epsilon epitaxy tool after the following surface preparation of the Si(001) substrates: (i) a "HF-Last" wet cleaning in a state-of-the-art DNS 3100 single wafer cleaning tool followed by (ii) an *in-situ* H_2 bake at 80 Torr for 225s at various temperatures : 750°C, 775°C, 800°C, 825°C and 850°C. Be it in the Dark Wide or Dark Narrow configuration of the SP2 metrology tool, we are faced with a very significant decrease of the haze (e.g. the light diffusion by the surface, which is proportional to the surface roughness [HAR 11a]) as soon as the bake temperature exceeds 775°C. Haze after 825°C–850°C bakes is indeed nearly 50 times lower than haze after 750°C–775°C H_2 bakes; it is close to values expected after a perfect Si epitaxy.

We have plotted in Figure 6 the interfacial O, C and F contamination (from Secondary Ion Mass Spectrometry) as a function of the H_2 bake temperatures in those samples. Be it at 825°C or 850°C, we are contamination-free (e.g. below the SIMS lower limits of detection of the elements probed: 10^{11} cm^{-2} for O and C $\Leftrightarrow$ 5×10^9 cm^{-2} for F). By contrast, we are faced with a significant interfacial C, O and F contamination at 750°C and 775°C, with notably an oxygen sheet concentration a third of the Si surface concentration ($2.4 \Leftrightarrow 6.8 \times 10^{14}$ cm^{-2}).

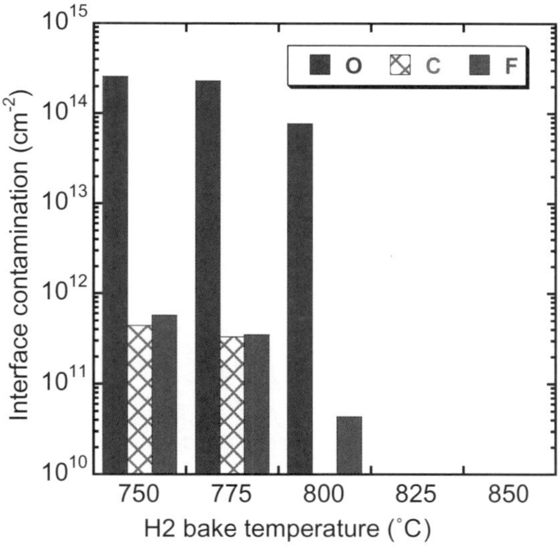

Fig. 6. Sheet concentration of O, C and F atoms at the interface between Si(001) substrates and 100 nm thick Si layers grown at 700°C after a "HF-Last" wet cleaning and 225s, 80 Torr H_2 bakes at various temperatures. 300 mm data. Y. Bogumilowicz *et al.*, unpublished results.

The situation is intermediary at 800°C, with O and F interfacial concentration several orders of magnitude lower than at 750°C and 775°C. The excellent correlation between Figure 5 haze evolution and Figure 6 interfacial contamination is to be highlighted.

SIMS has been used a while ago to profile the C, O and F atoms present in 50 nm thick Si layers grown at 650°C on Si (001) substrates (in LETI's 200 mm AMAT Epi Centura 5200 epitaxy tool) after a "HF-Last" wet cleaning followed by a two minutes, 20 Torr H_2 bake, either at 750°C or 775°C [ABB 04]. 750°C was too low to get rid of F and O contamination. Peaks were indeed present at the interface between the substrate and the epitaxial layer. No such peaks were observed at 775°C, however (same findings in [KOR 08]). These results are in full agreement with those of Brabant, who showed that 750°C H_2 bakes lasting as long as 10 to 20 minutes did not completely rid the Si(001) surface from O contamination [BRA 03]. Interfacial O contamination disappeared after 800°C H_2 bakes in [WOS 14b].

It is interesting to note that the H_2 bake threshold temperature around 775°C–800°C, above which C, O and F surface contamination is eliminated, does not seem to be influenced by the crystallographic orientation of the Si surface. A low interfacial O peak was shown by Destefanis [DES 08] after 20 Torr H_2 bakes at 775°C for two minutes (after "HF-Last" cleaning) of Si(110) substrates. This peak disappeared entirely at 800°C.

Similarly, changing, for (001) surfaces, from Si to SiGe or even pure Ge has little impact on the threshold temperature. O and F contamination peaks were indeed evidenced by Abbadie *et al.* [ABB 04] after 20 Torr H_2 bakes at 775°C for two minutes (post "HF-Last" cleaning) of SiGe 20% and 33% Strain Relaxed Buffers (SRBs) prior to their encapsulation with thin, tensily-strained Si layers. These contamination peaks disappeared at 800°C. With regards to pure Ge, there was no interfacial contamination peaks anymore after 20 Torr H_2 bakes at 750°C for two minutes [HAR 04a]; there was, however, a low O peak after bake at 700°C.

The H_2 bake pressure itself has also a definite impact on interfacial contamination; the higher it is, the lower the O removal efficiency will notably be. Hykavyy *et al.* were indeed faced, at 750°C, with a monotonous increase of the integrated O dose (at Si/Si interfaces) from $\sim 10^{13}$ atoms/cm^2 up to $\sim 7 \times 10^{13}$ cm^{-2} when the bake pressure increased from 1 Torr up to 150 Torr [HYK 16]. At 800°C, the integrated O dose increased from $\sim 5 \times 10^{12}$ cm^{-2} up to $\sim 3 \times 10^{13}$ cm^{-2} when switching from 20 to 150 Torr. Such findings are fully in line with those shown in Figure 5, where the threshold temperature mandatory to get rid of interfacial contamination is roughly 50°C higher at 80 Torr than at 20 Torr.

The temptation would then be, given those data, to carry out the H_2 bake at high temperatures (typically between 850°C and 900°C) and low pressures (20 Torr and less) for relatively long durations (two minutes or more), in order to be safe. This is however not always possible.

On ultra-thin SOI substrates with a mesa isolation, there is for instance a moat recess at the edges of the active Si zones and possibly an islanding of the Si film if the thermal budget of the *in-situ* H_2 bake is too high (top left image of Figure 7). The H_2 bake is then typically carried out at 650°C, 20 Torr for two minutes prior to the selective epitaxial growth of Si raised sources and drains on ultra-thin SOI (i.e. with a starting Si layer thickness of 3 nm, typically) [JAH 05].

Too high a H_2 bake temperature may also lead, for MOSFET transistors with etched gates, to a detrimental regrowth of the interfacial oxide layer present between the Si channel and the high permittivity gate dielectrics (such as HfO_2) or a detrimental diffusion of ions which had previously been implanted to dope the extensions. Because of re-flow, Si nanowires will change shape, with squared-based sections becoming circular (bottom left images of Figure 7) [TAC 09] and the contacts with the sources and drains' blocks becoming pinched-off and ultimately broken (right images of Figure 7) [DOR 07].

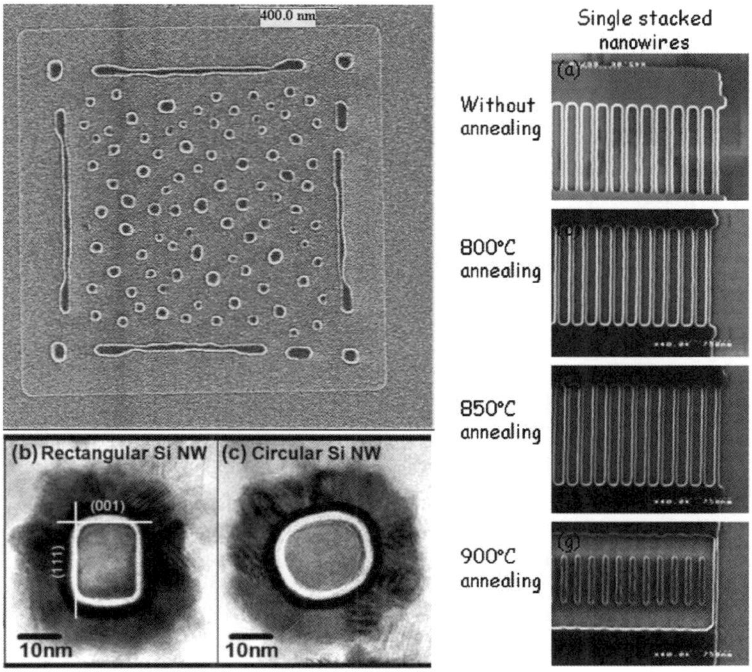

Fig. 7. (top left) top-view SEM image of a 8 nm thick patterned SOI(001) film after a 950°C, 2 min. H_2 bake: moat recess and islanding; (bottom left) cross-sectional TEM images of rectangular Si NWs becoming circular after 750°C, 2 min. H_2 bakes. (right) top-view SEM images of suspended Si NWs after various temperature H_2 bakes lasting 2 minutes at 20 Torr: pinch-off then breaking-off at the contact regions with the source/drain pads. 200 mm data. [JAH 05], [DOR 07], [TAC 09].

What about the structural and electronic properties of the epitaxial layers on partially contaminated Si surfaces? The impact of too low a H_2 bake temperature (between 650°C and 750°C) to remove all traces of O, C or F contamination at the interface is limited on (001) surfaces. The surface stays smooth after epitaxy [DES 08], Si/SiGe/Si stacks are of high crystalline quality in X-ray Diffraction, the electrical performances of Fully Depleted transistors built on top of SOI or SiGe/Si dual channels [LER 11] with Si [AND 07] or SiGe:B [BAU 10] raised sources and drains are excellent and so on. However, H_2 bakes at too low temperature (i.e. $T < 775$°C for two minutes at 20 Torr) have a catastrophic impact on the crystalline quality of epitaxial Si layers grown on Si(110) substrates [DES 08].

Whatever the surface, specific care must be taken during lithography and etch steps to remove all polymers, High-K dielectrics residues (e.g. oxides from transistor gates) and so on. Failure to do so delays growth, prevents epitaxy from taking place if the H_2 bake temperature is too low and so on [HAR 11b]. It may even lead to a significant deterioration of the layer crystalline quality [Liu 12].

Dry plasma etching based on NH_3, NF_3 and H_2 gases in a "Siconi" chamber connected to an epitaxy cluster tool can also be used in order to get rid at low temperature of native oxyde. Chemical reactions are then as follows [YAN 10]:

(i) Etchants are generated in a remote plasma cavity:

$$NF_3 + NH_3 \rightarrow NH_4F + NH_4F.HF \quad (1)$$

(ii) SiO_2 is transformed at ~30°C into a salt:

$$NH_4F \quad \text{or} \quad NH_4F.HF + SiO_2 \rightarrow (NH_4F)_2SiF_6\ (\text{solid}) + H_2O \quad (2)$$

(iii) This salt is sublimated at low temperature (100°C < T < 200°C)

$$(NH_4F)_2SiF_6\ (\text{solid}) \rightarrow SiF_4\ (\text{gas}) + 2NH_3\ (\text{gas}) + 2HF\ (\text{gas}) \quad (3)$$

Surfaces stay pristine afterwards as wafers are transferred under ultra-pure nitrogen from the "Siconi" chamber to the epitaxy chamber itself. Interfacial contamination after wet cleaning in single wafer tools is then avoided. Indeed, 300 mm wafers do not stay any more in Front Opening Unified Pods (FOUPs) before being loaded into inert gases load-locks.

Combinations of dry plasma and wet cleaning can be very useful in order to get rid of deleterious "residues" on patterned wafers. We have provided in Figure 8 some 3D Scanning Electron Microscopy (SEM) images of transistors after the use of a 650°C, 120s + 750°C, 30s hybrid H_2 bake followed by the 650°C selective epitaxy of 18 nm thick $Si_{0.7}Ge_{0.3}$:B raised sources and drains on each side of the gate (everything at 20 Torr). 4 different *ex-situ* surface preparation schemes were used beforehand:

(i) A "HF-Last" wet cleaning only;
(ii) The creation inside a remote "Siconi" chamber of the salt (e.g. reactions (1) and (2) above), this salt being sublimated afterwards in the epitaxy chamber;
(iii) A "HF-Last" wet cleaning followed by the creation inside a remote "Siconi" chamber of the salt (e.g. reactions (1) and (2) above), this salt being sublimated afterwards in the epitaxy chamber;
(iv) The full "Siconi" process (e.g. reactions (1) to (3)) followed by a "HF-Last" wet cleaning.

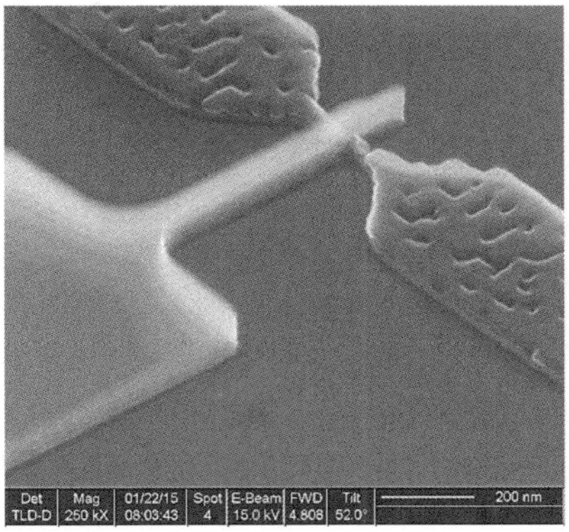

"HF-Last" wet cleaning only

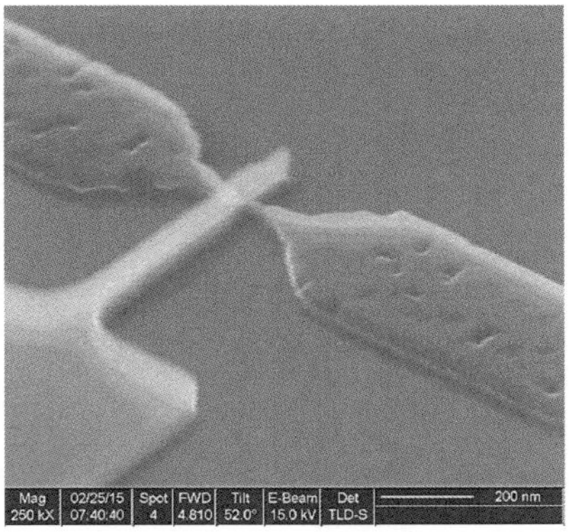

Salt creation in the "Siconi" chamber and sublimation in the epitaxy chamber

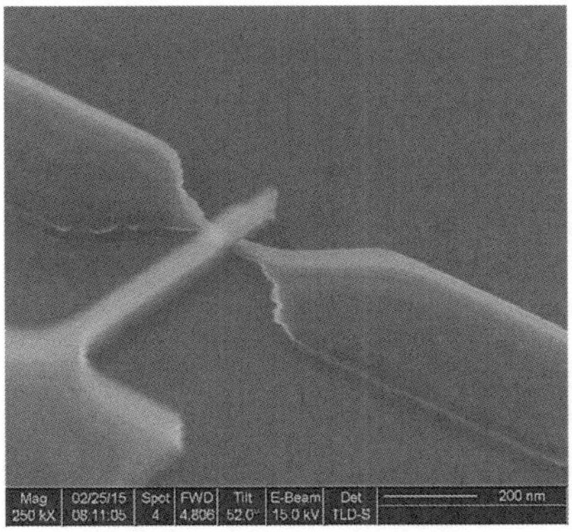

"HF-Last" wet cleaning followed by salt creation in the "Siconi" chamber and sublimation in the epitaxy chamber

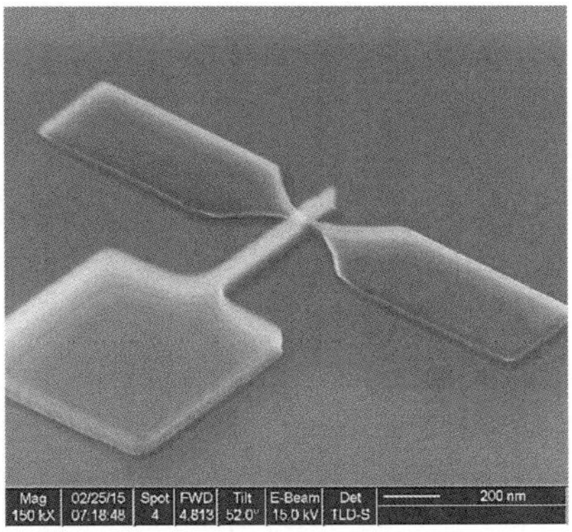

Full "Siconi" process followed by a "HF-Last" wet cleaning

Fig. 8. Benchmark of various surface preparation strategies in order to get rid of "residues" on the surface of patterned FD-SOI(001) wafers prior to the selective epitaxial growth of 18 nm thick SiGe:B raised sources and drains. 300 mm data [Lu 16].

While "HF-Last" wet cleanings or "Siconi" processes are not by themselves enough to get rid of "residues" (presence after SiGe:B SEG of numerous holes in the S/D regions, especially with a "HF-Last" wet cleaning), we do see that combining them yields 2D layers [Lu 16].

1.4. *Selective epitaxial growth of Si and SiGe with standard gaseous precursors*

Dichlorosilane (SiH_2Cl_2) is the silicon precursor of choice for the Selective Epitaxial Growth (SEG) of Si and SiGe in the Si windows of patterned Si substrates, the masking layers being dielectrics such as SiO_2 or SiN. This Si

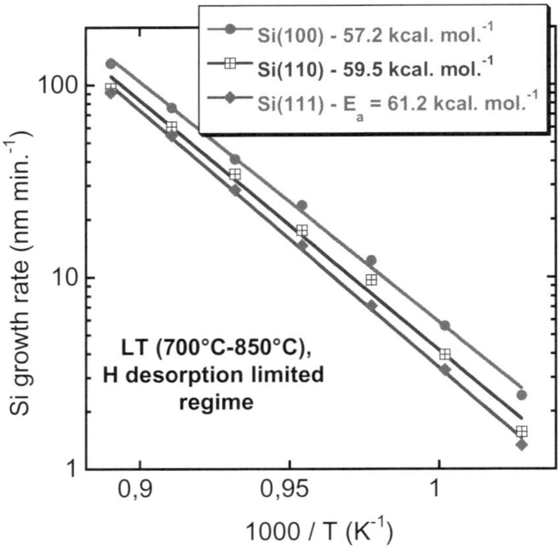

Fig. 9. Si growth rate at 20 Torr as a function of the reverse absolute growth temperature for (100), (110) and (111) surface orientations and a $F(SiH_2\,Cl_2)/F(H_2)$ mass-flow ratio equal to 0.01. 200 mm data. [HAR 06].

gaseous precursor is used for the growth of recessed or raised Si or SiGe:B sources and drains [HAR 04b, HAR 08, BAU 10, HAR 11b, BAR 13], SiGe/Si epitaxial bilayers which constitute the core of localized Silicon-On-Insulator type devices [MON 10] or the high mobility channels of advanced p-type MOSFET transistors [LER 11] and so on. Dichlorosilane does however have relatively low growth rates at low temperatures [HAR 07], as we will see later on.

The growth rate of Si at 20 Torr is shown in Figure 9 as a function of the reverse absolute temperature for three surface orientations: (100), (110) and (111) [HAR 06]. The first two correspond to the top and sidewalls of fins, respectively, while the third one might occur when facets form.

There is as expected an exponential increase of the Si growth rate with the temperature. We have extracted from those Arrhenius plots the activation energies associated with the (100), (110) and (111) surface orientations. They slightly increase with the Miller indexes of the surface: E_a 100 = 57 kcal. mol.$^{-1}$, versus E_a 110 = 59.5 kcal. mol.$^{-1}$ and E_a 111 = 61 kcal mol.$^{-1}$. Those values are in-between the Si-H bond energy, 47 kcal. mol.$^{-1}$ and the Si-Cl bond energy, 90 kcal. mol.$^{-1}$ (1 eV = 23.053 kcal mol.$^{-1}$). Such high activation energies are most probably due to increased surface coverages by Cl atoms (instead of almost exclusively H atoms) for high dichlorosilane mass-flows.

Dangling bond densities on (110) and on (111) surfaces are $1/\sqrt{2}$ and $1/\sqrt{3}$ times that on (100) surfaces, respectively. 29% (42%) *lower* Si growth rates on (110) (on (111)) than on (100) would then be expected based on such simple considerations (surface reconstructions notwithstanding). Those theoretical values are not far away from those experimentally obtained. Indeed, the Si growth rates on (110) (on (111)) are on average 25% (36%) lower than on (100) in the 700°C–850°C range. Such findings are in line with those reported in [PRI 11] and references therein.

The main features of the 20 Torr growth kinetics of SiGe using a gaseous mixture of dichlorosilane and germane (GeH_4) will be described in the coming figures. The SiGe growth rate and Ge concentration functions of the $F(GeH_4)/F(SiH_2Cl_2)$ Mass Flow Ratio (or MFR) are provided in Figure 10

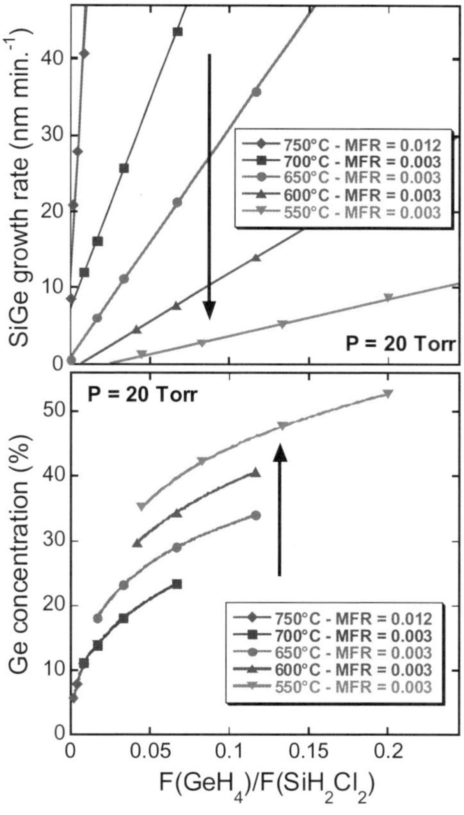

Fig. 10. SiGe growth rate (top) and Ge concentration (bottom) at 20 Torr functions of the $F(GeH_4)/F(SiH_2\ Cl_2)$ Mass-Flow Ratio for different temperatures between 550°C and 750°C ((100) surface orientation data). The $F(SiH_2Cl_2)/F(H_2)$ MFR, equal to 0.003 for $550°C \leq T \leq 700°C$, was multiplied by 4 at 750°C (i.e. 0.012) to reach low Ge contents. 300 mm data. [HAR 12].

for temperatures between 550°C and 750°C and a (100) surface orientation [HAR 12].

As previously shown, in another 200 mm epitaxial reactor [HAR 07] x, the concentration of Ge, increases sub-linearly with the GeH_4 flow. Such behavior is accurately described by a relationship of the type $x^2/(1-x) = n^*F(GeH_4)/F(SiH_2Cl_2)$, with $n = 0.98$ (700°C), 1.34 (650°C), 2.01 (600°C) and 2.52 (550°C).

The exponential increase of n as T (the absolute growth temperature) decreases may be modeled by $n = 5.2 \times 10^{-3} \exp(0.45\,eV/k_BT)$. The Suh and Lee model, which gives a physical description of the system and includes the $x^2/(1-x) = n^*\ F(GeH_4)/F(SiH_2Cl_2)$ formula, translates the fact that each surface Cl atom migrates towards a neighboring Ge atom, before being desorbed [SUH 00]. The SiGe growth rate otherwise increases linearly with the mass flow of GeH_4. This is due to the catalyzed desorption of surface H and Cl atoms (freeing dangling bonds for the adsorption of Si and Ge atoms) due to the presence of Ge atoms (Ge-H and Ge-Cl bonds, respectively 37 and 51 kcal. mol.$^{-1}$, are weaker than Si-H and Si-Cl bonds, 47 and 90 kcal. mol.$^{-1}$). However, the growth rate decreases and the Ge concentration increases when the growth temperature drops, making the 550°C selective epitaxy of SiGe layers rather complicated, at least with a $SiH_2Cl_2 + GeH_4$ mixture (prohibitively low growth rates for Ge contents below 35%).

The complexity of the situation increases when hydrochloric acid is added to the gaseous mixture, which is done, for instance, to promote selectivity versus SiN [HAR 07], [HAR 12]. At 650°C and 20 Torr, a growth rate drop of approximately 50% occurs with $F(HCl)/F(H_2) = 0.0015$ (top part of Figure 11). The SiGe growth rate increase with the germane flow, while being smaller, is also linear. Adding HCl leads to an increase in the concentration of Ge (bottom part of Figure 11).

In the gaseous phase, the added Cl atoms bond preferentially with Si, as the Si chlorides and chlorosilanes are more stable than the corresponding molecules formed with Ge (hence the Ge content increase). The increase of x with the $F(GeH_4)/F(SiH_2Cl_2)$ MFR is also sub-linear, with $n = 2.06$. Although the addition of B_2H_6 leads to a slight increase in the growth rate (no more than 25% here, with a $2 \times 10^{20}\,cm^{-3}$ B atomic concentration), it is difficult to selectively deposit SiGe:B RSDs below 650°C (as growth rates are then too low for Ge concentrations below 40%, [HAR 08], [HAR 11b]).

We have quantified in [DES 10] the impact of surface orientation (e.g. (100) and (110)) on the 20 Torr growth kinetics of SiGe at 600°C and 650°C. As expected, the SiGe growth rate decreases as the growth

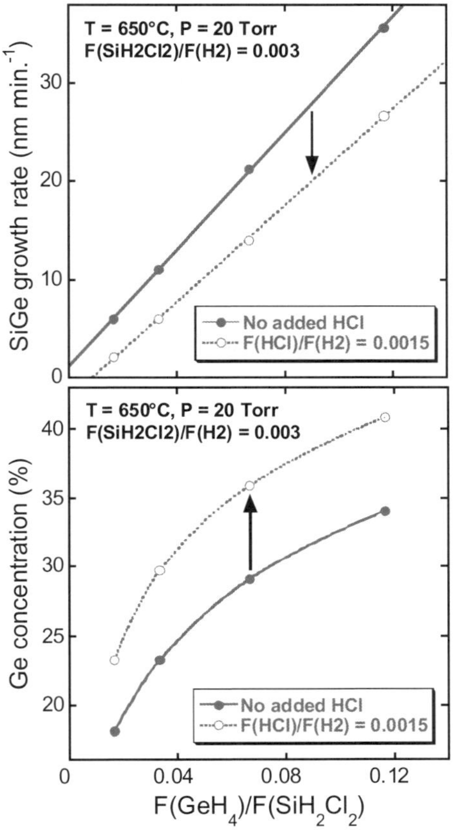

Fig. 11. SiGe growth rate (top) and Ge concentration (bottom) at 650°C and 20 Torr (versus the $F(GeH_4)/F(SiH_2Cl_2)$ Mass Flow Ratio) without or with HCl ($F(HCl)/F(H_2) = 0.0015$, then). The $F(SiH_2Cl_2)/F(H_2)$ MFR was equal to 0.003. 300 mm data, (100) surface orientation. [HAR 12].

temperature decreases from 650°C down to 600°C (irrespectively of the surface orientation). However, whereas the SiGe(100) growth rate increases linearly with the germane mass flow, the SiGe(110) growth rate saturates for high germane mass flows, reaching much lower values than on (100). The Ge concentration x dependence on the $F(GeH_4)/F(SiH_2Cl_2)$ mass flow ratio is sub-linear on (100) and well accounted for by $x^2/(1-x) = n \times F(GeH_4)/F(SiH_2Cl_2)$, with $n = 2.36$ at 600°C and 1.13 at 650°C. In contrast, x depends on (110) linearly on the $F(GeH_4)/F(SiH_2Cl_2)$ mass flow ratio, with a $x/(1-x) = m \times F(GeH_4)/F(SiH_2Cl_2)$ relationship linking the two ($m = 1.71$ at 600°C $\Leftrightarrow$ $m = 2.58$ at 650°C).

The discrepancy between the two surface orientations (which will come to the fore when growing SiGe on fins or trigate structures with (110)

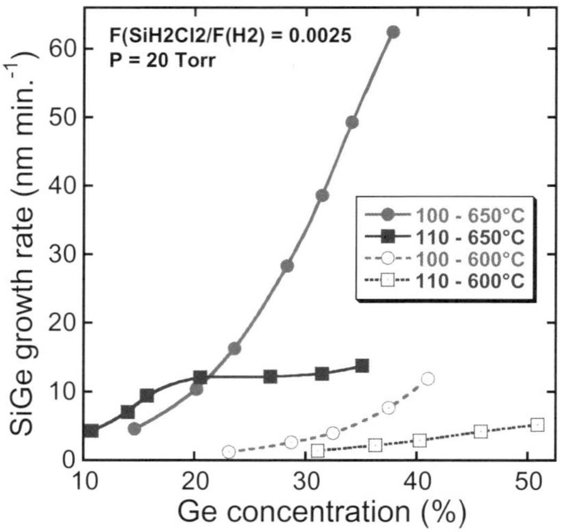

Fig. 12. SiGe growth rate at 20 Torr as a function of Ge concentration, this for two temperatures (e.g. 650°C and 600°C) and two surface orientations (e.g. (100) and (110)). The $F(SiH_2Cl_2)/F(H_2)$ mass-flow ratio was equal to 0.0025 while the GeH_4 mass-flow was varied. 200 mm data. [DES 10].

sidewalls and (100) tops), is well illustrated by Figure 12 plot of the SiGe growth rate as a function of the Ge content at 600°C and 650°C.

Growing stacks on substrates other than bulk Si can otherwise have a definite impact on growth kinetics. In order to illustrate this, we have deposited at 650°C, 20 Torr a 15 periods $\{Si_{0.8}Ge_{0.2}/Si\}$ superlattice on top of a {145 nm thick buried oxide/15 nm thick Si over-layer} SOI(100) substrate. We have plotted in Figure 13 the evolution with the deposited thickness of the Si and SiGe growth rates (top) and the Ge concentration (bottom). We are faced with growth rates which are at first 5% higher than bulk values. They drop down as the deposited thickness increases, reaching a minimum at around 100 nm (70% of the bulk growth rate). Growth rates then go back up. Values close to those on bulk Si(100) are reached for roughly 300 nm of deposited material. Growth rates then decrease once again. The evolution of the Ge concentration with the deposited thickness is the exact opposite of the SiGe growth rate. It reaches a maximum at around 100 nm (21.5%) and goes back to the bulk value (20.2%) for roughly 300 nm. Everything points out towards a surface temperature which is, because of the presence of the buried oxide, lower than the nominal one and which evolves with the deposited thickness. The same trend was evidenced (for the same growth settings) in another 200 mm tool with infra-red pyrometers instead of thermocouples as in Figure 13 [HAR 09]. The evolution of the

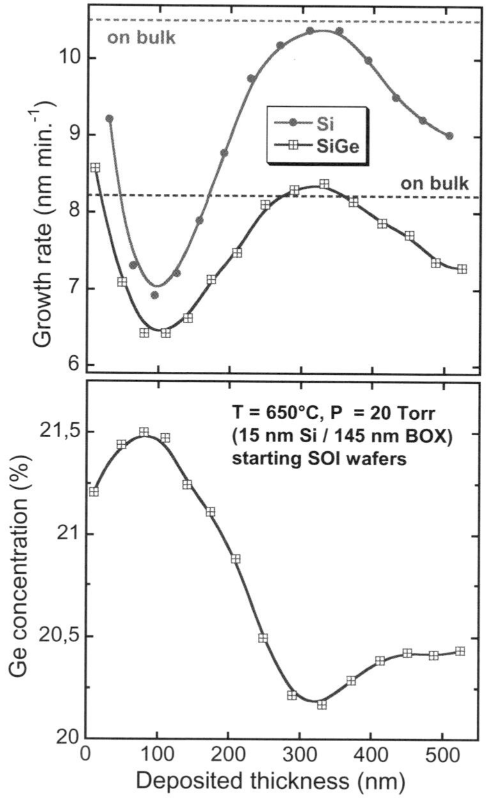

Fig. 13. (top) Si and SiGe growth rates together with (bottom) Ge content (bottom) as a function of deposited thickness on 300 mm blanket SOI substrates. Data coming from SIMS depth profiling of the Si and Ge contents in a $Si_{0.8}Ge_{0.2}$/Si superlattice grown at 650°C, 20 Torr with SiH_2Cl_2+GeH_4 (SiGe) or SiH_4(Si). The starting SOI(100) substrates consisted in 15 nm thick Si layers on top of 145 nm thick buried oxide. J. M. Hartmann *et al.*, unpublished results.

effective surface temperature was then as follows: from 651°C down to 638°C (100 nm), back up to 648°C (350–450 nm) and finally slowly decreasing to 646°C (for 800 nm). Growth kinetics also change with the buried oxide thickness [HAR 03a], [HAR 04b]; this will also have to be factored in when performing epitaxy on SOI wafers.

We are also faced, when working on patterned wafers, with global and local loading effects that are most troublesome in order to obtain SiGe layers of controlled thickness and composition [HAR 04b], [KOL 10], [KOL 11]. By "global" loading effects, we mean the definite growth rate increase and the slight Ge concentration increase that occurs when switching from blanket to patterned wafers. The amplitude of this phenomenon will be inversely proportional to the mask opening, e.g. the amount of the surface that is

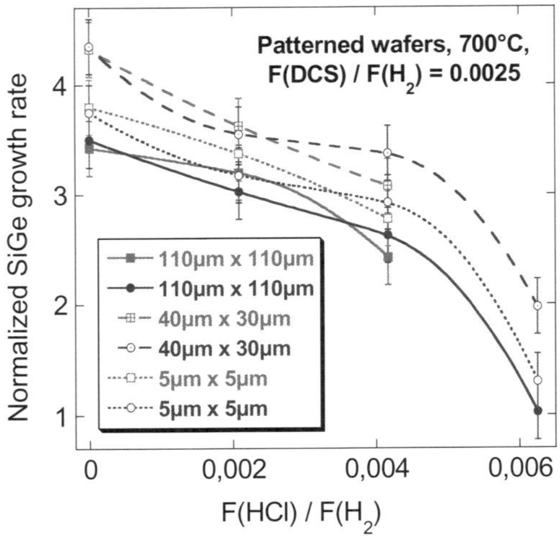

Fig. 14. Normalized SiGe growth rates at 700°C, 20 Torr in variable size Si windows functions of the F(HCl)/F(H_2) mass-flow ratio (F (SiH_2Cl_2)/F(H_2) = 0.0025 and F(GeH_4)/ F(H_2) = 1.25×10^{-4} (circles) or 1.67×10^{-4} (squares). Growth temperature: 700°C. The associated Ge concentrations are on blanket wafers in the 18%–32% range. 200 mm data, (100) surface orientation. [HAR 03b].

not covered by dielectrics. By "local" loading effects, we mean the growth rate and Ge concentration increase that occur when switching from large, closely packed Si windows to small, isolated ones. As an illustration, we have plotted in Figure 14 the normalized growth rates (e.g. the growth rate in a window divided by the growth rate on blanket wafers) at 700°C, 20 Torr for given SiH_2Cl_2 and GeH_4 flows and various HCl flows [HAR 03b]. Without any HCl, we have a factor 4 increase of the growth rate, which is due to a small mask opening: 8% only of the surface (the remainder is covered by SiO_2), versus 30%–40% in typical patterns. Adding HCl reduces the growth rate discrepancy, however. We otherwise see that the growth rate fluctuates with the window size.

1.5. *Use of low temperature SEG in actual devices*

We will describe in the following some successful uses of low temperature H_2 bakes and selective epitaxial growth for the fabrication of planar [LER 11] and nanowires-based [BAR 14] pMOSFET transistors with state of the art electrical performances. In order to modulate the threshold voltage V_T and have a high mobility channel for holes in [LER 11] devices, a {$Si_{0.8}Ge_{0.2}$3 nm/Si 2 nm} bilayer was grown at 650°C and 20 Torr with SiH_2Cl_2+GeH_4 (using Figure 10 data points) on a 300 mm Extra-thin (ET-)SOI substrate (starting Si thickness around 3 nm). A "HF-Last" wet cleaning followed by an *in-situ* H_2 bake at 650°C and 20 Torr for two minutes (in order to avoid any islanding of the very thin Si starting layer) was used to prepare the surface beforehand. A second epitaxy was then carried out later on in the process flow (after gate stack deposition and patterning) in order to selectively grow highly *in-situ* B doped $Si_{0.7}Ge_{0.3}$ raised sources and drains ([B] = 2×10^{20} cm^{-3}). This drastically reduces the access resistance of the transistors and to a certain extent, introduces a uniaxial compression in the SiGe channel, increasing the hole mobility. As with the channel, a temperature of 650°C was used for both the *in-situ* H_2 bake and the SiGe:B epitaxy (with a SiH_2Cl_2+GeH_4+HCl+B_2H_6 chemistry; cf. [HAR 13a]). Cross-sectional Transmission Electron Microscopy images of a pMOSFET transistor (gate length: 30 nm) and of the channel beneath its gate are shown in Figure 15. The {$Si_{0.8}Ge_{0.2}$/Si} bilayer has a perfect crystallinity (no extended defects; abrupt interfaces). The deposition process used for the $Si_{0.7}Ge_{0.3}$:B raised sources and drains is otherwise perfectly selective versus the Si_3N_4 spacers. {111} facets are however present at the boundaries between the source and drain regions and the gate.

Short gate length epitaxial SiGe/Si multi-(core/shell) p-type nanowire (NW) transistors were fabricated in [BAR 14]; a cross-sectional TEM picture

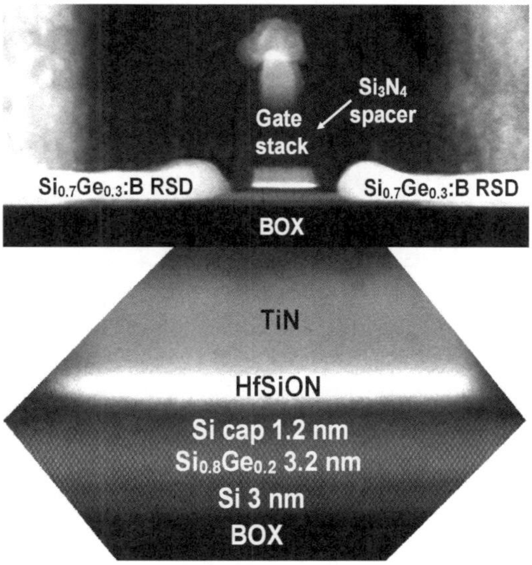

Fig. 15. Cross-sectional High Resolution TEM images of a pMOSFET transistor (of 30 nm gate length) with a SiGe/Si bilayer as channel for the holes and silicided, 18 nm thick $Si_{0.7}Ge_{0.3}$:B raised sources and drains. 300 mm data, (100) surface orientation for the starting ET-SOI substrate. [LER 11].

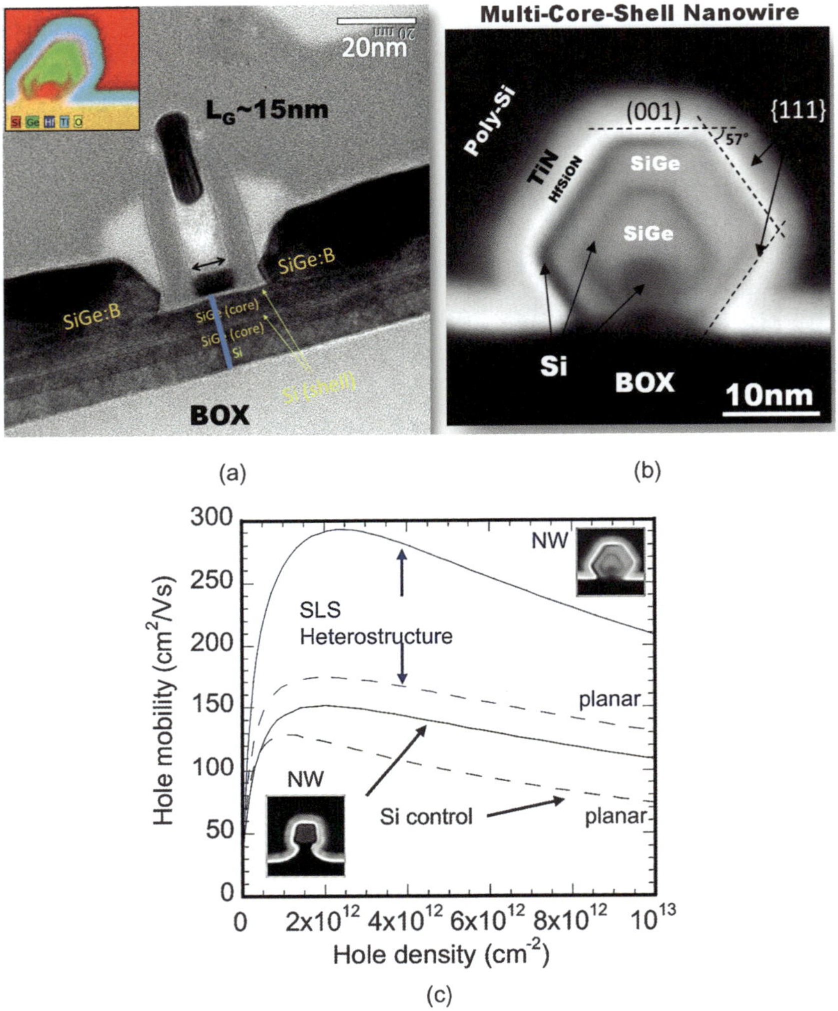

Fig. 16. (a) Cross sectional TEM image of a SiGe/Si multi-(Core/Shell) P-FET nanowire transistor with a 15 nm gate length (Inset: Energy Dispersive X-ray analysis of a core-shell NW with a HfSiON/TiN metal gate). (b) Cross-section of a hexagonal multi-(Core/Shell) nanowire with {111} facets. (c) Effective hole mobility in nanowire (full lines) and planar (dotted lines) MOSFETs, the channel being either Si or SiGe/Si Strained Layer Superlattices (SLS). Mobility is higher in the latter than in the former, all the more so in NW devices. 300 mm data. [BAR 14].

of such a device can be found in Figure 16(a). To that end, silicon nanowires with (110) sidewalls and (100) top surfaces were first of all patterned in (100) ET-SOI wafers by lithography and etching [BAR 12]. A "HF-Last" wet cleaning followed by a 2 min. H_2 bake at 650°C, 20 Torr was then used to remove native oxide and have Si NW surfaces fit for growth.

Dichlorosilane and germane were subsequently used for the 650°C, 20 Torr SEG of $Si_{0.7}Ge_{0.3}$/Si/$Si_{0.7}Ge_{0.3}$/Si shells around the Si NW cores (using once again Figure 10 data points). The formation of epitaxial SiGe facets along the {111} crystallographic planes is obvious in Figure 16(b). The access regions on each side of the {HfSiON/TiN/poly-Si} gates were later on thickened thanks to a 650°C, 20 Torr SEG of 18 nm thick $Si_{0.7}Ge_{0.3}$:B layers. Electrical transport measurements showed a hole mobility improvement up to 100% in SiGe/Si multi-(core/shell) NWs (70% in wide planar devices) compared to p-type Si p-MOSFETs; see Figure 16(c). A drive current enhancement of 60% compared to reference Si-channel devices was otherwise evidenced in multi-(core/shell) P-FET NWs scaled down to 15 nm gate length. The [BAR 14] paper on those results wined the 2014 IEEE EDS Paul Rappaport Award.

1.6. *Advanced low temperature cyclic deposition/etch processes for recessed and raised sources and drains*

We have recently shown that silane (SiH_4) and disilane (Si_2H_6) yielded vastly higher growth rates at low temperatures than dichlorosilane (SiH_2Cl_2) [HAR 12], [HAR 14a]. Such gases are intrinsically not selective versus dielectrics masks, however. This difficulty can be overcome with the so-called Cyclic Deposition/Etch (CDE) process, which was patented in 2006 by Bauer *et al.* [BAU 06], [BAU 07]. The general principle of this technique is as follows: a few nm of the desired material is deposited not selectively on patterned wafers, usually with high order silanes such as disilane (Si_2H_6) [LOU 12] or trisilane (Si_3H_8) [BAU 12a] (which delivers high growth rates at low temperatures [GOU 09], [TAK 10], [VIN 10]). Single-crystalline layers are then grown in the Si active area of those wafers, whereas poly-crystalline or amorphous layers are deposited on top of dielectrics. A selective etch of the latter type of layers is then performed using Cl_2 [BAU 12a], HCl + GeH_4 [BAU 12b] and so on. A thin single-crystalline layer is then obtained in the active area of patterned wafers, while dielectrics layers are free of deposits. The repetition of growth and etch steps (hence the "Cyclic Deposition/Etch" name) yields the active layer thickness aimed for, with full selectivity (see Figure 17 schematics inspired from [VER 07]).

CDE is usually conducted at low temperatures ($\leq$600°C) for several practical reasons: (i) as far as $Si_{1-y}C_y$:P layers are concerned, the lower the temperature and the higher the growth rate are, the higher the substitutional C content in those layers will be [HAR 4c], [BAU 07]; (ii) high Ge content

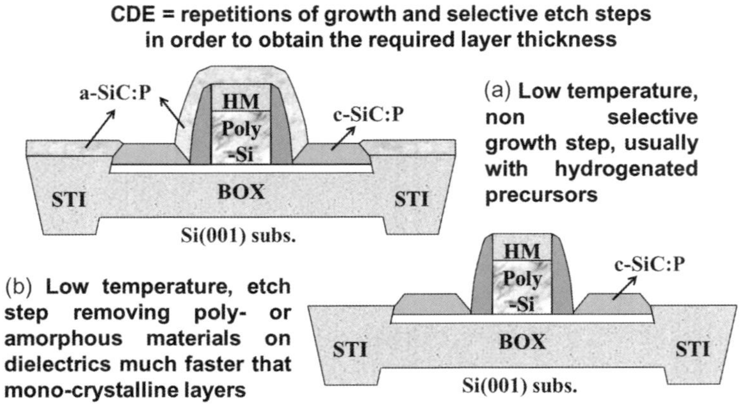

Fig. 17. Schematics of the individual steps of a Cyclic Deposition/Etch process (shown here for raised $Si_{1-y}C_y$:P sources and drains grown on each side of a *n*-type MOSFET built on top of a ET-SOI substrate. BOX stands for Buried Oxyde, STI for Shallow Trench Isolation and HM for Hard Mask. [HAR 13b]).

SiGe:B layers as in [HE 12] will remain fully compressively strained and (iii) amorphous layers deposited at low temperatures on dielectrics will be etched much faster (compared to single-crystalline layers) than higher temperature poly-crystalline layers [HAR 10], [BAU 12b].

Trisilane, one of the Si precursors of choice for CDE, has several drawbacks: (i) it is extremely expensive to produce with electronic-grade quality; (ii) being liquid, it requires a dedicated bubbler to be delivered into the growth chamber. Cl_2, the CDE etching gas advocated by Bauer *et al.*, needs to be injected in the growth chamber in dedicated supply lines, in order to avoid exothermic reactions with trisilane [BAU 12a]. A Si_2H_6+$SiCH_6$+PH_3 chemistry has recently been used by Loubet *et al.* to fabricate $Si_{1-y}C_y$:P RSDs with a CDE approach (using HCl as the etchant gas) [LOU 12]. There is however a scarcity of results in the literature concerning the use of disilane for the low temperature growth of group IV layers. We have therefore benchmarked SiH_4 and Si_2H_6 for the (001) growth of Si:P and $Si_{1-y}C_y$:P. We have also quantified the beneficial impact of adding GeH_4 to HCl in order to etch Si at low temperatures [HAR 13b]. Knowing the specifics of individual growth and etch steps (i.e. steps (a) and (b) of Figure 17) has enabled us to develop new 300 mm Si, Si:P and $Si_{1-y}C_y$:P CDE processes [HAR 13c], whose main features are summed up in the following.

Major differences between silane and disilane concerning the low temperature *in-situ* phosphorous doping of Si(100) have been identified, as shown in Figure 18. Phosphorous surface segregation has prevented us

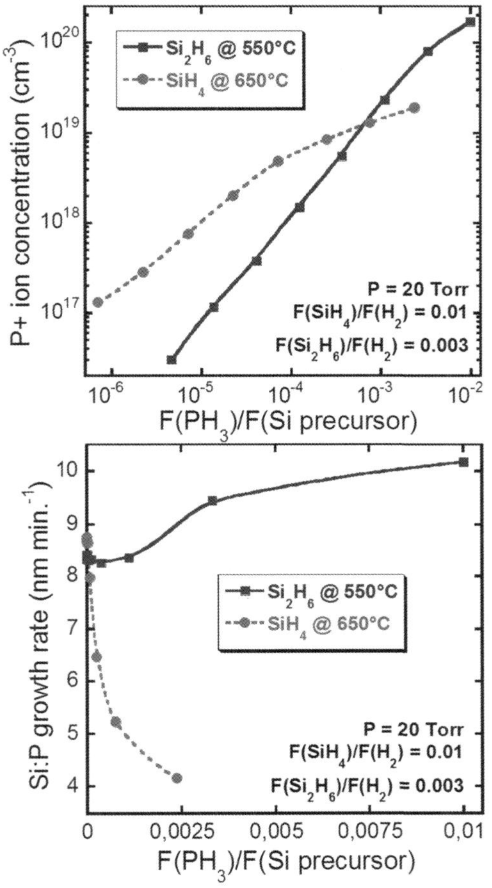

Fig. 18. log-log plot of the P^+ ion concentration (top) and linear plot of the Si:P growth rate (bottom) as a function of the $F(PH_3)/F$(Si precursor) Mass-Flow Ratio, for SiH_4 @ 650°C or Si_2H_6 @ 550°C ($P = 20$ Torr). Silane-based data were acquired a 200 mm epitaxy tool [GON 08], while disilane-based data are 300 mm [HAR 13b]. $F(SiH_4)/F(H_2)$ and $2F(Si_2H_6)/F(H_2)$ MFRs were similar (0.01 versus 0.006).

from reaching P^+ ion concentrations higher than a few 10^{19} cm^{-3} with SiH_4; the resulting surface "poisoning" led to a severe growth rate reduction (at 650°C, 20 Torr). Meanwhile, $[P^+]$ increased linearly with the phosphine flow when using Si_2H_6 as the Si precursor; values as high as 1.7×10^{20} cm^{-3} were achieved. The Si:P growth rate from Si_2H_6 was initially stable then increased as the PH_3 flow increased (at 550°C, 20 Torr).

Similar substitutional C concentrations $[C]_{subst.}$ were achieved in intrinsic $Si_{1-y}C_y$ layers grown at 550°C, 20 Torr with both Si gaseous precursors (up to 1.9%), although growth rates were approximately 6 times higher with Si_2H_6 than with SiH_4. (e.g. 3 instead of 0.5 nm min.$^{-1}$). Mono-methylsilane

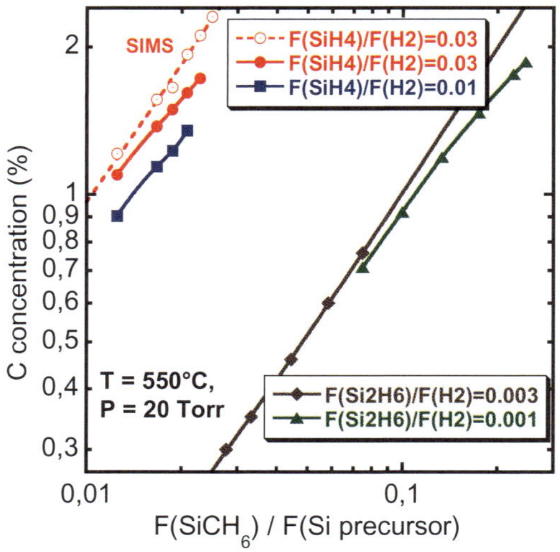

Fig. 19. log-log plot of the substitutional C concentration (full symbols, full lines) and, for one specific process condition, of the total C concentration (open circles, dotted line) in $Si_{1-y}C_y$ layers grown at 550°C, 20 Torr with either SiH_4 or Si_2H_6, this as a function of the $F(SiCH_6)/F(\text{Si flux})$ mass-flow ratio. Silane-based data were acquired in a 200 mm tool while disilane-based data are 300 mm. [HAR 13b].

flows 6.5 to 10 times higher were needed with Si_2H_6 than with SiH_4 to obtain the same $[C]_{subst.}$ on (100), as illustrated in Figure 19. Finally, ~30 nm thick $Si_{1-y}C_y$ layers became rough as the substitutional C content exceeded 1.6% (formation of numerous small size islands, as in [HAR 10], [SHI 12]).

We have then studied the structural and electrical properties of "low" and "high" C content $Si_{1-y}C_y$:P layers (~1.5 and 1.8%, respectively) grown at 550°C, 20 Torr with Si_2H_6. Adding significant amounts of PH_3 led to a reduction of the tensile strain in the films (still high crystalline quality in X-ray Diffraction). This was due to the incorporation of P atoms (at the expense of C atoms) in the substitutional sites of the Si matrix. $Si_{1-y}C_y$:P layers became rough as the PH_3 flow increased. Resistivities lower than 1 mOhm.cm were nevertheless associated with those $Si_{1-y}C_y$:P layers, with P atomic concentrations up to 4×10^{20} cm^{-3}.

We have otherwise quantified the etch rate increase occurring when adding comparatively small amounts of GeH_4 to HCl (etching of single-crystalline Si). Etch rates were almost the same for temperatures 1000°C and above. There was however a growing discrepancy at lower temperatures, with 12–36 times higher Si etch rates achieved at 20 Torr with HCl + GeH_4 than with pure HCl (the first atomic mono-layers became "SiGe-like" and

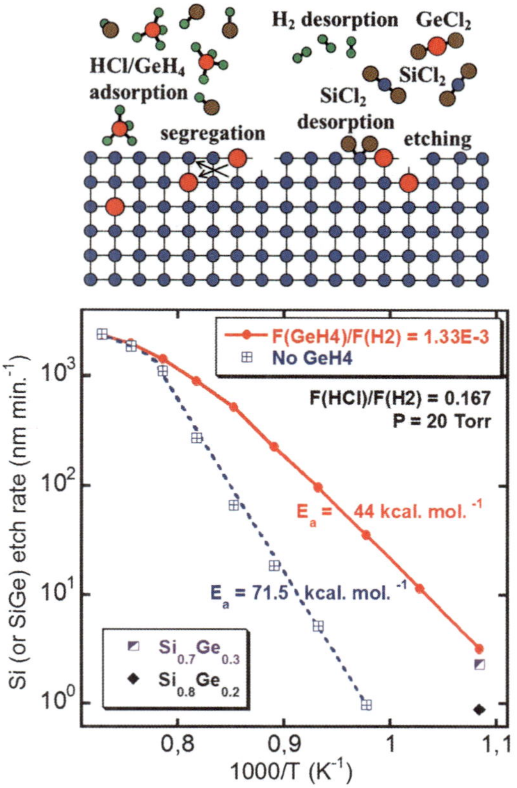

Fig. 20. Schematic showing the molecular processes occurring during the HCl+GeH_4 etch of single-crystalline Si. (bottom) Arrhenius plot of the 20 Torr etch rate of single-crystalline Si(100) with HCl or HCl+GeH_4. The 650°C, 20 Torr etch rates of single-crystalline $Si_{0.8}Ge_{0.2}$ and $Si_{0.7}Ge_{0.3}$ with pure HCl were added for comparisons purposes. The F(HCl/)/F(H_2) Mass-Flow Ratio was 0.167 (Si) or 0.1067 (SiGe), whereas the F(GeH_4)/F(H_2) MFR was equal to 1.33×10^{-3}. 300 mm data [HAR 13b].

were etched must faster). Workable etch rates (i.e. close to 1 nm min.$^{-1}$) were obtained at 600°C (versus 750°C for pure HCl), as shown in Figure 20. A chamber pressure increase otherwise helped us in reaching higher etch rates.

We have then evaluated various CDE processes in order to selectively grow at low temperatures Si, Si:P and $Si_{1-y}C_y$:P Raised Sources and Drains on patterned SOI(100) wafers. A Si_2H_6+PH_3+$SiCH_6$ chemistry was used for the 550°C growth steps. Meanwhile, the selective etch of layers on dielectrics was conducted at 600°C with HCl + GeH_4.

We have first of all studied the specifics of those isobaric (P = 20 Torr) CDE processes on bulk, blanket Si(001) substrates. High crystalline quality and smooth CDE-grown Si, Si:P and $Si_{1-y}C_y$(:P) layers were produced. However, 2–3% Ge was injected in the layers during the etch steps. These Ge atoms, being larger than Si and C atoms, reduced the tensile strain and thus the "apparent" substitutional C content $[C]_{subst.}$ in the $Si_{1-y}C_y$(:P) layers (from X-ray Diffraction). This $[C]_{subst.}$ reduction, close to 0.2%, occurred as soon as etch steps were used, with little or no dependence on the etch time per step. The two $SiCH_6$ mass-flows probed yielded $[C]_{subst.} = 1.3\%$ and 1.6% in our CDE-grown $Si_{1-y}C_y$ layers. Although process conditions were the same, $[C]_{subst.}$ was slightly lower in $Si_{1-y}C_y$:P layers (−0.1%). This was likely due to the incorporation of large amounts of P atoms at the expense of C atoms in the lattice. The atomic P concentration in our Si:P and $Si_{1-y}C_y$:P layers was indeed higher than 2×10^{20} cm^{-3}. The Si, Si:P and $Si_{1-y}C_y$(:P) thickness deposited per CDE cycle decreased linearly as the HCl+GeH_4 etch time increased. The "equivalent" etch rate (e.g. the slope of this linear decrease) was lower in intrinsic than in *in-situ* doped layers, however. The higher the C content was, the lower the etch rate otherwise was (0.53 nm $min.^{-1}$ for Si:P, versus 0.23 nm $min.^{-1}$ for "high" C content intrinsic $Si_{1-y}C_y$). A CDE strategy suppressed the islanding occurring for high C content, several tens of nm thick $Si_{1-y}C_y$:P layers grown in one step only (as in [SHI 12]). The resistivity of our $Si_{1-y}C_y$:P layers did not depend on the growth mode. It however increased with the C content, as in [LOU 12], from 0.48 mOhm.cm ($[C]_{subst.} = 0\%$) up to 0.87 and finally 1.06 mOhm.cm ($[C]_{subst.} = 1.3\%$ and 1.6%).

We have then calibrated, for 19–23 nm thick CDE-grown Si, Si:P and $Si_{1-y}C_y$:P RSDs, the HCl + GeH_4 etch time/step necessary to achieve full selectivity on patterned SOI(100) substrates. Selectivity was obtained for intrinsic Si as soon as 180s etch steps were used. Longer etch times were needed for Si:P and especially $Si_{1-y}C_y$:P (270 and 315s/CDE cycle, respectively).

The resulting S/D areas were rather smooth and slightly facetted, but as the poly-Si layers sitting on top of the gate stacks were unprotected with the integration scheme used here, they were completely removed with such etch times, as shown in Figure 21. Hard masks will thus have to be used in actual devices. Finally, we have compared the Si, Si:P and $Si_{1-y}C_y$:P layer thickness deposited per cycle on bulk, blanket Si and in the large metrology windows of patterned ET-SOI substrates. Similar values were obtained in all cases. Global loading effects are thus negligible.

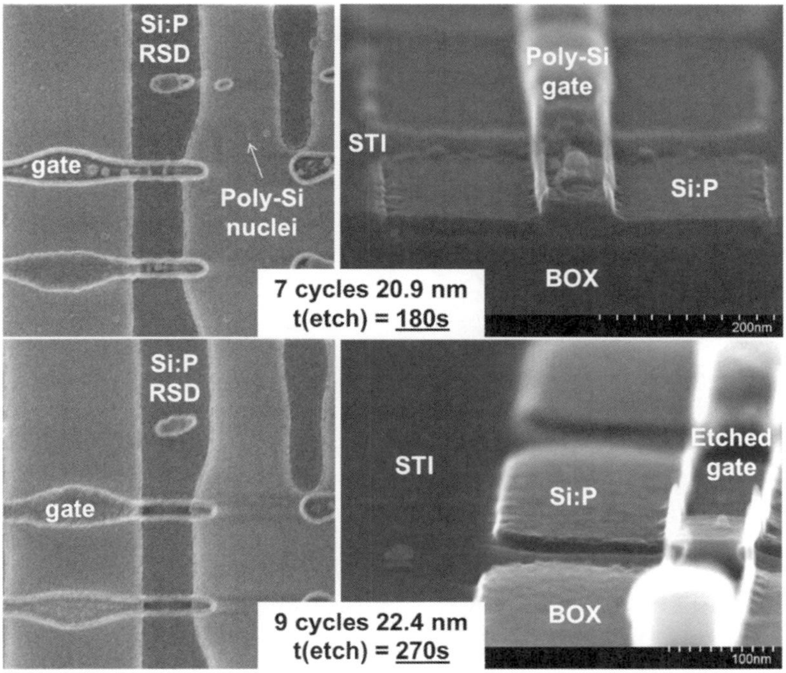

Fig. 21. Top view and cross-sectional SEM images of 300 mm ET-SOI transistors after the CDE of 21–22 nm thick Si:P layers in the source/drain regions. Growth steps: $T = 550°C$, $P = 20$ Torr, $F(Si_2H_6)/F(H_2) = 0.003 + F(PH_3)/F(H_2) = 6 \times 10^{-5}$, growth time = 30s/CDE cycle. Etch steps: $T = 600°C$, $P = 20$ Torr, $F(HCl)/F(H_2) = 0.167 + F(GeH_4)/F(H_2) = 1.33 \times 10^{-3}$, etch time = 180s or 270s/CDE cycle. Number of CDE cycles increased from 7 to 9 in order to have similar thicknesses. [HAR 13c].

We have otherwise leveraged our know-how on (i) disilane-based epitaxial growth and (ii) Cyclic Deposition/Etch strategies to develop an innovative 500°C process for the 300 mm selective deposition of heavily *in-situ* boron-doped SiGe:B RSDs on (100) ET-SOI wafers [HAR 14b] [Lu 16]. We have first of all evaluated the impact of diborane on the 500°C, 20 Torr growth kinetics of SiGe:B on blanket Si(001).

The growth rate significantly increased while the Ge content decreased as B_2H_6 was added to disilane + germane: from 3 up to 13 nm min.$^{-1}$ and from 45% down to 28%, respectively (see Figure 22). Layers were of excellent crystalline quality even for high substitutional B concentrations (from strain compensation: at most 5×10^{20} cm^{-3}). The lowest resistivity achieved, $\sim 4 \times 10^{-4}$ Ohm.cm, was a factor of two lower than our usual 650°C process (see Figure 23). Electrically active boron concentrations were otherwise as high as 7.5×10^{20} cm^{-3}.

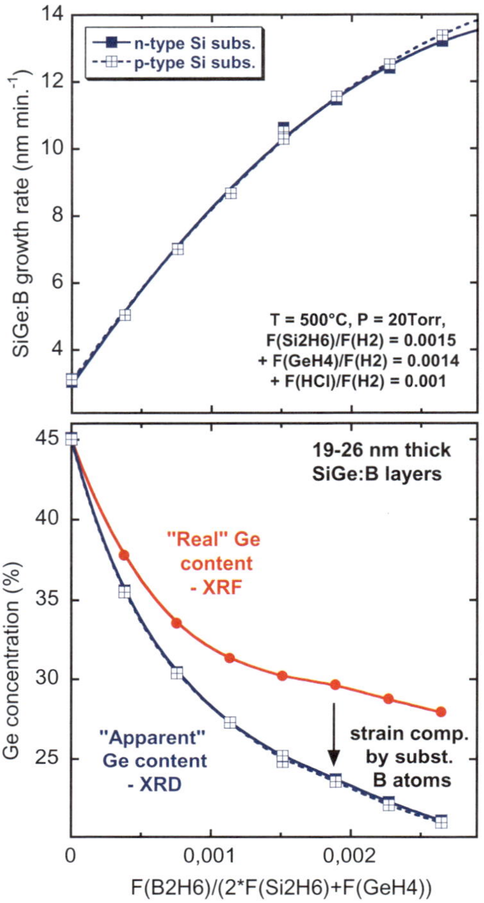

Fig. 22. SiGe:B growth rates (top) and real or "apparent" Ge concentrations (bottom) associated with SiGe:B layers grown at 500°C, 20 Torr with Si_2H_6+GeH_4+HCl and various amounts of B_2H_6. Strain compensation by B atoms, which are much smaller than Si and Ge atoms, explains why "Apparent Ge" contents from XRD are smaller than "Real" Ge contents from X-ray Fluorescence. [HAR 14b].

We have then studied whether or not straightforward SiGe:B SEG was feasible by adding HCl, an etchant gas that promotes selectivity on patterned wafers, to the gaseous mixture. As expected, the (100) SiGe:B growth rate decreased while the Ge content increased with the HCl flow. The poly-SiGe:B growth rate on SiO_2-covered Si wafers was close to the single-crystalline growth rate on blanket Si wafers, however (even for high HCl flows): see Figure 24. Co-flow SEG is thus not feasible at 500°C with Si_2H_6+GeH_4+HCl.

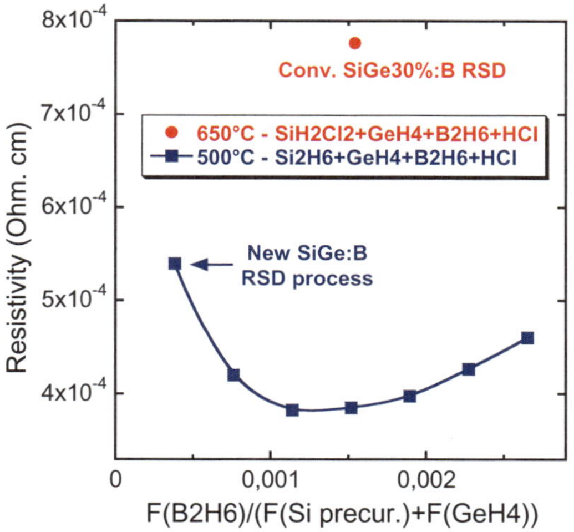

Fig. 23. Resistivity associated with the SiGe:B layers grown at 500°C, 20 Torr with Si_2H_6+GeH_4+HCl and various amounts of B_2H_6. Same layers as in Figure 22. 300 mm data, (100) surface orientation. [HAR 14b].

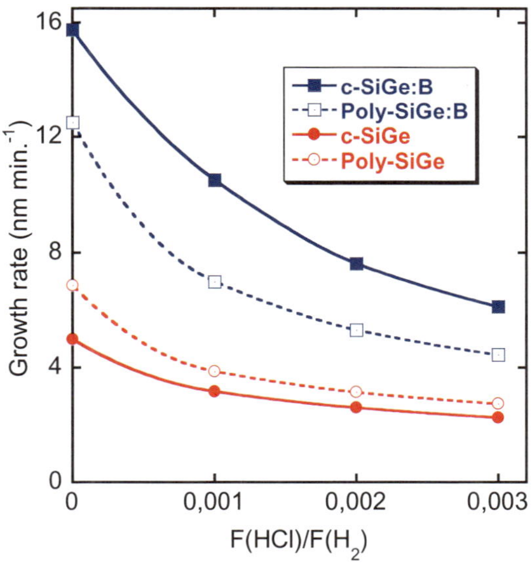

Fig. 24. Single crystalline and poly-crystalline SiGe(:B) growth rates at 500°C, 20 Torr for various F(HCl)/F(H_2) mass-flow ratios. The F(Si_2H_6)/F(H_2), F(GeH_4)/F(H_2) and F(B_2H_6)/F(H_2) MFRs were constant and equal to 0.0015, 0.0014 and 6.67×10^{-6}, respectively. 300 mm data. [HAR 14b].

We have then benchmarked various CDE strategies in order to obtain the selectivity aimed for at 500°C. The overall deposition time was constant at 432s, while various HCl partial pressures and total etch pressures were tested in simple Deposition/Etch (DE) processes or 12 cycles CDE processes. For cyclic processes, low HCl flows and etch pressures (20 Torr) resulted in almost nil poly-SiGeB etch rates (0.4 nm $min.^{-1}$ only). Significantly increasing both the HCl flow and the total etch pressure (80 Torr) delivered the poly-SiGe:B etch rates sought after (4 nm $min.^{-1}$). The etch selectivity (i.e. the poly-SiGe:B etch rate on SiO_2-covered substrates divided by the SiGe:B etch rate on blanket Si) was low, however (2.7 only). The resulting SiGe:B layers were otherwise 3D for long etch durations, making such CDE processes worthless (as for chlorinated chemistries at 650°C [HAR 13a]).

We have thus used a simpler 500°C DE process, with a single etch step at 740 Torr (atmospheric pressure) after the 20 Torr growth of the whole layer. Etch selectivity was improved when switching from high to medium HCl flows (from 2.6 up to 4.8), while still having high poly-SiGeB etch rates (24 nm $min.^{-1}$): see Figure 25. The resulting (100) SiGe:B layers were compressively-strained and of high crystalline quality (although the surface was roughened a bit). We have then switched to patterned SOI wafers with

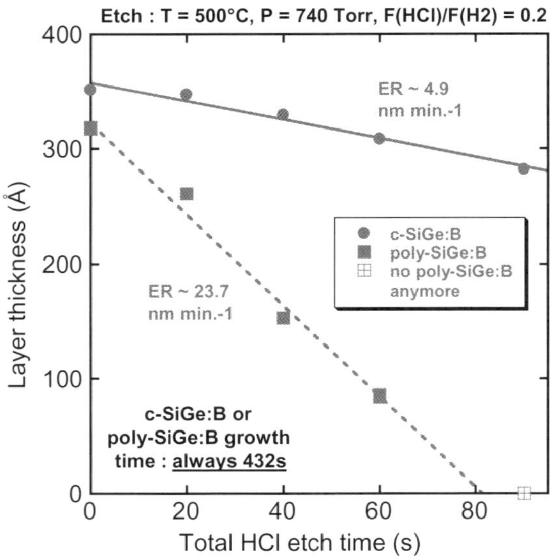

Fig. 25. Single crystalline and poly-crystalline SiGe:B layer thickness as a function of the HCl etch time for the optimum 500°C DE process. $Si_{0.65}Ge_{0.35}$:B deposition at 20 Torr lasted 432s with constant Si_2H_6, GeH_4 and B_2H_6 mass-flows. Meanwhile, the HCl etch step was conducted at P = 740 Torr with a low $F(HCl)/F(H_2)$ mass-flow ratio (0.2). 300 mm data.

gate stacks and studied the feasibility of such a DE process. Longer HCl etch times than the ones identified on blanket wafers were key in getting rid of poly-SiGe:B on top of dielectrics covered surfaces such as the SiO_2 isolation or the SiN spacers; rather smooth, facetted SiGe:B raised sources and drains were obtained in the end: see Figure 26.

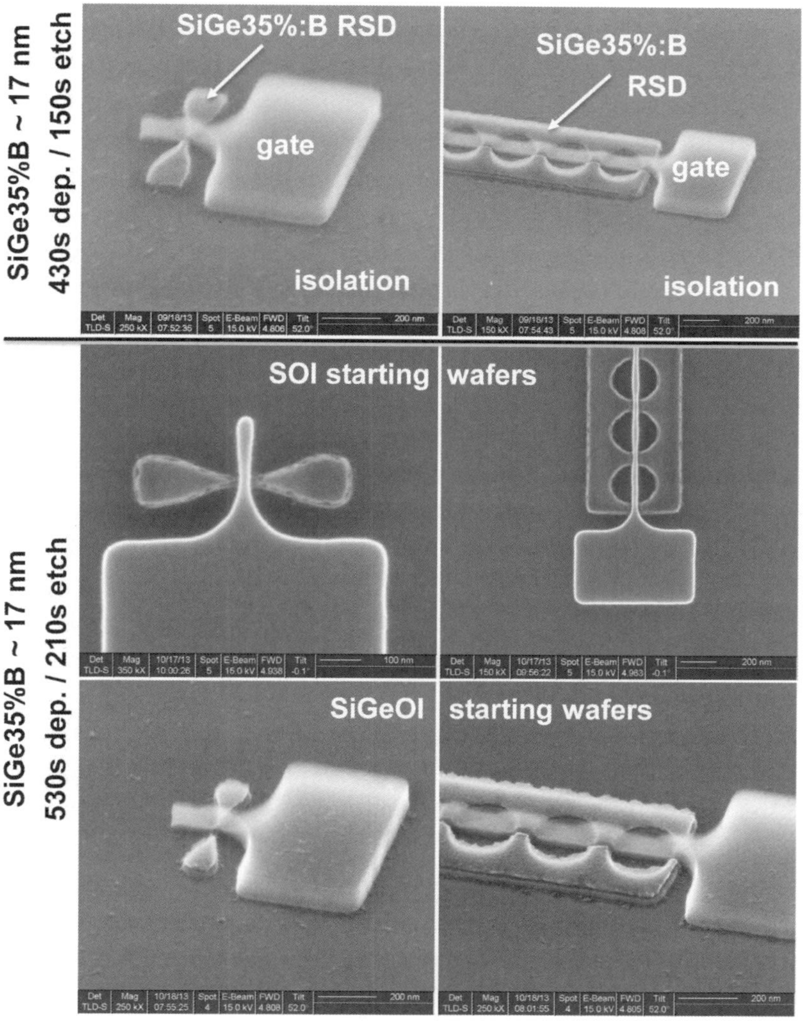

Fig. 26. 3 dimensional or top view SEM images of single and multi-fingered FETs on patterned ET-SOI or Ge-enriched ET-SiGeOI wafers ([GLO 14]) with gate stacks after the 500°C Deposition/Etch of SiGe:B RSD. $Si_{0.65}Ge_{0.35}$:B deposition occurred at 20 Torr with constant $F(Si_2H_6)/F(H_2)$, $F(GeH_4)/F(H_2)$, $F(HCl)/F(H_2)$ and $F(B_2H_6)/F(H_2)$ mass-flow ratios (0.0015, 0.0014, 0.001 and 1.67×10^{-6}, respectively). Meanwhile, the HCl etch steps were conducted at $P = 740$ Torr with a low $F(HCl)/F(H_2)$ MFR (0.2). 300 mm data, (100) surfaces. [HAR 14b].

1.7. *Conclusions and some short term perspectives*

In this chapter, we have presented some of the challenges one faces when selectively growing group-IV semiconductors on patterned wafers for nanoelectronics, namely: surface preparation, faceting, loading effects, surface temperature changes because of the presence of a buried oxide and so on. We have described some of the salient features of the chlorinated chemistry used for Selective Epitaxial Growth (e.g. $SiH_2Cl_2+GeH_4+HCl$) and shown two successful uses of that chemistry in advanced planar or core-shell based Metal Oxyde Semiconductor Field Effect Transistors (MOSFETs). Finally, we have detailed a new type of process that yields low temperature selectivity, e.g. the Cyclic Deposition/Etch (CDE) approach. It consists in growing Si (Si:P, SiC:P) or SiGe:B non selectively on patterned wafers, then selectively removing poly-crystalline materials (on dielectrics) while keeping some single-crystalline layers inside Si (or SiGe) windows.

Let us now discuss some current and near future group-IV epitaxy trends. Very low temperature (e.g. $T \leq 500°C$) epitaxy (with the use of CDE) has some major appeal, as it would enable monolithic 3D integration, e.g. the sequential processing of layers of ET-SOI transistors one upon the other (with metallic interconnections in-between) [FEN 14], [BAT 15], avoid 3D objects' (such as fins or tri-gates) shape changes and so on. Liquid Si precursors such as trisilane (Si_3H_8), tetrasilane (Si_4H_{10}, [KAN 93], [HAZ 16], [HART 16]) neopentasilane (Si_5H_{12}, [CHU 08]) and so on together with slightly more evolved Ge precursors such as digermane (Ge_2H_6, [WIR 13], [HAR 16]) will then be most handy (in combination with Cl_2 as the low temperature etchant gas).

The Ge content in the embedded or raised SiGe:B Sources and Drains (S/Ds) on each side of high performance p-type MOSFETs becomes higher and higher when moving from one technology node to the next one (50% for the 22 nm technology node [HYK 14]), in order to inject higher and higher amounts of uniaxial compressive strain in the channel of planar [DEN 12] and now fin-shaped devices. The low temperature selective epitaxy of Ge-rich SiGe layers (e.g. with Ge contents above 50%) will thus continue to be explored.

Such know-how will also be beneficial for the high Ge content SiGe cladding of Si finFETs, with therefore some superior hole mobilities [MER 14]. Several hundreds of nm thick Ge layers selectively grown in deep, narrow cavities (with a Si floor and SiO_2 sidewalls) can benefit from Aspect Ratio Trapping, resulting in rather low defect density, relaxed layers that can be used as the core of high mobility Ge FinFETs [VAN 14] (presence of Ge(110) planes, then). A variant of this approach is to selectively grow Ge_xSi_{1-x} buffers (with x typically in the 70%–85% range) instead of pure Ge in those cavities, followed by the epitaxy of thin, compressively-strained Ge layers, with therefore better transport properties than in unstrained Ge [MIT 14].

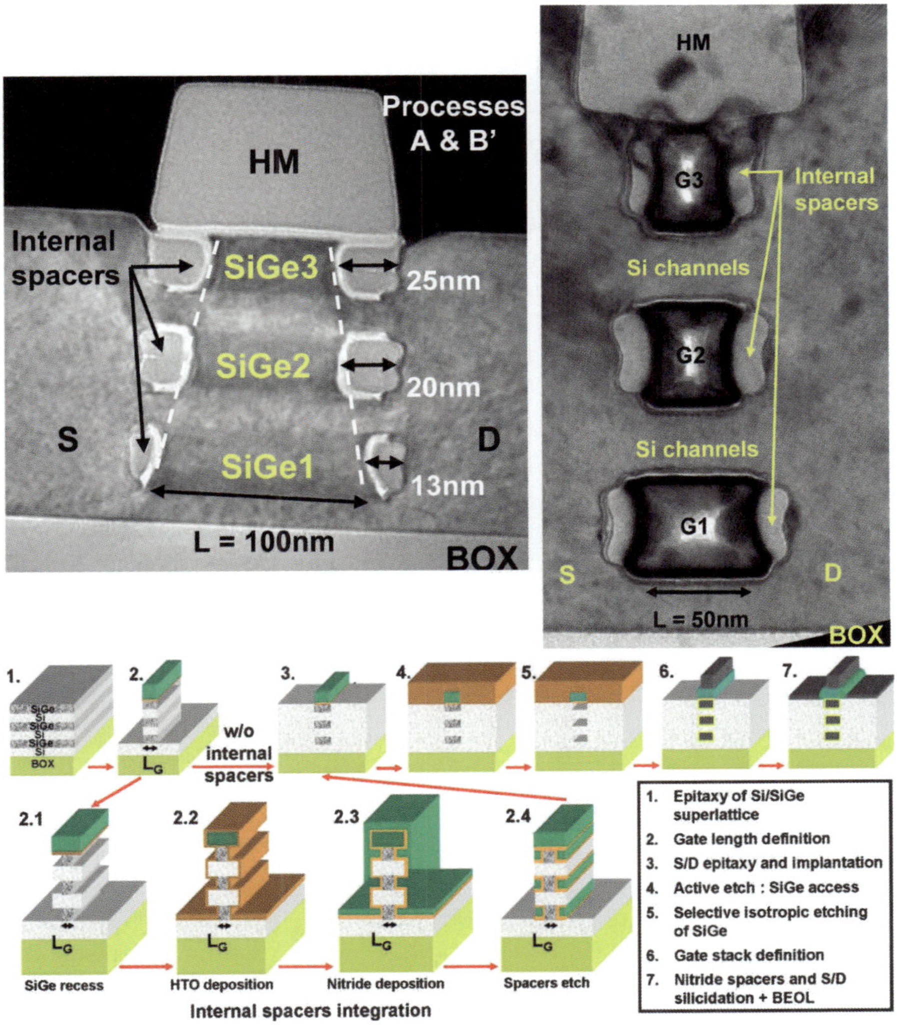

Fig. 27. (top) cross-sectional Transmission Electron Microscopy images of a Si multi-channel FET with SiN internal spacers and Si raised Sources and Drains [HAR 10]. The SiGe1, 2 and 3 layers still present in the left picture were selectively etched and the voids filled with {HfO_2/TiN/N^+ poly-Si} gate stacks (G1, G2 and G3 in the right picture). (bottom) Introduction of SiN internal spacers adds only four steps to the MCFET process flow sequence. 200 mm data. [BER 09].

One could otherwise use SiP binary alloys (instead of "regular" SiC:P films) in the S/D regions of n-MOSFETs. Although P atoms are not as small as C atoms ($a_P = 5.014$ Å instead of $a_C = 3.567$ Å and $a_{Si} = 5.431$ Å), the uniaxial tensile strain in short gate length devices will indeed be high if the P concentration is high (at most: 12% in [WEE 12]). This will most likely boost the mobility of electrons in short gate length n-MOSFETs.

Finally, using SiGe/Si multilayers together with sacrificial etching enables to fabricate stacked NWs with superior integration densities and device performances ([BER 09], [TAC 09], [BAR 15]), as shown in Figure 27. Fins with Ge fluctuations along the z axis are defined during the process (prior to the lateral selective etching of SiGe or Si). Ge enrichment (e.g. the high temperature thermal oxidation of SiGe, with a preferential oxidation of Si atoms and a "snowplough" effect for the Ge atoms) might also be handy during device fabrication [GLO 14], [NGU 14]. Such stacked NWs might therefore be the ultimate evolution (and the convergence point) of the finFET and the FD-SOI technologies.

References

Abbadie, A., Hartmann, J.-M., Holliger, P., Semeria, M. N., Besson, P., Gentile, P. (2004), Low thermal budget surface preparation of Si and SiGe, *Applied Surface Science*, **225**(1–4) p. 256–266.

Andrieu, F., Faynot, O., Rochette, F., Barbe, J.-C., Buj, C., Bogumilowicz, Y., Allain, F., Delaye, V., Lafond, D., Aussenac, F., Feruglio, S., Eymery, J., Akatsu, T., Maury, P., Brevard, L., Tosti, L., Dansas, H., Rouchouze, E., Hartmann J.-M., Vandroux, L., Casse, M., Fenouillet-Beranger, C., Boeuf, F., Brunier, F., Cayrefourcq, I., Mazure, C., Ghibaudo, G., Deleonibus, S. (2007), Impact of mobility boosters (XsSOI, CESL, TiN gate) on the performance of <100> or <110> oriented FDSOI cMOSFETs for the 32nm node, *Proceedings of the 2007 VLSI Conference*, p. 50–51.

Barraud, S., Coquand, R., Casse, M., Koyama, M., Hartmann, J. M., Maffini-Alvaro, V., Comboroure, C., Vizioz, C., Aussenac, F., Faynot, O., Poiroux, T. (2012), Performance of Omega-Shaped-Gate Silicon Nanowire MOSFET With Diameter Down to 8 nm, *IEEE Electron Device Letters*, **33**(11) p. 1526–1528.

Barraud, S., Coquand, R., Hartmann, J. M., Maffini-Alvaro, V., Samson M. P., Tosti, L., Allain, F. (2013), Enhanced Performance of P-FET Omega-Gate SoI Nanowire With Recessed-SiGe Source-Drain Down to 13-nm Gate Length, *IEEE Electron Device Letters*, **34**(9) p. 1103–1105.

Barraud, S., Hartmann, J. M., Maffini-Alvaro, V., Tosti, L., Delaye, V., Lafond, D. (2014), Top-Down Fabrication of Epitaxial SiGe/Si Multi-(Core/Shell) p-FET Nanowire Transistors, *IEEE Transactions on Electron Devices*, **61**(4) p. 953–956.

Barraud, S., Casse, M., Gaben, L., Nguyen, P., Hartmann, J. M., Samson M. P., Maffini-Alvaro, V., Tabone, C., Vizioz, C., Arvet, C., Pimenta-Barros, P.,

Glowacki, F., Bernier, N., Rozeau, O., Jaud, M. A., Martinie, S., Laccord, J., Allain, F., De Salvo, B., Vinet M. (2015), Opportunities and Challenges of Nanowire-Based CMOS Technologies, *Proceedings of the 2015 IEEE SOI-3D-Subthreshold Microelectronics Technology Unified Conference* (S3S), p. 1–3, Rohnert Park, California (USA), October.

Batude, P., Fenouillet-Bernager, C., Pasini, L., LU, V., Deprat, F., Brunet, L., Sklenard, B., Piegas-Luce, F., Casse, M., Mathieu, B., Billoint, O., Turkyilmaz, O., Sarhan, H., Thuries, S., Sollier, S., Widiez, J., Tabone, C., Samson M-P, Previtali, B., Rambal N., Ponthenier, F., Mazurier, J., Beneyton, R., Bidaud, M., Josse, E., Petirprez, E., Rozeau O., Rivoire, M., Euvard-Colnat, C., Seignard, A., Fournel, F., Benaissa, L., Hartmann, J. M., Besson, P., Kerdiles, S., Bout, C., Nemouchi, F., Royer, A., Ghibaudo, G., Signamarcheix, T., Haond, M., Clermidy, F., Faynot, O., Vinet, M. (2015), 3DVLSI with CoolCube process, an alternative path to scaling, *Proceedings of 2015 VLSI Technology Symposium*, p. 48–49.

Bauer, M., Arena, C., Bertram, R., Tomasini, P., Cody, N., Brabant, P., Italiano, J., Jacobson, P., Weeks, K. D. (2006), Selective deposition of silicon-containing films, US Patent 2006/0234504 A1.

Bauer, M., Machkaoustan, V., Arena, C. (2007), Highly tensile strained silicon–carbon alloys epitaxially grown into recessed source drain areas of NMOS devices, *Semiconductor Science and Technology*, **22** p. S183–S187.

Baudot, S., Andrieu, F., Weber, O., Perreau, P., Damlencourt, J. F., Barnola, S., Salvetat, T., Tosti, L., Brevard, L., Lafond, D., Eymery, J., Faynot, O. (2010), Fully depleted strained silicon-on-insulator p-MOSFETs with recessed and embedded silicon–germanium source/drain, *IEEE Electron Device Letters*, **31**(10) p. 1074–1076.

Baudrant, A. (editor) (2010), *Silicon technologies ion implantation and thermal treatment*, ISTE Ltd (UK) and John Wiley & Sons, Inc (USA), Chapter 4 by Hartmann, J. M. p. 209–331.

Bauer, M., Thomas, S. G. (2012a), Low temperature Si:C co-flow and hybrid process using Si_3H_8/Cl_2, *Thin Solid Films*, **520**(8) p. 3133–3138.

Bauer, M., Thomas, S. G. (2012b), Low temperature catalyst enhanced etch process with high etch selectivity for amorphous silicon based alloys over single-crystalline based alloys, *Thin Solid Films*, **520**(8) p. 3139–3143.

Bean, J. C., Feldman, L. C., Fiory, A. T., Nakahara, S., Robinson, I., K. (1984), Ge_xSi_{1-x}/Si strained layer superlattice grown by molecular beam epitaxy, *Journal of Vacuum Science and Technology A*, **2**(2) p. 436–440.

Bernard, E., Ernst, T., Guillaumot, B., Vulliet, N., Lim, T. C., Rozeau O., Danneville, F., Coronel, P., Skotnicki, T., Deleonibus, S., Faynot, O. (2009), First Internal Spacers' Introduction in Record High I_{ON}/I_{OFF} TiN/HfO_2 Gate Multichannel MOSFET Satisfying Both High-Performance and Low Standby Power Requirements, *IEEE Electron Device Letters*, **30** p. 148–151.

Brabant, P., Wen, J., Italiano, J., Landin, T., Cody, N., Haen L. (2003), Achieving a SiGe HBT epitaxial emitter with novel low thermal budget technique, *Applied Surface Science*, **224**(1–4) p. 347–349.

Bruel, M. (1995), Silicon on insulator material technology, *Electronics Letters*, **31**(14) p. 1201–1202.

Bühler, J., Steiner, F.-P., Baltes, H. (1997), Silicon dioxide sacrificial layer etching in surface micromachining, *Journal of Micromechanics and Microengineering*, **7**(1) p. R1–R13.

Chang, H.-R. (1985), Autodoping in silicon epitaxy, *Journal of the Electrochemical Society*, **132**(1) p. 219–224.

Chung, K. H., Yao, N., Benziger, J., Sturm, J. C., Singh, K. K., Carlson D., Kuppurao, S. (2008), Ultrahigh growth rate of epitaxial silicon by chemical vapor deposition at low temperature with neopentasilane, *Applied Physics Letters*, **92** 113506.1–113506.3.

Denneulin, T., Cooper, D., Hartmann, J. M., Rouviere, J. L. (2012), The addition of strain in uniaxially strained transistors by both SiN contact etch stop layers and recessed SiGe sources and drains, *Journal of Applied Physics* **112** 094314.1–094314.1.

Destefanis, V., Hartmann J-M., Hopstaken, M., Delaye, V., Bensahel, D. (2008), Low-thermal surface preparation, HCl etch and Si/SiGe selective epitaxy on (110) silicon surfaces, *Semiconductor Science and Technology*, **23**(10) p. 105018.1–105018.9.

Destefanis, V., Hartmann J.-M., Hugeni, J. L., Delaye, V., Samson, M. P., Boulitreau, P., Morand, Y., Brianceau, P., Arvet, C., Gautier, P., Monfray S., Skotnicki, T. (2010), Fabrication, structural and electrical properties of (110) localized silicon-on-insulator devices, *Semiconductor Science and Technology*, **25** p. 045014.1–045014.10.

Dornel, E., Ernst, T., Barbe, J. C., Hartmann, J. M., Delaye, V., Aussenac F., Vizioz, C., Borel, S., Maffini-Alvaro, V., Isheden, C., Foucher, J., (2007), Hydrogen annealing of arrays of planar and vertically stacked Si nanowires, *Applied Physics Letters* **91** 233502.

Dutartre, D., Seiss, B., Barge, D., Loubet, N. (2014), "Small Size" Effects in Si-based Epitaxies for Advanced CMOS Technologies, *ECS Transactions* **64**(6) p. 207–221.

Fenouillet-Beranger, C., Mathieu, B., Previtali, B., Samson M. P., Rambal, N., Benevent, V., Kerdiles, S., Barnes, J. P., Barge, D., Besson, P., Katchouli R., Casse, M., Garros, X., Laurent, A., Nemouchi, F., Huet, K., Toque-Tresonne, I., Lafond, D., Dansas, H., Aussenac, F., Druais, G., Perreau, P., Richard, E., Chhun, S., Petitprez, E., Guillot, N., Deprat, F., Pasini, L., Brunet, L., LU, V., Reita, C., Batude P., Vinet, M. (2014), New insights on bottom layer thermal stability and laser annealing promises for high performance 3D VLSI, *Proceedings of 2014 IEDM Conference*, p. 27.5.1–27.5.4.

Glowacki, F., Le Royer, C., Morand, Y., Pedini, J.-M., Denneulin, T., Cooper, D., Barnes, J.-P., Nguyen, P., Rouchon, D., Hartmann, J.-M., Gourhant, O., Baylac, E., Campidelli, Y., Barge, D., Bonnin, O., Schwarzenbach, W., (2014), Ultrathin (5 nm) SiGe-On-Insulator with high compressive strain (2 GPa): From fabrication (Ge enrichment process) to in-depth characterizations, *Solid-State Electronics*, **97** p. 82–87.

Gonzatti, F., Hartmann, J. M., Yckache, K., (2008), Low and High Temperature Boron and Phosphorus Doping of Si for Junctions and MEMS purposes, *ECS Transactions*, **16**(10) p. 485–493.

Gouye, A., Kermarrec, O., Halimaoui, A., Campidelli, Y., Rouchon, D., Burdin, M., Holliger, P., Bensahel, D. (2009), Low-temperature RPCVD of Si, SiGe alloy, and $Si_{1-y}C_y$ films on Si substrates using trisilane (Silcore®), *Journal of Crystal Growth*, **311**(13) p. 3522–3527.

Hart J., Hazbun R, Eldridge D., Hickey R., Fernando N., Adam T. N., Zollner S., Kolodzey J. (2016), Tetrasilane and digermane for the ultra-high vacuum chemical vapor deposition of SiGe alloys, *Thin Solid Films*, **604**, p. 23–27.

Hartmann J-M., Abbadie, A., Vinet, M., Clavelier, L., Holliger, P., Lafond, D., Semeria, M. N., Gentile, P. (2003a), Growth kinetics of Si on fullsheet, patterned and silicon-on-insulator substrates, *Journal of Crystal Growth* **257** p. 19–30.

Hartmann, J.-M., Bertin, F., Rolland, G., Laugier, F., Semeria, M. N. (2003b), Selective epitaxial growth of Si and SiGe for metal oxide semiconductor transistors, *Journal of Crystal Growth*, **259** p. 419–427.

Hartmann, J.-M., Abbadie, A., Papon, A. M., Holliger, P., Rolland, G., Billon, T., Fedeli, J. M., Rouviere, M., Vivien, L., Laval, S. (2004a), Reduced pressure chemical vapor deposition of Ge thick layers on Si(001) for 1.3–1.55 μm photodetection, *Journal of Applied Physics*, **95**(10) p. 5905–5913.

Hartmann, J.-M., Clavelier, L., Jahan, C., Holliger, P., Rolland, G., Billon, T., Defranoux, C. (2004b), Selective epitaxial growth of boron and phosphorous-doped Si and SiGe for raised sources and drains, *Journal of Crystal Growth*, **264** p. 36–47.

Hartmann, J. M., Ernst, T., Ducroquet, F., Rolland, G., Lafond, D., Papon, A. M., Truche, R., Holliger, P., Laugier, F., Semeria, M. N., Delonibus, S. (2004c), *High C content Si/$Si_{1-y}C_y$ heterostructures for n-type metal oxide semiconductor transistors*, Semiconductor Science and Technology **19** 593–601.

Hartmann, J. M., Burdin, M., Rolland, Billon, T. (2006), Growth kinetics of Si and SiGe on Si(100), Si(110) and Si(111) surfaces, *Journal of Crystal Growth* **294** p. 288–295.

Hartmann, J.-M. (2007), Low temperature growth kinetics of high Ge content SiGe in reduced pressure — chemical vapor deposition, *Journal of Crystal Growth* **305**(1) p. 113–121.

Hartmann, J-M., Gonzatti, F., Fillot, F., Billon, T. (2008), Growth kinetics and boron-doping of very high Ge content SiGe for source/drain engineering, *Journal of Crystal Growth*, **310** p. 62–70.

Hartmann, J.-M., Papon, A. M., Barnes, J. P., Billon, T. (2009), Growth kinetics of SiGe/Si superlattices on bulk and silicon-on-insulator substrates for multi-channel devices, *Journal of Crystal Growth*, **311** p. 3152–3157.

Hartmann, J. M., Py, M., Morel, P. H., Ernst, T., Previtali, B., Barnes J. P., Vulliet, N., Cherkashin, N., Reboh, S., Hytch, M., Paillard, V., (2010), Cyclic Deposition/Etch processes for the formation of Si raised sources and drains in advanced MOSFETs, *ECS Transactions*, **33**(7) p. 391–407.

Hartmann, J.-M., Abbadie, A., Favier, S. (2011a), Critical thickness for plastic relaxation of SiGe on Si(001) revisited, *Journal of Applied Physics* **110**(8) p. 083529_1 to 083529_8.

Hartmann, J.-M., Py, M., Barnes, J. P., Previtali, B., Batude, P., Billon T. (2011b), Low temperature boron and phosphorous doped SiGe for recessed and raised sources and drains, *Journal of Crystal Growth*, **327**(1) p. 68–77.

Hartmann, J.-M., Benevent, V., Damlencourt, J. F., Billon, T. (2012), A benchmarking of silane, disilane and dichlorosilane for the low temperature growth of group IV layers, *Thin Solid Films*, **520**(8) p. 3185–3189.

Hartmann, J. M., Benevent, V., Barnes, J. P., Veillerot, M., Lafond, D., Damlencourt, J. F., Morvan, S., Previtali, B., Andrieu, F., Loubet, N., Dutartre, D. (2013a), Mushroom-Free Selective Epitaxial Growth of Si, SiGe and SiGe:B Raised Sources and Drains, *Solid State Electronics*, **83** p. 10–17.

Hartmann, J. M., Benevent, V., Barnes, J. P., Veillerot M., Deguet, C. (2013b), Disilane-based Cyclic Deposition/Etch of Si, Si:P and Si1-yCy:P layers Part I: the elementary process steps, Semiconductor Science and Technology **28** p. 025017.1–025017.10.

Hartmann, J. M., Benevent, V., Barnes, J. P., Veillerot, M., Previtali B., Batude, P. (2013c), Disilane-based Cyclic Deposition/Etch of Si, Si:P and Si1-yCy:P layers Part II : the CDE features, Semiconductor Science and Technology **28** 025018.1–025018.11.

Hartmann, J. M., Benevent, V., Veillerot, M., Halimaoui A. (2014a), Potentialities of disilane for the low temperature epitaxy of intrinsic and boron-doped SiGe, *Thin Solid Films*, **557** p. 19–26.

Hartmann, J. M., Benevent, V., Andre, A., Sirisipanaporn, C., Veillerot M., Samson, M. P., Barraud, S., Essa, Z., Sermage, B. (2014b), Very Low Temperature (Cyclic) Deposition/Etch of *in-situ* Boron-Doped SiGe Raised Sources and Drains, *ECS Journal of Solid State Science and Technology*, **3**(11) p. 382–390.

Hartmann J. M., Aubin J., Barnes J. P. (2016), A Benchmark of Germane and Digermane for the Low Temperature Growth of Intrinsic and Heavily in situ Boron Doped SiGe, to be published in *ECS Transactions* (Proceedings of Symposium G05 of the PRiME/230th ECS Meeting, Honolulu (USA), October 2016).

Hashemi, P., Balakrishnan, K., Engelmann, S. U., Ott, J. A., Khakifirooz A., Baraskar, A., Hopstaken, M., Newbury, J. S., Chan, K. K., Leobandung, E., MO, R. T., Park D.-G. (2014), First Demonstration of High-Ge-Content Strained-$Si_{1-x}Ge_x$ (x = 0.5) on Insulator PMOS FinFETs with High Hole Mobility and Aggressively Scaled Fin Dimensions and Gate Lengths for High-Performance Applications, *Proceedings of 2014 IEDM Conference*, p. 16.1.1–16.1.4.

Hazbun R., Hart J., Hickey R., Ghosh A., Fernando N., Zollner S., Adam T.N., Kolodzey J. (2016), Silicon epitaxy using tetrasilane at low temperatures in ultra-high vacuum chemical vapor deposition, *Journal of Crystal Growth*, **444**, p. 21–27.

He, H., Brabant, P., Chung, K., Shinriki, M., Adam, T., Reznicek, A., Sadana, D., Hasaka, S., Francis, T. (2012), High strain embedded-SiGe via low temperature reduced pressure chemical vapor deposition, *Thin Solid Films*, **520**(8) p. 3175–3178.

Hersam, M. C., Guisinger, N. P., Lyding, J. W., Thomson, D. S., MOORE, J. S. (2001), Atomic-level study of the robustness of the Si (100)-2×1:H surface following exposure to ambient conditions, *Applied Physics Letters*, **78**(7), p. 886–888.

Houghton, D. C. (1991), Strain relaxation kinetics in $Si_{1-x}Ge_x/Si$ heterostructures, *Journal of Applied Physics*, **70**(4) p. 2136–2151.

Hykavyy, A., Rossel, E., Dhayalan, S., Witters, L., Mertens, H., Bender H., Favia, P., Loo, R. (2014), High Ge content SiGe thin films: growth, properties and integration, *ECS Transactions*, **64**(6) p. 831–839.

Hikavyy, A., Rossel, E., Kubicek, S., Mannaert, G., Favia, P., Bender H., Loo, R., Horiguchi, N. (2016), Integration of $Si_{0.30}Ge_{0.70}$ stressors in 14 nm fin-based p-MOS transistors, *Thin Solid Films*, **602**, p. 72–77.

Ishikawa, Y., Kumezawa, M., Nuryadi, R., Tabe, M. (2002), Effect of patterning on thermal agglomeration of ultrathin silicon-on-insulator layer, *Applied Surface Science*, **90**(1–4) p. 11–15.

Jahan, C., Faynot, O., Tosti, L., Hartmann, J.-M. (2005), Agglomeration control during the selective epitaxial growth of Si raised sources and drains on ultra-thin silicon-on-insulator substrates, *Journal of Crystal Growth*, **280**(3–4) p. 530–538.

Jerier, P., Dutartre, D. (1999), Boron autodoping in single-wafer epitaxy of silicon at reduced pressure, *Journal of the Electrochemical Society*, **146**(1) p. 331–335.

Kanoh, H., Sugiura, O., Matsumura, M. (1993), Chemical Vapor Deposition of Amorphous Silicon using Tetrasilane, *Japanese Journal of Applied Physics*, **32** p. 2613–2619.

Kolahdouz, M., Tabib Zadeh Adibi, P., Afshar Farniya, A., Shayestehamaminzadeh, S., Trybom, E., Di Benedetto, L., Radamson, H. H. (2010), Selective Growth of B- and C-Doped SiGe Layers in Unprocessed and Recessed Si Openings for p-type Metal-Oxide-Semiconductor Field-Effect Transistors Application, *Journal of the Electrochemical Society*, **157** p. H633–H637.

Kolahdouz, M., Maresca, L., Ghandi, R., Khatibi, A., Radamson H. H. (2011), Kinetic Model of SiGe Selective Epitaxial Growth using RPCVD Technique, *Journal of the Electrochemical Society*, **158** p. H457–H464.

Kormann, T., Garnier, P., Chabanne, G., Fortuin, A. (2008), *Ex-situ* wet clean and *in-situ* hydrogen clean for Si and SiGe epitaxy, *Thin Solid Films*, **517**(1) p. 269–271.

Le Royer, C., Villalon, A., Casse, M., Cooper, D., Mazurier, J., Previtali, B., Tabone, C., Perreau, P., Hartmann, J.-M., Scheiblein, P., Allain, F., Andrieu, F., Weber, O., Batude, P., Faynot, O., Poiroux, T. (2011), First demonstration of ultrathin body c-SiGe channel FDSOI pMOSFETs combined with SiGe(:B) RSD: drastic improvement of electrostatics (V_{th}, p tuning, DIBL) and transport (μ_0, I_{sat}) properties down to 23 nm gate length, *Proceedings of 2011 IEDM Conference*, p. 16.5.1–16.5.4.

Liu, J. F., Camacho-Aguilera, R., Bessette, J. T., Sun, X., Wang, X., Cai Y, Kimerling, L. C., Michel, J. (2012), Ge-on-Si optoelectronics *Thin Solid Films*, **520**(8) p. 3354–3360.

Liu, Q., De Salvo, B., Morin, P., Loubet, N., Pilorget, S., Chafik, F., Maitrejean, S., Augendre, E., Chanemougame, D., Guillaumet, S., Kothari, H., Allibert, F., Lherron, B., Liu, B., Escarabajal, Y., Cheng, K., Kuss, J., Wang,

M., Jung, R., Teehan, S., Levin T., Sankarapandian, M., Johnson, R., Kanyandekwe, J., He, H., Venigalla, R., Yamashita, T., Haran, B., Grenouillet, L., Vinet, M., Weber, O., Josse, E., Boeuf, F., Haond, M., Bataillon, J.-L., Kleemeier, W., Skotnicki, T., Khare, M., Faynot, O., Doris, B., Celik, M., Sampson, R. (2014), FDSOI CMOS Devices Featuring Dual Strained Channel and Thin BOX Extendable to the 10nm Node, *Proceedings of 2014 IEDM Conference*, p. 9.1.1–9.1.4.

Loubet, N., Adam, T., Raymond, M., Liu, Q. Cheng., Sreenivasan, R., Reznicek, A., Khare, P., Kleemeier, W., Paruchuri, V., Doris, B., Sampson R. (2012), Ultra-low resistivity *in-situ* phosphorous doped Si and SiC epitaxy for source/drain formation in advanced 20 nm n-type field effect transistor devices, *Thin Solid Films*, **520**(8) p. 3149–3154.

Lu C. M. V., Fenouillet-Beranger C., Hartmann J. M., Rodriguez P., Benevent V., Samson M. P., Previtali B., Tabone C., Casse M., Allain F., Romano G., Brunet L., Batude P., Skotnicki T., Vinet M. (2016), Integration of Low Temperature SiGe:B Raised Sources and Drains in p-type FDSOI Field Effect Transistors, to be published in *ECS Transactions* (Proceedings of Symposium G05 of the PRiME/230th ECS Meeting, Honolulu (USA), October 2016).

Mertens, H., Ritzenthaler, R., Hikavyy, A., Franco, J., Lee, J. W., Brunco, D. P., Eneman, G., Witters, L., Mitard, J., Kubicek, S., Devriendt, K., Tsvetanova, D., Milenin, A. P., Vrancken, C., Geypen, J., Bender, H., Groeseneken, G., Vandervorst, W., Barla K., Collaert, N., Horiguchi, N., Thean, A. V.-Y. (2014), Performance and Reliability of High-Mobility $Si_{0.55}Ge_{0.45}$ p-Channel FinFETs based on Epitaxial Cladding of Si Fins, *Proceedings of 2014 VLSI Technology Symposium*, p. 58–59.

Meyerson, B. S., Himpsel, F. J., Uram, K. J. (1990), Bistable conditions for low-temperature silicon epitaxy, *Applied Physics Letters*, **57**(10) p. 1034–1036.

Mitard, J., Witters, L., Loo, R., Lee, S. H., Sun, J. W., Franco, J., Ragnarsson, L.-Å., Brand, A., Ly, X., Yoshida, N., Eneman, G., Brunco, D. P., Vorderwestner M., Storck, P., Milenin, A. P., Hikavyy, A., Waldron, N., Favia, P., Vanhaeren, D., Vanderheyden A., Olivier, R., Mertens, H., Arimura, H., Sonja, S., Vrancken, C., Bender, H., Eyben, P., Barla K., Lee, S.-G., Horiguchi, N., Collaert, N., Thean, A. V.-Y. (2014), 15nm-W_{FIN} High-Performance Low-Defectivity Strained-Germanium pFinFETs With Low Temperature STI-Last Process, *Proceedings of 2014 VLSI Technology Symposium*, p. 138–139.

Monfray, S., Fenouillet-Berranger, C., Bidal, G., Boeuf, F., Denorme S., Huguenin, J. L., Samson, M. P., Loubet, N., Hartmann J-M., Campidelli, Y., Destefanis, V., Arvet, C., Benotmane, K., Clement, L., Faynot, O., Skotnicki, T. (2010), Thin-film devices for low power applications, *Solid State Electronics*, **54**(2) p. 90–96.

Nguyen, P., Barraud, S., Koyama, M., Casse, M., Andrieu, F., Tabone, C., Glowacki, F., Hartmann, J.-M., Maffini-Alvaro, V., Rouchon, D., Bernier, N., Lafond, D., Samson, M.-P., Allain, F., Vizioz, C., Delprat, D., Nguyen, B.-Y., Mazure, C., Faynot, O., Vinet, M., High Mobility W-Gate Nanowire P-FET on cSGOI Substrates Obtained by Ge Enrichment Technique, *Proceeding of the 2014 IEEE S3S Conference*, DOI: 10.1109/S3S.2014.7028187.

Pribat, C., Dutartre, D. (2011), Anisotropy effects during non-selective epitaxial growth of Si and SiGe materials, *Journal of Crystal Growth*, **334** p. 138–145.

Shinriki, M., Chung, K., Hasaka, S., Brabant, P., He, H., Adam, T. N., Sadana, D. (2012), Gas phase particle formation and elimination on Si(100) in low temperature reduced pressure chemical vapour deposition silicon-based epitaxial layers, *Thin Solid Films*, **520**(8) p. 3190–3194.

Suh, K. Y., Lee, H. H. (2000), Ge composition in $Si_{1-x}Ge_x$ films grown from SiH_2Cl_2/GeH_4 precursors, *Journal of Applied Physics* **88**(7) p. 4044–4047.

Tachi, K., Casse, M., Jang, D., Dupre, C., Hubert, A., Vulliet, N., Maffini-Alvaro, V., Vizioz, C., Carabasse, C., Delaye, V., Hartmann, J. M., Ghibaudo, G., Iwai H., Cristoloveanu, S., Faynot, O., and Ernst, T. (2009), Relationship between mobility and high-k interface properties in advanced Si and SiGe nanowires, *Proceedings of 2009 IEDM Conference*, 313.

Takeuchi, S., Nguyen, N. D., Goosens, J., Caymax, M., Loo, R., (2010), $Si_{1-x}Ge_x$ growth using Si_3H_8 by low temperature chemical vapor deposition, *Thin Solid Films*, **518**(6) p. S18–S22.

Trucks, G. W., Raghavachari, K., Higashi, G. S., Chabal, Y. J. (1990), Mechanism of HF etching of silicon surfaces: a theoretical understanding of hydrogen passivation, *Physical Review Letters*, **65**(4) p. 504–507.

Van Dal, M.,J. H., Vellianitis, G., Duriez, B., Doornhos, G., Hsieh, C. H.-H., Lee, B.-H., Yin, K.-M., Passlack, M. (2014), Germanium p-Channel FinFET Fabricated by Aspect Ratio Trapping, IEEE Transactions ON Electron Devices **61** p. 430–436.

Verheyen, P., Machkaoutsan, V., Bauer, M., Weeks, D., Kerner, C., Clemente, F., Bender, H., Shamiryan, D., Loo, R., Hoffmann, T., Absil, P., Biesemans, Thomas, S. G. (2008), Strain Enhanced nMOS Using *In-Situ* Doped Embedded $Si_{1-x}C_x$ S/D Stressors With up to 1.5% Substitutional Carbon Content Grown Using a Novel Deposition Process, *IEEE Electron Device Letters*, **29** p. 1206–1208.

Vincent, B., Loo, R., Vandervorst, W., Brammertz, G., Caymax M. (2010), Low temperature Si homo-epitaxy by reduced pressure chemical vapor deposition using dichlorosilane, silane and trisilane, *Journal of Crystal Growth*, **312**(19) p. 2671–2676.

Wada, K. (editor), Kimerling, L. C. (editor), *Photonics and Electronics with Germanium*, Wiley-VCH (Germany), July 2015. Chapter 3 by Hartmann, J. M. p. 49–76.

Weeks, K. D., Thomas, S. G., Dholabhai, P., Adams, J. (2012), Characterization and analysis of epitaxial silicon phosphorous allos for use on n-channel transistors, *Thin Solid Films*, **520** 3158–3162.

Wirths, S., Buca, D., B, Tiedmann, A. T., Bernardy, P., Hollander, B., Stoica, T., Mussler, G., Breuer, U., Mantl, S., (2013), Low temperature RPCVD epitaxial growth of Si1xGex using Si_2H_6 and Ge_2H_6, *Solid State Electronics* **83** p. 2–9.

Wostyn, K., Rondas, D., Kenis, K., Loo, R., Hikavyy, A., Douhard, B., Mertens, P. W., Holsteyns, F., De Gendt, S. S., D'urzo, L., Van Autryve L. (2014a), Use of a Purged FOUP to Improve H-Terminated Silicon Surface Stability prior to Epitaxial Growth, *ECS Transactions*, **64**(6) 669–673.

Wostyn, K., Kenis, K., Rondas, D., Loo, R., Hykavyy, A., Dhayalan, S., Douhard, B., Mertens, P., Holsteyns, F., De Gendt, S., Simpson, G., Bast G., Swaminathan, K. (2014b), Evaluation of the $Si_{0.8}Ge_{0.2}$-on-Si Epitaxial Quality By Inline Surface

Light Scattering: A Case Study On The Impact of Interfacial Oxygen, *ECS Transactions*, **64**(6) 989–995.

Yang, R., Su, N., Bonfanti, P., Nie, J., Ning, J., Li, T. T. (2010), Advanced *in-situ* pre-Ni silicide (Siconi) cleaning at 65nm to resolve defects in $NiSi_x$ modules , *Journal of Vacuum Science and Technology B*, **28**(1) p. 56–61.

Chapter 2

Silicon Carbide on Silicon: 3C-SiC Buffer Layers for GaN LEDs

J. Chai, L. Wang and S. Dimitrijev
Queensland Micro- and Nanotechnology Center,
Griffith University, Nathan, QLD 4111, Australia

1. Introduction

Dwindling resources are driving the worldwide momentum for more efficient energy usage through renewable energy, more efficient devices, and better sensing. In particular, general lighting has been targeted as it typically comprise of a substantial proportion of total electricity consumption. Solid-state lighting is a desirable replacement for the incumbent incandescent and fluorescence lighting technologies due to higher efficiency, potential for electronic controls, and longer lifespan. A common technique to produce white light, analogous to the room lights that we are used to, is via combining a blue light-emitting-diode (LED) with a suitable phosphor. Blue LEDs are made from the III-nitride family on heteroepitaxy substrates such as sapphire (Al_2O_3), silicon carbide (SiC), and silicon. The deposition of high quality silicon carbide (SiC) on Si substrates for subsequent nitride deposition will be the focus of this chapter.

SiC has many applications such as in LED fabrication, MEMS, and power devices due to a range of properties such as optical transparency, high thermal conductivity, high elastic modulus, and high breakdown field [Harris, 1995]. SiC exists in over 200 polytypes, with the polytype determined by its stacking sequence [Iwami, 2001], as shown in Fig. 1. The most commonly studied polytypes are the 4H, 6H, and 3C polytypes. The only polytype that can be epitaxially deposited on Si is 3C since both are of cubic crystal structure. Other polytypes such as 4H- and 6H- cannot

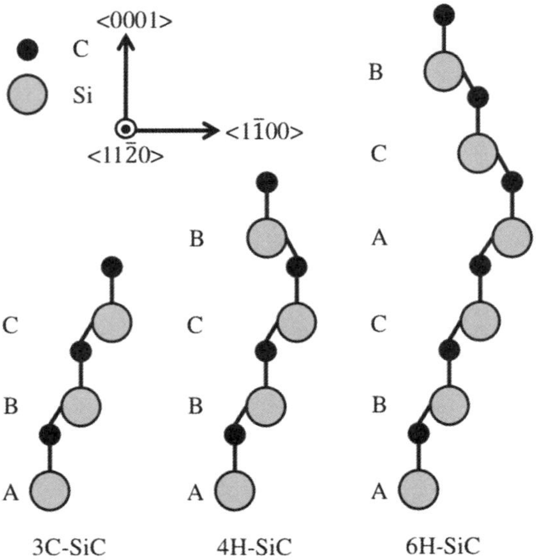

Fig. 1. Cross-sectional representation of the stacking sequences that define the main SiC polytypes [Iwami, 2001].

be epitaxially deposited on Si as they have non-cubic crystal structures and require deposition above the melting point of Si (~1400°C). The (111) orientation is preferable for subsequent nitride deposition as 3C-SiC(111) has a 3-fold symmetry [Dadgar *et al.*, 2007], which allows for epitaxial alignment with the GaN(0001) 6-fold symmetry [Komiyama *et al.*, 2006] (Fig. 2). 3C-SiC is the polytype referred to whenever SiC is mentioned herein, unless otherwise stated.

This chapter will explore the advantages of employing a thin 3C-SiC/Si template for nitride-based LED deposition, a typical low temperature 3C-SiC/Si template deposition procedure, effect of deposition conditions on the quality of 3C-SiC/Si templates, and effect of deposition conditions on 3C-SiC/Si film stress.

1.1. *Si as a low cost substrate for GaN LEDs*

A major stumbling block in widespread adoption of blue LEDs for general lighting applications is the key figure of merit of $ per watt. Homoepitaxy is the Holy Grail of epitaxy as the lattice and thermal coefficient match allows for high crystalline quality, and thus the highest performance devices. While GaN substrates have been in development for decades, they are currently only available in small wafer sizes (<2"), very expensive, and are still an immature technology. Therefore, heteroepitaxy of GaN

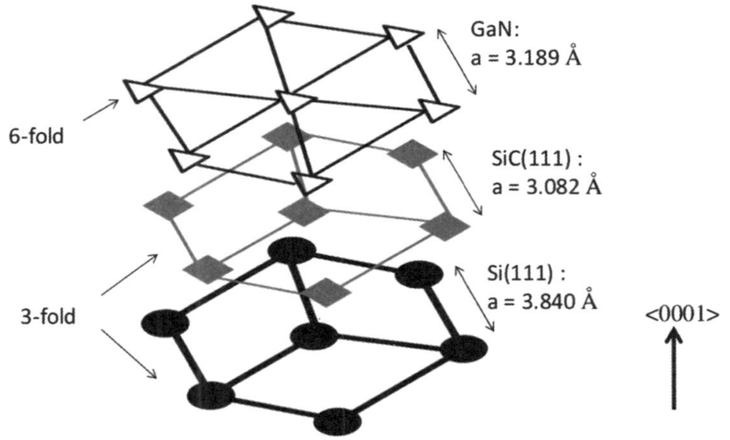

Fig. 2. Crystallographic alignment of a GaN/3C-SiC/Si(111) system.

LEDs are currently far more prevalent. High quality GaN deposited on Al_2O_3 and 4H-SiC substrates are already commercially available. However, the largest diameter, currently commercially available Al_2O_3 and 4H-SiC substrates, is 150 mm. These wafers can costs up to USD 1000's. On the other hand, 150 mm Si substrates are USD 10's, and high quality Si wafers are available up to 300 mm. Furthermore, the use of Si substrates enables direct integration of the LED structure with its control circuitry, utilization of existing Si fabs, and use of mature Si processing technologies.

The use of large area Si wafers can therefore greatly lower the cost per GaN device. In order to mitigate the issues with direct GaN-on-Si deposition as described in the next section, a thin 3C-SiC can be used as a buffer layer on Si, prior to GaN deposition.

1.2. *Minimising impact of GaN deposition onto Si via use of a 3C-SiC buffer layer*

Employing a 3C-SiC buffer layer on top of a Si substrate prior to GaN deposition has many advantages. A key issue in the early days of GaN-on-Si development was the issue of melt-back etching [Krost and Dadgar, 2002], whereby the reaction of Ga on Si resulted in Si substrate etching due to formation of Ga-rich Ga-Si alloy. Ammonia also reacts with the Si substrate to form an amorphous Si_3N_4 layer, which is detrimental for subsequent epitaxial growth. These issues have largely been eliminated using an AlN buffer layer under an initial TMAl pre-flow/pre-dose for direct GaN-on-Si deposition using MOCVD [Zhu *et al.*, 2013]. However, melt-back etching can still occur due to the presence of GaN deposits on the reactor architecture

Table 1. Comparisons between various buffer layers for deposition of GaN [Lahrèche *et al.*, 2000].

Substrate	Lattice mismatch with GaN (%)	Lattice mismatch with AlN (%)	In-plane thermal expansion coefficient ($\times 10^{-6}$ K^{-1})
Si	17	19	2.6
3C-SiC	3.9	0.95	3.9
GaN	—	2.5	5.6
AlN	2.5	—	4.2

that can outgas and subsequently react with the Si substrate in consecutive growths. There is therefore a need to clean reactor parts regularly (eg. after each run) to minimise this melt-back etching effect. With our CVD SiC-on-Si deposition process, a thin layer of 3C-SiC is simultaneously deposited on the Si substrate on both sides of the Si substrates. Such a 3C-SiC layer can completely eliminate melt-back etching on both sides of the Si substrates due to the low chemical reactivity of 3C-SiC, significantly reducing the need for additional reactor cleaning processes.

3C-SiC also has a smaller lattice mismatch than Si to both GaN and AlN [Lahrèche *et al.*, 2000], as shown in Table 1. This should allow for epilayers with a lower dislocation density and consequently higher efficiency LEDs and lifetime improvements [Uchida *et al.*, 2003]. Furthermore, 3C-SiC has a better thermal expansion coefficient match to GaN than Si, which has been shown to result in suppression of GaN film cracking [Komiyama *et al.*, 2006, Cordier *et al.*, 2008] and reduced wafer bow. Both reduced film cracking, and reduced wafer bow are critical in LED device processing stages to increase the yield, and thereby lower production costs. The ability to deposit crack-free thick GaN films is also especially critical in power device applications as the GaN film thickness determines the breakdown voltage achievable.

In addition, 3C-SiC is an effective chemical barrier due its chemical inertness. This allows the use of 3C-SiC as an etch-stop layer for post-deposition processing, removing the need for deposition of an additional etch stop layer. Moreover, the insertion of 3C-SiC between Si and GaN allows the use of 3C-SiC as a diffusion barrier, minimising potential impurity diffusion from the substrate through to the GaN overlayer.

Table 2 also shows that 3C-SiC has superior electrical and thermal properties compared to Si and Al_2O_3 substrates. 3C-SiC has far higher thermal conductivity than either Si or Al_2O_3. Consequently, this allows better heat extraction from the GaN LED to prolong its lifetime. In addition,

Table 2. Electrical and thermal properties of typical substrates used for GaN depositions [Liu and Edgar, 2002, Harris, 1995].

Substrate	Thermal conductivity (W/cm K)	Resistivity (Ω-cm)
Si	1.5	<0.001 to >30
Al_2O_3	0.23 to 0.25	$>10^{11}$
3C-SiC	3.2	0.001 to 10*

*measured on our 3C-SiC/Si(111) thin films.

the range of resistivity achievable in 3C-SiC, compared with insulating Al_2O_3, allows for the possibility of a vertically injected LED structure, compared with the conventional planar LED structure. Such a structure is preferable as it eliminates the need for expensive and time-consuming substrate removal/transfer. Substrate removal of large area wafers is also non-trivial. The ability to control the conductivity of substrate is also beneficial to increase the breakdown voltage in power devices. Hence it is desirable to retain the large area Si wafer to realise the cost savings. To minimise absorption from the Si substrate, a conductive distributed Bragg reflector (DBR) design can be used to enable vertical current injection while minimising optical losses. A DBR design for increasing the efficiency of planar devices on Si has already been reported [Mastro *et al.*, 2005].

2. 3C-SiC Deposition onto Si: Simultaneous and Alternating Supply Epitaxy

Using a LPCVD system [Wang *et al.*, 2011], highly uniform 3C-SiC thin films have been deposited on 150 mm Si(111) substrates (Fig. 3). The two main methods for deposition of SiC is using simultaneous supply epitaxy (SSE) and alternating supply epitaxy (ASE). In a SSE process, both the Si-containing and C-containing precursors are supplied at the same time throughout deposition. A SSE process typically operates at higher temperatures (>1250°C), requires a carrier gas such as H_2, and have high growth rates (several microns an hour). However, due to the high temperatures and high process pressures (>100 mbar) employed for SSE depositions, the SiC thin films typically suffers from large wafer bows and high thickness non-uniformity. The reader is directed to excellent reviews by Severino *et al.* [2012] and Ferro [2015] regarding SSE-deposited SiC on Si. While most SiC groups have deposited SSE SiC above 1250°C [Capano *et al.*, 2006, Chen *et al.*, 2010, Nagasawa *et al.*, 2006, Portail *et al.*, 2009, Severino *et al.*, 2010], we have chosen to instead study deposition at 1000°C

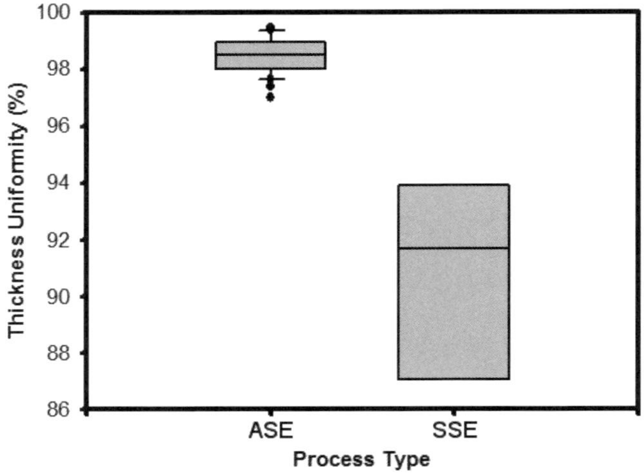

Fig. 3. High thickness uniformity of 3C-SiC deposited on 150 mm Si(111) using LPCVD at 1000°C. There was a variety of deposition conditions and 3C-SiC thicknesses, for both ASE and SSE processes. SSE processes were less studied due to the high thickness non-uniformity; hence less data points are available. The top and bottom whiskers represent the maximum and minimum data within 1.5 times of the interquartile range from the upper or lower quartiles, respectively. The top and bottom line of the box represents the upper and lower quartiles respectively. The center line of the box represents the median. The black dots represent any data outside the top and bottom whiskers.

to minimise stress due to thermal expansion coefficient mismatch and wafer bowing. In an ASE process, the Si-containing and C-containing precursors are supplied separately in alternating intervals (elaborated in the next section). Uniformity is generally poorer for the low temperature SSE process compared to the low temperature ASE process (Fig. 3). Hence the ASE process results in superior thickness uniformity, at least for the deposition temperatures of 1000°C~1200°C in an LPCVD reactor. In addition, our ASE process does not necessitate hydrogen carrier gas, which is required for typical SSE processes. This saves cost as there is no need for a hydrogen generation infrastructure.

2.1. *Alternating supply epitaxy: Typical deposition procedure*

Our 3C-SiC/Si(111) templates are deposited using a custom-built hot wall, low pressure chemical vapour deposition (LPCVD) [Wang *et al.*, 2011]. The schematic of the setup is shown in Fig. 4.

The deposition of 3C-SiC on Si(111) consists of several steps, as shown in Fig. 5. The Si(111) substrates are initially loaded into the reactor at 600°C under atmosphere without any prior chemical cleaning. After that, the reactor is pumped down to 0.001 mbar. The reactor temperature is

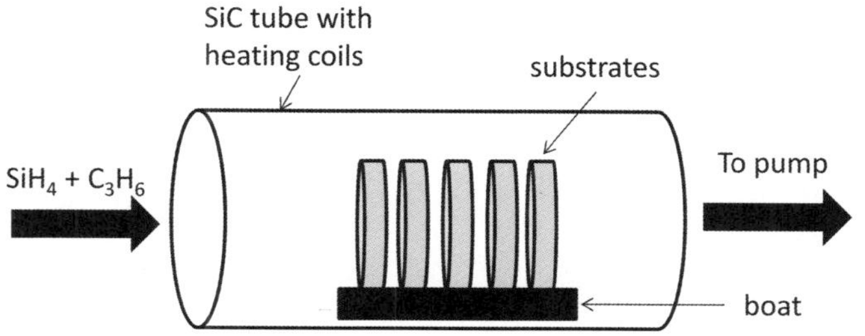

Fig. 4. Schematic of a LPCVD SiC reactor. Carbon sources can also be C_2H_2 or C_3H_8, instead of C_3H_6.

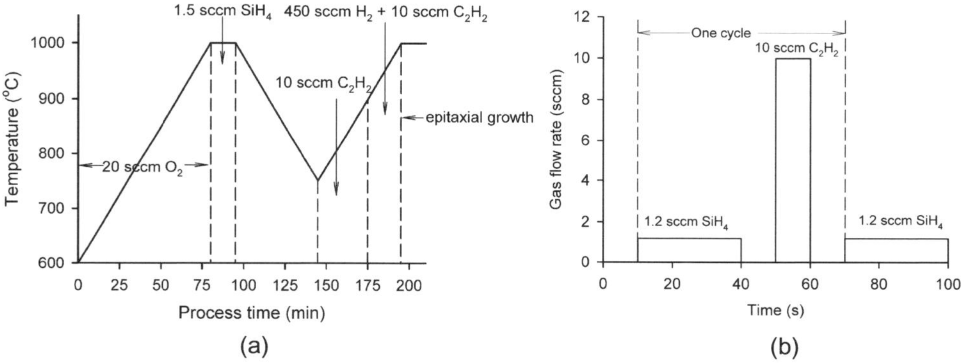

Fig. 5. Schematic diagrams of the gas supply procedure: (a) reactor temperature at various growth steps, (b) SiH_4 and C_2H_2 gas flow during each ASE cycle. Reprinted from [Wang *et al.*, 2011] with permission from Elsevier.

then ramped up from 600°C to 1000°C at a rate of 5°C/min with a flow of O_2 (20–100 sccm) in a pressure of 0.3–0.95 mbar. This oxidation step prevents carbon contamination of the Si surface via the thermal oxidation of a protective thin silicon dioxide layer. The carbon contamination may come from outgassing of the reactor walls during the temperature ramp-up. Once the reactor has reached 1000°C, 1.5 sccm SiH_4 is introduced for 15 minutes to remove the silicon dioxide and subsequently deposit a thin Si homoepitaxy layer.

After *in-situ* cleaning of the Si substrates, the reactor is cooled to 750°C in vacuum (<0.1 Pa). Then the temperature is ramped from 750°C to 1000°C at 5°C/min under 6.67 sccm C_3H_6. This carbonisation procedure has been found to be sufficient to suppress Si void formation and result in a carbonised surface with clear steps and low rms roughness (0.11 nm), as shown in Fig. 6. The preservation of step-like features after carbonisation indicates that the carbonised layer has not degraded the surface. Furthermore, XTEM

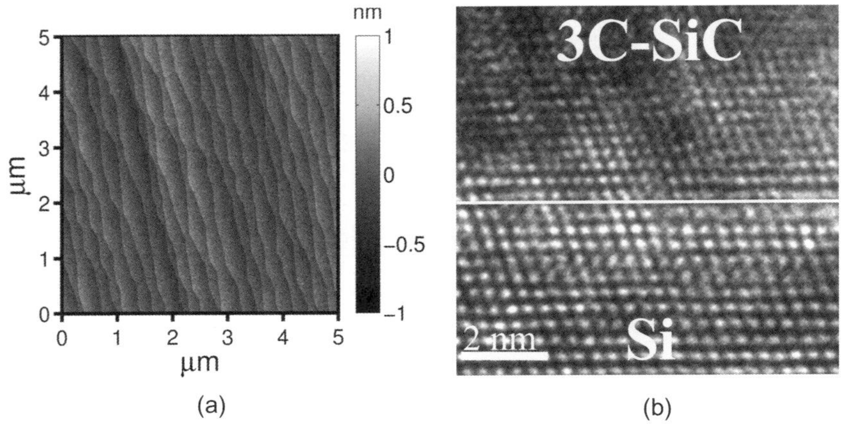

Fig. 6. (a) AFM of a carbonised (∼3 nm) on-axis Si(111) surface. (b) Cross-sectional TEM of a SiC/Si interface. Reprinted from [Wang *et al.*, 2014] with permission from Elsevier.

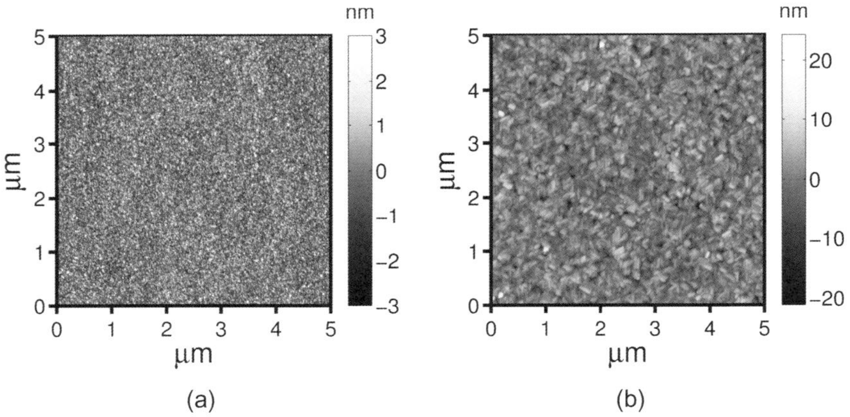

Fig. 7. $5 \times 5\,\mu$m AFM images after (a) ∼10 nm seed layer, (b) ∼250 nm epitaxial SiC layer on on-axis Si(111) substrate. Reprinted from [Wang *et al.*, 2014] with permission from Elsevier.

micrographs indicate a sharp SiC/Si interface, as well as the absence of any Si voids.

Approximately 10 nm SiC layer is then deposited on the carbonised layer via the SSE method, using simultaneous supply of 2 sccm SiH_4 and 12 sccm C_3H_6 at 1000°C. The step-like features of the carbonised layer is replaced by regular small grains in the SiC seed layer, with an rms roughness of 0.7 nm over a $5 \times 5\,\mu$m scan area (Fig. 7(a)). The conversion to granular morphology is due to the prevalence of island growth modes in high lattice mismatch heteroepitaxy, such as in this case of SiC on Si. Finally a thicker epitaxial layer (~250 nm) is deposited at 1000°C using an ASE method, with rms roughness ~3.5 nm. There are four main steps in each ASE cycle (Fig. 5(b)), depositing ~0.8 nm SiC per ASE cycle:

1. 30 s of 2.4 sccm SiH_4 with no C_3H_6 flow.
2. Pump out for 10 s with no gas flows.
3. 2 s of 6.67 sccm C_3H_6 with no SiH_4 flow.
4. Pump out for 10 s with no gas flow.

2.2. *SiC/Si colour chart*

High thickness uniformity SiC/Si (>99% across 150 mm wafer) allows for high wafer area utilization for subsequent devices. Using the previously described typical deposition process, thickness colour charts analogous to silicon oxide colour charts can also be defined [Wang *et al.*, 2013] as shown in Table 3.

3. Challenges in Depositing High Quality 3C-SiC on Si

3.1. *Si surface preparation prior to deposition*

Silicon surface preparation is extremely important for epitaxial film growth on Si substrates. The cleaning of the native oxide includes *ex-situ* and *in-situ* cleaning steps. Hydrofluoric acid is widely used as an *ex-situ* method to remove the oxide layer, but the deoxidised Si surface can be easily re-oxidized after loading the Si wafers into reactors, especially in our case where wafers are loaded under elevated temperatures (~600°C). However, even if clean Si wafers can be loaded under room temperature or under an inert purge gas (eg. N_2 purge gas), the Si wafer surface can still be contaminated by oxygen and carbon as a result of reactor furniture outgassing during the temperature ramping steps. In order to maintain the smoothness of the fresh Si surface and protect it from contamination, a protective oxide layer can be grown using chemical solutions before subsequent processing [Smith and Ghidini, 1982]. However a subsequent in-situ oxide removal process is still needed to get fresh Si surface prior to any epitaxial processes.

The *in-situ* thermal desorption of the thin silicon dioxide layer starts to occur at high temperatures (as a function of oxygen pressure) [Iraji-zad *et al.*, 2000]. However, this thermal desorption process is inhomogeneous/anisotropic, which cause roughening and faceting of the Si surface, with the effect more severe in the case of Si(111) compared to Si(100) [Iraji-zad *et al.*, 2000, Smith and Ghidini, 1982]. Hydrogen is typically

Table 3. Colour chart for SiC grown on Si substrates at magnification of 100 times when illuminated using light source colour temperature between 3200 to 3400 K. Thickness was measured using Nanometrics NanoSpec/AFT 210 and Rudolph AutoEL IV Ellipsometer using refractive index of 2.65. Reprinted from [Wang *et al.*, 2013] with permission from Trans Tech Publications Ltd.

Thickness [nm]	Image	Color Name	RGB Code [0-225]	Thickness [nm]	Image	Color Name	RGB Code [0-255]
Bare Si		Dim Gray	105-105-105	206	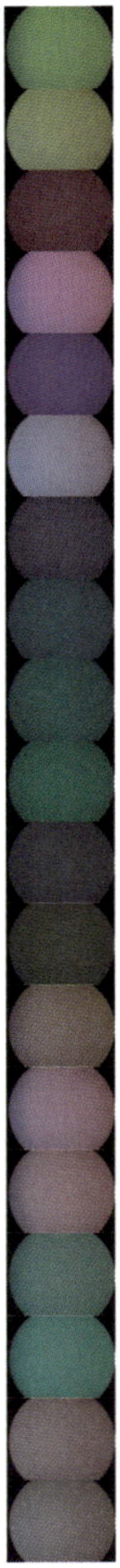	Green yellow	173-255-47
15		Dim Gray	105-105-105	218		Gold	255-215-0
40		Dark goldenrod	184-134-11	230		Brown	165-42-42
53		Blue violet	138-43-226	240		Violet	238-130-238
63		Royal blue	65-105-225	253		Blue violet	138-43-226
83		Ligh steel blue	176-196-222	264		Medium purple	147-112-219
95		Dark olive green	85-107-47	268		Light slate blue	132-112-255
110		Dark khaki	189-183-107	278		Medium aquamarine	102-205-170
120		Dark goldenrod	184-134-11	288		Medium sea green	60-179-113
127		Goldenrod	218-165-32	298		Forest green	34-139-34
132		Saddle brown	136-69-19	317		Dark olive green	85-107-47
140		Dark salmon	233-150-122	329		Olive drab	107-142-35
147		Medium orchid	186-85-211	335		Light coral	240-128-128
155		Purple	160-32-240	345		Orchid	218-112-214
159		Medium slate blue	123-104-238	357		Light pink	255-182-193
163		Slate blue	106-90-205	387		Medium sea green	60-179-113
174		Dodger blue	30-144-255	399		Medium spring green	0-250-154
188		Medium aquamarine	102-205-170	465		Rosy brown	188-143-143
196		Yellow green	154-205-50	487		Plum	221-160-221

employed to assist the removal of the oxide layer and maintain a smooth Si surface as atomic hydrogen can passivate the Si surface through formation of Si-H bonds. However no passivation can be achieved beyond 400°C as the Si-H is only stable below 400°C [Oda and Kiyota, 1996]. Nevertheless high temperature annealing under high hydrogen pressure (>100 mbar) has been effective for Si surface cleaning [Oda and Kiyota, 1996], possibly due to a balance between formation and breakage of Si-H bonds. In order for this technique to be effective, a quick ramp to high temperature is also required to ensure a clean and smooth Si surface. However, such high temperature annealing under hydrogen is not suitable when contamination sources are present, such as residual carbon in a SiC reactor. Furthermore, the fast high temperature ramp required is not practical for reactors with a slow heating rate.

In our custom-made low-pressure CVD reactor, the reactor has an upper pressure limit of ~100 Pa and a high temperature limit of 1300°C, with a targeted heating rate of 5°C/min. The reactor is also potentially contaminated by carbon due to the carbon-based precursors used in SiC depositions. A solely thermal decomposition of the native oxide will result in a rough surface, as shown in Fig. 9 [Wang *et al.*, 2014]. The surface has irregular terraces with a large density of particles pinned around the step edges, which are undesirable for subsequent epitaxial growth. Such detrimental defects occur due to the non-uniform desorption of silicon dioxide and the reaction of the Si surface with the residual carbon in the reactor.

Therefore a unique *in-situ* Si surface cleaning process has been developed to address the shortcomings of the more typical high temperature

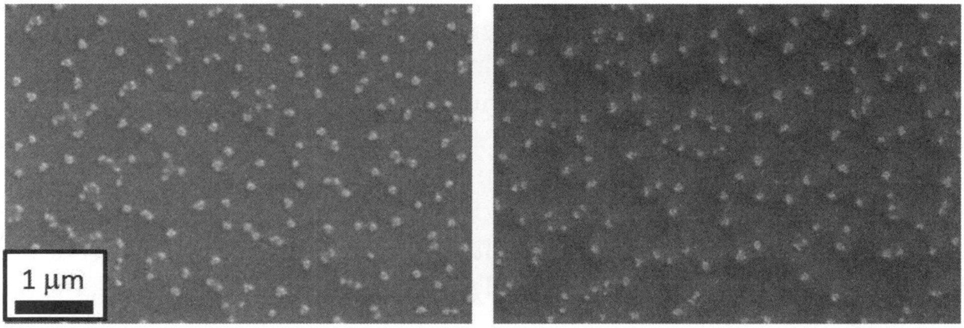

(a) (b)

Fig. 8. SEM images of (a) on- and (b) off-axis Si(111) wafers after ramping up from 600°C to 1000°C in vacuum without O_2 and SiH_4 flows [Wang *et al.*, 2014].

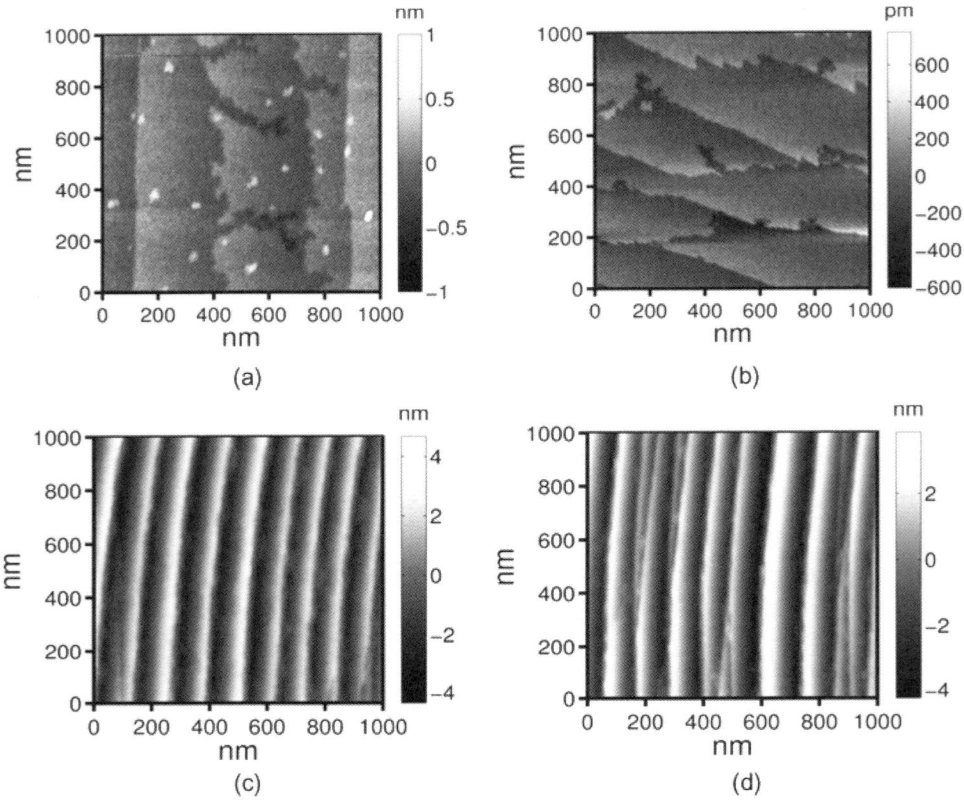

Fig. 9. Top-view AFM images of Si(111) surface after in-situ SiH_4 cleaning of (a) on-axis Si at 1000°C, (b) on-axis Si at 1200°C, (c) off-axis Si at 1000°C, and (d) off-axis Si at 1200°C. The scan area is 1 μm × 1 μm. Reprinted from [Wang *et al.*, 2014] with permission from Elsevier.

hydrogen cleaning used [Wang *et al.*, 2014]. The Si(111) substrate is initially ramped up to 1000°C under a high O_2 flow (∼95 Pa). The O_2 ramp prevents decomposition of the native oxide layer until the growth temperature is reached to prevent premature reactions between the Si surface and the residual carbon in the reactor. Upon reaching 1000°C, an isothermal bake is performed for 1.5 hours to minimise the risk of unwanted carbon-based reactions with the Si surface. After the isothermal bake, the O_2 flow is stopped, and a low pressure SiH_4 flow (1 sccm) is started. This allows the native oxide to decompose and subsequent deposition of a thin Si homoepitaxial buffer layer. Figures 9(a) and (b) show that the Si surface after deoxidation and Si deposition to have clear terraces free of particulates after cleaning at 1000°C and 1200°C for the off- and on-axis Si substrates, respectively. The rms roughness after cleaning at 1000°C for the on- and off-axis Si substrates are 0.17 nm and 1.55 nm, respectively. For the on-axis Si surface after Si cleaning and homoepitaxy deposition, the terrace width is approximately 155 to 322 nm with an average step height of 0.31 nm [Wang *et al.*, 2014], indicating that each step corresponds to double layers of Si atoms [Homma *et al.*, 1991]. The calculated off-cut angle is therefore less than 0.1°, consistent with the manufacturer's specifications. Tiny particulates are observed to evenly populate the Si surface, but these can be avoided by increasing the temperature to 1200°C, as seen in Fig. 9(b).

In contrast, the surface of an off-axis Si(111) wafer after the cleaning process (Fig. 9(c) and (d)) has much higher rms roughness (1.55 nm compared to 0.17 nm for on-axis). In addition, dense and highly parallel steps along $[\bar{1}10]/[1\bar{1}0]$ and normal to the primary wafer flat were observed. The step density is approximately twice as high as that of an on-axis wafer, with step widths between 82 and 117 nm, and step heights between 6.7 and 8.3 nm, which is more than 21 times higher compared to the step heights formed on on-axis Si wafers [Wang *et al.*, 2014]. Furthermore, particulate-free surfaces can be achieved at a lower temperature of 1000°C for off-axis Si, compared to the 1200°C required for on-axis Si.

3.2. *Buffer layer formation*

Heteroepitaxy of 3C-SiC on Si is complicated by the presence of ~20% and ~8% of lattice and thermal expansion coefficient mismatch, respectively. The first step involves deoxidation of the Si substrate through a SiH_4 cleaning process, as described previously in Section 3.1.

The second step in SiC/Si heteroepitaxy typically involves low temperature carbonisation of the Si surface into a 3C-SiC-like buffer layer using a carbon-based precursor such as acetylene (C_2H_2), propane (C_3H_8) or propylene (C_3H_6). We have chosen to use primarily C_3H_6 and C_2H_2 as our carbon precursor (because they have better reactivity at the low temperatures to be used for depositions). Unoptimised carbonisation will result in Si out-diffusion from the Si substrate, forming interface voids. The voids are triangular and square for carbonisation on (111) and (100) Si, respectively (Fig. 10). While the presence of voids may be desirable in terms of stress compliance, it can result in rougher SiC epilayer growth and therefore unsuitable as a template for subsequent nitride deposition.

The third step and final step in the buffer layer process is the deposition of a thin (~10 nm) seed layer using the SSE process (Fig. 7(a)). The SSE process is favoured over the ASE process for the seed layer to ensure an initial uniform SiC layer across the substrate. If the seed layer was

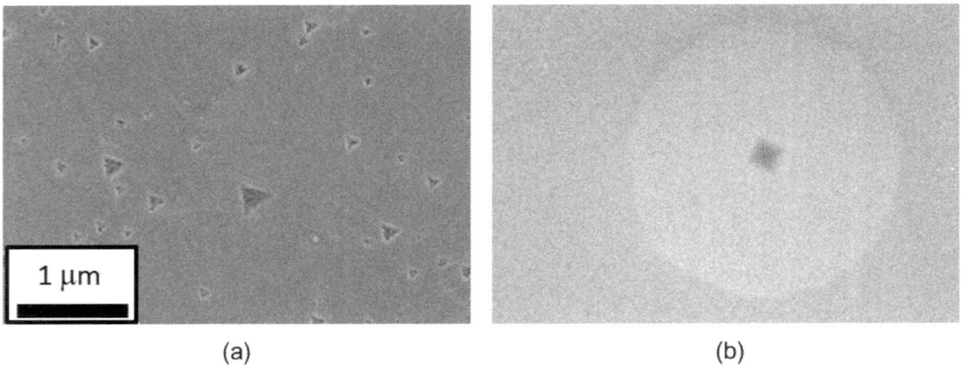

Fig. 10. SEM of Si void formation due to unoptimised carbonisation on (a) Si(111) and (b) Si(100) substrates.

deposited using the ASE process, then there can be voids formed in the seed layer, which is detrimental for subsequent deposition. The void formation mechanism is still under investigation and is postulated to be due to outdiffusion of Si atoms from the substrate during the pump out time between supply of each precursor.

3.3. *Si(111) substrate miscut dependence*

After converting the Si(111) surface into SiC by carbonisation (Fig. 11(a)), the surface of the on-axis wafer surface is quite smooth with a RMS roughness around 0.11 nm (similar to *in-situ* cleaned Si surface) and the step-like features are still visible. The presence of steps after carbonisation indicates that the chemical reaction that forms SiC during the carbonisation does not significantly change the initial surface morphology. This SiC layer thickness was measured to be around 3 nm using ellipsometry.

Figure 12(a) shows that the step-like feature is also visible for SiC deposited on off-axis Si substrates. However, unlike the on-axis case where the RMS roughness was similar before and after carbonisation, the RMS roughness after carbonisation on off-axis Si substrate is increased to 3.08 nm, which is a significant change from the RMS roughness of 1.55 nm of the cleaned Si surface. This indicates that the Si surface and thin carbonised film (~3 nm) may not be stable during the cool down process — Si atoms might easily diffuse through the thin SiC layer and cause the formation of Si voids: the higher stress of the SiC grown on off-axis substrate may contribute to the formation of Si voids, which are absent in the SiC grown on on-axis substrate with similar thickness; the mismatch in thermal expansion coefficients between Si and SiC also introduces stress during the cool down

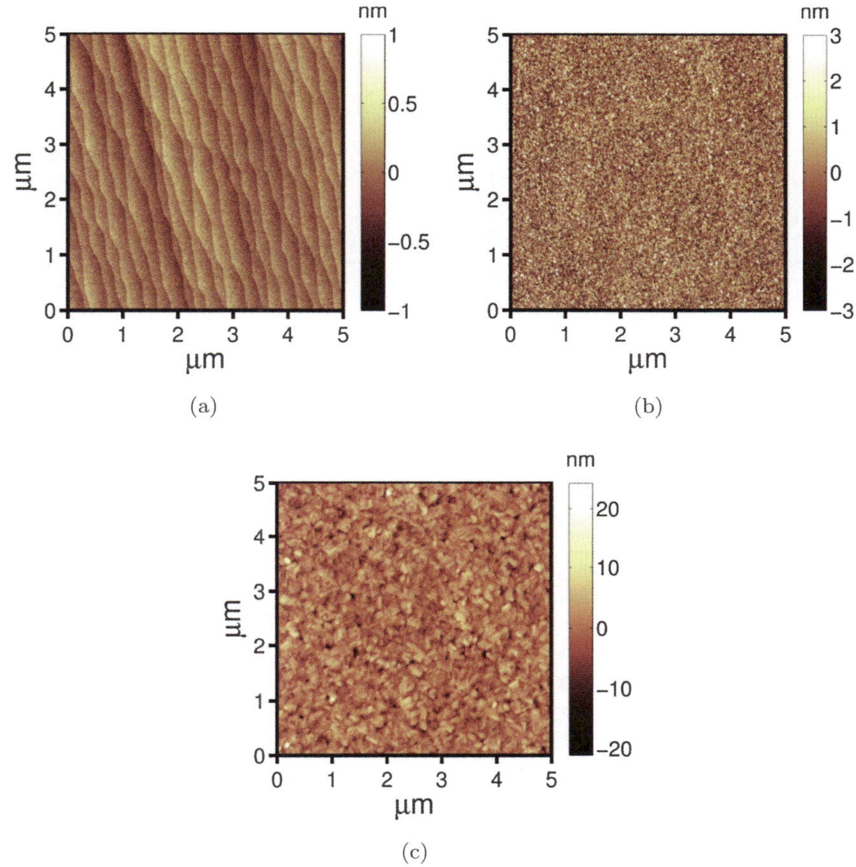

Fig. 11. Top-view surface morphology of SiC films grown on on-axis Si(111) substrate at various process stages: (a) after carbonisation (∼3 nm), (b) after 8 ± 2 nm SiC growth, and (c) after 240 ± 5 nm SiC deposition. The scan area is $5\,\mu\text{m} \times 5\,\mu\text{m}$. Reprinted from [Wang *et al*., 2014] with permission from Elsevier.

process, which can also contribute to the formation of Si voids. This gives rise to an increase in roughness that is not seen with the thicker SiC films, which could be due to formation of Si voids in the case of the carbonised film (due to the much larger step height, the carbonisation conditions may need to be specifically optimized for off-axis wafers). With growth of thicker SiC films, the film become more robust and stable during the cool down process, so they retain their roughness characteristics. From TEM evaluation of thicker SiC film grown on off-axis.Si substrate, there are no Si voids observed at the SiC/Si interface, which supports the assumption that Si voids only occurred in thin carbonised film (∼3 nm) if it is cooled down to room temperature.

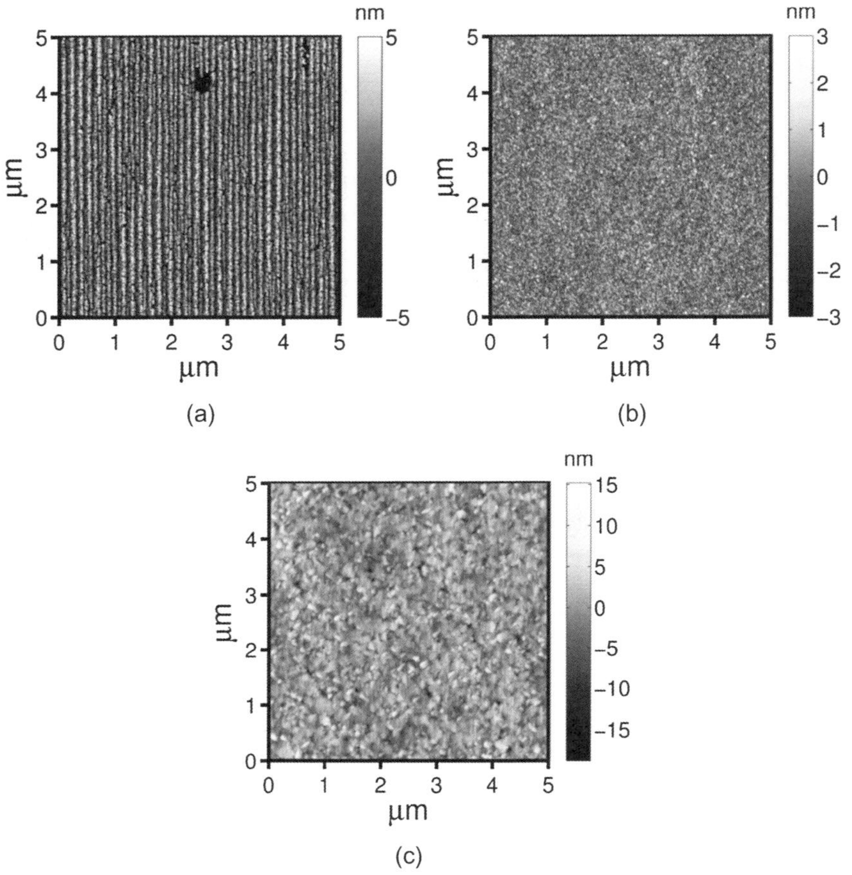

Fig. 12. Top-view surface morphology of SiC films grown on ~4.0° off-axis Si(111) substrate at various process stages: (a) after carbonisation (~3 nm), (b) after 8 ± 2 nm SiC growth, and (c) after 240 ± 5 nm SiC deposition. The scan area is $5\,\mu m \times 5\,\mu m$. Reprinted from [Wang *et al.*, 2014] with permission from Elsevier.

Due to the island growth mode of SiC at the subsequent film growth stage, the step-like features of the carbonised layer disappeared and the surface became rougher with increasing film thickness. After 8 ± 2 nm SiC deposition, the SiC surface is composed of fine grains and no clear pattern can be recognized in the AFM image in Figs. 11(b) and 12(b). No obvious difference can be observed in surface morphology for SiC film grown on on-axis and off-axis substrates. This indicates that the previously observed Si voids in the case of off-oriented substrates are either not formed or covered up by thin SiC growth. The trend of RMS roughness of Si substrates and deposited SiC film at different stages is shown in Fig. 13. The RMS roughness for the 8 ± 2 nm SiC layer is around 0.98 nm, for SiC films

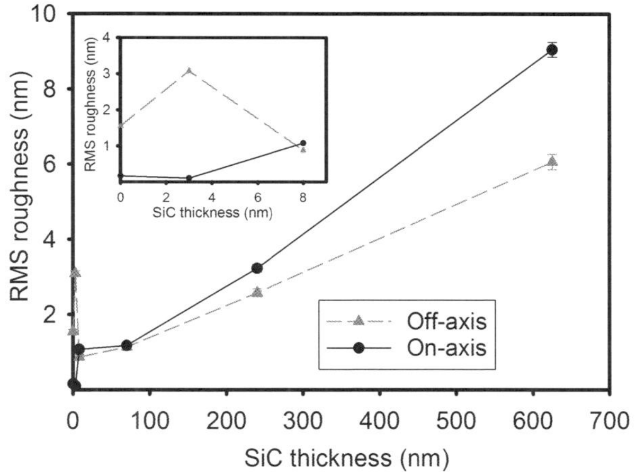

Fig. 13. RMS roughness of Si substrates and SiC films at various thicknesses. The inset is an enlargement of RMS roughness for thin SiC films (<10 nm). Reprinted from [Wang *et al.*, 2014] with permission from Elsevier.

deposited on both on- and off-axis substrates. For 70 nm SiC film, the RMS roughness data are also comparable at a value of ~1.16 nm. With the further increase in thickness, triangular features were observed on SiC with a thickness of 240 ± 5 nm when using on-axis Si substrate (Fig. 11(c)). In contrast, the SiC grown on off-axis substrate (shown in Fig. 12(c)) had a vastly different morphology; it is anisotropic, with lines predominantly developed in the directions parallel to the initial Si surface steps. The RMS roughness is slightly smaller for 240 ± 5 nm SiC grown on off-axis substrate (2.57 nm) than that on on-axis substrate (3.22 nm), even though both had comparable roughness values after 9 nm SiC deposition. The impact of substrate orientation on RMS roughness was further exemplified by increasing the deposited SiC film thickness, for a ~625 nm SiC film, the RMS roughness reached a value of 6.06 nm for SiC grown on off-axis Si, compared with a value of 9.05 nm for SiC grown on on-axis Si. It seems to indicate that our growth conditions may be better suited for SiC growth on off-axis Si substrates, where the denser steps on off-axis substrate enhanced step-flow growth mode rather than island growth. Despite the vastly different growth conditions, this observation is in agreement with the observation by Severino *et al.* [2010] that a smoother film was deposited on off-axis Si than on-axis Si, the root cause being attributed to enhanced lateral growth. Although the roughness is smaller when using off-axis substrate, the XRD evaluation (discussed in detail in Sec. 3.4) shows a significant tilting of the grown SiC layer relative to the substrate. The tilting of the grown layer might also contribute to the smaller roughness.

3.4. *Epitaxial alignment to Si(111)*

Heteroepitaxy can also result in epitaxial misalignment due to substrate miscut. The orientation and crystalline quality of the SiC films on Si(111) can be evaluated using high resolution XRD. The SiC films are confirmed to be oriented in the [111] direction using a 2θ-ω scan.

Rocking curve (RC) scans at different azimuthal angles from 0° to 360° (relative to the wafer primary flat) with an increment of 30° were performed to quantify the degree of tilt between the epilayer and the substrate. [Pesek *et al.*, 1991] The difference in omega position between Si(111) and SiC(111) peak can be used to evaluate the degree of tilt between the grown epitaxial layer and the substrate using Eq. (1):

$$(\omega_{\mathrm{SiC}} - \omega_{\mathrm{Si}})_{\mathrm{maximum}} - (\omega_{\mathrm{SiC}} - \omega_{\mathrm{Si}})_{\mathrm{minimum}} = 2\Delta\alpha_0, \quad (1)$$

In Eq. (1), ω_{SiC} is the omega angle of SiC(111) peak at certain azimuthal angle, ω_{Si} is the omega angle of Si(111) peak at the same azimuthal angle, and $\Delta\alpha_0$ is the relative titling angle of the epilayer to the substrate. Figures 14 and 15 show that the relative tilt is minimal (0.05°) for ~250 nm SiC grown on on-axis substrate. In contrast, SiC grown on 4° off-cut substrate was positively tilted/inclined 1.66° from the Si substrate, which means that the grown SiC is around 5.60° tilted from the nominal [111] direction [Wang *et al.*, 2014]. If

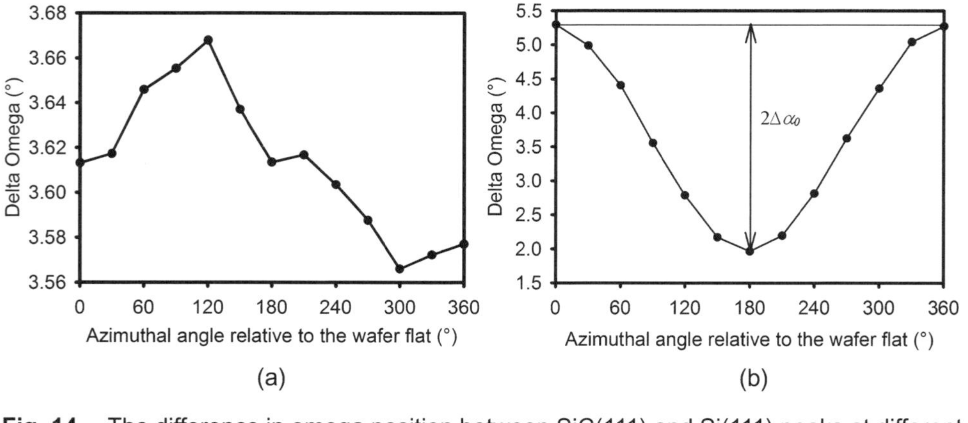

Fig. 14. The difference in omega position between SiC(111) and Si(111) peaks at different azimuthal angle for (a) on-axis Si(111) and (b) 4°off-cut Si(111). Delta omega is equal to $(\omega_{\mathrm{SiC}} - \omega_{\mathrm{Si}})$ at a certain azimuthal angle. Reprinted from [Wang *et al.*, 2014] with permission from Elsevier.

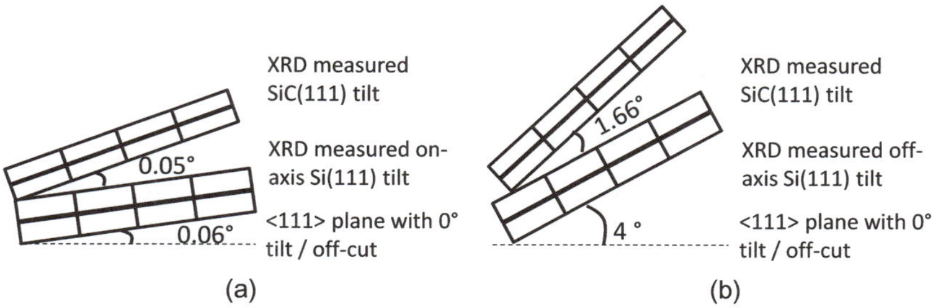

Fig. 15. Illustration of alignment between SiC and Si substrates with different off-cuts. (a) on-axis Si, (b) off-axis Si. Note, figure is not drawn to scale and is for illustration purposes only.

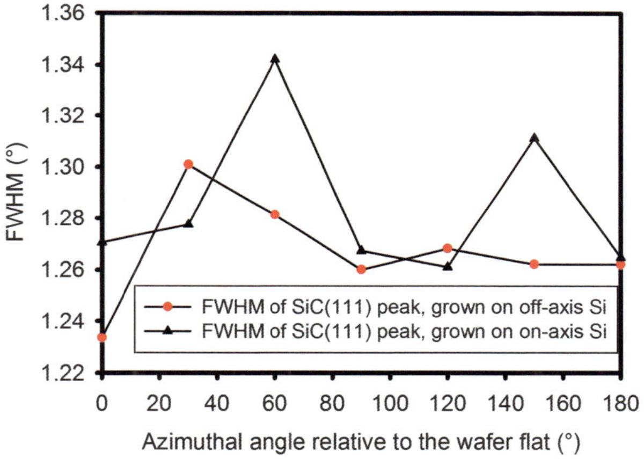

Fig. 16. FWHM of RC scans for SiC films (245 ± 5 nm) at different azimuthal angles. Reprinted from [Wang *et al.*, 2014] with permission from Elsevier.

the goal is to minimise the final tilt in the SiC film, then on-axis substrates are preferred.

While there is no clear relationship between SiC quality parallel or normal to the surface steps, the SiC FWHM is generally lower when grown on off-axis Si, compared to that grown on on-axis Si (Fig. 16).

3.5. *Effect of doping*

SiC can be doped n-type and p-type using N and Al, respectively. SiC is n-type as grown, likely due to intrinsic defects or residual impurity incorporation. A conductive SiC layer in conjunction with a conductive substrate is desirable to allow integration of a DBR structure into an LED to enable vertical current injection. This enables the use of a substrate as an electrode and therefore negates the additional costs induced by the

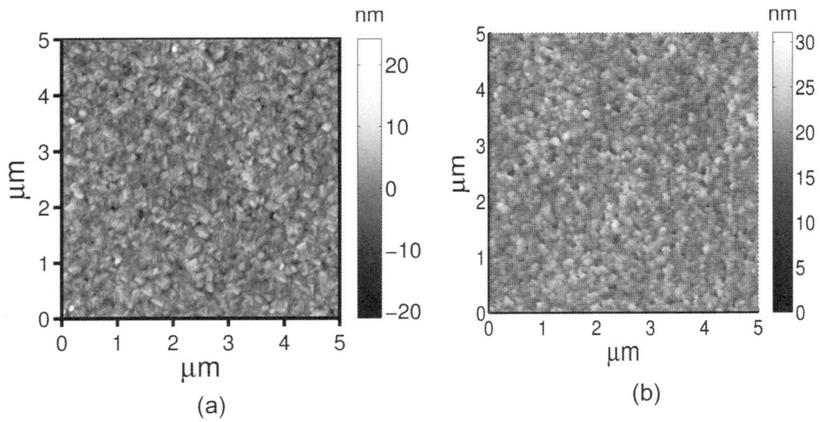

Fig. 17. AFM images of (a) undoped SiC (same as Fig. 7(b)), and (b) N-doped SiC to resistivity of ~0.001 Ω-cm.

substrate transfer processes required for current planar devices grown on Al_2O_3 substrates.

N-type doping is typically performed using nitrogen, which is normally supplied using ammonia and substitutes in C sites. The defect level is quite shallow at around 54 meV [Harris, 1995] in 3C-SiC. A high concentration of N can be incorporated to achieve resistivity in the 0.001 Ω.cm range and electron concentration in the 10^{20} cm^{-3} range, as determined by Hall effect measurements. The electrical properties of N-doped SiC degrade with N doping, with the Hall mobility reduced to ~1 cm^2/Vs at the highest N doping, whereas the Hall mobility for undoped SiC of comparative thickness (~250 nm) is usually an order of magnitude higher. The surface morphology can also degrade with N doping, especially at high levels of N incorporation (Fig. 17).

For LED applications, n-type doped SiC is preferred for used as an electrode. However, in HEMT applications, an insulating substrate, and consequently insulating template, is preferred due to the high breakdown voltage required from those devices. A p-type dopant is required in this case to counteract the naturally n-type carriers, and is typically performed using B or Al, with Al being preferred as it has a shallower defect level (160–260 meV) [Harris, 1995] than B (735 meV) [Harris, 1995]. Al substitutes for Si when incorporated as p-type dopant [Larkin, 1997]. Each cycle in the LPCVD epitaxial deposition process is as follows:

1. Supply SiH4 for 10s at 2 sccm.
2. Pump out for 30s.

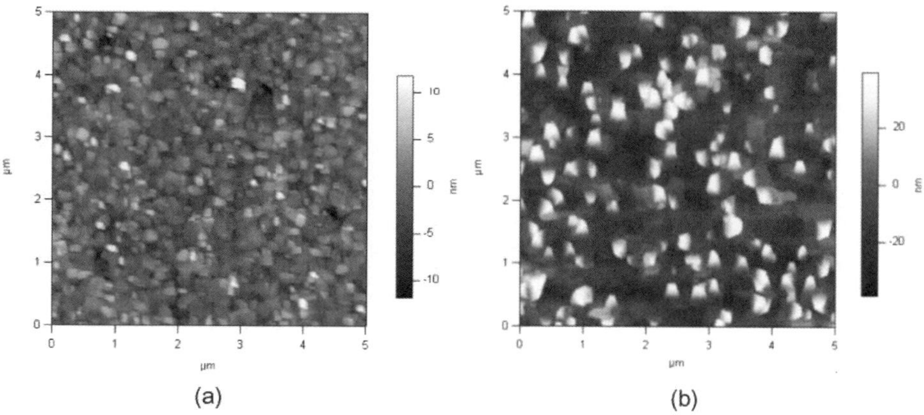

Fig. 18. AFM image of ~250 nm (a) undoped SiC, (b) Al-doped SiC grown on Si(100).

3. Supply TMAl for 10–20s at 0.1 sccm.
4. Supply 1.5 sccm C_3H_6 and 0.1 sccm TMAl for 10s.
5. Pump out for 30s.

We studied the impact of Al doping on SiC grown on Si(100). The SiC surface morphology, after doping with mid-10^{19} cm^{-3} Al atoms [Wang *et al.*, 2011], significantly degraded compared to its undoped counterpart, as shown in Fig. 18. The rms roughnesses of the SiC before and after Al doping are 2.6 nm and 8.4 nm, respectively.

$\theta - 2\theta$ scans of Al-doped SiC/Si(100) clearly show single crystalline SiC films as only the 3C-SiC(200) and 3C-SiC(400) peaks were observed (Fig. 19). Table 4 shows that the FWHM of the 3C-SiC(200) improves with increasing thickness. Interestingly, unlike many other compound semiconductors where doping degrades crystal quality, a fairly high incorporation of Al in the SiC/Si(100) does not appear to significantly degrade the crystal quality as the undoped rocking curve FWHM [Wang *et al.*, 2011] of 0.7°is similar to that measured for Al-doped SiC [Wang *et al.*, 2011] of 0.68°.

The TEM and electron diffraction pattern of the SiC/Si interface show that Al-doped SiC/Si(100) has a smooth interface devoid of etch pits and is clearly single-crystalline. The TEM shows that the defect density reduces away from the interface, consistent with improvement in the rocking curve FWHM for thicker films. This clearly demonstrates that high quality, high Al atomic concentration SiC/Si can be readily achieved.

Hot-probe and Hall effect measurements confirm that the conductivity of Al-doped SiC/Si(100) films are p-type, with a hole concentration of 1.6×10^{17} cm^{-3} at Al atomic concentration of mid-10^{19} cm^{-3}. Four-point

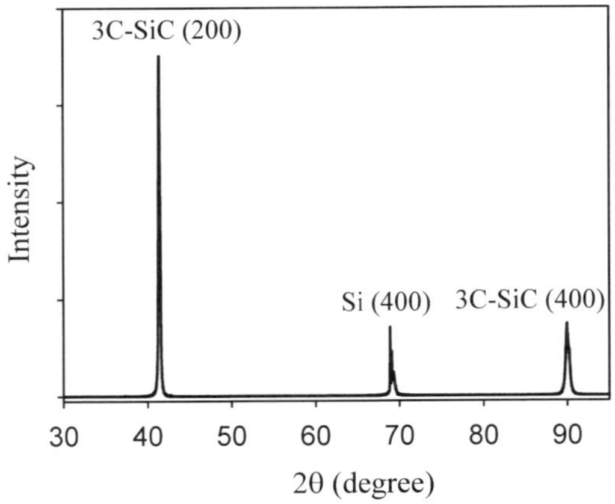

Fig. 19. XRD θ-2θ scan of 990 nm Al-doped SiC/Si(100). This film was deposited using C_2H_2 as the C-precursor, instead of C_3H_6. Reprinted from [Wang *et al.*, 2011] with permission from Elsevier.

Table 4. XRD measurements of Al-doped SiC/Si(100) grown to various thicknesses. The FWHM listed are measured for the 3C-SiC(200) peak. Al atomic concentration is in the mid-10^{19} cm^{-3} region. This film was deposited using C_2H_2 as the C-precursor, instead of C_3H_6. [Wang *et al.*, 2011].

Thickness	$\theta - 2\theta$ FWHM (°)	Rocking curve FWHM (°)
290	0.31	1.20
620	0.24	0.72
990	0.22	0.68

probe measurements yielded a resistivity of 0.31 Ω-cm. Based on the measured values of resistivity and hole concentration, the mobility is approximately 126 cm^2/Vs. This value is higher than typically reported in literature [Wagner *et al.*, 2009, Yamanaka *et al.*, 1987] and attest to the high crystal quality of the Al-doped layers.

At first glance, the low measured hole concentration at the high Al atom concentration is expected to be due to incomplete ionizations of the deep Al acceptors, which is the case for Mg in GaN depositions. However, temperature dependent conductivity measurements show that the film conductivity decreases with increasing temperature, which is inconsistent with deep acceptor behaviours. The linear relationship observed in the log-log plot in Fig. 21 implies that phonon scattering is dominant, consistent with p-type layers on Si. High compensating n-type dopants from background impurities or structural defects may also contribute to the low Al dopant activation.

(a) (b)

Fig. 20. (a) Cross-sectional TEM of Al-doped SiC/Si(100) along the [110] direction of electron incidence. The SiC film delaminated from the Si substrate during TEM sample preparation. (b) SAED of the same film as observed along the [110] electron incidence. This film was deposited using C_2H_2 as the C-precursor, instead of C_3H_6. Reprinted from [Wang *et al.*, 2011] with permission from Elsevier.

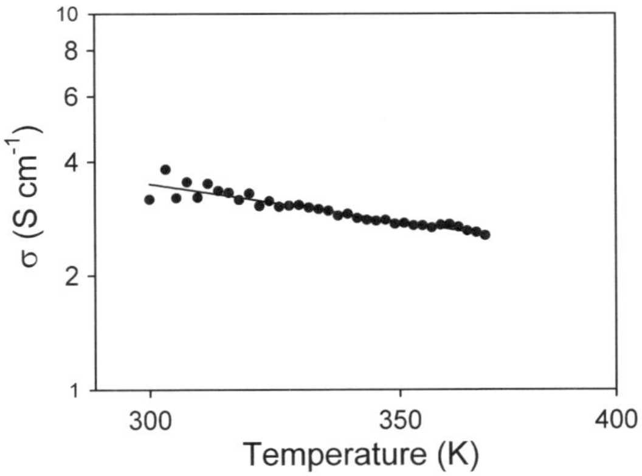

Fig. 21. Temperature dependence of conductivity of SiC/Si(100) doped with mid-10^{19} cm^{-3} Al atoms. This film was deposited using C_2H_2 as the C-precursor, instead of C_3H_6. Reprinted from [Wang *et al.*, 2011] with permission from Elsevier.

4. 3C-SiC on Si: Stress Engineering

4.1. *Orientation dependence: Si(100) vs. Si(111)*

SiC deposited on Si(100) and Si(111) has very different stress characteristics. SiC/Si(100) is expected to have an extrinsic stress of around 300 MPa based on a deposition temperature of 1000°C, which matches the experimentally measured stress as shown in Fig. 22. However, the SiC/Si(111) residual

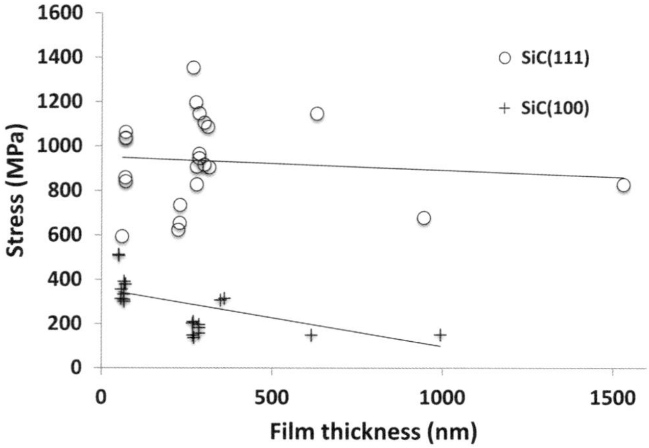

Fig. 22. Residual stress from wafer curvature measurements for SiC on Si(100) and Si(111) substrates. Reprinted with permission from [Iacopi *et al.*, 2013]. Copyright [2013], AIP Publishing LLC.

stress is significantly higher, at around four times that of SiC/Si(100). The thermal and lattice mismatches are expected to be similar and hence the differences arises from the differences in relaxation mechanisms between the two orientations. Furthermore, the stress in SiC/Si(100) films decreases much faster with increasing thicknesses, compared to SiC/Si(111) films.

TEM micrographs in Fig. 23(a) show that the stacking faults (SFs) decrease away from the SiC/Si(100) interface. The SF inclination is around 55° from (100), corresponding to the angle between the {100} and {111} planes. In contrast, Fig. 23(b) show that the SFs do not appear to decrease much away from the SiC/Si(111) interface. The density of SFs for the SiC(111) film also appear to be much higher than for a SiC(100) film. The SF inclination is ~70° from (111), corresponding to the angle between {111} and {−111} planes. As such, SF annihilation via two intersecting SF are less likely to occur in SiC(111) than for SiC(100). In addition, the SFs in SiC(100) is approximately 40% more efficient at relieving residual biaxial stress than in SiC(111). Consequently, a thicker or higher SF density SiC(111) film will be required to achieve the same strain relaxation as a SiC(100) film. [Iacopi *et al.*, 2013]

4.2. *Process dependence*

By varying the carbonisation temperature or the growth temperature, the residual stress in the SiC(111) films will also vary (Fig. 24). Higher carbonisation and higher growth temperatures both lead to higher residual stress. [Iacopi *et al.*, 2013].

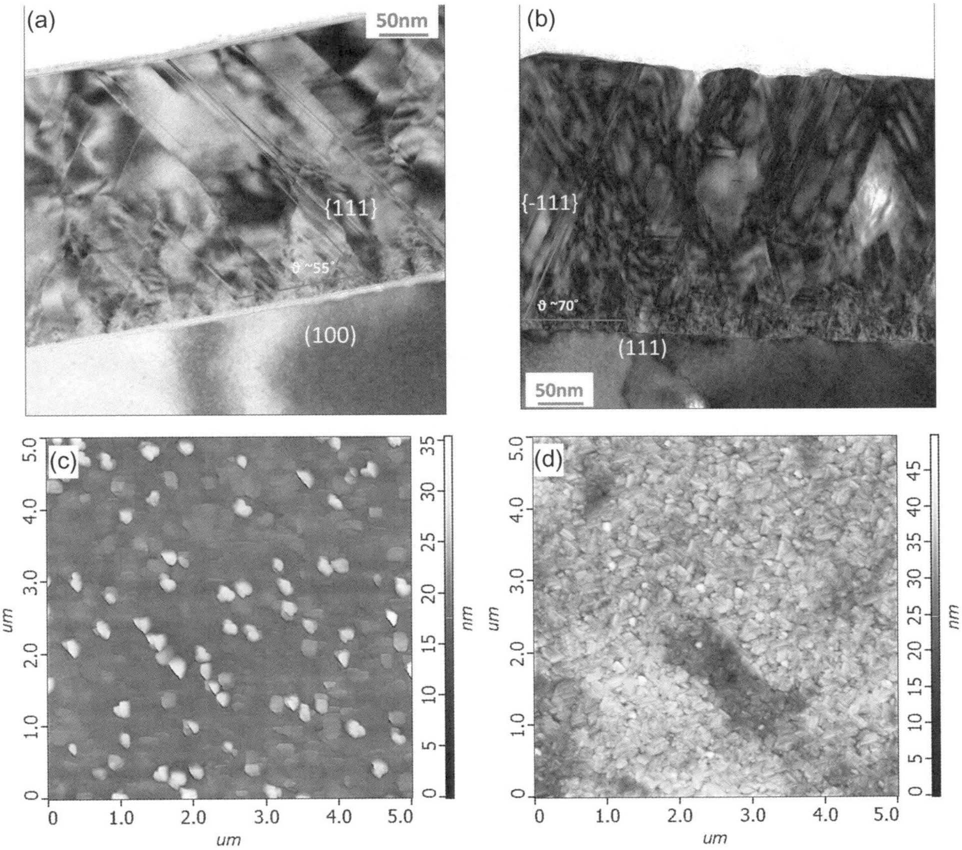

Fig. 23. Cross-sectional TEM micrographs of 300 nm SiC on (a) Si(100) and (b) Si(111). AFM images in (c) and (d) show the corresponding surface images of (a) and (b), respectively. Reprinted with permission from [Iacopi *et al.*, 2013]. Copyright [2013], AIP Publishing LLC.

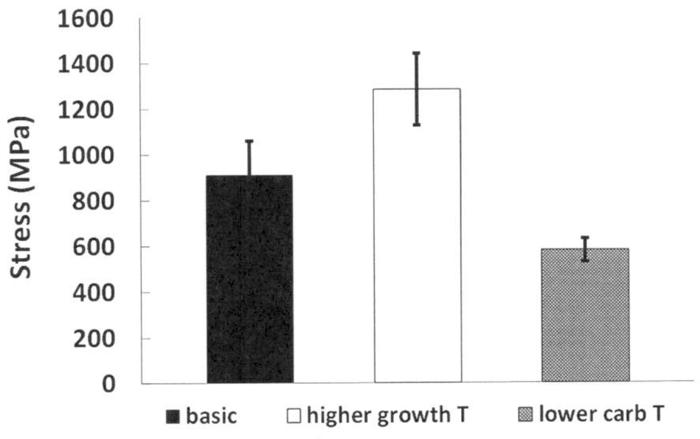

Fig. 24. Average stresses of 300 nm thick SiC(111) grown with a baseline process, higher growth temperature (1050°C), and lower carbonisation temperature (750°C). Reprinted with permission from [Iacopi *et al.*, 2013]. Copyright [2013], AIP Publishing LLC.

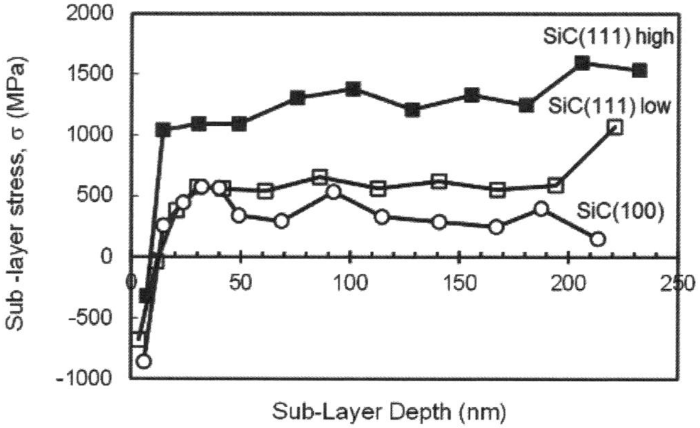

Fig. 25. Residual stress depth profile of SiC on Si films from wafer curvature measurements. Reprinted from [Iacopi *et al.*, 2013] with permission from Elsevier.

Table 5. Thickness and total residual stress of the epitaxial films used for the detailed stress analysis shown in Fig. 26. Reprinted from [Iacopi *et al.*, 2013] with permission from Elsevier.

Film	Thickness (nm)	Total film stress (MPa)
SiC(100)	271 ± 1	270 ± 15
SiC(111) low stress	234 ± 1	500 ± 25
SiC(111) high stress	245 ± 1	1300 ± 65

It should also be noted that the residual stress of a SiC/Si film is non-uniform across its thickness. As shown in Fig. 25, all three types of films are under highly compressive states below ~10 nm of the SiC/Si interface. Above 10 nm, the high stress SiC(111) rapidly reach high (~1000 MPa) tensile stress, whereas the low stress SiC(111) stabilises at ~500 MPa after about 40 nm. In contrast, the SiC(100) gradually decrease in stress, and end up at approximately 150 MPa tensile stress at the surface. [Iacopi *et al.*, 2013] The differences in stress behaviour is correlated with the reduction rate of the stacking faults for the different orientations, as previously discussed. [Iacopi *et al.*, 2013]

4.3. *Si(111) substrate miscut dependence*

As discussed in Sec. 3.4, SiC(111) films grown on on-axis and off-axis Si(111) substrates have different overall tilt and also different surface morphology. The residual stress, measured using a wafer curvature measurement system at different azimuth angles, is shown in Fig. 26. The stress is fairly uniform

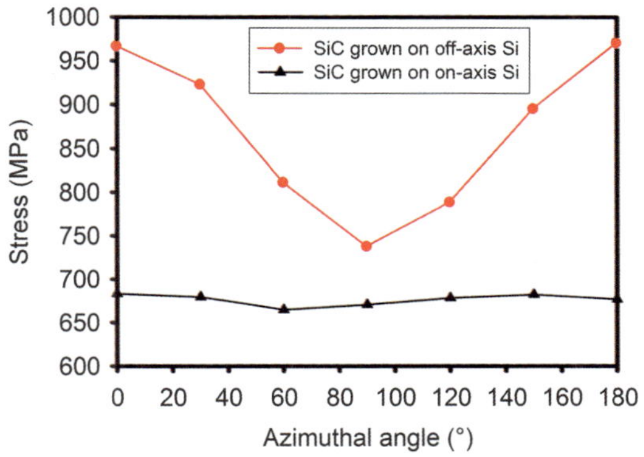

Fig. 26. Residual stress distribution in ~240 nm SiC films grown on on-axis and off-axis Si(111) substrates. 0° corresponds to the wafer flat. Reprinted from [Wang *et al.*, 2014] with permission from Elsevier.

for SiC(111) grown on on-axis Si(111), with an average tensile stress of 676 MPa. However, there is a clear difference in the SiC(111) stress at different azimuth angles. At azimuth angle of 0°, the curvature scan direction is normal to the steps. Whereas at azimuth angle of 90°, the curvature scan direction is parallel to the steps. Hence Fig. 26 shows that residual stress is influenced by the surface steps. The step density for SiC deposited on off-axis Si(111) is nearly doubled compared to SiC deposited on on-axis Si(111). Due to the lattice mismatch between SiC and Si, the stress normal to the growth direction has to be accommodated by epilayer deformation. Hence denser steps will accumulate more stress, as observed in the case of SiC on off-axis Si(111).

5. Demonstration of Blue LED on 3C-SiC/Si(111)

An LED structure consisting of 20 nm AlN and 2.7 μm GaN with five InGaN/GaN quantum wells have been deposited on a 250 nm SiC/Si(111) [Han *et al.*, 2011]. High resolution rocking curve of the GaN(0002) peak gives a FWHM of 730 arcsec. Figures 27(a) and (b) show that the thin SiC layer on the Si(111) substrate is quite defective, with a high density of stacking faults. While visible defects are observed in the n-GaN layer, the density is far less than for the SiC layer. Crucially, Figs. 27(b) and (c) clearly show that the stacking faults in SiC layer does not appear to propagate to the subsequent AlN layer. Some of the n-GaN defects instead appear to originate from the AlN/n-GaN interface, with the origin of other n-GaN defects being less clear. Upon deposition of contacts on the structure and applying current, emission of blue light (wavelength peak at 453 ± 24 nm) was observed. This

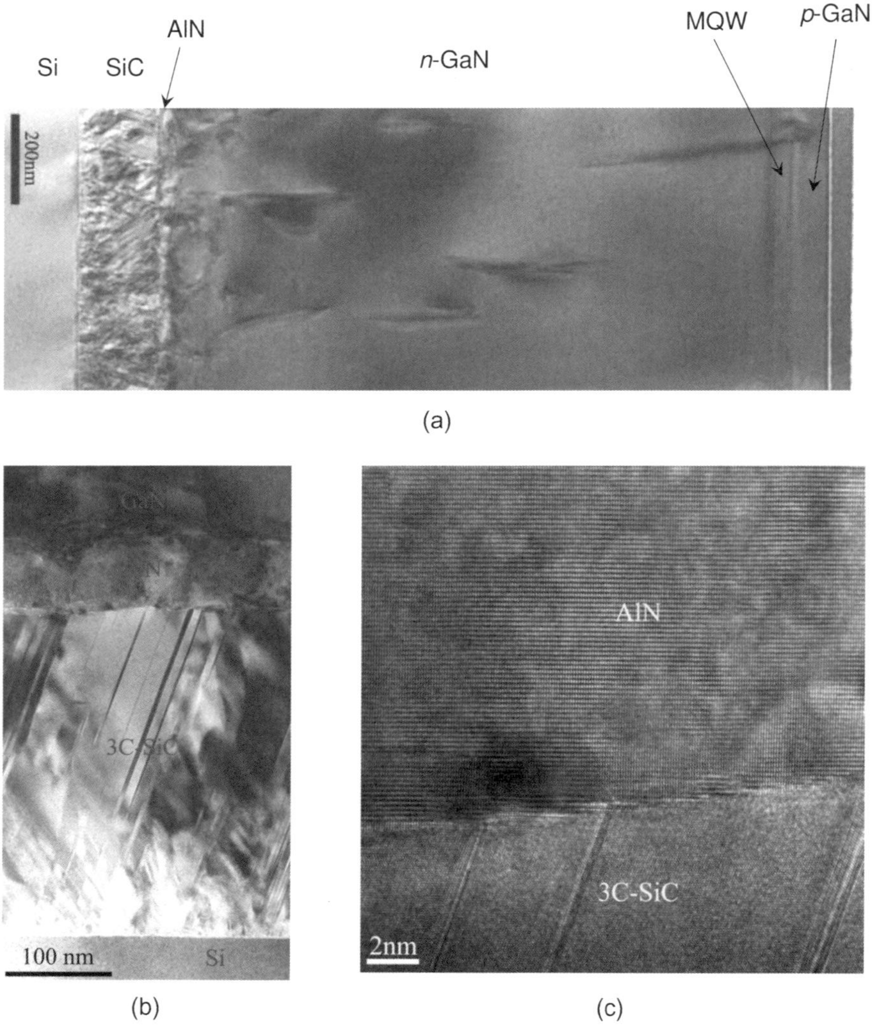

Fig. 27. Cross-section TEM of a blue LED structure deposited on a SiC/Si(111) template showing (a) the full structure, (b) the SiC/AlN/GaN layers, (c) high resolution micrograph of the SiC/AlN interface. Reprinted from [Han *et al.*, 2011] with permission from Trans Tech Publications Ltd.

clearly demonstrates the capability of using SiC/Si templates for blue LED deposition.

6. Summary

Thin 3C-SiC templates on Si substrates are a promising technology for GaN deposition, in particular for LED applications. The use of these

SiC/Si templates prevents melt-back etching and provides better lattice and thermal match for subsequent GaN deposition in comparison to bare Si substrates. The usage of Si substrates also enables monolithic integration of the optical device with its control circuitry. Conductive SiC/Si templates can be used for vertical current injection while avoiding substrate removal processes that are used in current technologies. Highly uniform, large area SiC/Si templates have been demonstrated.

A typical low temperature CVD SiC-on-Si process involves several basic stages: Si wafer cleaning, low temperature carbonisation, seed layer deposition, and then epilayer deposition. As with any large lattice mismatch heteroepitaxy, SiC-on-Si is no different. Inadequate Si cleaning will result in void formation at the interface, instead of smooth, step-like Si surfaces. Proper optimization of the low temperature carbonisation is also required to avoid interface void formation. Substrate miscut angle also affects the surface morphology. SiC/Si templates deposited on on-axis substrates typically have higher rms roughness than those deposited on off-axis substrates. However epilayers deposited on off-axis substrates have a higher relative tilt and a higher residual stress than those deposited on on-axis substrates. The higher stress for off-axis templates is attributed to the denser steps on off-axis substrates.

In summary, various process conditions that affect various properties of a SiC layer deposited on Si have been presented in this chapter.

References

Capano, M. A., Kim, B. C., Smith, A. R., Kvam, E. P., Tsoi, S. and Ramdas, A. K. (2006), *J. Appl. Phys.*, **100**(8) p. 083514.

Chen, W.-Y., Wang, W.-L., Liu, J.-M., Chen, C.-C., Hwang, J.-C., Huang, C.-F. and Chang, L. (2010), *J. Electrochem. Soc.*, **157**(3) p. H377.

Cordier, Y., Portail, M., Chenot, S., Tottereau, O., Zielinski, M. and Chassagne, T. (2008), *J. Cryst. Growth*, **310**(20) p. 4417.

Dadgar, A., Schulze, F., Wienecke, M., Gadanecz, A., Bläsing, J., Veit, P., Hempel, T., Diez, A., Christen, J. and Krost, A. (2007), *New J. Phys.*, **9**(10) p. 389.

Ferro, G. (2015), *Crit. Rev. Solid State Mater. Sci.*, **40**(1) p. 56.

Han, J. S., Dimitrjiev, S., Wang, L., Iacopi, A., Shuang, Q. and Xu, X. G. (2011), 8th European Conference on Silicon Carbide and Related Materials (ECSCRM 2010), Material Science Forum, p. 801.

Harris, G. L. (1995). *Properties of Silicon Carbide* (INSPEC, the Institution of Electrical Engineers, United Kingdom).

Homma, Y., Tomita, M. and Hayashi, T. (1991), *Surf. Sci.*, **258**(1–3) p. 147.

Iacopi, F., Brock, R. E., Iacopi, A., Hold, L. and Dauskardt, R. H. (2013), *Acta Mater.*, **61**(17) p. 6533.

Iacopi, F., Walker, G., Wang, L., Malesys, L., Ma, S., Cunning, B. V. and Iacopi, A. (2013), *Appl. Phys. Lett.*, **102**(1) p. 011908.

Iraji-Zad, A., Taghavinia, N., Ahadian, M. and Mashaei, A. (2000), *Semicond. Sci. Technol.*, **15**(2) p. 160.

Iwami, M. (2001), *Nucl. Instrum. Methods Phys. Res. Sect. A-Accel. Spectrom. Dect. Assoc. Equip.*, **466**(2) p. 406.

Komiyama, J., Abe, Y., Suzuki, S. and Nakanishi, H. (2006), *J. Appl. Phys.*, **100**(3) p. 033519.

Komiyama, J., Abe, Y., Suzuki, S. and Nakanishi, H. (2006), *Appl. Phys. Lett.*, **88**(9) p. 091901.

Krost, A. and Dadgar, A. (2002), *Mater. Sci. Eng., B*, **93**(1–3) p. 77.

Lahrèche, H., Vennégùes, P., Tottereau, O., Laügt, M., Lorenzini, P., Leroux, M., Beaumont, B. and Gibart, P. (2000), *J. Cryst. Growth*, **217**(1–2) p. 13.

Larkin, D. J. (1997), *Physica Status Solidi (b)*, **202**(1) p. 305.

Liu, L. and Edgar, J. H. (2002), *Mat. Sci. Eng. R*, **37**(3) p. 61.

Mastro, M. A., Holm, R. T., Bassim, N. D., Eddy, C. R., Gaskill, D. K., Henry, R. L. and Twigg, M. E. (2005), *Appl. Phys. Lett.*, **87**(24).

Nagasawa, H., Yagi, K., Kawahara, T., Hatta, N. and Abe, M. (2006), *Microelectron. Eng.*, **83**(1) p. 185.

Oda, K. and Kiyota, Y. (1996), *J. Electrochem. Soc.*, **143**(7) p. 2361.

Pesek, A., Hingerl, K., Riesz, F. and Lischka, K. (1991), *Semicond. Sci. Technol.*, **6**(7) p. 705.

Portail, M., Zielinski, M., Chassagne, T., Roy, S. and Nemoz, M. (2009), *J. Appl. Phys.*, **105**(8) p. 083505.

Severino, A., Bongiorno, C., Piluso, N., Italia, M., Camarda, M., Mauceri, M., Condorelli, G., Di Stefano, M. A., Cafra, B., La Magna, A. and La Via, F. (2010), *Thin Solid Films*, **518** p. S165.

Severino, A. (2012) *Silicon Carbide Epitaxy*, ed. Via, F. L, Chapter 7 "3C-SiC epitaxial growth on large area silicon: thin films," (Research Signpost, India) p. 145.

Smith, F. and Ghidini, G. (1982), *J. Electrochem. Soc.*, **129**(6) p. 1300.

Uchida, S., Takeya, M., Ikeda, S., Mizuno, T., Fujimoto, T., Matsumoto, O., Goto, S., Tojyo, T. and Ikeda, M. (2003), *IEEE J. Sel. Top. Quantum Electron.*, **9**(5) p. 1252.

Wagner, G., Schmidbauer, M., Irmscher, K., Tanner, P. and Fornari, R. (2009) *Mater. Sci. Forum*, p. 165.

Wang, L., Dimitrijev, S., Han, J. S., Iacopi, A., Hold, L., Tanner, P. and Harrison, H. B. (2011), *Thin Solid Films*, **519**(19) p. 6443.

Wang, L., Dimitrijev, S., Han, J. S., Tanner, P., Iacopi, A. and Hold, L. (2011), *J. Cryst. Growth*, **329**(1) p. 67.

Wang, L., Dimitrijev, S., Iacopi, A., Hold, L., Walker, G., Chai, J. and Massoubre, D. (2014) European Conference on Silicon Carbide and Related Materials (ECSCRM 2014), Materials Science Forum, *in press*.

Wang, L., Dimitrijev, S., Walker, G., Han, J. S., Iacopi, A., Tanner, P., Hold, L., Zhao, Y. and Iacopi, F. (2013) 9th European Conference on Silicon Carbide and Related Materials (ECSCRM 2012), Materials Science Forum, p. 279.

Wang, L., Iacopi, A., Dimitrijev, S., Walker, G., Fernandes, A., Hold, L. and Chai, J. (2014), *Thin Solid Films*, **564** p. 39.
Yamanaka, M., Daimon, H., Sakuma, E., Misawa, S. and Yoshida, S. (1987), *J. Appl. Phys.*, **61**(2) p. 599.
Zhu, D., Wallis, D. J. and Humphreys, C. J. (2013), *Rep. Prog. Phys.*, **76**(10) p. 106501.

Chapter 3

Epitaxy of GaN on Silicon

Yu Cao
HRL Laboratories, CA

Oleg Laboutin and Wayne Johnson
IQE LC LLC, Taunton, MA, USA

Satyaki Ganguly
Intel, OR

Grace (Huili) Xing and Debdeep Jena
Cornel University, NY

1. Introduction

This chapter covers the epitaxy of GaN semiconductor heterostructures on Silicon substrates from an applications perspective. In section 2, the major challenges encountered in the epitaxy of GaN on Si are discussed. These include the choice of the substrate type and orientation, nucleation, the various effects of lattice and thermal expansion mismatch during and after epitaxy, and practical issues such as system maintenance, cleanliness, reproducibility, and throughput. Section 3 provides details of growth by MOCVD, emphasizing the importance of proper system design and *in-situ* tools for monitoring and controlling wafer bow and curvature during and after epitaxy. In section 4, the MBE growth technique is discussed. In Section 5 GaN based devices on Silicon and their CMOS compatibility is discussed. Finally Section 6 includes a commentary on the future perspectives on epitaxy of GaN on Silicon.

2. Challenges in Epitaxy of GaN on Silicon

2.1. *Substrate choices*

There are few suitable substrates for GaN epitaxial growth. Though the cost of native bulk GaN substrates is steadily decreasing, it remains cost-prohibitive for high throughput applications compared to several other choices. The commercially available wafer size is also limited to 2-inch diameter, which is not cost-effective for industrial production. Sapphire and SiC are currently the most commonly used substrates. Mature growth techniques have been developed on both of them. SiC substrates offer excellent thermal conductivity and small lattice mismatch to GaN (~3.4%). This makes SiC ideal for both GaN electronic and optoelectronic devices that require efficient heat dissipation. However, the high cost of SiC substrates and limited substrate size limit its cost-effectiveness for mass production of GaN devices. Sapphire has a larger lattice mismatch to GaN (13%). Large size sapphire substrates have been commercialized with 8-inch diameter already available for research and development. Due to its optical transparency and low cost, high volumes of sapphire substrates have been widely used in the LED market. However, the poor thermal conductivity (~0.4 W/cm-K), which is more than one order of magnitude lower than nitride materials, limits sapphire from widespread use in RF and power devices.

Silicon, as the most studied semiconductor material, has the lowest cost among all suitable substrates for nitride growth. Unlike most CMOS devices, the hexagonal crystal structure of the nitrides requires silicon substrates of (111) orientation. Depending on the substrate resistivity requirements, silicon substrates are produced by different methods such as float-zone (FZ), Czochralski (CZ), magnetic Czochralski (MCZ) process, etc. FZ silicon is very pure, and resistivities more than 10,000 ohm-cm can be obtained in single crystals. The largest commercially available FZ silicon wafers have a diameter of 200 mm. However, the price of this substrate is higher than the same size of sapphire wafers. CZ silicon can be very conductive using intentional boron or phosphorus doping. Arsenic or antimony can also be used as the dopants for CZ silicon. CZ substrates are inexpensive and support large size with the (111) substrate diameter up to 450 mm. This greatly helps to reduce the device cost in industrial production. The resistivity and cost of MCZ silicon is between what is achievable from FZ and CZ. The thermal conductivity of silicon is 1.3 W/cm-K, which is about the same as GaN. Thus, silicon substrate is not a thermal bottleneck for several applications. More attractively, growing nitride layers on silicon makes it possible for the wafers to be processed in commercial silicon foundries. The option of removing silicon substrates by widely used chemical-mechanical polishing (CMP) and wet etching makes GaN on silicon promising for sensor applications, high breakdown power devices, as well as for devices with other advanced architectures.

2.2. *Epitaxy challenges*

The benefits of GaN MOCVD technology employing Si substrates can be reaped if certain challenges in its implementation are appropriately addressed. One of the difficulties of MOCVD of nitrides on silicon is caused by high growth temperatures commonly used in this technique. Silicon has a relatively low thermal and chemical stability which makes it vulnerable to parasitic chemical reactions, intermixing with nitrides and melt-back etching.[1] To mitigate such undesirable effects, AlN is widely used as the nucleation layer. However aluminum is a substitutional shallow p-type dopant in silicon. It can create a conductive layer at the AlN/silicon interface. If high resistivity silicon like FZ Si is selected as the substrate for GaN HEMTs, this parasitic path could lower the breakdown voltage and introduce switching loss during transistor operation.

Multiple challenges exist for crystal growers who want to produce high quality nitride layers on silicon substrates. First of all the lattice constant mismatch between the c-plane GaN and the (111)-plane Si is high: 16.9%.[2] This indicates that large amount of misfits will exist at the epitaxy interface if GaN is directly grown on Si. A high density of threading dislocations is inevitably formed in the nucleation layer and most of them extend into the subsequent buffer layers. These defects degrade the nitride film quality and can cause charge trapping or buffer leakage in electronic devices, or low internal quantum efficiency in optoelectronic devices.

Thermal expansion mismatch is another major challenge for GaN on silicon epitaxy. The difference of the thermal expansion coefficients between GaN and Si is about 57%.[2] This creates significant challenges for epitaxy development and production. During the growth at a constant temperature, the bulk nitride film is free of thermal stress. However when the wafers cool down at the end of growth, the GaN layer contracts more in the c-plane compared to silicon due to its larger thermal expansion coefficient. This results in concave shaped wafers. If a wafer bows too much, cracks are generated at the edge of the wafer to release excess stress during the cool-down process.[3] Wafers with large bowing also exhibit high warp values defined as the difference between the maximum and minimum distances of the median surface of a free and unclamped wafer from a reference plane.[4] High wafer bow and warp indicate a large amount of residual stress, which makes the wafer susceptible to breaking during device processing, thus lowering the production yield. Bow and warp introduce particular difficulties during lithography steps, when uniform focus depth is desired across the wafer.

Melt back etching is another issue when GaN is grown directly on silicon due to the Ga-Si eutectic reaction at the growth interface.[5] The GaN layer in the area where the reaction occurs cannot coalesce well and results in holes in the film surface after growth. To prevent such eutectic formation, AlN is widely used as the nucleation layer. However, aluminum is a p-type dopant in silicon. As mentioned above, it creates a hole-rich conductive layer at the AlN/Si interface on the Si side. Such parasitic conductive layer is harmful for electronic devices. It can increase switching loss during device operation, and introduce other instabilities in circuits.

Single-crystal bulk silicon is known to experience a brittle-to-ductile transition (BDT) at a temperature of $\sim$550°C,[6] which is below nitride growth temperatures by either MOCVD or MBE. Thus silicon substrates are compliant, and inclined to bow if any external force is applied. Using AlN as the nucleation layer, compressive stress is always introduced in any Al(Ga)N or GaN layers grown on top due to the lattice mismatch. In return, stress with same strength forces the silicon substrate to a convex bow during growth. Such bow is the source of difficulty of achieving uniform wafer temperature across large size wafers. The center of a convex wafer is always cooler than the edge. Such temperature variation results in differences in the growth rate, ternary or quaternary alloy composition, and stress distribution across the wafer. This affects the sheet resistance in HEMTs, or wavelength uniformity in LEDs. When silicon is ductile at high temperatures, plastic deformation can occur if the substrate bows too much and the internal stress reaches the yield point. Once plastic deformation occurs, instead of becoming more concave driven by thermal mismatch during the cooling process, the wafer will remain convex even after cool-down. Such high internal stress in substrates formed during growth can even lead to wafer breakage inside the reactor after the plastic deformation occurs.

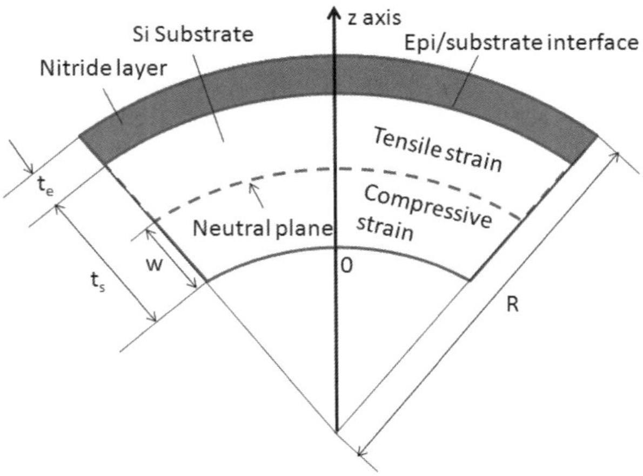

Fig. 1. Schematic of strained nitride-on-silicon heterointerface.

Based on an analysis of wafer curvature in heterostructures,[7] Figure 1 represents a generalized model of a strained nitride/Si interface. From this model, it can be shown that the strain at any location in the Si substrate is given by $\varepsilon = \frac{t_s}{2R-t_s} \sim \frac{t_s}{2R} = \frac{t_s k}{2}$ with wafer curvature defined as $k = 1/R$. Thus the strain in the silicon substrate is correlated with the substrate thickness and the curvature.

As the stress increases, the deformation becomes plastic. Then Hollomon's equation can be used to express the plastic stress $\sigma = K\varepsilon^n$, where K is the material strength index and n is the strain hardening exponent. At high temperature, n is a function of the shear modulus, dislocation density, Burger's vector, and other factors. The upper limit for wafer curvature depends on the ultimate strength of silicon, beyond which cracks and fractures will occur. Using the model in Figure 1 and Hollomon's equation, it is easy to see the upper limit for curvature is roughly proportional to the inverse of substrate thickness, which is consistent with upper limits we experimentally observe during the MOCVD growth of GaN on Si. Since the parameters n and K are unknown for our substrates, the maximum curvature k was estimated using the elastic strain relationship $\sigma = \frac{E\varepsilon}{1-\nu}$, where E is the Young's modulus and ν is Poisson's ratio. In this way, the maximum stress in the silicon substrate is actually defined by the substrate thickness and the curvature. The yield stress in silicon has been reported to be between 10–20 MPa at ~1000°C.[8] Using this stress range, Figure 2 shows the theoretical curvature limit for semi-standard Si wafers with the diameter between 100 mm and 200 mm. It can be seen that larger wafers has lower tolerance for wafer bow during growth, which is in accordance with experimental data.

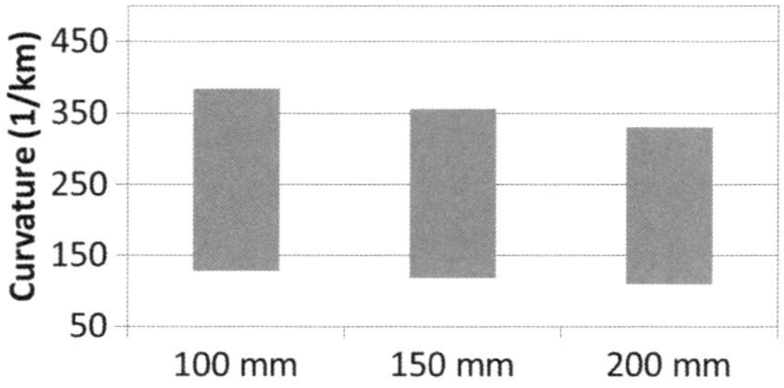

Fig. 2. Theoretical maximum wafer curvature during GaN on silicon growth.

Another difficulty for the MOCVD growth of nitrides on silicon is caused by some deposition on the reactor parts in close proximity to the growth zone. The gas entrance, reactor walls and wafer carrier (susceptor) are the most common parts that accumulate significant deposits. Typically, these deposits contain high concentrations of carbon and group III elements in a metallic form, and are susceptible to oxygen and moisture contamination. These materials can be easily released into the gas phase and transported to the substrate surface prior to or during nitride deposition, damaging epitaxial films.

3. MOCVD Growth of GaN on Silicon

3.1. *MOCVD reactor requirements*

As mentioned earlier, the parasitic deposition of the growth materials inside a MOCVD reactor creates a challenge for the epitaxy of nitrides on silicon. In order to keep reactor clean, several methods can be used in the growth recipes, reactor design and maintenance. The reactor design and growth conditions are set up so that laminar flow conditions exist between the reactor entrance and the growth zone. The return and recirculation flows carrying byproducts of the synthesis reactions from the growth zone and reactor exhaust back to the reactor entrance is thus avoided. Maintaining a low temperature in the gas phase, and employing low reactor pressure can also eliminate premature chemical reactions in the gas phase. Especially important for vertical type reactors, the reactor gas entrance is maintained at a low temperature using a temperature-controlled water-cooling system. If the cleanliness of the reactor entrance cannot be sustained during a deposition process, its design often employs some hardware components that can be easily replaced after each deposition process. *In-situ* cleaning procedures are also used for some MOCVD reactors.

Particular attention must be paid to the wafer carriers (susceptors). It is important to remove all, or most of the nitride deposits on a wafer carrier accumulated during a growth process prior to starting the next one. If the nitrides are not removed, they can migrate onto the Si substrate during the growth preparation steps, making the substrate unsuitable for high quality epitaxy. It is a common practice in the MOCVD of nitrides to bake out the wafer carriers after each run. One way to bake out a wafer carrier is to do that inside the reactor after each growth process. However, this reduces the reactor time available for growth. The other way to clean wafer carriers is to bake them out after each growth process in a dedicated external oven operating at high temperatures. A typical bake out temperature is about 1300–1400°C for the effective removal of nitride deposits.

The growth temperature control in modern MOCVD reactors is realized via the emissivity-corrected optical thermometry, often integrated with laser deflectometry in the *in-situ* control tools. The *in-situ* tools are usually mounted on the reactor view ports located above the epitaxial wafers. One large-area monitoring tool, or a few point-of-monitor tools can be installed to cover the entire growth zone. If the growth of nitrides on silicon is carried out using susceptor temperature control, the wafer temperature can drift with time due to the wafer bow, producing a gap between the wafer and carrier. Moreover, the width of the gap is uneven across the wafer leading to temperature non-uniformity. Wafer temperature control is needed in the nitride-on-silicon epitaxy to sustain constant and uniform growth temperatures. A difficulty in wafer temperature control is that the wafer emissivity can be compromised by destructive optical interference in the nitride films accompanied by severe wafer bow.

3.2. *Importance of in-situ characterization tools*

A decades ago, when *in-situ* monitoring tools were not widely available, crystal growers relied solely on post-growth characterization for the development of each epi layer. Therefore multiple growths were required for either SEM or X-ray measurements to determine only one parameter such as the growth rate. The availability of *in-situ* monitoring tools provided a solution to this problem. As a powerful tool for epitaxy development and diagnosis, *in-situ* monitoring offers functions such as monitoring epi surface smoothness, calibrating growth rate, measuring wafer temperature, as well as the wafer curvature in real time during the growth. Further advanced techniqes even allow the measurement of the bandgap of the material directly during the growth. Reflectance measurements have been widely used in GaN growth on traditional substrates like sapphire or SiC. The laser or LED beams are incident on the wafers through the reactor viewports. The reflected beams are collected by photodetectors through the same viewport. By measuring the intensity of the reflected beams, growers are able to (a) examine the surface roughness when the growth is stopped, and (b) measure the grown film thickness and extract the growth rate from the growth time used. A good example for the monitoring of the surface roughness is during the GaN growth on sapphire substrates. Typically a low-temperature GaN is used as a nucleation layer. The nucleation layer growth is followed by a ramp-up in the growth temperature. The nucleation layer decomposes at high temperatures, roughening the surface and decreasing the thickness. During the temperature ramp up, the sapphire surface can be exposed if GaN islands are formed. This process is accompanied by a significant drop in the *in-situ* reflectance intensity. When the growth of GaN is resumed at a high temperature, the reflectance recovers due to the transition of the surface morphology from three-dimensional islands to planar. The reflectance recovery indicates a smooth surface of GaN film. Monitoring the reflectance change closely and starting high temperature GaN growth at the right time helps to achieve high quality GaN films on sapphire substrates.

To perform reflectance-based growth-rate calibration, relatively thick films are required. Figure 3 shows a typical reflectance profile for the growth of a nitride HEMT structure. In this example, AlN is used as the nucleation layer, followed by the GaN buffer and the channel. The

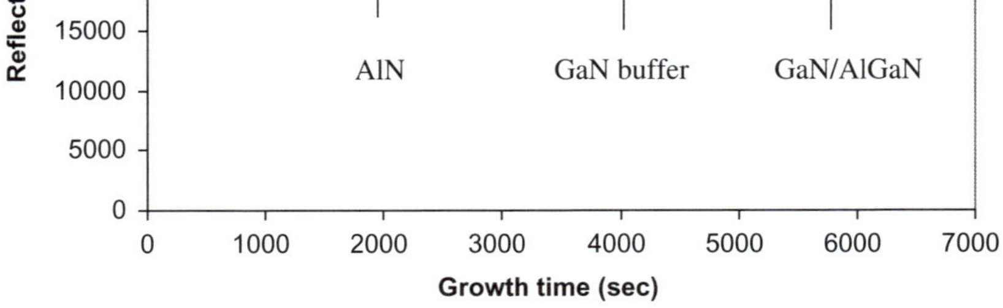

Fig. 3. *In-situ* reflectance evolution of a typical HEMT structure.

structure is capped with an AlGaN barrier. The oscillations seen in the reflectance profile are due to the interference of light undergoing multiple reflections between the epi/substrate interface and the epi surface cavity. The growth rate and thickness of each layer in the structure can be calculated using the appropriate refractive index. In the nitride on silicon growth, multiple AlGaN layers with different Al compositions are often used as the buffer layer. The *in-situ* reflectance measurements provide a convenient way to tune the growth rates for all layers in one calibration run while extracting the Al composition from post-growth X-ray.

Wafer temperature measurement is another useful feature enabled by *in-situ* tools. Infrared pyrometry may be used to check the temperature of the growth surface if corrected to the surface reflectivity variations. To measure the temperature of transparent substrates and films, techniques utilizing the temperature-dependent optical absorption edge are used. This technique also enables low temperature monitoring, a capability beyond the scope of traditional pyrometers. In most reactor designs, the wafer carrier rotates during the growth. Thus the temperature across the whole wafer can be monitored. During nitride growth on silicon, wafers often bow due to the accumulated stress. If wafers bow in a concave shape, the wafer edge is cooler than its center. The wafer temperature difference caused by bowing could be as large as more than 10°C in 150 mm wafers and even larger when the wafer size scales up to 200 mm and beyond. This leads to non-uniformity in the film thickness and alloy composition. In InGaN quantum-well based LEDs, a 10°C or more difference is very large and can greatly affect the LED wavelength in the final heterostructure. In multiple-wafer production type reactors, it is also important to make sure all the wafers are grown under the same condition. Thus keeping the growth temperature for each wafer within a certain spec window is highly preferred. Abnormal temperature changes during growth from run to run also indicate that there could be issues with hardware, so that engineers can schedule maintenance.

Monitoring the wafer curvature during growth of nitrides on Si substrates is very important: more so than for any other substrate. The bowed wafer surface changes the direction of the reflected beam and causes a change in the spot position in the CCD detector. The wafer curvature can be calculated based on this change. Knowing the wafer curvature, one can also quantify the strain introduced into the nitride film as the growth evolves. When the wafer bow is too high and the internal stress exceeds the mechanical limit, the wafer can break. The bowed wafer can also be tossed inside the reactor, since the pocket of the wafer carrier can be no longer deep enough to hold the wafer. *In-situ* curvature monitoring thus provides a chance to stop the growth before the wafer bow exceeds the safe range.

3.3. *Growth schemes for stress management and defect control*

Thick GaN device layers that are 2 μm and beyond on Si substrate are required to realize the potential of nitride materials.[9] Due to the large lattice and thermal mismatch between nitride epitaxial layers and the Si substrate, large mechanical stress is built into the structures. Special growth techniques must be employed to manage this stress. Three stress-management techniques enabling the growth of thick crack-free nitride epitaxial films on Si substrates have been frequently reported in the literature. They are: (1) the growth using AlGaN compositionally graded transition layers (TL),[10] (2) low temperature AlN insertion layers[3,11–13] AlN/GaN short-period multiple layer structures or super-lattices (SL).[14,15]

The AlGaN TL-based scheme employs one or more AlGaN layers. Grown on a AlN NL, the AlGaN has a composition graded from high to low Al concentration or all the way from AlN down to GaN. The gradient can be formed continuously within one layer or using few steps/layers each of constant composition. The important transition from tensile to compressive stress in nitride film occurs during the growth of the AlGaN TLs.

The stress evolution during the growth of nitride films can be observed using *in-situ* laser deflectometer that is installed over the quartz window viewport in the MOCVD reactor to record the curvature of the epitaxial wafer. Typical wafer curvature evolution during a growth of the nitride structure on silicon employing AlGaN TLs is shown in Figure 4 (blue curve). The nitride structure consists of the AlN nucleation layer, two AlGaN TLs and a GaN layer. The Al concentration in the first and second TLs is about 60 and 30 atomic %. The thicknesses of AlN, first transition, second transition and GaN layers are about 100, 400, 250 and 800 nm, respectively.

One can see in Figure 4 that the wafer curvature decreases, or trends negatively during the AlN NL growth (region II) and increases or trends positively during the AlGaN TL (region III) and GaN layer (region IV) growth. The negative and positive trends in the wafer curvature evolution correspond to concave and convex wafer bowing due to tensile and compressive stress accumulation in nitride films respectively. The tensile stress in an AlN NL grown on the silicon substrate arises from the lattice mismatch between nitride and silicon, and the compressive stress from the lattice mismatch between the AlN, AlGaN TLs and GaN.

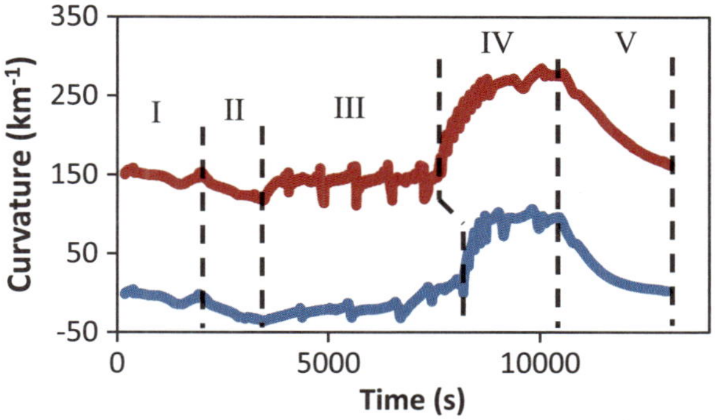

Fig. 4. Wafer curvature evolution during deposition of two nitride structures on Si substrate, AlGaN TL-based (blue curve) and SL-based structure (red curve). Major steps of deposition process are enumerated, substrate heating and cleaning (I), AlN NL growth (II), AlGaN TL for TL-based and SL growth for SL-based structure (III), GaN growth (IV) and cool-down (V).

The thicknesses of all nitride layers in the AlGaN TL-based structure illustrated in Figure 4 are so that it is safe to assume a high degree of stress relaxation in each layer. However, it has been shown that significant residual stress still remains in the nitride films long after the onset of stress relaxation, gradually decreasing with increasing thickness of the nitride film.[16] The layer sequence starting from AlN and finishing with the GaN layer enables a sustainable injection of compressive stress throughout the structure, resulting in substantial compressive stress (convex wafer bow) at the end of GaN growth. The compressive stress accumulated during the nitride growth balances the tensile stress arising from the structure cool-down process when the wafer bows concave (see Figure 4, region V). It is well known that the tensile stress accumulated during the cool-down process occurs due to the thermal mismatch between the nitride structure and substrate.

The transition from AlN NL to GaN layer can be also realized using AlN/GaN multiple layer structure or and AlN/GaN superlattice. Similar to the AlGaN layers in the TL-based scheme, the AlN/GaN SL helps to build a compressive stress in the nitride structure and to balance the effect of thermal mismatch between nitrides and silicon. The wafer curvature evolution during the growth of the nitride structure consisting of AlN, AlN/GaN SL and GaN layer is illustrated in Figure 4 (red curve). The thicknesses of AlN NL, AlN/GaN SL and GaN layer are 100, 700 and 800 nm, respectively. One can see that similar to the AlGaN TL-based scheme, the wafer curvature trends concave during AlN growth (region II) and convex during AlN/GaN SL growth (region III). Compressive stress is consistently injected into the nitride films throughout the SL growth owing to the compressive stress in the GaN grown over the AlN in each pair of the SL structure. The compressive stress further increases during the growth of the GaN layer over the SL (region IV), counter-balancing the tensile stress arising from the cool-down process at the end of the growth (region V).

If a thick GaN layer is grown over an AlN NL, the stress in the GaN induced by the lattice mismatch tends to relax gradually with increasing layer thickness. The stress relaxation process is accelerated by grain boundaries and dislocations abundantly present in the NL and at the beginning of the GaN growth. In order to sustain the compressive stress accumulation in thick GaN films, thin AlN layers grown at low temperature can be inserted into the GaN. A thin AlN layer grown over GaN at a high temperature is strained by underlying GaN and cannot introduce a compressive stress into the subsequently overgrown GaN. A low growth temperature facilitates quick strain relaxation in thin AlN layers making them a suitable template for the subsequent growth of compressively strained GaN. A few AlN insertion layers can be employed in thick GaN films.

3.4. *Control of wafer warp and bow*

Unlike the growth on sapphire and SiC substrates, severe tensile stress is accumulated during the growth of nitrides on Si substrates, leading to post-growth wafer bow and warping. *In-situ* monitoring of the wafer curvature, temperature and reflectance is essential to control the stress in the nitride-on-Si structures and to obtain essentially flat wafers. The concave and convex wafer bows correlate with tensile and compressive stresses respectively. Adjustments in the nitride structure design, and in the growth conditions enable a measurable tuning of the wafer shape.

Figure 5 illustrates the wafer shape adjustment exemplified by the nitride structure employing AlN/GaN SL to transition from an AlN NL to a GaN layer. This figure shows the wafer curvature as a function of the growth time for two structures with different SLs. The thickness of the GaN layers in the SL is adjusted from 21 to 28 nm while the thickness of the AlN layers and number of the AlN-GaN pairs in SL are kept constant, about 4 nm and 35, respectively. One can see that the slope of the wafer curvature evolution during the SL growth becomes significantly steeper with increasing thickness of the GaN layers. That corresponds to increasingly larger compressive stress in the structure and increasingly larger convex bow of the wafer. Using the thickness of GaN layers in the

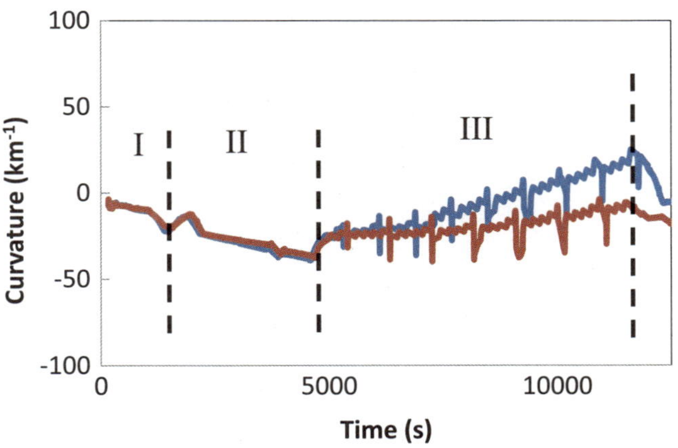

Fig. 5. Wafer curvature evolution during deposition of two SL-based nitride structures on Si substrate. The thickness of GaN layers in the SL is 21 nm (red curve) and 28 nm (blue curve) while the thickness of AlN layers and number of AlN-GaN pairs are constant, about 4 nm and 35, respectively.

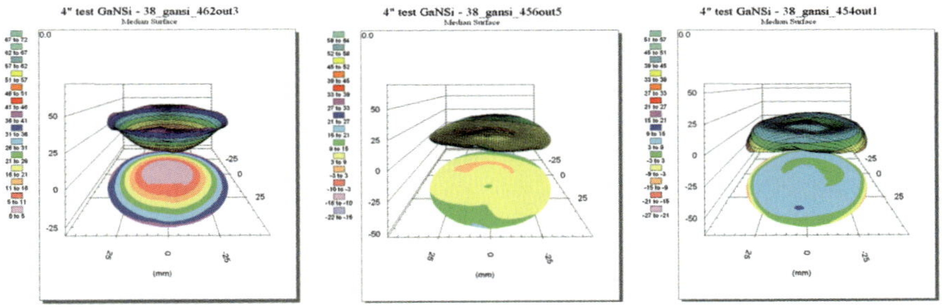

(a) (b) (c)

Fig. 6. Median wafer surface of nitride-on-Si structures employing SL-based transition scheme. Wafer shape is adjusted from concave, $-34\,\mu$m bow (a) to nearly flat, $-3\,\mu$m bow (b) and further to convex, $15\,\mu$m bow (c) by increasing thickness of GaN layers in SL. Median wafer surfaces were obtained using MTI Instrument Proforma 200 SA tool.

SL, the post-growth wafer shape can be changed from concave to nearly flat and further to convex as illustrated in Figure 6.

3.5. *Results: surface morphologies, structural, transport properties, photoluminescence*

Post-growth wafer characterization is necessary to assess material properties and evaluate potential device performance of epitaxial structures. The evaluation of the uniformity of material properties is an important aspect of wafer production. Basic wafer material characterization includes evaluation of the surface morphology using Nomarski contrast optical microscopy

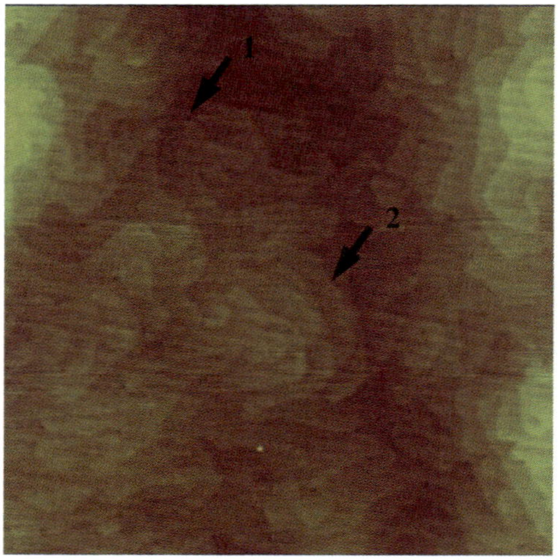

Fig. 7. $2 \times 2\,\mu m^2$ AFM image on GaN-based HEMT structure on Si substrate. Maximum Z height is 5 nm. Arrows 1 and 2 point to surface pit and atomic step, respectively.

and atomic force microscopy (AFM), crystal quality evaluation using transmission electron microscopy (TEM) and high-resolution X-ray diffraction (XRD), electron transport properties evaluation employing the Hall-effect, and optical properties using photoluminescence (PL) measurements.

A typical surface morphology of GaN-based high electron mobility transistor (HEMT) structure grown on Si substrate is shown in Figure 7. The $2 \times 2\,\mu m^2$ area of the wafer surface was imaged using an AFM instrument operating in a tapping mode. The root mean square (rms) surface roughness is about 0.3 nm. One can see the characteristic features of the nitride surface such as atomic steps and pits. The atomic steps have a roundish appearance and are often terminated by pits of about 10–100 nm in size. The origin of the pits on the GaN surface is generally understood as the result of intersection of threading dislocations with the surface. The pit density can be used to estimate the dislocation density in GaN, which is usually in the range of 10^9–10^{11} cm^{-2}.

In nitride-based high electron mobility transistor (HEMT) structures, a two-dimensional electron gas (2DEG) is formed in the channel layer (usually GaN) near a heterojunction barrier (usually AlGaN). Basic electron transport properties of the HEMT structure can be assessed using Hall effect measurements at room temperature. A typical GaN-based HEMT structure grown on Si substrate exhibits electron mobility in the range of about 1500–1800 $cm^2/V\,s$ with associated sheet charge density in the range of about 5×10^{12}–1×10^{13} cm^{-2}.

Room temperature PL measurements of the HEMT structures results in a signal from the channel layer, in close vicinity of the 2DEG, in addition to the thick underlying layer. If the GaN is used as a channel layer material, the PL signal consists of a sharp near band-edge emission peak centered at about 363 nm that dominates over the blue band emission centered at about 450 nm and a yellow band centered around 570 nm. Both blue and yellow PL emissions are usually associated with various defects in GaN. A dominant near band-edge emission peak is indicative of good material quality of the device layers.

4. MBE Growth of GaN on Silicon

4.1. *Challenges in MBE growth*

Driven by the market demand for photonic and electronic devices, the epitaxy of GaN on Si by MOCVD has prospered and matured owing to the availability of large size and low cost Si wafers, coupled with the high throughput of epitaxy. The high growth temperature necessary in MOCVD poses some thermal constraints for potential introduction of nitride epitaxy in a CMOS process line.[17] Typical MBE growth temperatures for epitaxy of GaN on Si are much lower, in the 700C range, allowing a favorable thermal budget. MBE has lagged in maturity, primarily because of far lower volume of research and development effort devoted to this technique compared to MOCVD. The traditional benefits of MBE such as the availability of *in-situ* characterization tools, extremely abrupt doping profiles and heterojunctions, little or no carbon and hydrogen contamination, and high mobility carrier channels proven in other III-V semiconductors carries over to the epitaxy of III-Nitrides. MBE allows a far wider range of growth regimes — ranging from effectively metal-rich conditions in plasma-MBE mimicking liquid-phase epitaxy, to almost MOCVD-like conditions without carbon in NH_3-MBE. The metal-rich LPE-type growth conditions in plasma-MBE have enabled the successful demonstration of lateral regrowth in many forms. For example, very low resistance source and drain ohmic contacts to reduce delays and losses, and to improve device reliability in HEMTs have been demonstrated by MBE,[18] and is now available as a service. Till date, the most abrupt and high quality AlN/GaN binary interfaces are still produced by MBE.

Like in MOCVD, the high defect density due to the lattice mismatch is the major challenge in MBE of GaN on Si. The large thermal mismatch causes cracking of the GaN films during the cool-down after high-temperature epitaxy. The critical thickness for cracking depends on the growth temperature, in addition to the layer and substrate thicknesses. MBE grown samples allow a larger critical thickness than MOCVD owing to the lower growth temperature. However, cracking of the epi layer remains a pressing issue.

Si melt back etching is also a severe problem in MBE epitaxy of GaN on Si. At high growth temperatures, Ga and Si react strongly to form an alloy,[19] making it difficult to obtain epitaxial layers of high crystalline quality. The strong out-diffusion of Si through the epi layer results in a poor surface morphology, when GaN is directly grown on Si. This issue is common to MOCVD and MBE. To prevent melt-back etching, when AlN nucleation layers are used, the thickness, flux, and substrate temperature must be precisely controlled to prevent significant degradation of the GaN buffer layer. An example of a degraded epitaxial layer surface morphology is shown in Figure 8(a). In addition to highly crystalline AlN nucleation layers, a very flat surface is also a prerequisite to achieve the epitaxy of smooth layers of GaN. The steps and terraces formed during growth promote inversion domain boundaries[20,21] and develop cubic GaN inclusions embedded into the hexagonal matrix.[22,23] Figure 8(b) shows such cubic inclusions[24] which are present in AlN.

In some other reports, substrate step-induced rotation of the AlN planes have been demonstrated,[25] which may contribute to the generation of threading dislocations that propagate into the buffer layers. Thus the choice of a proper nucleation layer and the subsequent growth conditions are key to realizing high quality MBE GaN buffer layers on Si. The quality

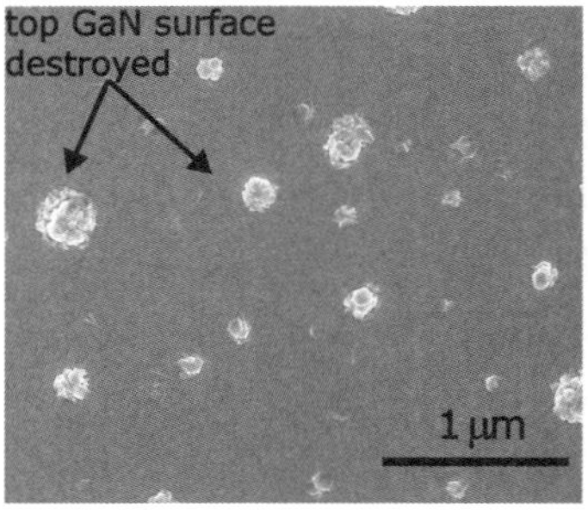

(a)

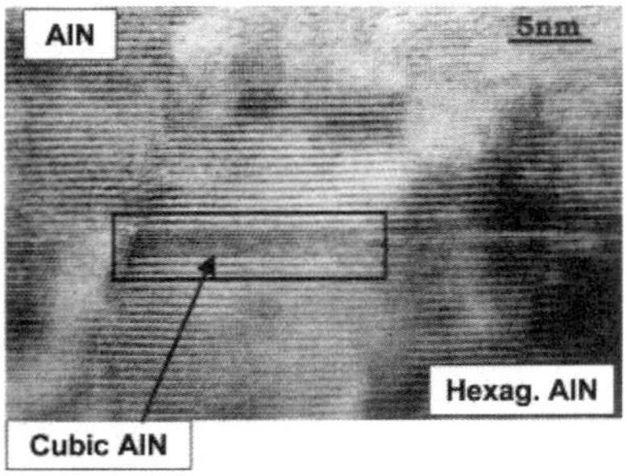

(b)

Fig. 8. (a) Showing top view SEM image of the MBE grown GaN surfaces getting destroyed in a large scale due to un-optimized growth condition. (b) HRTEM micrograph indicating the presence of cubic inclusions embedded in hexagonal matrix for AlN layers.[24] Reprinted from E. Calleja *et al.*, Growth of III-nitrides on Si(111) by molecular beam epitaxy Doping, Optical and electrical properties (Fig. 3), Journal of Crystal Growth, Vol 201–202, pp. 296–317. Copyright 1999, with permission from Elsevier.

of subsequent heterostructures hinges on the success of the MBE growth of the nucleation layer on Si, and the subsequent buffer layer.

4.2. *Growth schemes for stress management and defect control*

When GaN is grown directly on a clean Si (111) surface by RF plasma-MBE, an amorphous SiNx layer forms at the interface because of the reaction Si with the active nitrogen species. The reaction

$$Si + 4/3GaN \rightarrow 1/3Si_3N_4 + 4/3Ga, \Delta G^{1000K} = -134.7\,kJ, \tag{4.1}$$

shows that the formation of SiNx is favored at typical plasma-MBE growth temperatures.[26] The formation of such an amorphous layer was reported by Nakada *et al.*[27] and was used to grow crystalline GaN buffer on top as shown in Figure 9.

However, several reports suggest that the GaN layers grown on top of the amorphous SiNx is not of a high crystalline quality. The X-ray FWHM values range between 70 and 100 arcmin. Alternative interlayers such as SiC,[28] Al_2O_3,[29] and GaAs[30] were explored. Eventually, AlN has emerged as most attractive for the subsequent growth of high crystalline quality GaN buffer layers on Si by MBE. The role of the AlN nucleation layer is to decouple the top GaN layer from the underlying Si substrate: mechanically, chemically, thermally, and electronically. The AlN layer

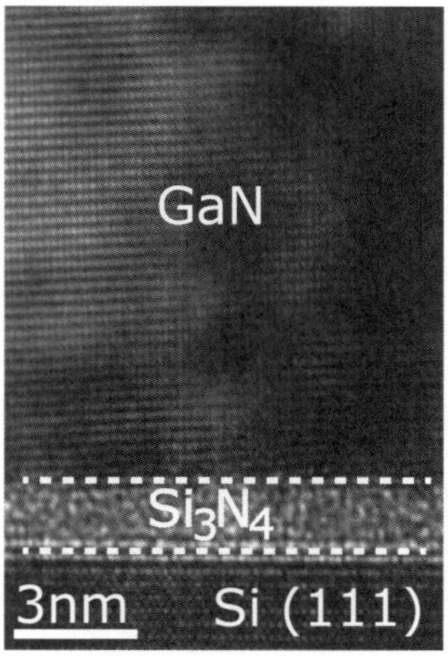

Fig. 9. A high resolution electron micrograph of a cross section of a GaN lattice structure grown on the silicon nitride formed on Si(111) substrate.[27] The silicon nitride layer shown here has some crystalline character. This figure has been reproduced with permission from Y. Nakada, I. Aksenov and H. Okumura, GaN heteroepitaxial growth on silicon nitride buffer layers formed on Si(111) surfaces by plasma-assisted molecular beam epitaxy (Fig. 5), Appl. Phys. Lett., Vol 73, No 6, pp. 827–829, 1998. Copyright 1998, AIP Publishing.

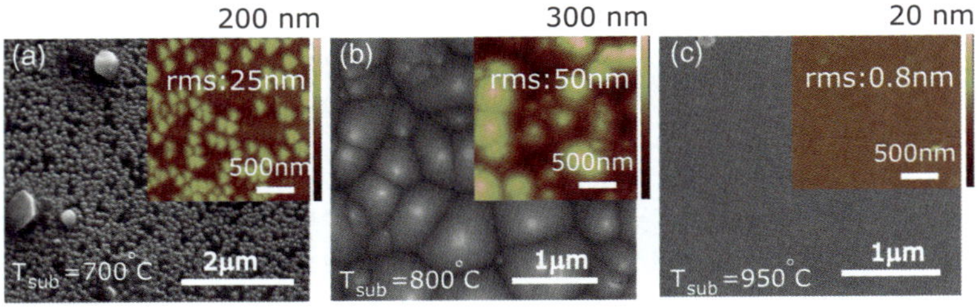

Fig. 10. SEM images of the surface of AlN nucleation layers grown by MBE on Si(111) at various substrate temperatures. Higher growth temperatures led to a smooth two dimensional surface morphology. The corresponding AFM scans are shown in the inset.[33]

induces a compressive stress in the subsequently grown GaN layers, partially counterbalancing both the growth as well as thermally induced tensile stress, avoiding cracking. It was also observed that similar to MOCVD process, a thin low temperature (LT) AlN seed layer followed by a high temperature (HT) AlN nucleation layer[31] is sometimes useful in growing high quality GaN crystal on Si by MBE.[32]

The growth temperature and stoichiometry of the AlN nucleation layer have to be carefully controlled for subsequent GaN epitaxy. Typically for substrate thermocouple temperature $T_{sub} \sim 900°C$ or above, at RF power ~400W and for a near-stoichiometric Al flux of $\sim 1 \times 10^{-7}$Torr, a smooth two dimensional (2D) AlN surface is obtained.[24] A lower T_{sub} results in 3-D island growth. The evolution of the surface morphology of the AlN nucleation layer with T_{sub} is shown in Figures 10(a-c).[33] A high-temperature AlN nucleation layer results in the smoothest surface morphology.

Similar to RF-MBE, NH_3-MBE also relies on high-temperature AlN nucleation layers grown at ~1100°C.[34] Prior to the AlN growth on Si, Al pre deposition[35,36] is carried out sometimes to completely overrule the formation of amorphous SiNx. However, the polarity of the grown film is strongly dependent on the Al pre-deposition time. Recently Dasgupta *et al.* have reported[37] that Al pre-deposition can cause N-polar AlN growth. Al pre-deposition can saturate the dangling bonds of Si and then subsequent exposure to N plasma will result in N-polar AlN growth as shown in Figure 11. Also, since Al acts as a p-type dopant in Si, excess Al can diffuse into Si, doping it with holes. The resulting conductive layer in Si is confirmed by Hall-effect measurements.[38] The conductive layer is undesirable for many electronic device applications since it forms a parasitic conduction path. Thus, proper choice of Al-flux and T_{sub} are crucial for subsequent GaN buffer growth on Si.

On top of the AlN nucleation layer, a high quality GaN crystal has been reported by MBE following a two-step growth method.[24] In this method,

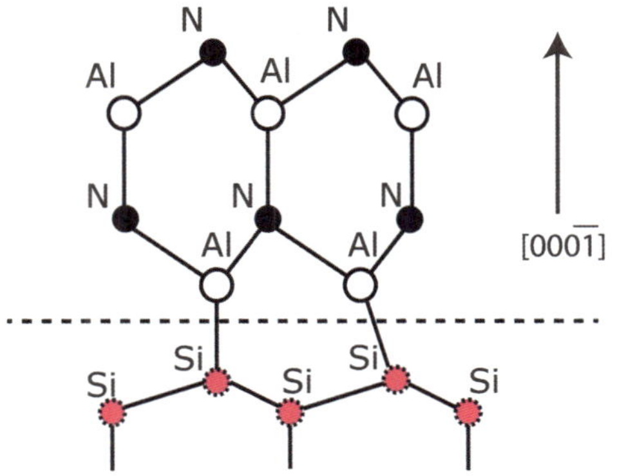

Fig. 11. Al atoms saturate the Si dangling bonds, resulting in N-polar growth with subsequent exposure to N plasma.

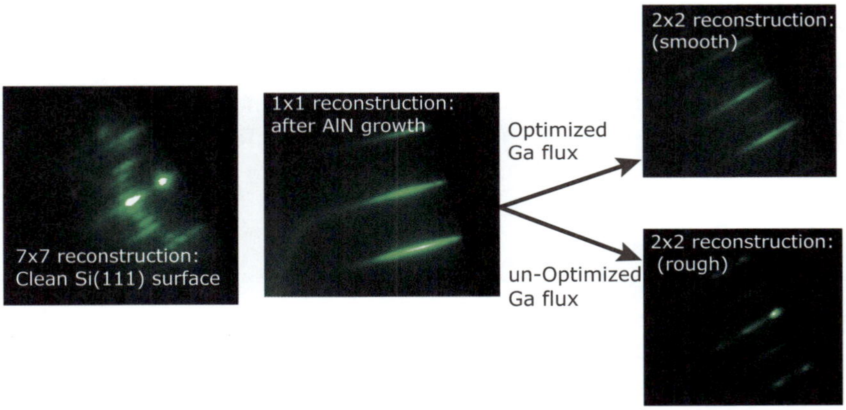

Fig. 12. Evolution of RHEED images during III-nitride growth on Si(111) by MBE has been shown.

initial GaN growth is performed at a slow growth rate to promote the coalescence of 3D islands. Then a fast growth rate ∼500 nm/hr enhances the lateral growth and suppresses defect propagation. The *in-situ* monitoring of RHEED patterns during MBE growth helps track the surface morphology during growth.

The evolution of RHEED images from the beginning of the growth to the end is shown in Figure 12. The clean Si (111) substrate surface has unstable dangling bonds reconstruct into a form called the dimer-adatom-stacking-fault (DAS) model.[39] The unit mesh of the surface structure becomes 7×7 times larger than the underlying bulk structure. Figure 12(a) shows this 7×7 reconstruction for a pristine Si (111) surface. If an amorphous SiNx layer is formed, the RHEED image turns diffuse. This imaging is very useful in preventing the formation of the amorphous layer. After AlN nucleation, a clear 1×1 reconstruction is observed. After the growth at the end of the cool down cycle, a 2×2 reconstruction is observed. This is a typical signature of Ga-polar growths. Proper choice of the metal fluxes determines the surface morphology of the epi layers. *In-situ* RHEED images is thus useful to guide the flux control during the growth. Streaky-spotty RHEED patterns as shown in Figure 12 indicate rough, or 3-D GaN surfaces grown on Si. On the other hand, smooth 2-D GaN layer produce streaky RHEED as shown in the same figure.

4.3. *Results: Surface morphologies, Structural, Transport properties, Photoluminescence*

To evaluate the crystal quality of the MBE grown GaN, XRD and PL measurements are carried out frequently, in addition to transport characterization. Calleja *et al.* reported[24] that by employing optimized AlN nucleation layer and a two-step growth, very strong and narrow linewidth (~15 meV) low-temperature PL spectra can be observed. Figure 13(a) shows the comparison of PL spectra for the GaN grown by RF MBE on Si employing

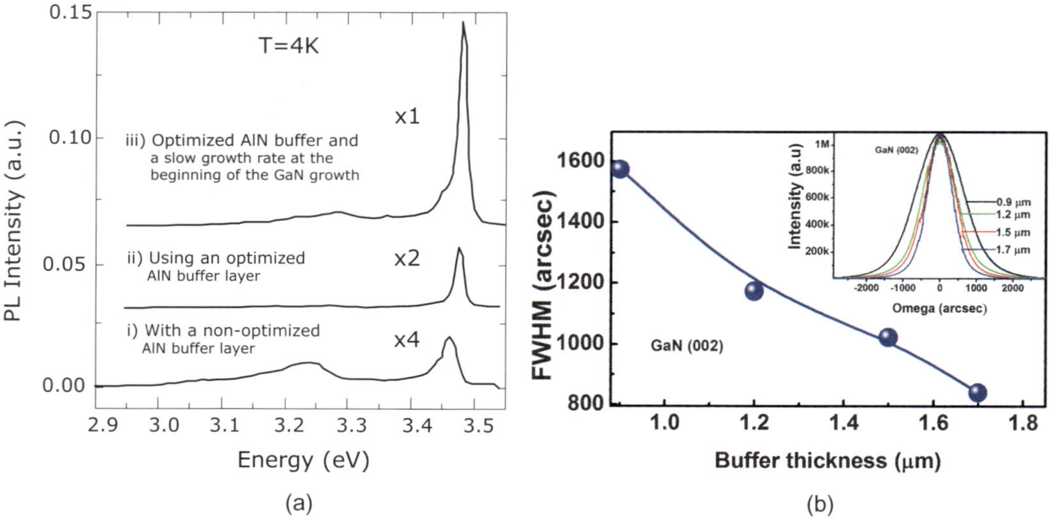

Fig. 13. (a) Low temperature PL spectra of GaN layers after different optimization processes: (i) grown on top of a non-optimized AlN buffer, (ii) using an optimized AlN buffer, and (iii) using an optimized AlN buffer layer and a two-step GaN growth mode.[24] Reprinted with permission from E. Calleja et al., Growth of III-nitrides on Si (111) by molecular beam epitaxy Doping, Optical and electrical properties (Fig. 11(b)), Journal of Crystal Growth, Vol 201-202, pp. 296–317. Copyright 1999, Elsevier. (b) HR-XRD measured FWHM of the GaN (002) rocking curve as a function of GaN buffer thickness. Inset shows the rocking curves.[40] Copyright 2012, The Japan Society of Applied Physics.

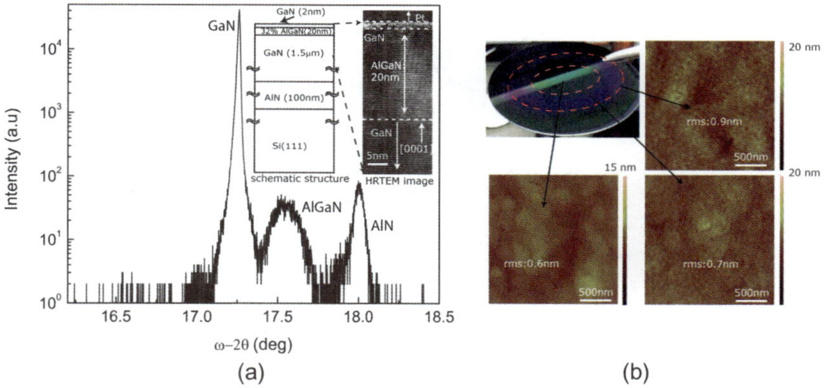

Fig. 14. (a) X-ray diffraction pattern of (002) ω-2θ scan for GaN(2nm)/AlGaN(20 nm)/GaN(1.5 μm)/AlN(100 nm) HEMT epitaxial structure. The schematic layer structure and the HRTEM image confirming the thicknesses are shown in the inset. (b) The AFM image map of the grown AlGaN HEMTs on 3″ HR-Si showing smooth surface morphologies. Very little variation of *rms* roughness is suggestive of high uniformity across the 3″ wafer.[33]

different nucleation and buffer growth conditions. Figure 13(b) shows the XRD rocking curve FWHM of the GaN (0002) peak as a function of GaN buffer thickness grown by ammonia MBE.[40] The FWHM decreases with increasing GaN buffer thickness. This is a signature of the improvement of the crystalline quality as the buffer thickness is increased. Recently Hoke *et al.*[17] have reported 1.7 μm thick crack free RF-MBE grown GaN HEMTs on 4″ Si(111) with GaN (002) FWHM ~360 arcsec, which is very attractive. The AFM scans, HRTEM images and $\omega - 2\theta$ XRD scans of a typical AlGaN HEMTs structure consisting of [2 nm GaN/20 nm $Al_{0.32}Ga_{0.68}N$/1.5 μm GaN/100 nm AlN/Si (111)] grown on 3″ wafers by RF MBE is shown in Figure 14.[33]

One of the key challenges to achieve high performance DC and RF GaN devices on Si is to ensure low-leakage (or high resistivity) GaN buffer layers. Electronic conductivity of the GaN buffer layer in turn depends on the growth conditions of the AlN nucleation layer. If a significant difference between the 300 K and 77 K charge density is observed in Hall effect measurements, it indicates the freezeout of shallow dopants in a parallel conduction path through the buffer layer. Figure 15(a) shows a secondary ion mass spectrometry (SIMS) depth profile of an AlGaN/GaN HEMT structure grown on Si(111) by RF MBE with a slightly metal-rich AlN nucleation layer. It illustrates significant Si incorporation into the AlN and the subsequent GaN layers. Although no silicon flux was used during the growth, a silicon concentration of $\sim 2 \times 10^{20}\,cm^{-3}$ in the AlN

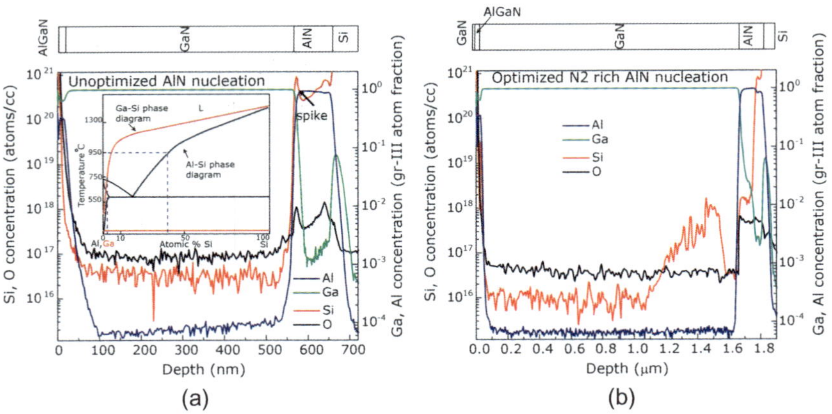

Fig. 15. (a) SIMS depth profiles of Al, Ga, Si, and O concentrations of the GaN HEMTs grown by plasma MBE on HR-Si using metal rich AlN nucleation layer. Si concentration in excess of 10^{20} cm^{-3} is seen in the AlN nucleation layer. A large Si concentration spike is seen at the AlN /GaN interface. The Al-Si and Ga-Si schematic phase diagram is shown in the inset. (b) SIMS depth profile of the HEMTs with N_2 rich AlN nucleation shows less Si in the AlN layer, and the absence of the sharp Si spike at the AlN/GaN interface.[33]

layer and large spikes at the GaN/AlN interface can be seen. This Si is due to out-diffusion from the Si substrate; its incorporation in GaN makes it conductive enough for noticeable difference in the 300 K and 77 K charge density. This conductive layer is responsible for conductive 'buffer' leakage[41] currents in III-nitride HEMTs, affecting the speed, performance, and on/off ratio of the transistors significantly.

The Al-Si schematic phase diagram[42] shown in the inset of Figure 15(a)[33] indicates that at such a temperature in excess of 40% atomic weight of Si can be dissolved in Al. This can be expressed by the following equation,[43]

$$\text{Excess Al (liquid)} + \text{Si} \rightarrow \text{Al} + \text{Si (molten mix)} \quad \Delta G < 0. \qquad (4.2)$$

The Si dissolved in excess Al thus propagates ('floats') and reaches the AlN/GaN interface. At this point the Al shutter is closed and Ga shutter is opened. The excess Al gets consumed by active N* preferentially over Ga and Si. The solubility of Si in Ga at 740°C (typical GaN buffer growth temperature) is much less as compared to Al and is shown in the schematic Ga-Si phase diagram[44] in Figure 15(a). Then Si cannot propagate anymore and gets precipitated, resulting in a large concentration spike [26] at the AlN/GaN interface. The transport of Si through AlN layer can be reduced significantly by incorporating a slightly N_2-rich AlN layer. This was reported for GaN growth on SiC substrates.[43] In N_2 rich AlN the absence of excess Al prevents the formation of Al + Si molten mix. As a result, Si cannot effectively propagate to form the spike at the interface. Figure 15(b) shows that using such a growth condition the Si incorporation in AlN can be significantly reduced. More crucially, the removal of the Si spike at the AlN/GaN heterointerface by employing N_2 rich nucleation eliminates the parallel conducting channel, resulting in all-around superior RF performance.[45] Although the exact breakdown mechanism for GaN HEMTs on Si still remains poorly understood, the elimination of this parallel path helps boost the breakdown voltages of GaN HEMTs on Si. A higher ON/OFF ratio can prove beneficial for digital electronics using GaN on Si as well.

5. GaN on Si Device Applications

Initial GaN on Silicon efforts were directed towards optical devices-primarily for solid-state lighting applications. GaN LEDs on Silicon have been successfully realized, and several companies are trying to commercialize this technology. The low-cost advantage is not so severe compared to Silicon in this arena. Silicon absorbs visible photons, which requires several tricks to reflect the light back into the nitride layers where they are generated. In this aspect, sapphire is preferable because of its optical transparency. For very high thermal loads, SiC has also been found useful. There is hope that the integration of GaN photonics with Silicon in the future can lead to unconventional applications related to on-chip photonic circuits. Clearly because of the much shorter wavelength of visible light compared to IR, one can bend it over smaller diameters, and make a denser integration of photonic circuits. This is an area that is still under investigation. GaN on Silicon LEDs have reached appreciable performance levels, but not lasers, because of the defect problem, compounded by the thermal conductivity problem.

Typical GaN-based HEMT structures with AlGaN barriers grown on Si substrate by MBE/MOCVD exhibit electron mobilities in the range of about $\sim 1600\,\text{cm}^2/\text{V s}$ with associated sheet charge density in the range of about 0.8×10^{13}–$1.4 \times 10^{13}\,\text{cm}^{-2}$.[17,32,46] By exploiting MBE regrown ohmic contact AlGaN HEMTs have been demonstrated with peak current gain cutoff frequency $f_T = 153\,\text{GHz}$, saturation drain current density in excess of 1.3 A/mm and R_{ON} as low as $1\,\Omega\,\text{mm}$.[46] The low on-resistance is attractive for reducing losses in high power and high frequency microwave applications, and mitigating the problem of the low- thermal conductivity of Si compared to SiC. Since metal-rich growth conditions in plasma-MBE enable the wetting of sidewalls, very effective regrowth is obtained. This technique has been crucial to push the RF performance of GaN HEMTs to record levels, and is expected to be exploited for more device applications in the future. With thinner barriers and further device scaling, integration of regrown contacts with GaN on Si HEMT platform thus offers pathways to boost the performance and efficiency of high-power and high-speed devices while harvesting the many advantages of being on a Si platform.

Apart from attractive RF characteristics, improved breakdown characteristics of GaN HEMTs on Si have also been reported,[47] making this GaN on Si technology highly attractive for the power electronics industry in addition to the RF electronics market. The utilization of blocking voltage boosting technology (BVB) where by the aid of selective ion implantation the depletion region in the Si substrate has been widened thus resulting in an off-state breakdown voltage up to 1340 V with the total epitaxial thickness of GaN as thin as 1.4 μm. Following these significant advances of GaN on Si electronics the material quality and microwave loss characteristics have reached the elevated level previously achievable only by GaN on SiC devices. A maximum power-added efficiency (PAE) in excess of 57% with a power performance of 4 W/mm at 10 GHz has been reported[48] for GaN MISHEMTs as shown in Figure 16. Moreover, large periphery CPW S-band high performing MMICs have been demonstrated where DC stress testing and RF life testing have been performed to ensure stable operation.[48]

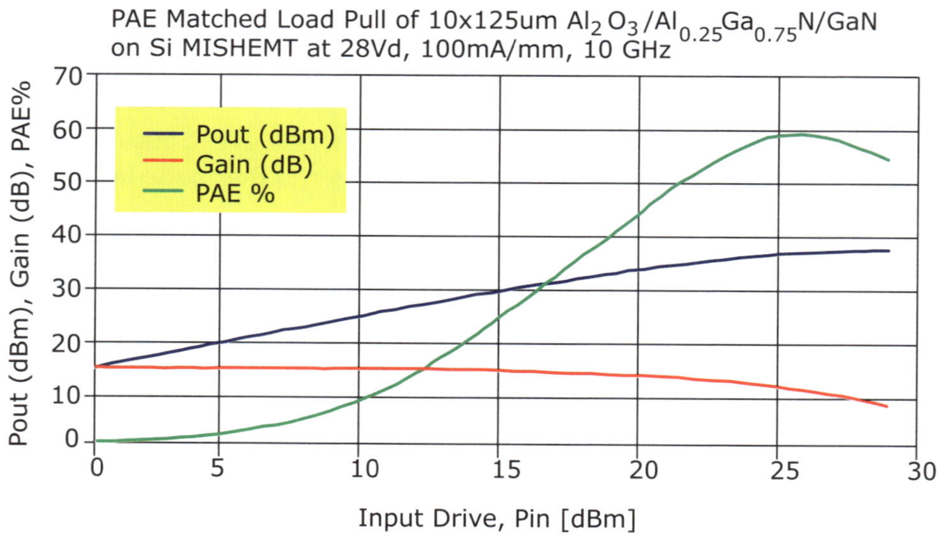

Fig. 16. PAE match load pull data for 10 μm × 125 μm Al_2O_3/AlGaN/GaN/Si MISHEMT with 250 nm gamma gate.[48] Copyright 2013 CS MANTECH Conference Figure 2, page 345–348, May 2013.

There is a long cherished dream of the heterogeneous integration of III-Vs on Silicon to take advantage of superior analog and mixed signal performances of III-Vs integrated with the unmatched digital logic functionality of Si VLSI. The first step towards that goal was the successful demonstration[49] of GaN HEMTs and Si MOSFET on the same Si (111) substrate. This was a commendable effort, but suffered from the poor CMOS performance owing to the high density of traps in Si (111) compared to the conventional Si (100). Later a double wafer bonding approach was undertaken and high performance Si (100) MOSFET and GaN HEMTs were both reported on the same wafer.[50] The real demonstration[51] in terms of integrated circuit came from Raytheon. Similar to that of InP HBTs and Si CMOS integration[52] as reported earlier, modified SOI wafers were used for this purpose. In this case (100) orientation was chosen for CMOS fabrication whereas GaN was grown on the selectively etched (111) windows of the handle substrate. This was the first successful demonstration of a fully functional heterogeneously integrated GaN HEMT — Si CMOS circuit consisting of three silicon PMOS transistors and four GaN HEMTs shown in Figure 17.

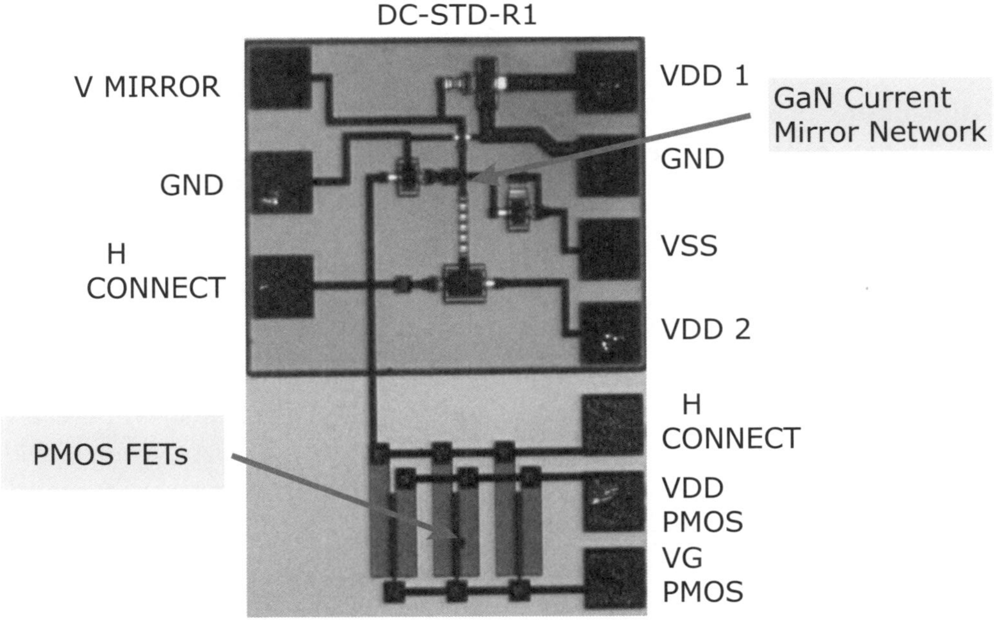

Fig. 17. 0.95 mm × 0.95 mm current mirror integrated circuit consisting of three Si PMOS transistors and four GaN HEMTs.[51] This figure has been reproduced with permission from W. E. Hoke *et al.*, Monolithic integration of silicon CMOS and GaN transistors in a current mirror circuit (Fig. 8a), J. Vac. Sci. Technol. B Vol 30, No 2, 2012, p02B101-1(-6). Copyright 2012, American Vacuum Society.

These types of circuits are particularly useful since Si transistors provide a stable and precise reference current which is otherwise difficult to achieve in GaN HEMTs owing to the instability arising from the temperature and threshold voltage variation. This heterogeneous integration and first circuit level demonstration will open up possibilities for a new class of devices which include advanced RF, mixed signal and power electronics circuits, high power digital-analog converters, RF transceivers on a chip[53] etc. Recently manufacturing of 200 mm GaN-on-Si process in an Au free CMOS fab compatible environment has been reported both by MBE[53] and MOVPE.[54] Using standard CMOS tool set and successfully eliminating Ga contamination- fully functional AlGaN MISHEMT power devices on 8-inch Si wafer has been demonstrated. The happy marriage of GaN and Si thus promises an exciting future ahead.

The most significant inroad of GaN on Silicon technology has been in the arena of power electronics. Electronic switches for high voltages with Silicon start becoming highly energy inefficient as the voltage requirement increases. This is a direct consequence of the 1.1 eV bandgap of Silicon, which results in a low critical breakdown electric field. One can increase the voltage handling capacity of Silicon transistors in various ways — such as the IGBT, the superjunction, the CoolMOS, etc by making the carrier drift regions longer, or by other tricks of field-balancing. The large physical dimensions of such Silicon high-power transistors results in significant additional resistances, reducing the energy efficiency of switching, and making the switching process slow as well. The most significant change going from Silicon to GaN or SiC is the far wider bandgap, which leads to a significantly higher breakdown electric field. Thus, the device size can be shrunk back to recover much of the resistive losses. For this reason, SiC and GaN are destined to replace Si devices in the few-hundred Volt electronic switching applications. Going beyond into kV and few 10s of kVs switching, SiC and GaN will dominate the marketspace, and will create several application arenas where Silicon transistors cannot perform. GaN on Silicon HEMT technology is currently competing for an market adoption in the $\sim$600 V power switching window. Compared to Silicon, it can deliver much higher switching efficiency, and compared to SiC, it offers a lower cost by virtue of its marriage with Silicon. In addition to several semiconductor companies jostling for this piece of the pie, most significantly, in December of 2014 at the IEDM conference, the largest Silicon foundry TSMC announced[55] the opening of a GaN on Silicon HEMT foundry service, which is a testament to the success, the importance, and the very bright future of this technology.

6. Future Prospects

Cost is a major driving forces to develop GaN on silicon. Although epitaxy techniques for GaN on silicon are much more mature today compared to just a few years ago, there are still quite a few challenges for future development. To further reduce the cost, substrates sizes more than 8-inch will be adopted. Larger wafers provide the chance to save both epitaxial and processing resources, and to reduce the unit cost of each device. As for SEMI standard Si substrates, thickness increases with the wafer diameter. This helps reducing the wafer bow during and after growth. Some growers even use non-standard thick Si substrates, 1 mm or thicker. However, that could bring up an incompatibility issue with the stepper lithography processes in silicon foundries.

The wafer size will likely keep increasing and eventually go beyond 200 mm in the future. This approach also requires the reactor capacity to be expanded accordingly to accommodate as many and as large wafers as possible to be manufactured at the same time. During a MOCVD growth, the edge of the wafer carrier always loses heat quicker compared to its central area. Thus, the heaters need to be designed to compensate this heat loss and to provide uniform temperature profile across whole growth zone. This is particularly difficult as the reactor circumference becomes larger. In MOCVD, the laminar flow at growth interface is necessary for high quality film growth. To meet this requirement in large capacity commercial reactors, reactor apparatus has to rely on increasingly elaborate designs of gas entrance and exhaust. More control possibilities are also required to adjust the flow pattern so that all wafers can be grown under identical conditions. All of these require MOCVD manufacturers to continuously improve and further optimize their existing system designs.

As the GaN epitaxy volume has been ramping up in recent years and is expected continue growing in the near future, it is a good time to think about how to improve production throughput. Higher growth rates are preferable for industrial production since it saves the time of each production run. Due to higher growth rates, MOCVD is more commonly used than MBE to produce thick nitride film on silicon. As the reactor size increases, the loading and unloading time becomes also important in the throughput optimization. The wafer carriers used in MOCVD are normally made of graphite and coated by SiC. The carrier thickness increases with the diameter to provide enough robustness for the wafer loading process. Therefore, larger heat capacity is expected in the carriers of large reactors. This significantly slows down the wafer cooling at the end of process and makes the whole process effectively longer compared to the growths in smaller chambers. One way to solve this problem and to improve the throughput is to build a buffer chamber to store fresh and grown wafers and rely on robot technology that can load and unload the wafer carriers at relative high temperatures.

Cleaning the contamination source of Ga left over from the previous run in MOCVD growth chambers is another factor for delay. To prevent the melt-back etching effect, MOCVD chamber cleaning and internal parts replacement are required before each run starts. Different MOCVD manufacturers have adopted both *in-situ* and *ex-situ* cleaning methods, while tradeoffs also exist. On one hand, *in-situ* cleaning introduces more tool downtime. On the other hand, *ex-situ* cleaning requires additional investment into facilities and equipment to support high temperature bakes of wafer carriers.

There are also other ways to optimize the production efficiency. For example, cluster tools have also been designed with more than one growth chamber and one common transfer chamber. The growth chambers can be designed to perform the same type of epi process or to deposit different parts of one epi structure at the same time. If one chamber can be used for AlN nucleation growth only and the other for growing the following Ga-containing layers, there is no need to perform wafer carrier cleaning after every run. Source saving is another important topic to reduce production cost. A good example is the growth of AlN/GaN superlattices. At the beginning of the AlN layer growth, the aluminum MO source flows into the growth chamber while the gallium MO source flows into the vent line. The gallium and aluminum sources are swapped between growth chamber and vent line when the AlN layer is completed and the GaN layer starts. The gallium and aluminum sources are thus wasted when flowing into vent line. Such a waste increases with increasing number of layers and thickness of superlattices. Source saving requires a reduction in the MO source flow when the source flows into the vent line. Such technique can also extend the lifetime of the reactor exhaust components thus making maintenance cycle longer and increasing reactor throughput.

Besides the improvements in hardware, new epitaxy techniques have been also developed to enhance GaN manufacturability. Wafer regrowth (secondary growth) is one of such techniques. In the past, the regrowth was considered as an expensive approach since more wafer preparation at the device level was needed. However, when wafer size scales up, the additional cost per device may not be so prohibitive. The regrowth technique was most successfully applied to form ohmic contacts for lateral transistor devices. Very low contact resistivity has been demonstrated, helping to improve high frequency device characteristics. In vertical devices, regrowth has been also demonstrated and may reach commercialization in the near future.

Patterned growth techniques could open new opportunities for device applications. Due to the difficulty in p-type doping in GaN and the low mobility of holes, p-channel GaN based HEMT technology is not as mature as those using n-channels. This limits applications of nitrides in CMOS-type logic. But monolithic heterogeneous integration of nitride devices with silicon-based architectures is still highly desirable for high-power and low-loss switching applications. Though the mass transport occurring at high temperature does not allow devices to have the best performance using MOCVD-regrown ohmic contacts, owing to high growth rates and good uniformity, MOCVD could be advantageous for large area selective regrowth, especially when the growth temperature is compatible with Si processing temperatures. This may provide the opportunity for fabricating microelectronics for a wide range of applications at low cost, and could open an entirely new chapter for III-V semiconductors.

References

Shu C. K. *et al.*, (1998), *Appl. Phys. Lett.* **73**, 641.

Egawa, T. and Shuhaimi, A. (2010), *J. Physics D: Appl.* Physics **43**, 354008.

ASTM F534 3.1.2.

ASTM F1390.

Sunkara, M. K., Sharma, S., Miranda, R., Lian, G. and Dickey, E. C. (2001), *Appl. Phys. Lett.* **79**, 1546.

St. John, C. (1975), *Phil. Mag.* **32**, 1193.

Dieing, T. and Usher, B. F. (2003), *Phys. Rev. B*, **67**, 054108.

Rabier, J. and Demenet, J. L. (1982), *Proc. IEEE*, **70**, 420.

Selvaraj, S. V., Suzue T. and Egawa, T. (2009), *IEEE Electron Device Lett.* **30**, 6, 587.

Nitronex, Patent US 6649287B1.

Reiher, A., Blasing, J., Dadgar, A., Diez, A. and Krost, A. (2003), *J. Cryst. Growth* **248**, 563.

Zhang, B. S., Wu, M., Liu, J. P., Chen, J., Zhu, J. J., Shen, X. M., Feng, G., Zhao, D. G., Wang, Y. T., Yang, H. and Boyd, A. R. (2004), *J. Cryst. Growth* **270**, 316.

Amano, H., Iwaya, M., Kashima, T., Katsuragawa, M., Akasaki, I., Han, J., Hearne, S., Floro, J. A., Chason, E. and Figiel, J. (1998), *Jpn. J. Appl. Phys.* **37** (Part 2), L1540.

Feltin, E., Beaumont, B., Laugt, M., de Mierry, P., Vennegues, P., Lahreche, H., Leroux, M. and Gibart, P. (2001), *Appl. Phys. Lett.* **79**, 3230.

Selvaraj, S. L., Suzue, T. and Egawa, T. (2009), *IEEE Electron Device Lett.* **30**, 587.

Raghavan, S. and Redwing, J. M. (2005), *J. Appl. Physics* **98**, 023514.
Hoke, W. E., Kennedy, T. D., Mosca, J. J., Kerr, A. J., Torabi, A., Davis-Hearns, S. and LaRoche, J. R. (2011), *J. Vac. Sci. Technol. B*, **29**, 03C107-1.
Shinohara, K., Corrion, A., Regan, D., Milosavljevic, I., Brown, D., Burnham, S., Willadsen, P. J., Butler, C., Schmitz, A., Wheeler, D., Fung, A. and Micovic, M. (2010), *IEDM Tech. Dig.*, 30.1.1.
Sunkara, M. K., Sharma, S., Miranda, R., Lian, G. and Dickey, E. C. (2001), *Appl. Phys. Lett.*, **79**, 1546.
Sverdlov, B. N., Martin, G. A., Morkoc, H. and Smith, D. J. (1995), *Appl. Phys. Lett.*, **67**, 2063.
Tanaka, S., Scott Kern, R. and Davis, R. F. (1995), *Appl. Phys. Lett.*, **66**, 37.
Wu, X. H., Fini, P., Keller, S., Tarsa, E. J., Heying, B., Mishra, U. K., DenBaars, S. P. and Speck, J. S. (1996), *Jpn. J. Appl. Phys.*, **35**, L1648.
Lei, T., Ludwig, K. F. and Moustakas, T. D. (1993), *J. Appl. Phys.*, **74**, 4430.
Calleja, E., Sanchez-Garcia, M. A., Sanchez, F. J., Calle, F., Naranjo, F. B., Munoz, E., Molina, S. I., Sanchez, A. M., Pacheco, F. J. and Garcia, R. (1999), *J. Cryst. Growth*, **201/202**, 296.
Bourret, A., Barski, A., Rouviere, J. L., Renaud, G. and Barbier, A. (1998), *J. Appl. Phys.*, **83**, 2003.
Hoke, W. E., Torabi, A., Mosca, J. J. and Kennedy, T. D. (2007), *J. Vac. Sci. Technol. B*, **25**, 978.
Nakada, Y., Aksenov, I. and Okumura, H. (1998), *Appl. Phys. Lett.*, **73**, 827.
Steckl, A. J., Devrajan, J., Tran, C. and Stall, R. A. (1996), *Appl. Phys. Lett.*, **69**, 2264.
Kobayashi, N. P., Kobayashi, J. T., Dapkus, P. D., Choi, W. J., Bond, E., Zhang, X. and Rich, D. H. (1997), *Appl. Phys. Lett.*, **71**, 3569.
Yang, J. W., Sun, C. J., Chen, Q., Anwar, M. Z., Khan, M. A., Nikishin, S. A., Seryogin, G. A., Osinsky, A. V., Chernyak, L., Temkin, H., Hu, C. and Mahajant, S. (1996), *Appl. Phys. Lett.*, **69**, 3566.
Semond, F., Cordier, Y., Grandjean, N., Natali, F., Damilano, B., Vezian, S. and Massies, J. (2001), *Phys. Stat. Sol (A)*, **188**, 501.
Radhakrishnan, K., Dharmarasu, N., Sun, Z., Arulkumaran, S. and Ng, G. I. (2010), *Appl. Phys. Lett.*, **97**, 232107.
Ganguly, S., Verma, J., Xing, H. G. and Jena, D. (2014), *Appl. Phys. Exp.*, **7**, 105501.
Nikishin, S. A., Faleev, N. N., Antipov, V. G., Francoeur, S., Peralta, L. G. D., Seryogin, G. A., Temkin, H., Prokofyeva, T. I., Holtz, M. and Chu, S. N. G. (1999), *Appl. Phys. Lett.*, **75**, 2073.
Stevens, K. S., Ohtani, A., Kinniburgh, M. and Beresford, R. (1994), *Appl. Phys. Lett.*, **65**, 321.
Calleja, E., Sanchez-Garcia, M. A., Monroy, E., Sanchez, F. J., Munoz, E., Sanz-Hervas, A., Villar, C. and Aguilar, M. (1997), *J. Appl. Phys.*, **82**, 4681.
Dasgupta, S., Wu, F., Speck, J. S., Mishra, U. K., (2009), *Appl. Phys. Lett.*, **94**, 151906.
Dharmarasu, N., Radhakrishnan, K., Sun, Z. Z. and Agrawal, M. (2011), *Phys. Stat. Sol. (C).*, **8**, 2075.
Takayanagi, K., Tanishiro, Y., Takahashi, M. and Takahashi, S. (1985), *J. Vac. Sci. Technol. A*, **3** 1502.

Dharmarasu, N., Radhakrishnan, K., Agrawal, M., Ravikiran, L., Arulkumaran, S., Lee, K. E. and Ing, N. G. (2012), *Appl. Phys. Exp.*, **5**, 091003.

Cao, Y., Zimmermann, T., Xing, H. and Jena, D. (2010), *Appl. Phys. Lett.*, **96**, 042102.

Murray, J. L. and McAlister, A. J. (1984), *Bull. Alloy Phase Diagrams*, **5**, 74.

Hoke, W. E., Torabi, A., Hallock, R. B., Mosca, J. J. and Kennedy, T. D. (2006), *J. Vac. Sci. Technol. B*, **24**, 1500.

Olesinski, R. W., Kanani, N. and Abbaschian, G. J. (1985), *Bull. Alloy Phase Diagrams*, **6**, 362.

Sze, M. and Ng, K. K. *Physics of Semiconductor Devices*. New York: Wiley, 3rd ed., pp. 396–397.

Ganguly, S., Song, B., Hwang, W. S., Hu, Z., Zhu, M., Verma, J., Xing, H. G. and Jena, D. (2014), *Phys. Stat. Sol. (C)*, **11**, 887.

Umeda, H., Suzuki, A., Anda, Y., Ishida, M., Ueda, T., Tanaka, T. and Ueda, D. (2010), *IEDM Tech. Dig.*, 20.5.1.

LaRoche, J., Hoke, W., Altman, D., McClymonds, J., Alcorn P., Smith, K., Chumbes, E., Letaw, J. and Kazior, T. *CS MANTECH Conference*, 345, 2013.

Cyurlia, P. N., Semond, F., Lester, T., Bardwell, J. A., Rolfe, S., Tang, H. and Tarr, N. G. (2010), *Electron. Lett.*, **46**, 253.

Chung, J. W., Lee, J.-K., Piner, E. L. and Palacios, T. (2009), *IEEE Electron Device Lett.*, 30, 1015.

Hoke, W. E., Chelakara, R. V., Bettencourt, J. P., Kazior, T. E., LaRoche, J. R., Kennedy, T. D., Mosca, J. J., Torabi, A. and Kerr, A. J. (2012), *J. Vac. Sci. Technol. B*, 30, 02B101.

Kazior, T. E., LaRoche, J. R., Lubyshev, D., Fastenau, J. M., Liu, W. K., Bulsara, M. T., Fitzgerald, E. A., Urtega, M., Bergman, J., Choe, M. J., Lee, K. J., Seong, T., Seo, M., Yen, A., Smith, D., Clark, D., Thompson, R., Drazek, C. and Guiot, E. (2010), *IEEE Compound Semiconductor IC Symposium (CSICS)*, 1–4.

Kazior, T. E., LaRoche, J. R. and Hoke, W. E. (2013), *IEEE Compound Semiconductor IC Symposium (CSICS)*, 1–4.

Marcon, D., Jaeger, B. D., Halder, S., Vranckx, N., Mannaert, G., Hove, M. V. and Decoutere, S. (2013), *IEEE Trans. Semicond. Manufacturing*, **26**, 361.

Kwan, M. H., Wong, K.-Y., Lin, Y. S., Yao, F. W., Tsai, M. W., Chang, Y.-C., Chen, P. C., Su, R. Y., Wu, C.-H., Yu, J. L., Yang, F. J., Lansbergen, G. P., Wu, H.-Y., Lin, M.-C., Wu, C. B., Lai, Y.-A., Hsiung, C.-W., Liu, P.-C., Chiu, H.-C., Chen, C.-M., Yu, C. Y., Lin, H. S., Chang, M.-H., Wang, S.-P., Chen, L. C., Tsai, J. L., Tuan, H.C. and Alex Kalnitsky, (2014), *IEEE IEDM Tech. Digest*, 17.6.1.

Chapter 4

III-V Semiconductors on Si: Arsenides and Phosphides for QWFET and BJT

Dmitri Lubyshev
IQE Inc., Bethlehem, PA.

Early experiments on the growth of III-V compound semiconductor epitaxial layers on silicon substrates started in the middle of the 1980's and were focused on obtaining low defect density epitaxial material, mainly GaAs layers, with applications in discrete high-frequency diodes and transistors. The primary motivation behind this work at the time was cost reduction via the substitution of expensive GaAs substrates by much cheaper Si substrates. The last three decades witnessed an exponential increase in the integration level of traditional silicon technology with >1 billion transistors/chip and scaling down of transistor characteristic size to 11 nm. Aggressive scaling of Si devices improved the frequency characteristics, achieving F_{max} of ~450 GHz for 32 nm Si System-on-Chip technology.[1] Likewise, performance of III-V transistors, both quantum well field effect transistors (QWFETs) and heterojunction bipolar transistors (HBTs) improved significantly as well, with F_t and F_{max} exceeding 1 THz.[2,3] Evolution of both technologies caused changes in scope for III-V device integration on Si. Currently this could be exemplified by monolithic integration of high-speed and optoelectronic devices with existing CMOS,

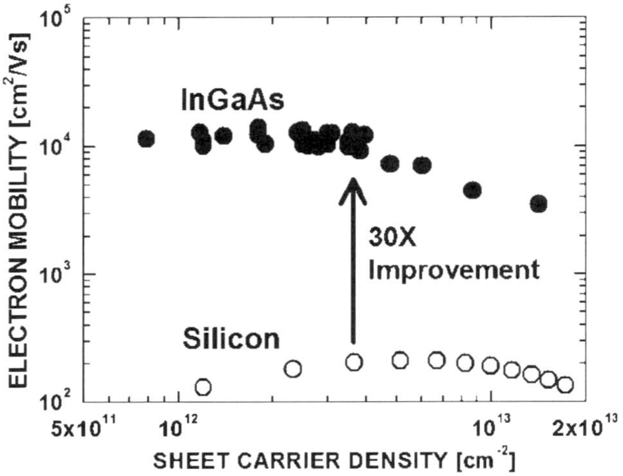

Fig. 1. Comparison of 300 K electron mobility for Si MOSFET with InGaAs/InAlAs QWFET.[4]

or even complete substitution of Si-based logic by high performance III-V components. The main catalyst for such integration is the III-V fabrication compatibility with existing Si fabrication process, maintaining the same integration level and advantage of large-diameter Si substrates (currently 8–12″, with prospective wafer diameter up to 18″).

Our choice of III-V materials for active device layers was focused on InP and lattice matched or elastically strained InAlAs and InGaAs ternary alloys. Such a material system offers excellent band-gap engineering and transport properties for both QWFETs and HBTs, and can be used as a platform for optoelectronic devices. The major advantage of n-type channel III-V QW FETs over Si-based FETs is a factor of 30 × increase in electron mobility as shown on Figure 1.[4] For HBTs, the same material system offers efficient band offset for the emitter-base heterojunction, InGaAs base material p-type doping up to 1×10^{20} cm^{-3},[5] low impact ionization rate for InP collector layers, and n-type doping in contact layers up to 5×10^{19} cm^{-3}. These material system advantages coupled with optimization of HBT topology and device processing has enabled HBT devices with F_t and F_{max} up to 0.52 and 1.1 THz.[3]

In this work, two different approaches will be discussed for III-V monolithic integration on Si:

(1) QWFETs grown directly on Si using a composite metamorphic buffer layer
(2) HBTs grown on Silicon-on-lattice-engineered-substrates (SOLES) developed by MIT and now commercially available.

We will begin by discussing general issues common to both device integration approaches. Then, unique growth aspects for each case, along with their experimental data will be presented separately. Material development discussed in this work was based primarily on growth by solid source molecular beam epitaxy (MBE). However, successful growth of III-V QWFET structures has also been demonstrated by metal-organic vapor phase epitaxy (MOVPE).

1. General Challenges in Direct Integration of III-V Semiconductors on Si, Ge, and SOLES Substrates

Direct growth of III-V compound semiconductors, like GaAs, GaP, and InP, on Si and Ge substrates were extensively researched and evaluated for over three decades. Although most attention was focused on GaAs growth on Si, observed phenomena are common and applicable to other combinations of mono-atomic and compound semiconductors.

Major challenges for the growth of GaAs on Si have been identified and summarized in the literature (e.g., review by Fang *et al.*[6] and more recently by Bolkhovityanov and Pchelyakov[7]) and are briefly summarized here:

(1) Formation of anti-phase boundaries (APBs) when switching from non-polar semiconductor (e.g. Si) to compound semiconductor (e.g. GaAs).
(2) Difference in thermal expansion coefficients ($2.3 \times 10^{-6}\,K^{-1}$ for Si versus $6.63 \times 10^{-6}\,K^{-1}$ for GaAs), which can lead to additional dislocation generation or even cracking of epitaxial layers during cool down after growth.
(3) 4% lattice mismatch between Si substrate and GaAs epilayer resulting in the formation of high density of misfit dislocations (up to 10^9–$10^{10}\,cm^{-2}$).
(4) Low surface wetting and poor chemical reactivity between Ga adatoms and Si substrate that can result in the formation of 3D GaAs islands instead of a continuous layer during the GaAs-on-Si nucleation process.
(5) Additional defect generation during island coalescence.

The formation of anti-phase GaAs domains in the growth on vicinal Si surfaces is related to the mono-atomic steps on a clean Si surface. The Si atoms in the (100) plane are arranged as dimers with the orientation of each successive layer rotated by 90° relative to the steps directly above and below. In contrast, the (100) surface of III-V compound semiconductor consists of bi-atomic steps due to the bonding nature of the binary compound. Thus

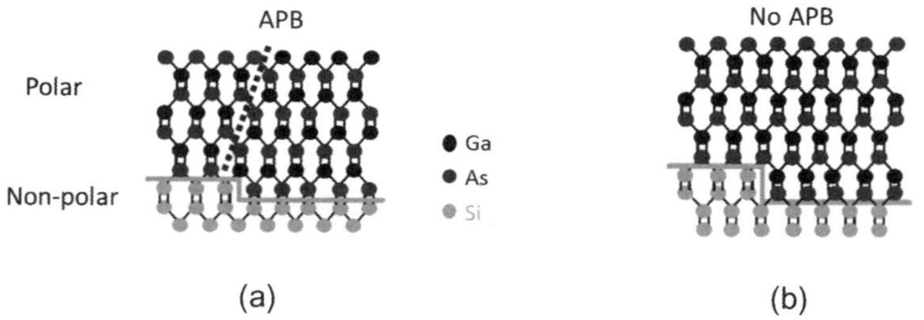

Fig. 2. (a) Schematic of anti-phase boundaries formation for compound semiconductors on the surface of mono-atomic semiconductors; (b) APB-free growth on Si surface with bi-atomic steps.

its growth on top of the mono-atomic step semiconductor (Si or Ge) will cause formation of boundaries shifted vertically by half of the GaAs lattice parameter (see Figure 2(a)), also known as anti-phase boundaries (APBs). This problem can be solved by using off-axis Si (100) substrates with 4 – 6° mis-cut angle toward the <110> direction.[8] Such atomically-clean Si surface can be converted to a predominately bi-atomic step array surface by thermal treatment at 900 – 1000°C in ultra-high vacuum.[9] Nucleation of compound semiconductor on these surfaces can produce APB-free epitaxial layers, as shown in Figure 2(b). We have evaluated GaAs growths on Si substrates with mis-cut angles 2, 3, 4 and 6°. Growth on substrates with 2° and 3° mis-cut resulted in GaAs layers with APBs as evident by the (4 × 4) reflection high energy electron diffraction (RHEED) patterns observed during growth at substrate temperatures (Ts) of 520°C. Growth on both 4° and 6° mis-cut substrates resulted in APB-free layers with (2 × 4) RHEED patterns, but the latter tends to produce rougher surfaces due to the formation of hillock-shaped defects. Based on these experiment results, we concluded that 4° mis-cut is optimal for growth of III-V (GaAs) material on Si under our growth conditions.

An atomically clean surface, free from oxide and carbon contaminations, is another critical condition for APB-free nucleation of GaAs on Si. In small amounts, these contaminants can inhibit the desired conversion from mono-atomic to bi-atomic step arrays during thermal treatment. In large amounts, carbon contaminants can easily react with Si, forming SiC islands which generate clusters of dislocations, stacking faults, or micro twins that are very difficult to eliminate in subsequent growth layers (see transmission electron microscopy (TEM) images in Figure 3).

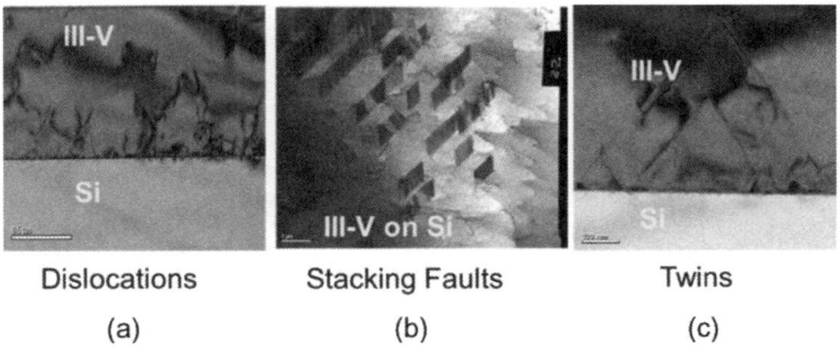

Fig. 3. TEM images of typical defects related to non-optimal GaAs nucleation on Si: (a) dislocations; (b) stacking faults (plan-view TEM); (c) micro twin defects.[23]

2. Composite Metamorphic Buffer Design

As discussed in the previous section, the APB challenge can be overcome by using appropriate substrate orientation. Addressing the other four challenges is not as straightforward, and relies more on the specific design and growth process for the III-V layers themselves. For example, the difference in thermal expansion coefficient between the substrate and the III-V epilayers is a fundamental parameter of the materials involved, and it places a limit on the maximum epilayer thickness that can be grown without cracking during cool down after growth. However, even below this thickness the residual strain can cause dislocation generation and accelerated dislocation gliding. This problem cannot be solved completely, but can be reduced by decreasing the total of epilayer thickness (which becomes a design goal below) and by thermal cycling of the buffer layer during growth.

Our goal is to design and implement an appropriate metamorphic buffer (M-buffer) template to transition from the Si or Ge substrate to the InP-based materials used to create our device structures of interest. The M-buffer has four main requirements:

(1) Effective spanning of the lattice parameter from Si to InP
(2) Efficient dislocation filtering to a level acceptable for devices
(3) Low surface roughness necessary for device processing
(4) Minimal total buffer thickness

For direct III-V integration of InP-based device structures on Si, we choose a composite metamorphic buffer consisting of GaAs ($\sim$4% lattice

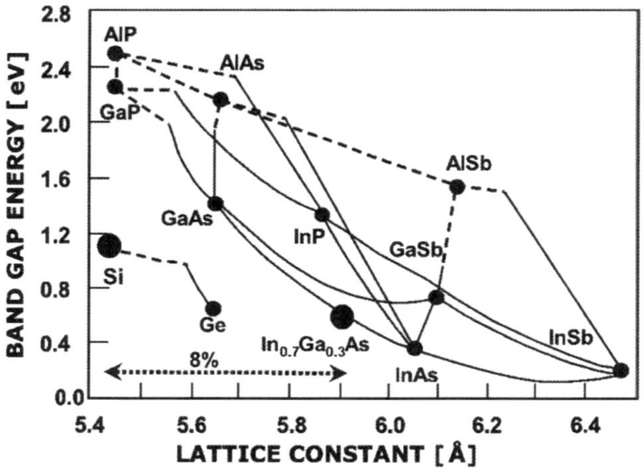

Fig. 4. Semiconductor band gap vs. lattice parameter; 8% difference in lattice constant between Si substrate and $In_{0.7}Ga_{0.3}As$ active layer need to accommodate in composite metamorphic buffer.

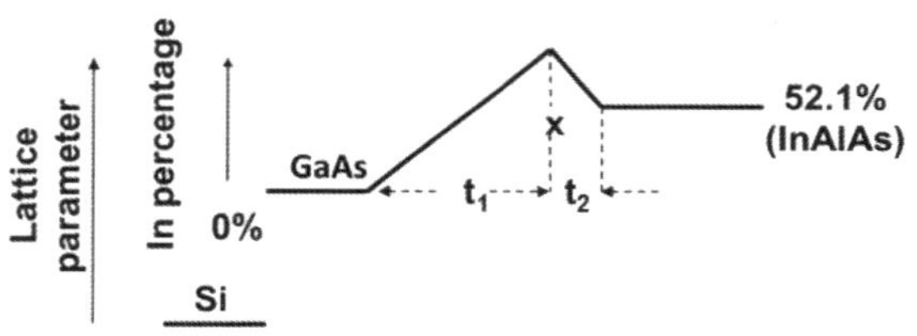

Fig. 5. Schematic of composite buffer design. Parameters x, t_1 and t_2 for graded InAlAs layer are chosen for complete compensation of residual strain in M-buffers.[10] InAlAs composition at the end of the M-buffer is lattice matched to InP.

mismatch with Si) and graded InAlAs (compensating for the additional ~4% lattice mismatch between the active layers and GaAs). The classic III-V band gap energy diagram (Figure 4) demonstrates the lattice parameter mismatch with in Si. A schematic of a composite buffer is shown in Figure 5. The GaAs layer is typically 0.5–1.0 μm thick and the graded InAlAs layer typically has a total thickness of 0.5–1.5 μm. The wide band gap InAlAs helps electrically insulate the III-V device layers from the underlying substrates, and can help reduce buffer leakage and back gating effects in fabricated QWFET devices. The design of this layer is generic and applicable to the growth of both QWFET on Si and HBT on SOLES substrates. In order to obtain the InP lattice parameter on top of the M-buffer, the indium mole fraction was ramped from ~0.03 to ~0.60 and then back to 0.52. This inverse graded step was used to compensate for residual strain in the epilayer. Buffer design with a continuously graded InAlAs layer and inverse graded step was initially proposed by Cordier[10] and later successfully adopted for InP-based metamorphic high electron mobility transistors (M-HEMTs) and M-HBTs grown on GaAs substrates.[11–14] Such graded M-buffers on GaAs with thickness ~1.1 μm demonstrated effective dislocation filtering resulting in dislocation density around 1×10^7 cm^{-2} and cross-hatched surface morphology with measured atomic force microscopy (AFM) root-mean-square (rms) roughness of around 15–35 Å in the device active layers. Device performance of M-HEMT grown on GaAs substrates were very close to reference devices grown on InP. As minority carrier devices, MHBTs show higher sensitivity to the presence of residual defects in the base area and exhibit figures-of-merit (FOM), such as current gain and breakdown voltage, about 85% from reference devices on InP. The remainder of this chapter will build upon the baseline issues just discussed. First, we will look at the growth of QWFET structures directly on Si substrates, and then we will discuss HBTs grown on SOLES substrates. Both device architectures will require an M-buffer to transition from the group IV surface to the InP-based III-V compound semiconductor alloys. While there are many similarities, the specific nature of the growth process is quite unique for each of these two device structures.

3. Quantum Well Field Effect Transistors (QWFETs) Grown Directly on Si Substrates

Development of III-V based QWFETs is targeting digital logic application on Si, and thus needs to satisfy some specific and strict performance requirements. One key advantage of III-V materials for this application is their superior channel mobility compared to Si. The following discussion will focus on $In_{0.7}Ga_{0.3}As$-based QW layers pseudomorpically strained to the M-buffer, although the same approach is also applicable to other similar alloys. The low electron effective mass of InGaAs, together with its large band offset with both InP and latticematched InAlAs, offer flexibility in QWFET design with good carrier quantum confinement in the QW for a wide range of carrier concentrations. In addition, any structure proposed should be suitable for aggressive device scaling to match the ultra-small device size achieved for Si CMOS. This would require a reduced composite buffer thickness without degradation of device characteristics caused by the presence of residual defects and dislocations. The possibility to confine misfit dislocations in very thin layers as close as possible to the substrate and realize rapid reduction in defect density across the composite M-buffer becomes a key point for this work. We will describe critical aspects in the growth procedure and will present data from material and device characterization in the following sections.

3.1. *Buffer design and growth optimization*

The growth process for QWFET structure on Si began with surface preparation and deoxidation of the Si substrate. Native SiO_2 oxide typically sublimes in ultra-high vacuum at temperatures around 1000 to 1050°C. However, running the substrate heater at such high temperatures can generate heat load on the chamber walls and cryoshrouds of the MBE reactor, which can negatively impact background pressure, thus risking substrate surface contamination immediately following the oxide desorption. In general, Si wafer de-oxidation on large production tools are best accomplished in a separate, clean chamber. For this work, using a Veeco GEN-III single wafer MBE system, we evaluated various types of Si surface treatments prior to GaAs deposition. An ultraviolet ozone process helps reduce surface carbon contaminations of the Si wafer, and thus limits formation of SiC during the desorption procedure, but does not help lower the de-oxidation temperature. Implementation of a group-III assisted desorption process, where the surface is exposed to a Ga or In molecular flux, can lower the de-oxidation temperature to ~850°C, but the process window is very narrow and the results are not easily reproducible in terms of surface quality. A reproducible and clean Si surface could be obtained with the pre-growth cleaning procedure described in Ref. [15]. Following some experimentation, we were able to establish a modified wet-chemical wafer cleaning procedure that was simpler to implement immediately prior to loading the Si substrate into the MBE chamber. This process includes subjecting the substrate to SC1 solution ($NH_4OH:H_2O_2:H_2O = 1:1:5$) for 10 minutes, de-ionized water rinse, SC2 solution ($HCl:H_2O_2:H_2O = 1:1:6$) for 10 minutes, and a final dip in buffered HF solution. This last step keeps the Si surface oxide-free and hydrophobic with the interfacial Si bonds saturated by hydrogen. Reduced oxide desorption temperature (600–800°C) with minimum SiC contaminations on the substrate surface was consistently obtained with this pre-growth cleaning process.

The growth process for GaAs buffer growth on Si was evaluated *in-situ* by monitoring the RHEED patterns in the [1,1,0] and [1,−1,0] directions as illustrated in Figure 6. The RHEED patterns from freshly prepared Si substrates exhibited diffused background from a thin oxide layer as well as crystalline diffraction patterns from Si (Figure 6(a)). After

(a)
(b)
(c)
(d)
(e)

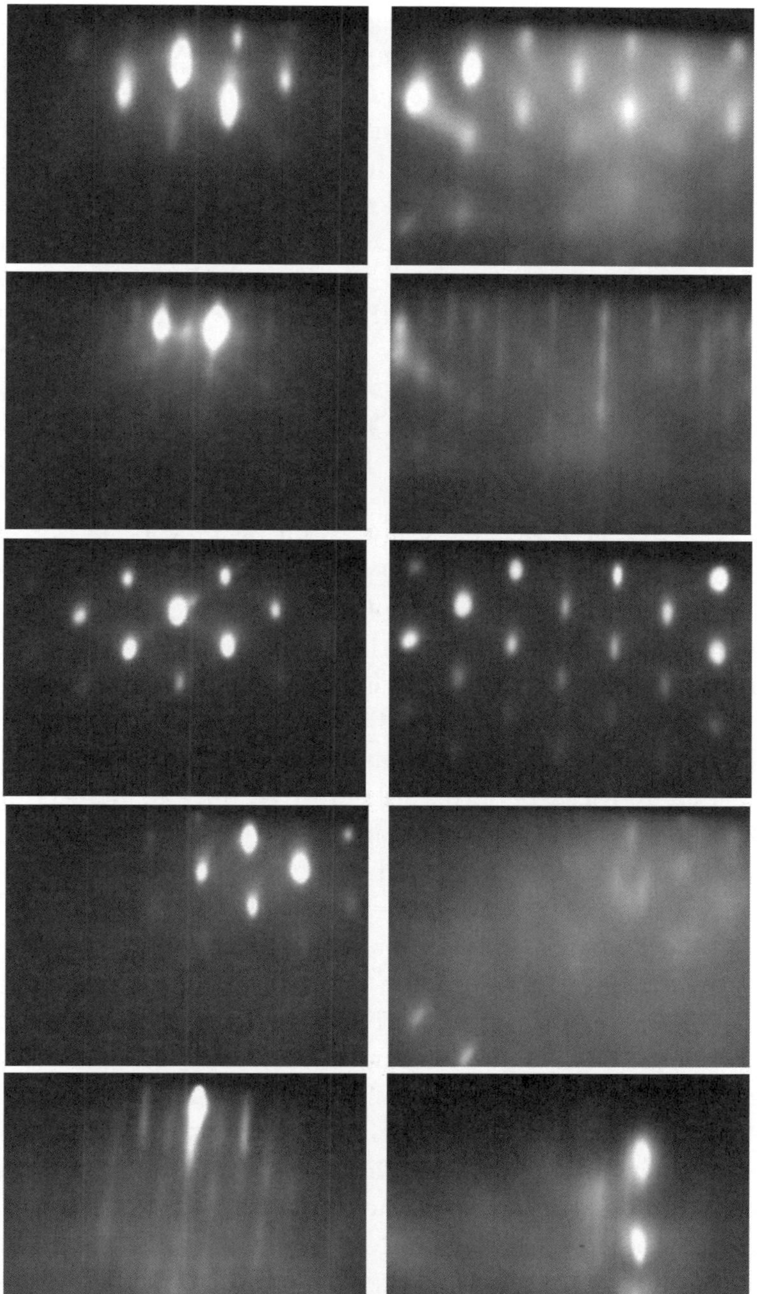

Fig. 6. RHEED pattern evolution during growth of GaAs buffer on Si: (a) Si substrate with oxide after transfer to growth chamber; (b) after de-oxidation at 900°C; (c) after deposition of 10 ML of GaAs at reduced Ts; (d) after growth of 0.2 μm GaAs; (e) after growth of 0.5 μm GaAs.

desorbing the surface oxide, further annealing at ~850°C for 20 minutes promoted the coalescence of mono-atomic steps into desirable bi-atomic steps required for APB-free GaAs nucleation. This is evident by the presence of a (2×2) superstructure in the RHEED patterns shown in Figure 6(b). Initial nucleation of GaAs on Si caused the formation of 3D islands as indicated by the spotty RHEED patterns taken after deposition of 10 MLs of GaAs (Figure 6(c)). The presence of micro-twin crystal defects was also evident at this stage. Subsequent growth resulted in the transformation of the spotty patterns into elongated RHEED streaks, elimination of micro-twin defect features, and the appearance of a (2×4) superstructure (Figure 6(d)). The presence of GaAs (2×4) superstructure on Si suggested the growth of primarily APB-free material, where otherwise (4×4) RHEED patterns would appear. This stage of growth, corresponding to island coalescence and elimination of crystal defects and faults related to nucleation, was the most crucial step for successful buffer growth. If surface roughening exceeded a certain critical level, 3D morphology would develop and would not support the uniform growth of the thick M-buffer and device layers. Under optimized conditions, material crystal quality and surface roughness should improve with thickness of the GaAs buffer layer. RHEED images after deposition of 0.5 μm of GaAs, Figure 6(e), showed bright (2×4) patterns with long steaks typical of an atomically-smooth surface. It is important to note that RHEED patterns are sensitive only to defects and surface roughness on a small scale comparable to the electron coherence length, and Nomarski microscopy and AFM were used to further assess surface morphology. Typical AFM rms roughness values of 5 to 25 Å can be obtained from 0.5–2.0 μm thick GaAs buffer layers grown on Si by MBE.

For the initial GaAs nucleation on Si, challenges posed by low surface wetting and poor chemical reactivity between Ga adatoms and the Si substrate must be overcome. Our solution involved the use of a two-step growth procedure starting with a low GaAs nucleation temperature of 200–300°C to obtain full surface coverage by GaAs, followed by increased growth temperature for the remainder of the GaAs buffer. The low nucleation temperature limits surface mobility of Ga adatoms, and provides high density of small atomic clusters that helps reduce defect generation during island coalescence and form a smooth surface. Alternative methods reported to improve surface coverage and wetting while compensating the lattice mismatch difference include the insertion of non-III-V buffer layer materials. For example, the chemical properties of CaF_2 and SrTiO layers allow them to be compliantly grown on Si with low dislocations and help bridge the difference in lattice constants, but they provide limited advantage due to difficulties associated with subsequent III-V material nucleation on them.[16–20] On the other hand, Ge and SiGe surfaces are more suitable for nucleation and growth of III-V compounds (especially GaAs), but they present their own challenges to obtain thin, low dislocation density buffer layers on Si[21] as will be discussed in more details in the HBT section below.

Typically, ~0.2 μm of GaAs was grown using the 2-stage growth process. The remainder of the GaAs buffer was then accomplished with progressive growth and annealing cycles. This method is similar to the cyclic anneal procedures developed for MOVPE.[22] The growth and annealing temperatures were optimized for our MBE growth process with the maximum limited to 570°C to prevent surface roughening. The *in-situ* annealing cycles are found to be effective in annihilating defects, filtering dislocations, and accommodating stress related to the difference in thermal expansion coefficient between Si and GaAs.

The reduction of dislocation density as a function of annealing temperature and GaAs buffer thickness was evaluated by comparing the full-width at half-maximum (FWHM) values of GaAs layer peaks in high resolution X-ray diffraction (HRXRD) measurements. It is not trivial to quantify dislocation density in thin layers especially when they are changing with layer thickness. This will require the use of sophisticated measurements such as X-ray reciprocal space map (RSM) and transmission electron microscopy (TEM). RSM can provide data on dislocation density, uniformity of plastic relaxation within the substrate plane, residual strain, and possibly epi-layer tilt. TEM analysis is helpful in determining defect types and distribution as well as dislocation filtering mechanisms. Due to the complexity of these techniques, we used a simplified approach for GaAs buffer optimization that would allow us to track the change in dislocation density with growth conditions. Quick, simplified semi-quantitative dislocation density estimation in the GaAs buffer layer was made using the equation:

$$D = \beta^2/9b^2$$

where D is the dislocation density, β is the FWHM of HRXRD GaAs peak in radians, and $b = \sqrt{2}a$ for 60° dislocations with $a = 5.65$ Å for the GaAs lattice constant.

The GaAs buffer growth procedure described above yielded layers with low FWHM values of 350, 320, and 260 arcsec at buffer thickness values of 0.7, 1.2, and 3.5 μm, respectively. Estimated average dislocation densities in

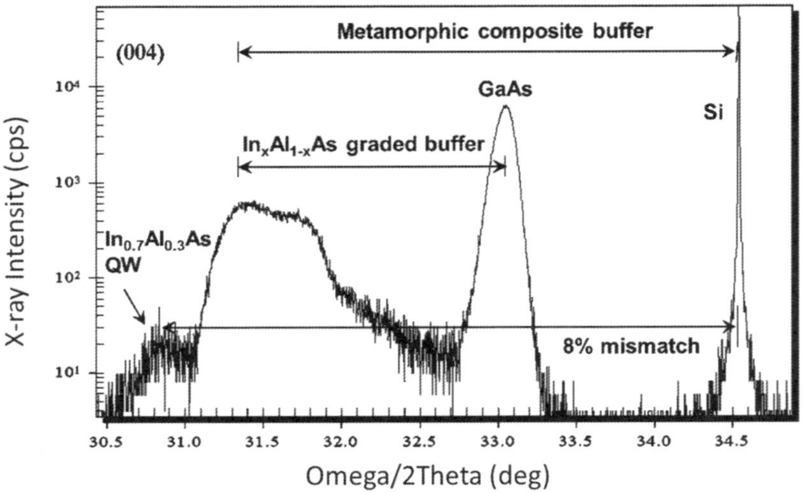

Fig. 7. HRXRD scan of composite metamorphic buffer on Si.[23,24]

these GaAs layers were 5.0×10^7, 4.2×10^7 and 2.8×10^7 cm^{-2}, respectively. Reduction of buffer thickness below 0.7 μm caused rapid increase in FWHM values with more scattered results.

A graded InAlAs M-buffer was used to transition the lattice constant from GaAs to InP and was grown around 370 to 400°C. As shown in Figure 7, diffraction peaks from the GaAs buffer, the InAlAs graded M-buffer with inverted step, and the strained QW active layer on top are all clearly visible in the HRXRD spectrum.[23,24]

Determination of defect density in the composite M-buffer may be complicated due to non-uniform dislocation distribution in both the GaAs portion of the buffer and in the linearly-graded InAlAs layer. If dislocation density rapidly decreases with thickness, FWHM measurement cannot be used alone to assess material quality since it is sensitive to the average density throughout the entire layer. In addition to the rapid change in dislocation density with buffer thickness, difference in dislocation generation and filtering was observed on composite M-buffers in the two orthogonal in plane directions resulting in significant changes in FWHM measured along the [1,1,0] and [1,−1,0] directions. This issue can be partially assessed by X-ray RSM measurements which enable the evaluation of plastic relaxation anisotropy, residual strain between layers, and tilt between epilayer and substrate, if present. On the other hand, cross-section (X-) and plan-view (PV-) TEM are helpful in determining the dislocation distribution vertically through the layer as well as on the surface, with detectivity limit around

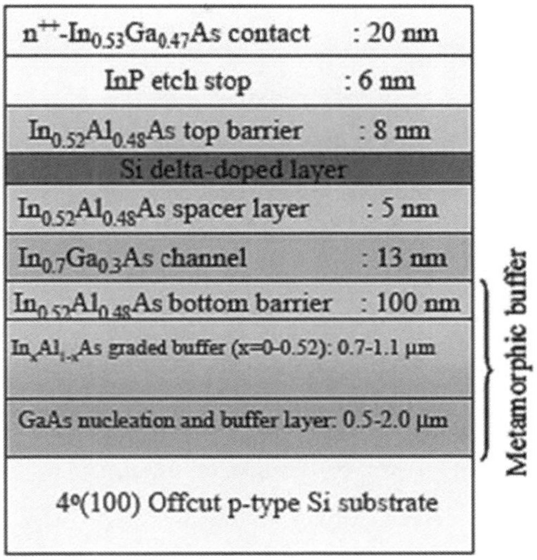

Fig. 8. Heterogeneous integration of $In_{0.7}Ga_{0.3}As$ QWFET on Si substrate using metamorphic composite buffer architecture with total buffer thickness from 3.2 to 1.3 μm.[23]

1×10^7 cm^{-2}. Additional data on material quality can be obtained by comparing material transport properties of the QWFET structures grown on InP and on Si substrates. Ultimate proof of the viability of our heterogeneous growth approach will come from device performance comparison with growth on native substrates.

During the course of this work, the possibility to reduce the composite M-buffer thickness without significant degradation of material performance was explored.[23] For material quality evaluation, an M-HEMT structure with layer design shown in Figure 8 was used. The device active layers consisted of $In_{0.7}Ga_{0.3}As$ QW, $In_{0.52}Al_{0.48}As$ spacer and barrier layers with Si delta doping, InP etch stop layer, and $I_{0.53}Ga_{0.47}As$ contact layer. The QW layer was pseudomorphically strained to the buffer layer. The higher indium mole fraction in the QW layer provided improved transport properties compared to lattice-matched InGaAs due to the combination of reduced electron effective mass and larger band gap discontinuity, hence allowing a higher charge concentration in the QW without thermionic escape of the carriers back to the barrier layer.

Dislocation distribution and material structural quality were evaluated. TEM images of the wafers with composite M-buffer thickness of 2.0, 1.5, and 1.3 μm are shown in Figures 9(a–c). The GaAs buffer layer showed high defect density at the GaAs/Si interface, but significant reduction toward the GaAs/InAlAs graded M-buffer interface. However, threading dislocations

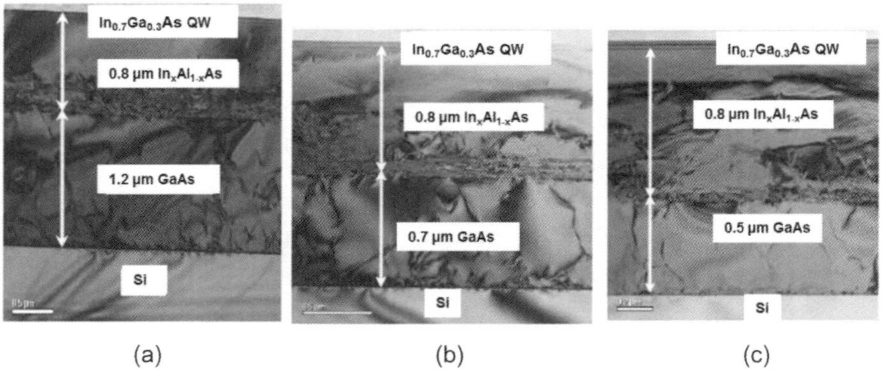

Fig. 9. Cross-sectional TEM images of $In_{0.7}Ga_{0.3}As$ QWFET on Si using composite buffers with total buffer thickness: (a) 2.0 μm; (b) 1.5 μm; (c) 1.3 μm.[23]

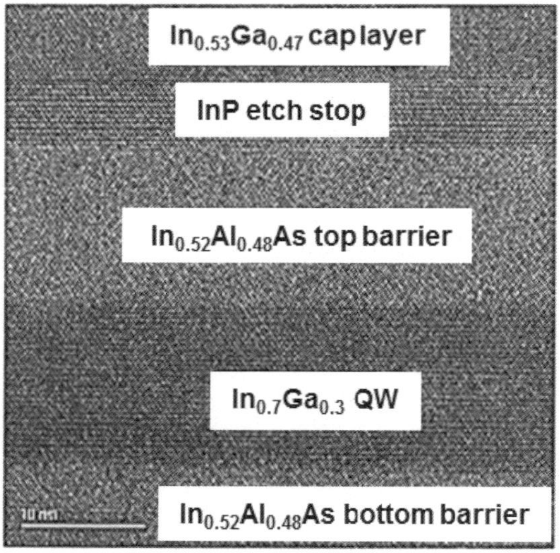

Fig. 10. High resolution X-TEM of QW device area.[23]

were observed to propagate into the top of GaAs layers in all cases. Another area with high dislocation and defect density was observed at the beginning of the graded InAlAs layer. Dislocation density rapidly decreased after approximately 0.4 μm of InAlAs deposition, leaving the device layers virtually dislocation-free. We speculate that the strain field created by the InAlAs grading help blocked some of the dislocations propagating from the GaAs layer (see GaAs/InAlAs interface in Figure 9(c)). High resolution TEM image, Figure 10, confirmed that the device active layers were essentially defect-free with sharp interfaces between all layers.

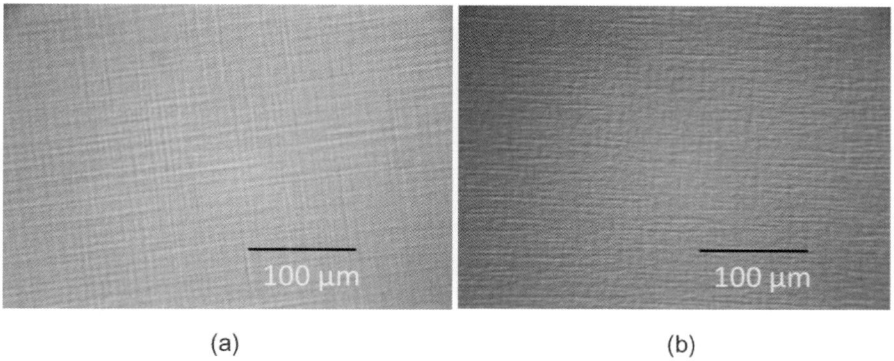

Fig. 11. Nomarski images of device structure grown on: (a) GaAs and (b) Si. Observation of cross-hatch pattern on Si indicates high quality of M-buffer on top of GaAs buffer.[23]

Strain relaxation during lattice mismatch growth can occur not only by dislocations or defect generation, but also through the development of a 3D growth surface, and sometimes by epilayer tilt relative to the substrate. Residual stress applied to the substrate itself can cause wafer bowing. Island coalescence during nucleation can roughen the GaAs buffer surface, resulting in an orange-peel-like morphology. Growth of linearly-graded layers on GaAs substrates typically results in the development of a cross-hatch pattern[10] as evidence on a reference QWFET structure grown on GaAs (Figure 11(a)). A superposition of cross-hatch pattern with orange-peel background was observed for a full stack structure grown on Si (Figure 11(b)). The rms surface roughness measured on a $5\times5\ \mu m^2$ AFM scan was in the range of 30 to 40 Å (Figure 12), similar to that of an $In_{0.7}Ga_{0.3}As$ QWFET grown on GaAs.[14]

Electron transport properties were evaluated by comparing nominally the same QWFET structures grown on InP substrate (reference), on GaAs substrate with InAlAs M-buffer, and on Si substrate with composite M-buffer. The presence of dislocations and defects in the active layer was typically inferred by degradation of electron mobility and sometimes by change in Si dopant activation. A summary plot, Figure 13, compared data for such structures with room temperature (RT) electron charge density nominally in the 1.9 to $4.4\times10^{12}\ cm^{-2}$ range. Room temperature mobility for QW structures grown on all three types of substrates was in the 9,000 to 10,200 cm^2/Vs range, close to the mobility limit for such structure design. Minimal or no RT mobility degradation compared to reference sample grown on InP was reported for structures grown on Si, suggesting low defect density in the active layers due to effective dislocation filtering within

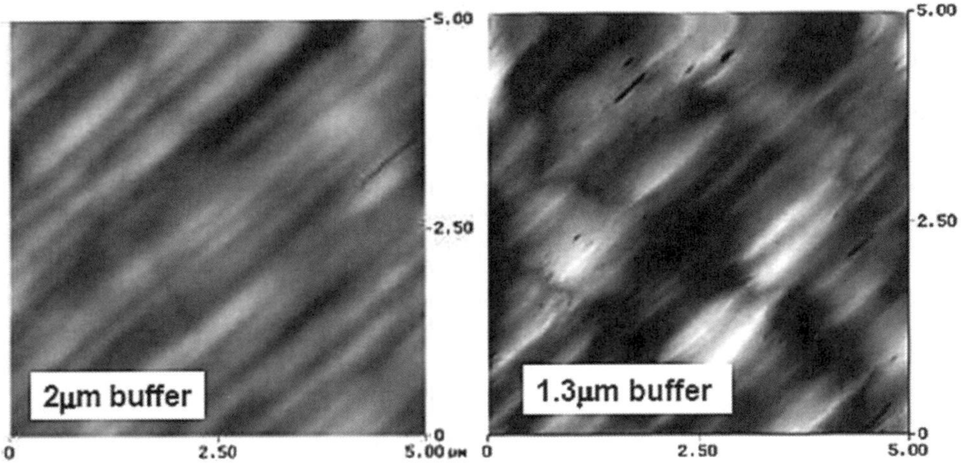

Fig. 12. AFM images of device structure on Si with composite buffer thickness (a): 2 μm, rms = 30 Å and (b): 1.3 μm, rms = 39 Å.[23]

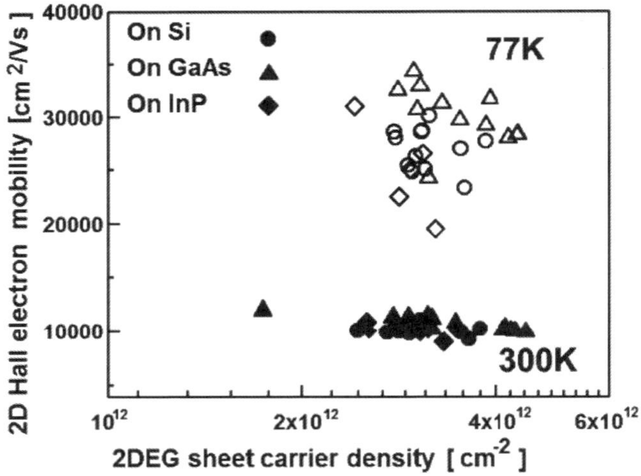

Fig. 13. Electron mobility versus sheet carrier density in $In_{0.7}Ga_{0.3}As$ QW grown using composite buffer on Si, metamorphic buffer on GaAs and lattice-matched growth on InP (reference structures).[23]

the composite M-buffer. The scattering of 77 K mobility values, observed for all three types of structures, was related to the optimization of active layer design; more specifically to the spacer layer and InP etch stop layer thickness. The effect of total composite buffer thickness on QW mobility was evaluated by measuring a set of structures with the same active layer design and doping, grown with different composite buffer thicknesses as shown

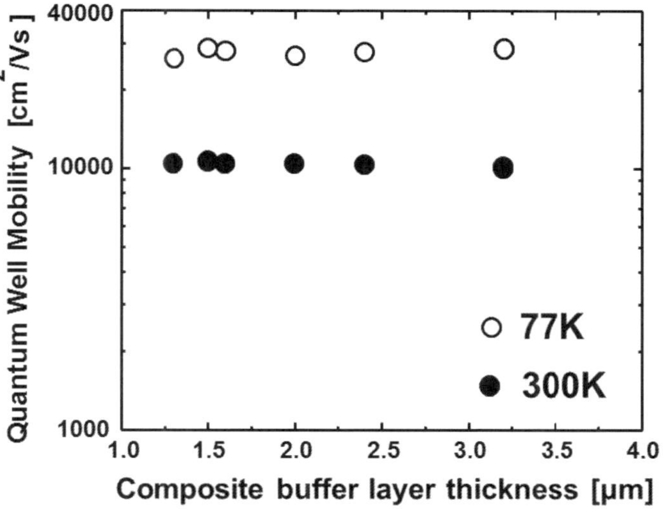

Fig. 14. $In_{0.7}Ga_{0.3}As$ QW mobility versus composite buffer thickness on Si at 300 K and 77 K.[23]

in Figure 14. No mobility degradation was observed for buffer thickness ranging from 1.5 to 3.2 μm, with a minimal reduction for the thinnest, 1.3 μm design.

The majority of the data III-V Si integration presented here was obtained from samples grown by MBE, which offers excellent process control. However, MBE growth is nonselective and line-of-sight, which may pose challenges for process integration and nonplanar 3D devices.[24] In contrast, an alternative epitaxial growth method, metal-organic vapor phase epitaxy (MOVPE), offers several advantages such as selective area growth, conformal growth on 3-D structures, and lower surface particle density. Prior to this work, there have been few studies[25] on the feasibility of MOVPE-grown III-V films on Si for logic device applications, and no direct comparisons with MBE materials has been reported. Here, we compared MOVPE and MBE growth processes, demonstrating for the first time that MOVPE III-V material quality can match that of MBE wafers.[26] Composite M-buffer thickness for both MBE and MOVPE was aggressively reduced to 1.3 μm and consisted of a 0.5 μm GaAs layer grown directly on Si followed by a 0.7 μm graded InAlAs layer. GaAs buffer layers grown by both methods demonstrated good surface morphology (Figures 15(a, b)) with rms values of 20 and 24 Å for MOVPE and MBE, respectively. The MOVPE-grown buffer shows slightly better structural quality as evidenced not only by the slightly lower rms value, but also by the absence of

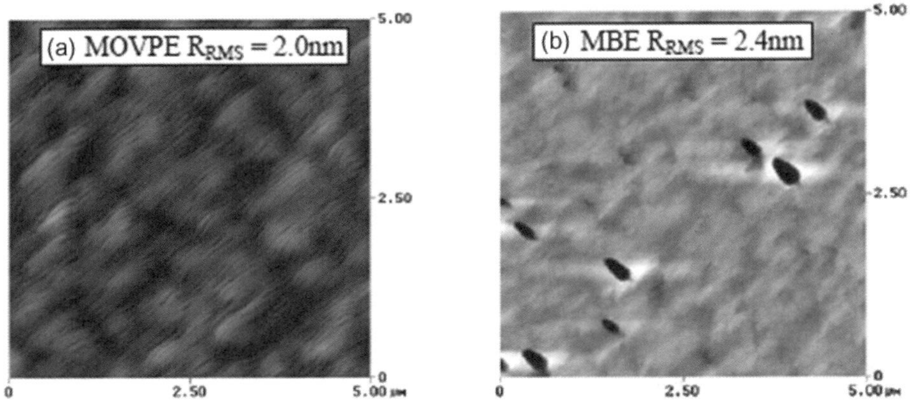

Fig. 15. Typical AFM scans of GaAs buffer with thickness 0.5 μm grown on Si by (a) MOCVD and (b) by MBE.[26]

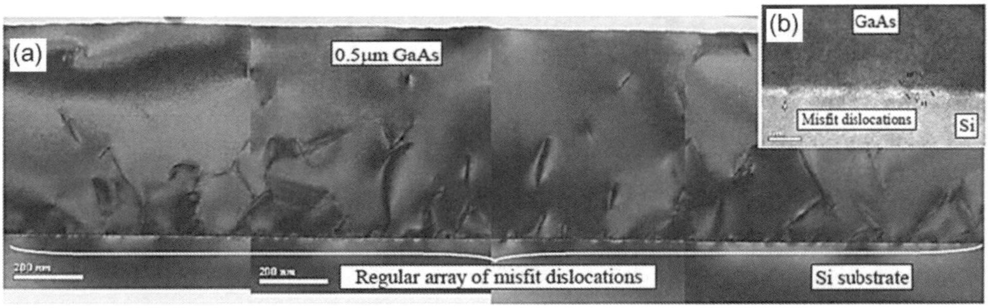

Fig. 16. Cross-sectional TEM of the 0.5 μm GaAs buffer on Si grown by MOVPE: (a) large area composite image and (b) high resolution TEM, showing a regular array of misfit dislocations at the Si-GaAs interface.[26]

defects propagated through the layer (black spots on MBE wafer shown in Figure 15(b), and by lower (004) ω-FWHM values (600 arcsec for MOVPE versus 1000 arcsec for MBE). The nature of MOVPE quality improvement was studied by analyzing the X-TEM image of the GaAs buffer depicted in Figure 16. The large area composite image shows the presence of threading dislocations, as well as the formation of a regular array of edge misfit dislocations at the GaAs/Si interface (see high resolution image in the inset). The presence of these dislocation arrays are highly desirable as it can accommodate a significant portion of the strain directly at the GaAs/Si interface, thereby reducing the density of dislocations threading into the GaAs layer. This mechanism of plastic relaxation was not evident in our MBE-grown material. However, artificial generation of interfacial misfit dislocation arrays has been successfully employed[27–29] to grow high quality

GaSb and AlSb layers on GaAs and Si substrates by MBE. As reported in Ref. [28], low dislocation density was obtained by establishing a planar growth mode after depositing only ∼3 MLs of GaSb on GaAs. In this case, strain relaxation occurred through the formation of 90° in-plane dislocation network, as opposed to the generation of large-size nucleation islands coalescing via the merging of {111} planes which cause 60° misfits. We speculated that the reduced dislocation density in our MOCVD-grown GaAs material is related to the formation of continuous flat layer at the early stage of growth due to higher nucleation island density and/or higher lateral growth rate.

Surface morphology and defect density of full QWFET device stacks with buffer thickness 1.3 μm was evaluated by AFM, TEM, and RSM.[26] Both MOVPE and MBE surfaces appeared uniform and featureless, but roughness of the former was somewhat higher than the latter (60 Å versus 44 Å as shown in Figure 17). PV-TEM images of MOVPE sample revealed the presence of defects with size $\sim 50 \times 100\,\text{nm}^2$, suggesting incomplete defect filtering in the composite buffer as compared to defect-free structures grown lattice-matched on InP (Figure 18). Defect density estimated by TEM was about $\sim 2 \times 10^9\,\text{cm}^{-2}$. A typical (004) X-ray RSM, Figure 19, showed diffraction peaks from the Si substrate, GaAs buffer, graded InAlAs M-buffer, and the QW device layers. The ω-FWHM of peaks from the 0.1 μm bottom barrier layer and the QW (in this case with the same lattice parameter) were 1600 and 1800 arcsec for MOVPE and MBE, respectively. These results demonstrated that the active layer material quality of MOVPE was similar to that of MBE.

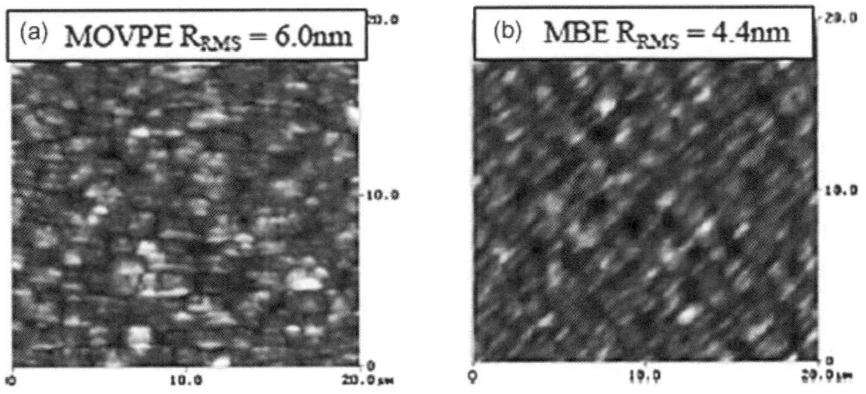

Fig. 17. Typical AFM scans from the entire device stack with n+ cap, delta-doped layer, InP layer and QW layer: (a) MOCVD grown material: (b) MBE grown material.[26]

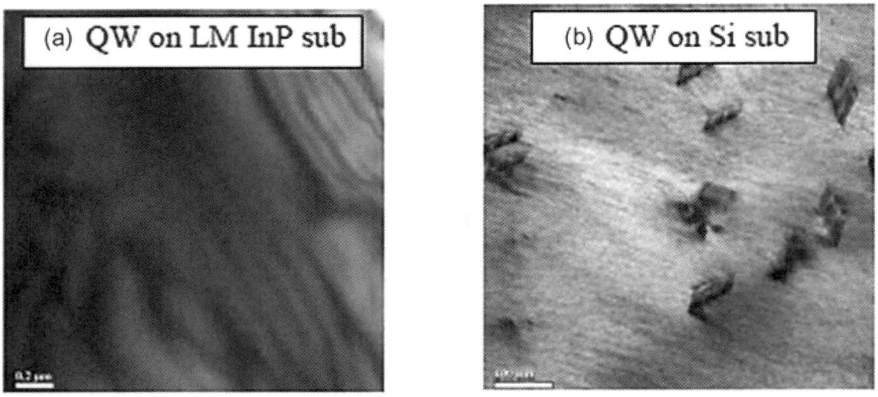

Fig. 18. Plan-view TEM images from 50 nm thick InGaAs QWs: (a) reference structure grown on InP and (b) grown on Si by MOCVD.[26]

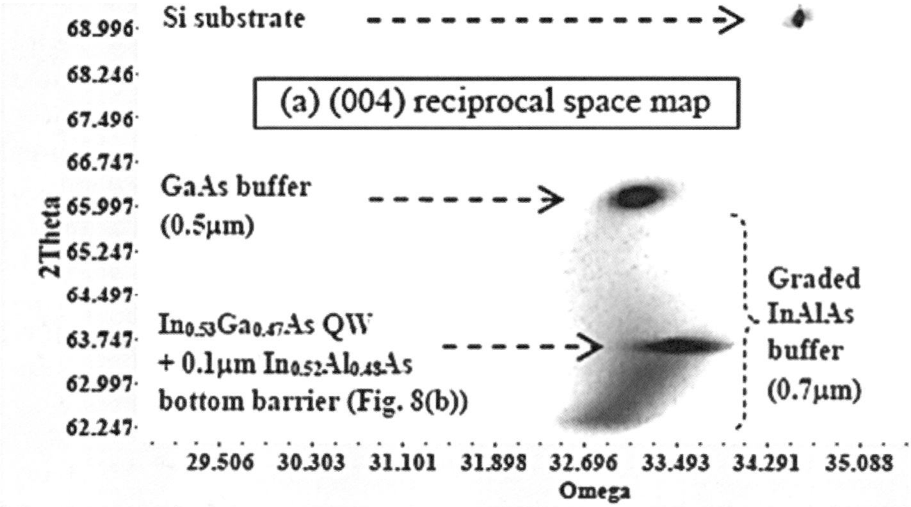

Fig. 19. (004) X-ray reciprocal space map from device stack structure with 1.3 μm composite buffer layer grown on Si by MOVPE.[26]

Transport properties of structures grown by MOVPE were compared to material grown by MBE on InP (reference structure) and Si using Hall Effect measurements. The MOVPE $In_{0.53}Ga_{0.47}As$ QW structure grown on Si exhibited high Hall mobility, matching that obtained by MBE structures grown on lattice-matched InP at 300 K (~8000 cm^{-2}/Vs) and 77 K (>22000 $cm^{-}2$/Vs). These values well-exceeded the 6180 and 16500 cm^{-2}/Vs reported in Ref. [25], which to the best of our knowledge is the only other MOVPE data for III-V material grown on Si published to-date. Both MBE and MOVPE grown materials were examined for the formation of multiple transport paths with parasitic parallel conduction. Quantitative mobility spectrum analysis data for the device stack showed the presence of only one *n-type* conductivity peak, which was attributed to electrons contained inside the QW, thus confirming the absence of parallel conduction.

The high-quality Hall mobility values measured on the III-V/Si structures cannot be a stand-alone FOM characteristic to ensure device quality. The previous paragraph demonstrated a high defect density $\sim 2 \times 10^9$ cm^{-2}, evaluated by RSM and X-TEM for these structures with composite buffer thickness 1.3 μm, which at first appears contradictory to the high Hall mobility values. The Hall measurement is taken over a larger surface area, where individual dislocations may be of small consequence. However, these structural defects might cause significant dispersion in electron transport on the microscopic scale and lead to the difference in electrical characteristics and reliability of neighboring small area transistors on chip. Thus, we cannot rely on either high Hall mobility values or on dislocation density values alone to determine overall device material quality. Instead, these must be evaluated in combination with other evaluation techniques more sensitive to device degradation, such as structural analysis and electrical characteristics of small area devices to determine the overall performance of these III-V devices on Si.

In summary, to successfully integrate a III-V QWFET structure grown directly on Si substrates, we optimized the design and growth conditions of a composite M-buffer to simultaneously satisfy the competing requirements of reduced total thickness, effective dislocation filtering, low surface roughness, and excellent transport properties. Dislocation density in the active device area varied from $\sim 2 \times 10^9$ to $\sim 5 \times 10^7$ cm^{-2} for composite M-buffer thickness ranging from 1.3 to 3.0 μm for MBE-grown material. TEM images indicated that the top part of the GaAs buffer was not dislocation free, but contained residual threading dislocations that were partially filtered by the subsequent growth of the graded InAlAs portion of the M-buffer. Further reduction in dislocation density in the GaAs layer required optimization of the GaAs nucleation procedure or development of an alternative buffer, since the GaAs/Si interface was a major source of defect generation. The surfaces of III-V QWFET-on-Si wafers typically exhibit cross-hatch patterns, developed during graded InAlAs M-buffer growth, superimposed on a uniform "orange-peel" background originating from the GaAs layer. Surface roughness was determined by AFM to be in the 35 to 60 Å range, which was acceptable for device fabrication. Electron mobility in QWFET structures grown on Si reached 85–100% of the values measured on reference structures grown lattice-matched on InP, and were typically in the range of 8,500 to 10,000 cm^{-2}/ Vs for a channel charge range of 2.0 to 4.5 $\times 10^{12}$ cm^{-2}.

3.2. *Transistor fabrication and device results*

Device optimization for monolithically integrated III-V QWFET on Si was performed in parallel with epitaxial growth optimization and passed through several interactions. The goal for the first generation QWFET device was to check feasibility and to obtain baseline device characteristics for comparison with standard gate length (L_G) Si *n*-MOSFET transistors.[23] Epitaxial structure design used for device fabrication is shown in Figure 8. Mesa device structure was fabricated by a combination of wet-chemical and reactive ion etches to recess the gate towards the channel as well as Pt/Au Schottky gates to enable enhancement-mode (e-mode) operation and improve short channel performance of the device (Figure 20). The I_D-V_{DS} characteristics of the e-mode L_G = 80 nm QWFET on Si with 1.3 μm composite M-buffer is shown in Figure 21. The device exhibits good transistor characteristics and achieves threshold voltage (V_T) of +0.11 V, I_{Dsat} = 0.32 mA/μm and I_{ON}/I_{OFF} = 2150 at V_{DS} = 0.5 V with 0.5 V V_G swing. Figure 22 shows V_T as a function of L_G for e-mode and depletion mode (d-mode) operation. A positive V_T shift of ~600 mV from d-mode

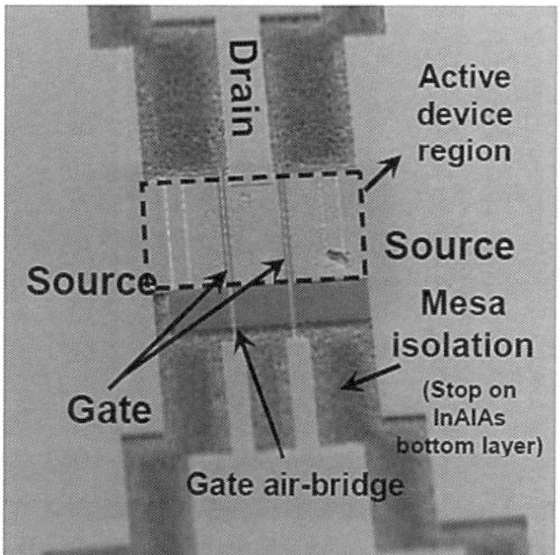

Fig. 20. SEM micrograph of a two-finger $In_{0.7}Ga_{0.3}As$ QWFET on Si with Pt/Au gate air-bridge at the mesa edge and Ti/Pt/Au source/drain metals.[23]

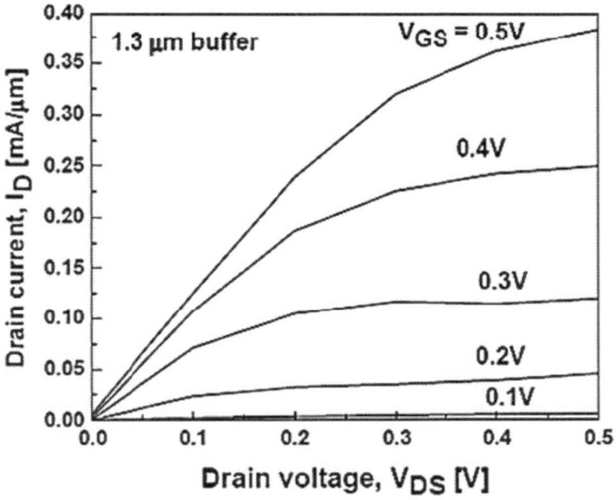

Fig. 21. I_D-V_{DS} characteristics of enhancement-mode $L_G = 80\,\text{nm}$ $In_{0.7}Ga_{0.3}As$ QWFET on Si with 1.3 μm buffer.[23]

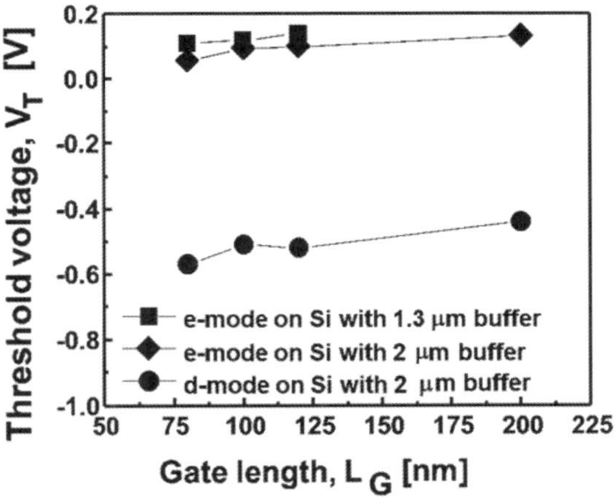

Fig. 22. Threshold voltage as a function of L_G for enhancement-mode (e-mode) and depletion-mode (d-mode) $In_{0.7}Ga_{0.3}As$ QWFET on Si.[23]

to e-mode operation was accomplished through gate recess etch. Figure 23 shows the transconductance (G_m) characteristics of e-mode and d-mode $L_G = 80\,\text{nm}$ QWFET on Si with different composite buffer thicknesses. The improved G_m values of e-mode devices with 1.3 and 1.5 μm buffer layers compared to that of 2 μm buffer device were due to improved transistor fabrication process. Figure 24 shows the current gain (h_{21}) dependence on frequency for an e-mode device with an estimated F_t of 302 GHz. Figure 25

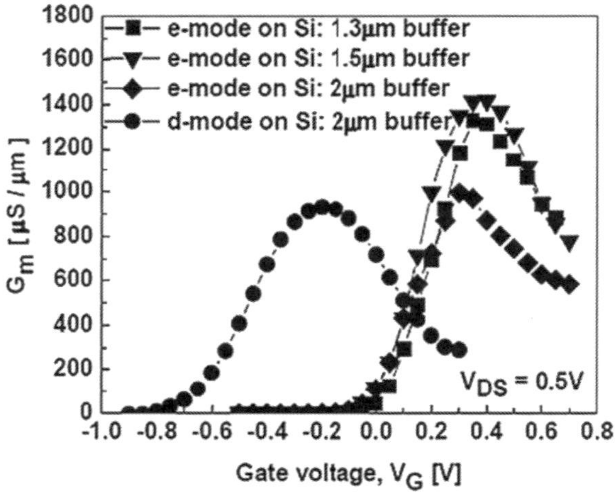

Fig. 23. Transconductance, G_m characteristics of e-mode and d-mode L_G $In_{0.7}Ga_{0.3}As$ QWFET on Si with different composite buffer thickness at $V_{DS} = 0.5\,V$.[23]

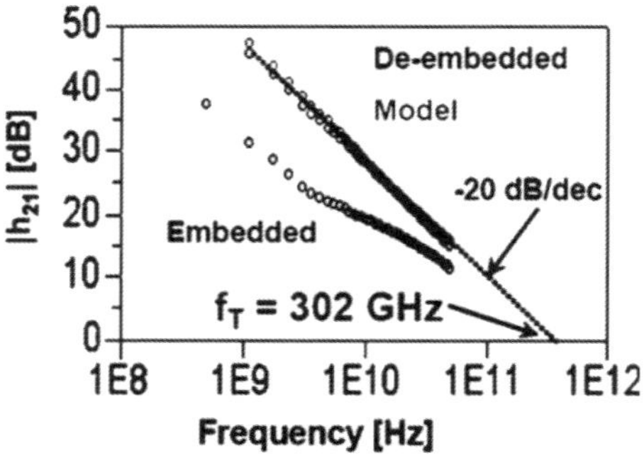

Fig. 24. Current gain, h_{21}, versus frequency for the 80 nm L_G e-mode $In_{0.7}Ga_{0.3}As$ QWFET on Si with 1.3 μm composite buffer, showing embedded data and de-embedded data at $V_{DC} = 0.5\,V$.[23]

shows cut-off frequency versus dc power dissipation for an e-mode L_G = 80 nm QWFET on Si with 1.5 μm composite M-buffer. Compared to standard L_G = 60 nm Si MOSFET, the e-mode QWFET monolithically integrated on Si exhibits 20×reduction in dc power dissipation for the same speed performance or 2×gain in speed performance for the same power.

The second iteration of device optimization was focused on reduction of large gate leakage (J_G) with scaling of the upper InAlAs barrier thickness above the QW.[30] For future transistor scaling, there are significant challenges in identifying a suitable gate oxide material with high dielectric constant (K) for integration with III-V transistors. The ideal high-K gate dielectric will enable the reduction of J_G and electrical gate oxide thickness

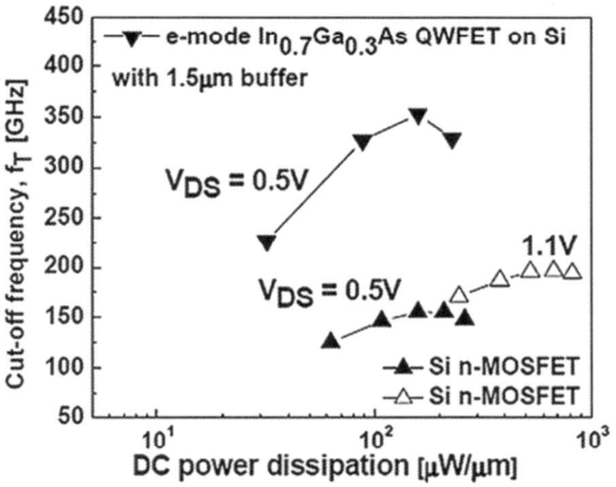

Fig. 25. Cut-off frequency as function of DC power dissipation for the e-mode $L_G = 80$ nm $In_{0.7}Ga_{0.3}As$ QWFET on Si with 1.5 μm composite buffer.[23]

(t_{OXE}), while maintaining good interface properties with low surface states density and minimal degradation of carrier mobility in the channel. A composite high-K $TaSiO_x/InP$ gate stack was developed and high-performance short-channel QWFET on Si structures with significantly reduced t_{OXE} and J_G was realized.

The device structure was further modified by incorporating an upper barrier InP layer to minimize the distance between gate and channel. Band diagram obtained using Schrodinger-Poisson simulation indicated carrier wave function confinement in the InGaAs QW layer (Figure 26(b)). The effectiveness of the 2 nm InP layer as an upper barrier was confirmed by achieving Hall electron mobility of 10,000 cm^2/Vs at 300 K with no parallel conduction. TEM image in Figure 27 shows a QWFET structure with a physical gate length of 75 nm and a 40 Å $TaSiO_x$/20 Å InP composite gate stack. Figure 28 compares the J_G dependence on V_G for QWFET structures with composite $TaSiO_x/InP$ gate stack versus Schottky gate. Incorporation of the high-K gate dielectric into the InGaAs QWFET structure reduced t_{OXE} by 11 Å while dropping J_G by a factor of >1000. The fabricated device exhibited sub-threshold slope of 85 mV/decade and drain induced barrier lowering (DIBL) of 35 mV/V at $V_{DC} = 0.5$ V. The drive current and peak G_m at the same conditions were 0.49 mA/μm and 1750 μS/μm, respectively, and the measured effective electron velocity (V_{eff}) in the high-K QWFET structure was ~4× higher than that in strained Si MOSFET (see Fig. 29).[30] These results demonstrated further enhancement in device properties for InGaAs QWFET with composite high-K gate over strained Si MOSFET.

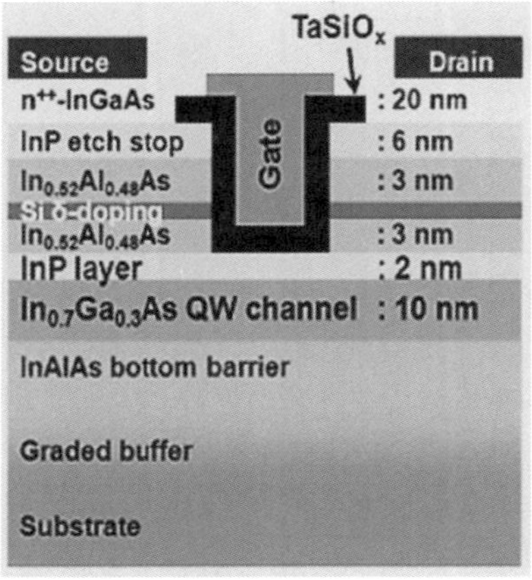

(a)

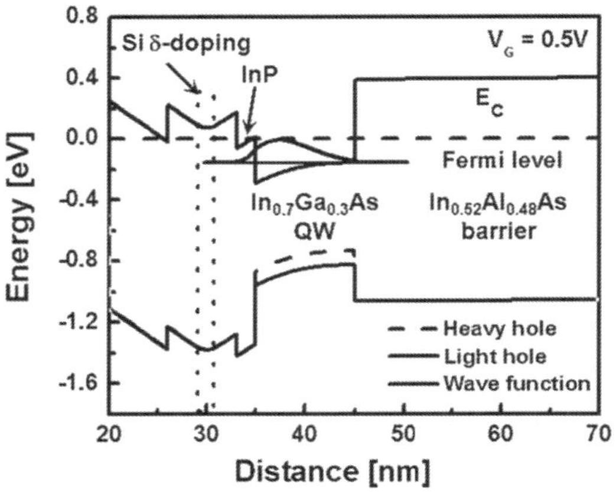

(b)

Fig. 26. (a): Schematic of $In_{0.7}Ga_{0.3}As$ QWFET on Si with 2 nm InP upper barrier and a 4 nm $TaSiO_X$ and (b): Band gap diagram obtained using Schrodinger-Poisson simulation, indicating carrier confinement in QW layer.[30]

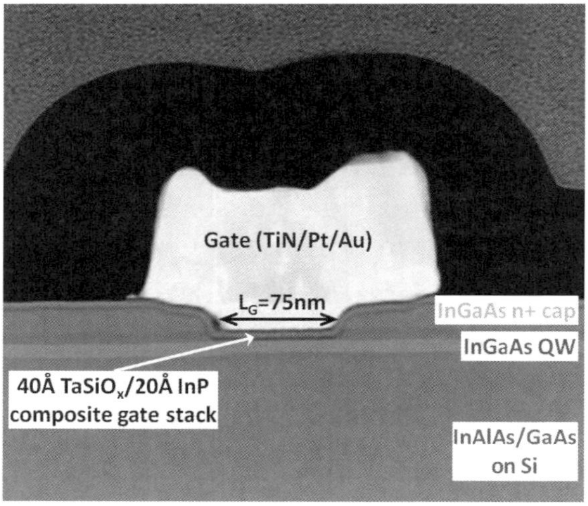

Fig. 27. TEM micrograph of QWFET with $L_G = 75$ nm showing the 2 nm InP upper barrier layer and 4 nm $TaSiO_X$ high-K gate dielectric, which form a composite $TaSiO_X$-InP gate stack.[30]

4. Direct Monolithic Integration of III-V Devices and Si CMOS on Silicon Substrate

Heterogeneous integration of high-speed III-V devices with Si CMOS circuitry has attracted significant interest for its potential in next generation

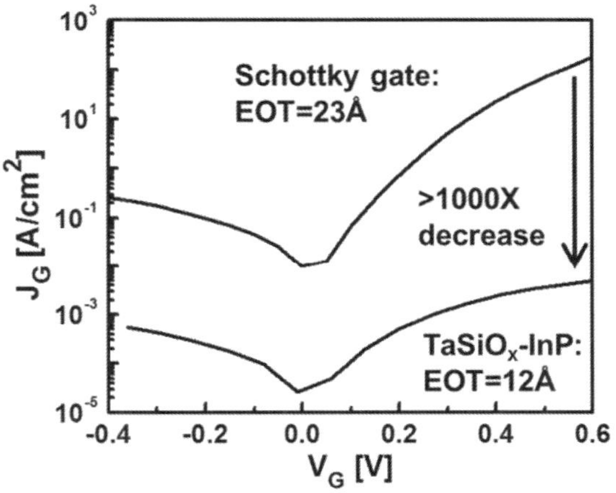

Fig. 28. J_G versus V_G of $In_{0.7}Ga_{0.3}As$ QWFET with composite 4 nm composite $TaSiO_x$-InP gate stack and Schottky gate electrode on 6 nm $In_{0.52}Al_{0.48}As$ barrier.[30]

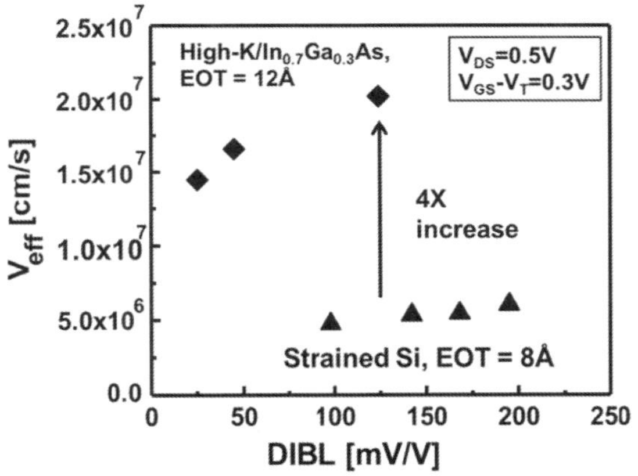

Fig. 29. Measured effective electron velocity (V_{eff}) versus DILB comparing $In_{0.7}Ga_{0.3}As$ QWFET with composite 4 nm composite $TaSiO_x$-InP gate stack and strained Si MOSFET at $V_{DC} = 0.5$ V and gate overdrive of $(V_{GS} - V_t) = 0.3$ V.[30]

monolithic hybrid integrated circuits (ICs) for analog and digital applications. Such integration schemes can provide circuit designers with the absolute technological flexibility of the "best junction for the function" without compromising the yield or scalability of CMOS, or the speed and breakdown characteristics of the III-V devices. This level of technological flexibility have been demonstrated by a variety of integration approaches such as traditional hybrid chip-and-wire or flip chip on a multilayer board[31]; chip-on-chip bonding[32,33]; III-V epilayer transfer onto Si ICs[34]; and wafer bonding to realize 3D hybrid ICs.[35,36] Our effort in this area used a direct monolithic integration approach with fabrication of III-V device clusters and CMOS circuitry on the same Si wafer.[37–49] To provide flexibility to circuit designers, our growth processes were initially developed for both InP-based metamorphic high electron mobility transistor (M-HEMT) and heterojunction bipolar transistor (M-HBT) structures for digital and analog devices. The M-HBT structure, which actually includes a double-heterojunction design around the base, was used for circuit demonstration.

Our approach is based on the use of Silicone-on-Lattice Engineered Substrate (SOLES), developed at MIT[50,51] and manufactured by SOITEC using their Smart-Cut™ process.[52,53] From the top down, the SOLES wafer comprises of the following layers: Si, SiO_2, Ge, SiO_2, and a Si substrate. The topmost silicon layer is used for SOI CMOS fabrication to realize an integrated CMOS/HBT process. The schematic of a SOLES substrate and X-TEM for device layers are shown in Figure 30. Essential features of these substrates include compatibility with Si CMOS fabrication processes, easy electrical isolation, and a lattice constant (Ge) suitable for growth and integration of III-V devices. The top Si layer of the substrate enables fabrication of CMOS circuitry using a standard process. The embedded Ge layer enables direct growth of high quality III-V epitaxial material inside patterned windows on the Si substrate. The Si and Ge layers are crystallographically independent from the substrate and are electrically insulated by the SiO_2 layer. The crystal orientation of the thin embedded Ge layer was (001) with a 6° mis-cut toward the <111>, which promoted the growth of the III-V epilayer with minimal anti-phase boundary formation.[54,55]

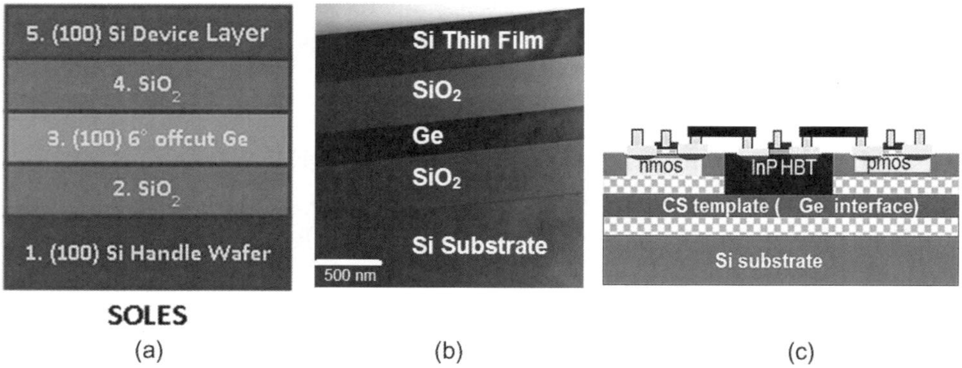

(a) (b) (c)

Fig. 30. (a): Schematic of the SOLES substrate; (b): cross-sectional TEM of the layers, and (c): the integration of Si CMOS with III-V devices grown on the exposed Ge layer in a window opening.[39]

The process flow for integrating III-V devices with Si CMOS involved the following steps: CMOS circuitry processing; substrate patterning to expose the embedded Ge in windows; III-V epitaxial growth; removal of polycrystalline material deposited outside of the windows; III-V device processing; and interconnect fabrication between III-V devices clusters and Si CMOS. The integration layout is imaged in Figure 30(c). A borophosphosilicate glass (BPSG) layer was initially deposited on the SOLES substrates to protect the Si CMOS layer during the patterning process and help facilitate the subsequent removal of polycrystalline III-V material surrounding the windows. Growth windows with a wide range of sizes were patterned by etching through the BPSG, the topmost Si layer, and the first internal oxide layer to expose the embedded Ge layer. Multiple iterations were required to perfect the window etching and cleaning process to optimize the sidewall profile and minimize damage to the Ge layer to support high-quality III-V epitaxial growth inside the windows. For this integration scheme to work, III-V epitaxial growth had to be realized in small windows with low aspect ratio between the window size and the epitaxial layer thickness. Challenges related to MBE growth on a patterned surface are discussed below.

4.1. *Composite buffer design and III-V material growth on SOLES*

A key advantage of using SOLES for III-V material integration is having an embedded Ge layer as a growth template for the GaAs buffer layer. Germanium has a lattice parameter very close to that of GaAs which helps minimize the formation of misfit dislocations. Additionally, Ge surface wetting is more favorable for GaAs nucleation. The thermal expansion coefficient difference between Ge ($5.8\times10^{-6}\ K^{-1}$) and GaAs ($6.63\times10^{-6}\ K^{-1}$) is significantly less than that between Si and GaAs, which helps to reduce residual strain in the grown structure after cool down. By optimizing the mis-cut angle of the Ge layer (6° in our case) and the substrate temperature profile for the oxide desorption, we were able to minimize the APB formation during the GaAs on Ge nucleation process. The combination of these favorable factors enabled the growth of smooth and low defect density GaAs layers on Ge with an estimated dislocation density of $\sim10^4\ cm^{-2}$ in a more reproducible and manufacturable process.[41]

The metamorphic buffer design on SOLES is similar to that used for direct growth on a Si substrate as previously shown in Figure 5, with the key exception of having Ge as the starting surface. Other differences in buffer designs are related to the total buffer thickness required for device planarization, and a smaller lattice parameter change between the substrate template layer and the device layers (4% for growth on Ge compared to 8% on Si). The final buffer design fixed the total thickness as 1.7 μm, with 0.6 μm of GaAs and 1.1 μm of graded InAlAs. In addition, for the growth of III-V material on patterned substrates in small windows, the epilayer surface should remain flat with minimum distortions and defect generation at the edges of the windows.

The optimization of GaAs growth conditions were performed on blank Ge substrates. The optimized process was then transferred to SOLES. De-oxidation of the Ge surface before MBE growth was performed at a low substrate temperature of ~500°C. High temperature treatment for formation of bi-atomic step arrays was achieved by substrate annealing at 650°C. The growth of GaAs on Ge was performed in a single-stage process at $T_s = 600$°C.[41] The evolution of surface structure was monitored by RHEED and shown in Figure 31. The RHEED patterns from properly de-oxidized Ge substrate revealed a well-defined (2×2) surface structure. Nucleation of the GaAs layer did not lead to a spotty pattern that signifies the formation of 3D islands as was observed for GaAs nucleation on Si. After depositing 50 Å of GaAs, the surface displayed a (2×1) superstructure with minor faceting or surface interface roughening (Figure 31(b)). The standard GaAs (2×4) RHEED patterns were observed after the deposition of 200 Å GaAs, and an atomically smooth surface, similar to that of homo-epitaxial growth on GaAs, was obtained after 1000 Å of growths (see Figures 31(c) and (d)).

The growth procedure for the InAlAs graded M-buffer was similar to the one used for direct growth on Si described earlier. A minor adjustment of indium composition on the inverse graded step was made to provide strain compensation for the device layers on top, and to obtain optimal surface smoothness and minimal device parameter degradation for the HEMT and HBT structures. Prior to the growth on SOLES, full stack device layers with InAlAs graded M-buffer were grown on GaAs, Ge, and Ge-on-insulator-on-Si (GeOI/Si) substrates. The difference in substrate material properties such as thermal expansion properties and surface conditions can affect the strain relaxation in the M-buffer. This is especially important for growth on GeOI/Si substrates due to the complex nature of possible residual strain in the thin Ge interacting with the oxide layer and the M-buffer.

GeOI/Si substrates are essentially truncated SOLES without the top-most SiO_2 insulating layer and Si layer.[39] Schematic diagrams of the M-HEMT and M-HBT structures are shown in Figures 32(a) and (b). The single-pulse-doped M-HEMT structure included a 200 Å $In_{0.53}Ga_{0.47}As$ channel and 50 Å $In_{0.52}Al_{0.48}As$ spacer layer. The pulse doping was

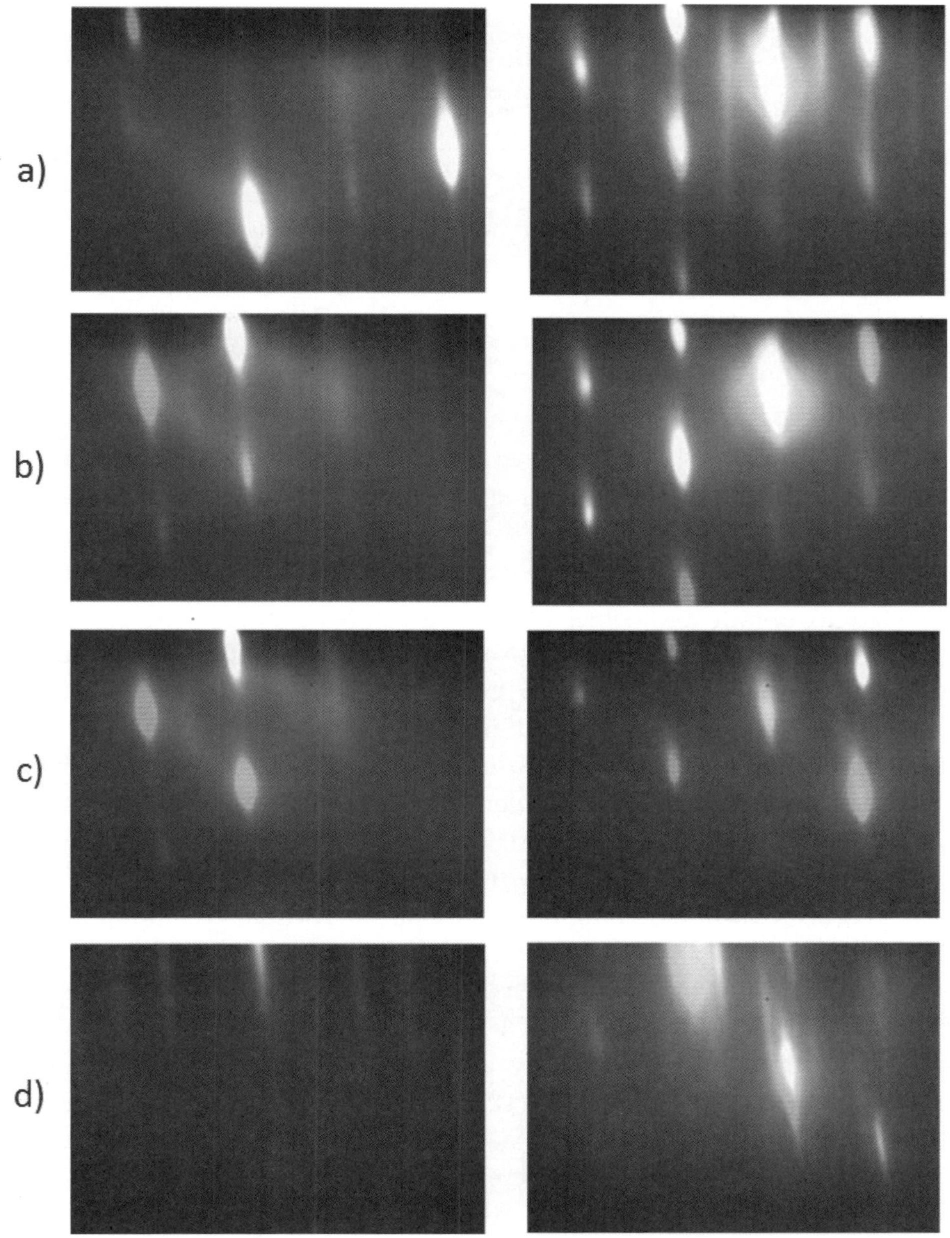

Fig. 31. RHEED patterns demonstrating the evolution of growth of GaAs on Ge substrate: (a) Ge substrate after desorb; (b) after 50 Å of GaAs deposition; (c) after 200 Å GaAs deposition; and (d) after 1000 Å GaAs deposition.

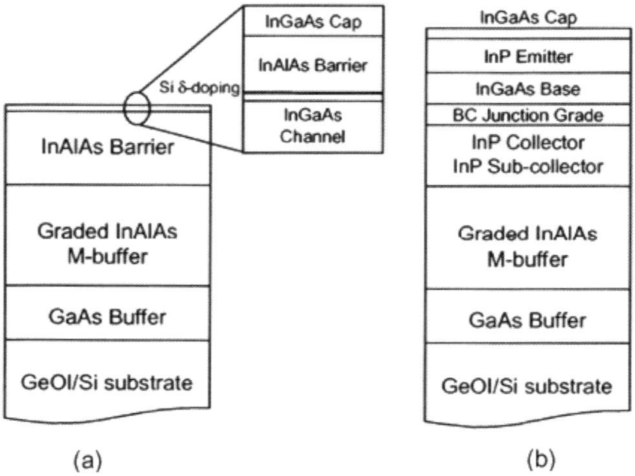

Fig. 32. Schematic diagram of (a) M-HEMT and (b) M-HBT structures on GeOI/Si substrates.[41]

adjusted to obtain a sheet charge carrier density around $3 \times 10^{12}\,cm^{-2}$. The generic M-HBT structure with type-I band alignment employed InP collector and emitter layers and a 350 Å thick InGaAs base layer that was C-doped at $\sim 4 \times 10^{19}\,cm^{-3}$. A digital alloy grade was incorporated to suppress current blocking at the base-collector junction.

The surface morphology of the 0.5 μm GaAs buffer layer grown on GeOI/Ge substrate was evaluated. Nomarski and AFM images revealed high surface quality with an AFM rms roughness of ~9 Å on a $5 \times 5\,\mu m^2$ scale (Figure 33(a)). The single step GaAs nucleation, described above, also minimized Ge out-diffusion, and consequently the unintentional background doping in GaAs grown on Ge and GeOI/Si substrates. Electrochemical capacitance-voltage (E-*CV*) profiles across the optimized GaAs buffer show the layer to be fully depleted with no discernable charge buildup at the GaAs–Ge interface (Figure 34). X-TEM analysis (Figure 35) shows a very clean GaAs/Ge interface with no rampant APBs or dislocation nucleation and propagation. Close examination of the GaAs/Ge interface does show limited formation of APBs in some regions, which quickly annihilate within the first 100 Å of the film growth. The dislocation density in the top GaAs layer stayed below the X-TEM detection limit ($<10^7/cm^2$). HRXRD spectrum showed a well-defined, sharp HRXRD peak from the fully relaxed GaAs buffer layer with a FWHM of 36 arcsec. Based upon the small lattice mismatch between GaAs and Ge, we estimated that the threading dislocation density in the GaAs buffer was very low, in the $10^4\,cm^{-2}$ range.

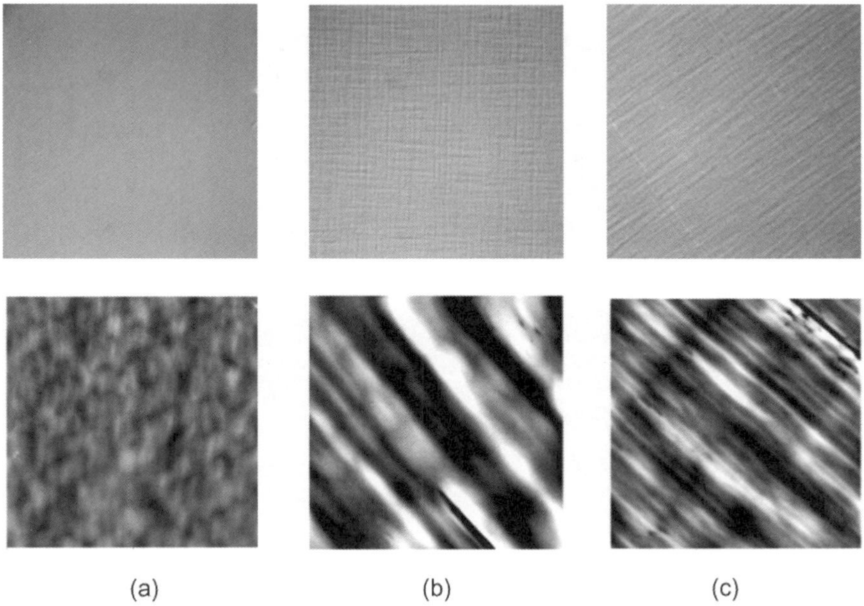

Fig. 33. Nomarski optical microscope (top), $100 \times 100\,\mu$m^2 and AFM (bottom, $5 \times 5\,\mu$m^2 scan) images. (a): $0.5\,\mu$m GaAs layer on GeOI/Si substrate; (b): reference M-HEMT on GaAs; (c): M-HEMT on GeOI/Si substrate. The AFM rms roughness values are 9, 34, and 29 Å for (a), (b), and (c), respectively (see Fig. 47).[41]

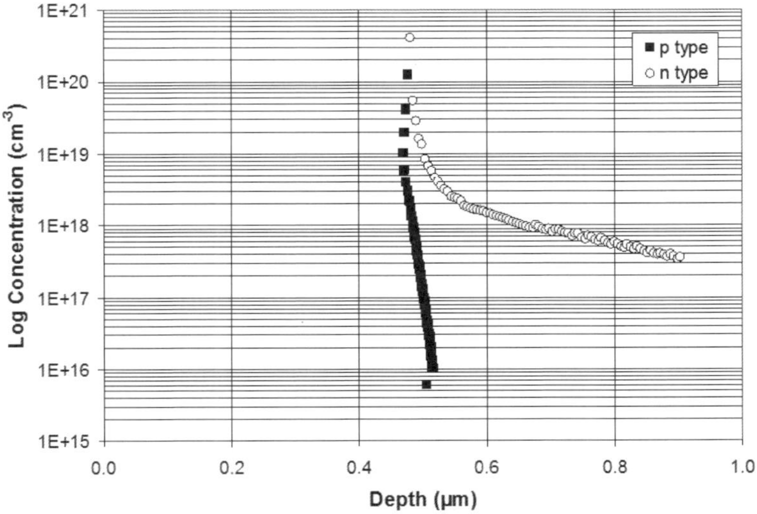

Fig. 34. E-CV plot of a GaAs buffer grown on Ge substrate, demonstrating the unintentionally doped GaAs buffer to be fully depleted.[41]

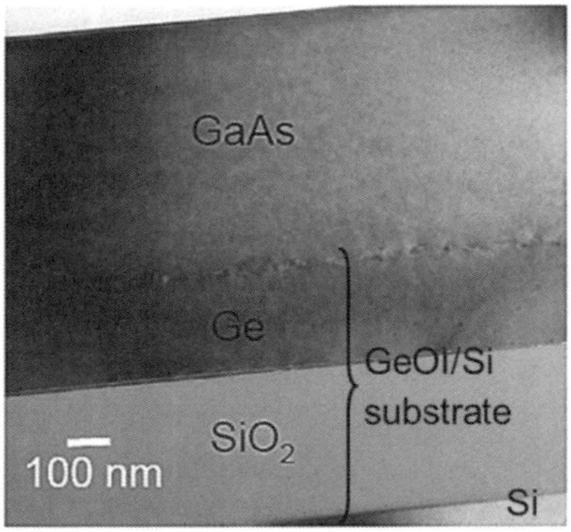

Fig. 35. TEM images of GaAs layer grown on GeOI/Si substrate.[41]

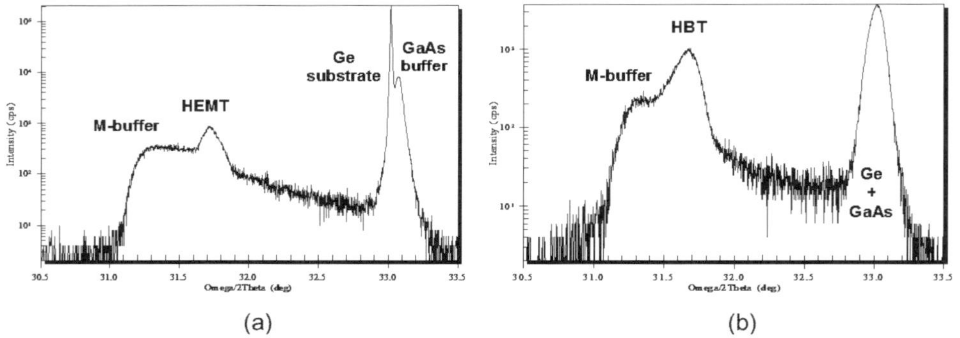

Fig. 36. High resolution XRD (004) spectra of (a): M-HEMT structure grown on a Ge substrate and (b): M-HBT structure grown on a GeOI/Si substrate.[41]

Typical HRXRD spectra for M-HEMT grown on Ge and for M-HBT on GeOI/Si are shown in Figure 36. Since the thickness of Ge layer in the GeOI substrate is only 1000 Å, it appeared as a single broadened peak merged with the GaAs buffer in the HRXRD spectrum (Figure 36(b)). The grading of the M-buffer composition results in the continuous signal detection extending to the left of the substrate peak. The peaks protruding out of the M-buffer signal are the device active layers, which are nominally matched to the InP lattice constant. The amplitude difference between the M-HEMT peak and the M-HBT peak is related to the difference in their total layer thickness. The distribution of threading dislocations depicted in the X-TEM images of a M-HBT structure grown on GeOI/Si (Figure 37) is typical for this graded M-buffer design, with high densities localized in the lower portion of the graded InAlAs buffer and efficient filtering toward the upper portion of

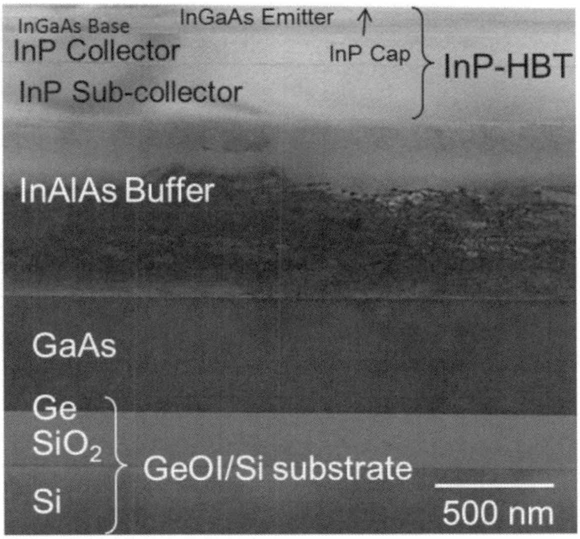

Fig. 37. Cross-section TEM image for M-HBT structure grown on GeOI/Si substrate.[44]

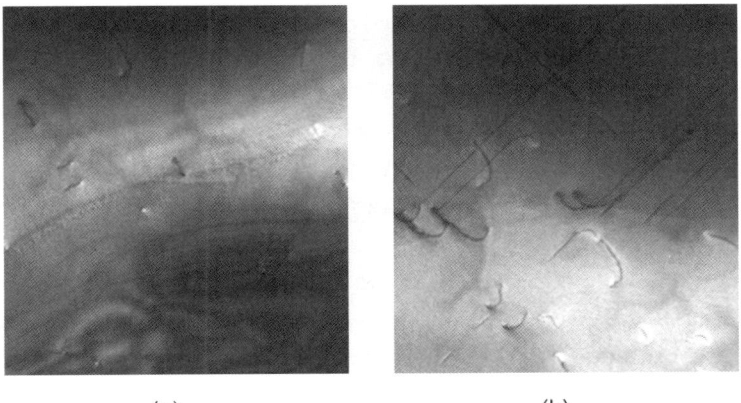

Fig. 38. Plan-view TEM evaluation of residual dislocation density at the top of an InP-HBT structure on un-patterned GeOI/Si substrate. Images from two representative regions are shown.[44]

the buffer and device layers. The dislocation density in the M-HBT base layer was below the detectivity limit of X-TEM and was estimated to be $\sim 3 \times 10^7\ \text{cm}^{-2}$ by PV-TEM (Figure 38). Similar distribution of threading dislocation was also observed for the M-HEMT structure (see Figure 39).

The effectiveness of our composite M-buffer was evaluated by comparing the figures-of-merit of M-HEMTs and M-HBTs grown on different substrates with reference lattice-matched structures grown on InP. Table 1

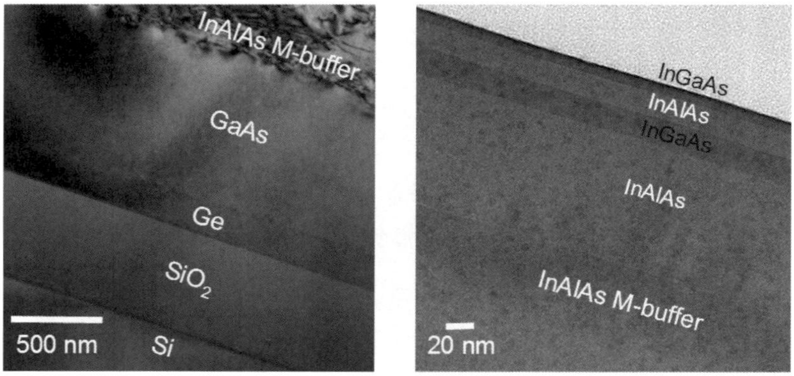

(a) (b)

Fig. 39. X-TEM images of a M-HEMT structure grown on a GeOI/Si substrate showing defect redistribution on: (a) GeOi/Si- M-buffer interface and (b) in the active device layers.[41]

summarizes the values for 300 K and 77 K Hall electron mobilities, charge-mobility product, and surface roughness for M-HEMT structures. In all cases, the value of the 300 K Hall electron mobility exceeded 9,000 cm^2/Vs and matched well with the value from the reference InP structure. Despite using identical M-buffer design and growth conditions such as substrate temperature and III-V ratios for this set of growths, we observed elevated rms roughness for structures grown on Ge and GeOi/Si, 48 Å and 50 Å, respectively, compared to 34 Å for the structure grown on GaAs. We speculated that the difference in surface morphology was related to the difference in thermal expansion coefficients between these types of substrates. This disparity might have produced different strain compensation in the graded InAlAs layer. Further optimization of growth temperature and maximum indium percentage in the InAlAs layer reduced the rms roughness of M-HEMTs structures on GeOI/Si substrates to below 35 Å. The optimized setting was applied to subsequent growth of M-HBTs structures, resulting in rms roughness of 31 Å and 36 Å on Ge and GeOi/Si structures, respectively (Table 2).

The growth of high quality M-HBT structures is more challenging compared to M-HEMTs. As minority-carrier devices, HBTs are widely believed to be more sensitive to crystal defects. Threading dislocations in the thin base layer and at the base-emitter and base-collector junctions can reduce carrier lifetimes and increase junction leakage currents that cause reduction in the current gain (β) and base-collector breakdown voltage.[14] Large area device dc parameters from HBT structures grown on Ge, GeOI/Si, and InP substrates are summarized in Table 2. The difference in performance of HBTs

Table 1. Summary of Hall and AFM data for M-HEMT structures grown on InP, GaAs, Ge, and GeOI/Si substrates [41].

Substrates	n_S at 300 K (cm^{-2})	μ at 300 K ($cm^2/V\ s$)	$n_S \times \mu$ at 300 K (1/V s)	n_S at 77 K (cm^{-2})	μ at 77 K ($cm^2/V\ s$)	AFM rms roughness (A)
InP	2.98E + 12	10252	3.06E + 16	2.90E + 12	36107	1.5
GaAs	2.96E + 12	10115	2.99E + 16	2.74E + 12	29138	34
Ge	3.07E + 12	9399	2.89E + 16	3.13E + 12	23582	48
GeOI/SI	3.04E + 12	10851	3.30E + 16	2.91E + 12	35445	55

Table 2. Summary of large area device dc characteristics and AFM data for M-HBT structures grown on InP, Ge, and GeOI/Si substrates.[41]

	Large-area device DC measurements							
Substrates	Gain (β) at Ib = 1.2 mA	Base Rsh (Ω)	V_{offset} (V)	BV_{ceo} (V)	B-E junction V_t/V_r (V)	B-C junction V_t/V_r (V)	Ideality factors (n_c/n_b)	Afm rms roughness (A)
InP	60	647	0.10	5.0	0.51/2.7	0.42/7.0	1.08/1.29	4.0
Ge	60	695	0.11	5.3	0.52/1.6	0.41/7.3	1.10/1.33	31
GeOI/Si	55	644	0.12	4.9	0.52/2.4	0.42/6.5	1.10/1.28	36

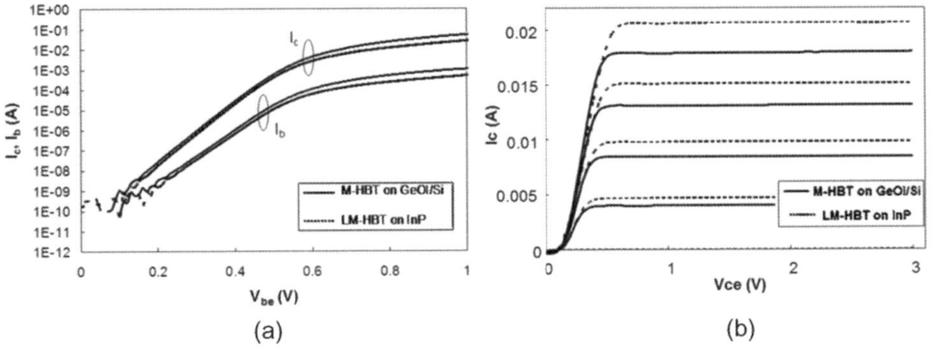

Fig. 40. (a): Gummel plot and (b): common emitter characteristics for large area HBT devices fabricated on GeOI/Si and InP substrates.[41]

on InP and GeOI/Si substrates is illustrated by Gummel plots and common emitter *I-V* characteristics shown in Figures 40(a) and (b), respectively. The dc parameters of M-HBT structures grown on Ge and GeOI/Si substrates compared well with baseline structures grown on InP substrates, with only ~10% degradation in the current gain. It is particularly encouraging that the breakdown voltages were essentially the same for all three structures.

The excellent electrical properties and low surface roughness of M-HEMT and M-HBT structures grown on GeOI/Si confirm the effectiveness of our composite M-buffer design and represent the critical step for III-V device integration with Si CMOS.

4.2. *Material growth and characterization on patterned SOLES*

Epitaxial growth by MBE on patterned SOLES substrates poses several specific challenges. First, unlike MOVPE, MBE is a non-selective growth method. Due to the high sticking coefficients of group-III atoms, growth proceeds across the entire wafer, resulting in polycrystalline deposition on the BPSG mask material that requires a patterned, dry etch process for its removal (see Figure 41). Problems can arise in future processing of the wafers if this polycrystalline material is too rough. We also needed to optimize our growth conditions to produce a high filling factor within the windows opened through the mask, while maintaining good quality GaAs on Ge nucleation and InP device growth on the Ge surface inside these windows. Potentially, epitaxial growth on the limited area within the Ge windows could reduce defect density by limiting misfit dislocation length and by reducing dislocation interaction and multiplication.[56] Small-area epitaxy within the windows may also enhance dislocation filtering within the M-buffer and hence help reduce the threading dislocation density in the device layers.[56] Our goal was to achieve a filling factor in the Ge windows that will allow placement of the active device junctions to within $\sim 5\,\mu$m of the edge of the window opening for intimate heterogeneous integration.

During growth optimization, we found that Ga droplet formation on exposed Si in the sidewall of the windows can result in preferential, wire-like growth that can protrude into the window openings, thus shrinking the area for high quality III-V growth. Optimizing the window etching and

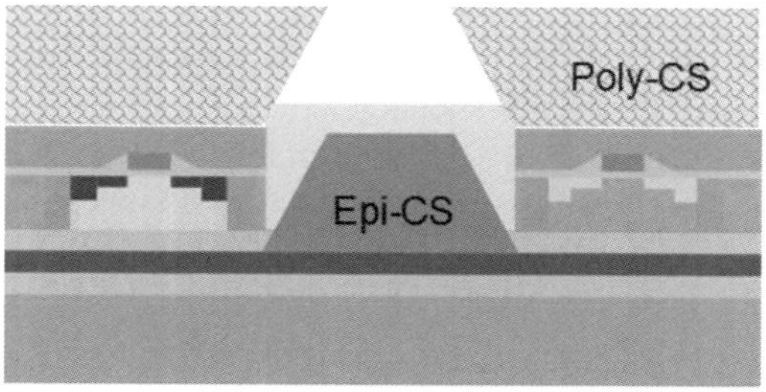

Fig. 41. MBE growth on patterned SOLES substrate: single crystal material grown in the window on the Ge nucleation layer; polycrystalline material formed on top of BPSG protective mask.

(a) (b) (c)

Fig. 42. SEM images of polycrystalline defects formed on the protective BPSG layer, related to non-optimal growth conditions: (a) rough surface and wire-like defects[45]; (b) and (c) wire-like defects related to Ga droplets on the edge of exposed Si layer.

epitaxial growth processes were key to achieving high quality device layers. To minimize the roughness of the polycrystalline material and nucleation along the window sidewalls, growth conditions previously optimized for un-patterned growth had to be modified for patterned substrates. Instead of using a 600°C single-stage growth, the GaAs buffer layer was grown at reduced temperature to minimize Ga surface migration and inhibit droplet formation and wire nucleation. Growth conditions for the initial GaAs nucleation on Ge, graded InAlAs M-buffer, and InP-HBT structure remained the same for both patterned and un-patterned substrates. As depicted in Figure 42, scanning electron microscope (SEM) images of M-HBT epitaxial material grown on SOLES with an un-optimized process revealed rough surface, poor edge definition, and formation of nanowire-like materials near the window edges. After careful optimization, however, high-quality InP-HBT epitaxy in windows as small as $15 \times 15\,\mu m^2$ with well-defined edges was reproducibly realized across 100 mm diameter wafers with good uniformity (Figure 43).[45] Upon removal of the polycrystalline material from the BPSG protective mask, post-MBE patterned SOLES wafers (Figure 43(c)) exhibit well-defined window edges with a gap of only $\sim 2\,\mu m$ between the sidewall of the window and the smooth MBE-grown region inside.[44] Nomarski and AFM images (Figure 44) revealed cross-hatch patterns inside the windows which are characteristic of high-quality metamorphic growth, as is the low rms roughness of 13 Å from a $20\times 20\ \mu m^2$ AFM scan. Additionally, the polycrystalline III-V material on the BPSG outside the windows appears uniform and of manageable roughness. The structural quality of M-HBT structures grown on patterned SOLES was confirmed by HRXRD to be qualitatively the same as those obtained on blanket GaAs substrates (Figure 45). Well-defined peaks from the GaAs buffer layer on

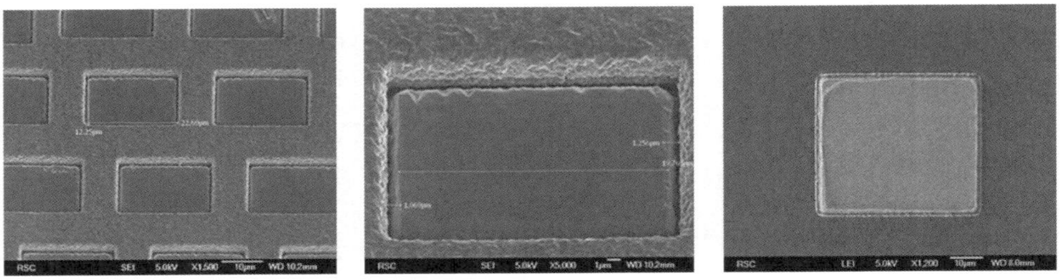

(a) (b) (c)

Fig. 43. Surface morphology (SEM) of epilayers on patterned surfaces grown under optimized growth conditions: (a) and (b) single crystal material grown in windows and polycrystalline material on top of BPSG mask; (c) after polycrystalline material removal from BPSG prior to HBT fabrication. The exclusion area on window edges $<2\,\mu$m.[44]

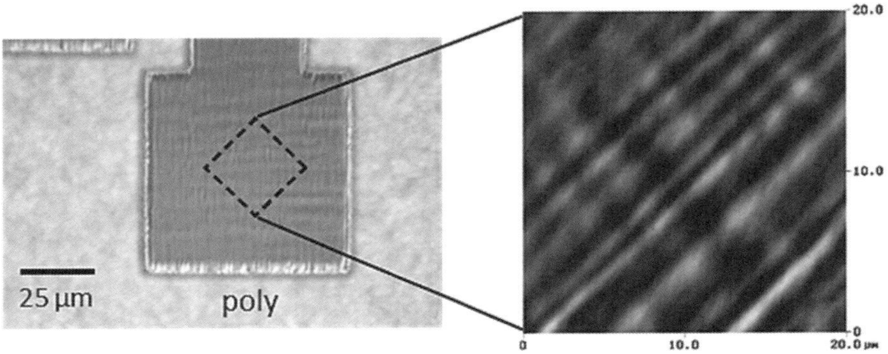

(a) (b)

Fig. 44. Evaluation of surface roughness of HBT material grown in windows: (a) Nomarski image and (b) AFM image from central area of the window; AFM rms roughness is 13 Å.[44]

Ge, the linearly-graded InAlAs M-buffer with an inverse step, and the InP-HBT layers on top can be clearly seen in the HRXRD spectrum. The FWHM for the GaAs and HBT layers were close to the values measured for layers grown on blanket substrates, indicating comparable structural quality.

To evaluate the electrical properties of M-HBT structures grown on patterned SOLES, small-area devices from various window sizes were fabricated. An SEM image of an array of M-HBT devices prior to final device passivation is shown in Figure 46. High device yield and reproducibility of dc characteristics were obtained using an optimized fabrication process. The distance between the active junction area and the window edge is $<5\,\mu$m, meeting the heterogeneous interconnect requirements between the III–V devices and the CMOS circuit. Good dc and RF device characteristics were

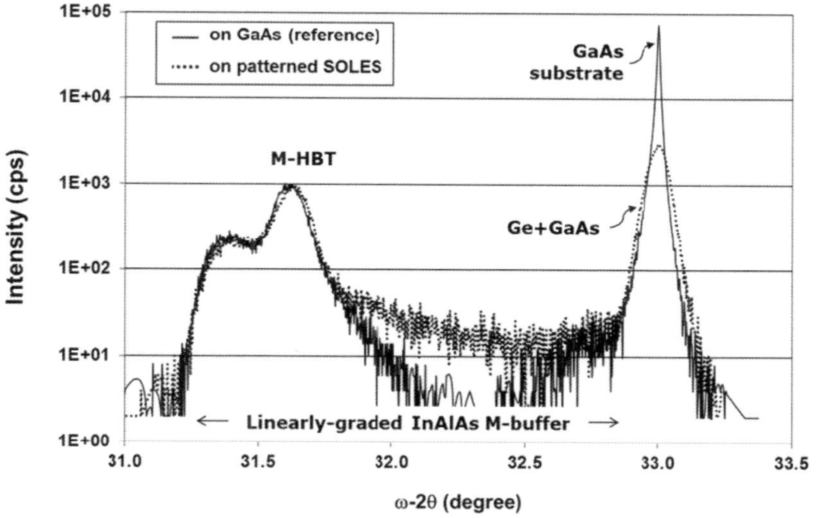

Fig. 45. InP HBT HRXRD (004) spectra measured inside the growth window of a patterned SOLES substrate. Reference spectrum from the same structure grown on GaAs substrate is also shown.[44]

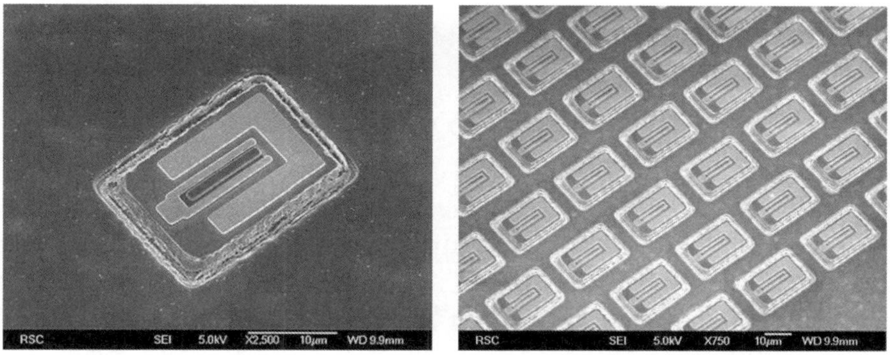

(a) (b)

Fig. 46. SEM images of InP HBT devices: (a) single device in a $15 \times 15\,\mu m^2$ window on SOLES; (b) device array.[44]

measured for devices in windows as small as $15 \times 15\,\mu m^2$. $1 \times 3\,\mu m^2$ emitter devices exhibit β of 40 and a common-emitter breakdown voltage (B_{VCEO}) of 6 V at a current density of $10\,mA/mm^2$. An f_t of 190 GHz was extrapolated from the RF plots, consistent with previous results demonstrated on GeOI/Si substrates and comparable to what one would expect from a lattice-matched InP-HBT with the same device dimensions.[41] Further device

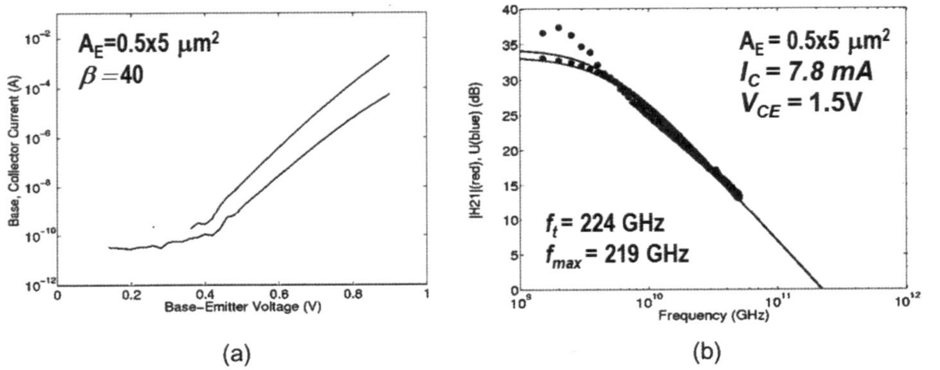

Fig. 47. (a): Measured Gummel characteristics and (b) RF gains of a $0.5 \times 5\,\mu m^2$ InP HBTs grown on a SOLES substrate.[44]

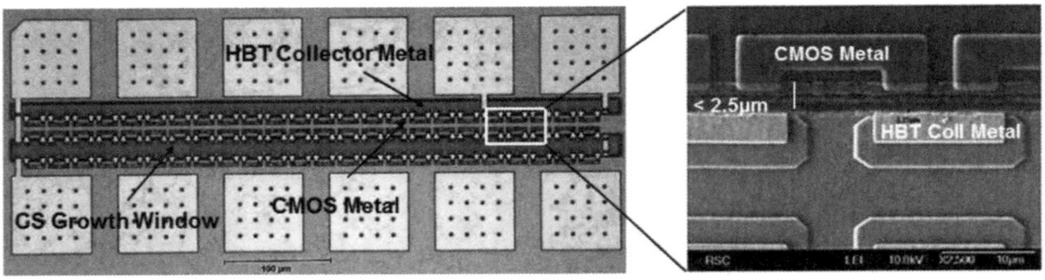

Fig. 48. CMOS and HBT device layout before interconnect fabrication.[45]

scaling to $0.5 \times 5\,\mu m^2$ emitter improved f_t and f_{max} to 224 and 219 GHz, respectively (see Fig. 47.).[45]

Interconnect fabrication between Si CMOS and III-V M-HBT circuitry is another important step in heterogeneous integration. To facilitate the interconnect process, the thickness of the epitaxial layers and depth of the windows are optimized such that the III-V M-HBTs and CMOS transistors are planar. Obtaining well-defined window edges with minimal roughness, high epitaxial filling factor in the windows, and low defect density are all critical factors for high yield interconnect fabrication. With this truly planar approach, interconnect length as small $2.5\,\mu m$ was fabricated. Dedicated test structure with M-HBTs and CMOS contact pads was used to evaluate quality and yield of interconnects. The layout of these structures before metallization and the same area after interconnect fabrication is shown in Figures 48 and 49, respectively. This approach resulted in 100% interconnect yield.

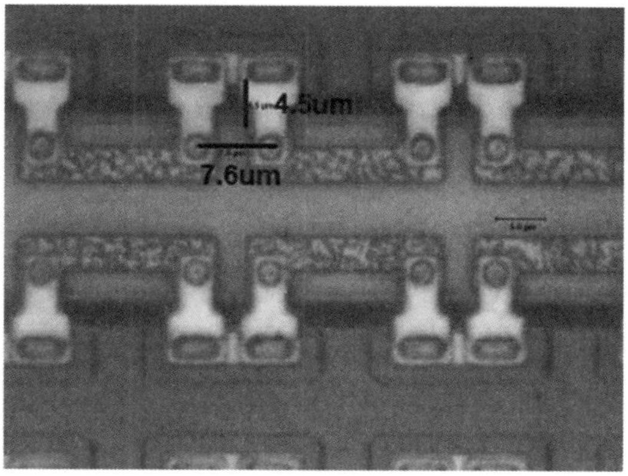

Fig. 49. Optical image of interconnects fabricated between CMOS and HBTs.

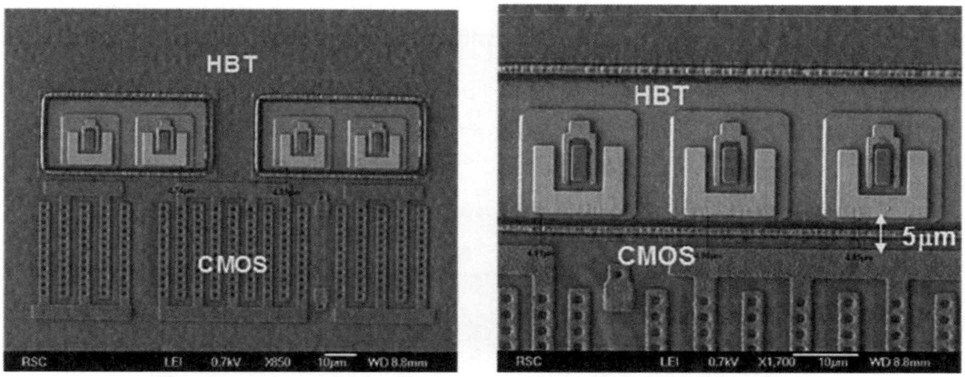

Fig. 50. SEM images of HBT adjacent to Si CMOS on SOLES before interconnect metallization.[45]

A differential amplifier test vehicle was designed and fabricated based on the heterogeneous integration approach described above. The layout of III-V HBTs clusters with Si CMOS for core differential amplifier before metallization is shown in Figure 50. Figure 51 shows the lowfrequency gain of the core differential amplifier. A peak low frequency gain of 454 V/V was measured. At lower frequencies, the gain is observed to decrease slightly. We believe this is due to device self-heating (increased output conductance of HBT).[45] The high frequency gain measurements are extracted using a similar scalar approach for determining the core amplifier characteristics. A unity-gain frequency of >25 GHz was extracted from high frequency test data and the DC-gain∗unity gain bandwidth product is measured to be 1.1×10^4 V/V GHz.[45] Figure 52 shows the measured output waveform of the amplifier when driven to saturation. A peak output swing of

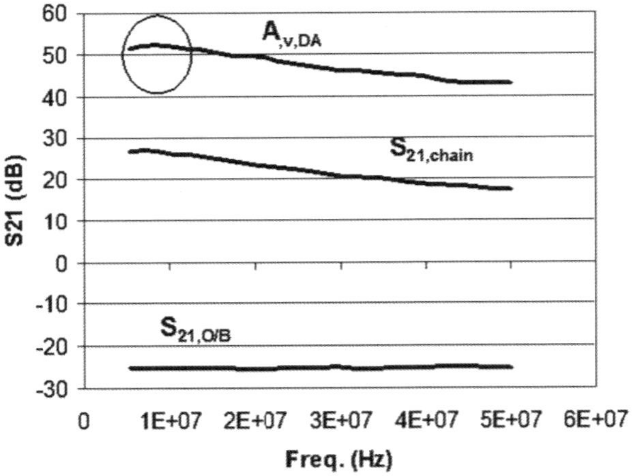

Fig. 51. Measured S21 of amplifier gain and output buffer test circuits at low frequencies. Low frequency gain AV, DA is given by AV, DA = S21, chain –S21,O/B. Core diff amp utilized three $2 \times 5\ \mu m^2$ HBTs in each diff-amp branch with 6-finger 2 μm PMOS devices. Iss = 10 mA, Vss = -6 V.[45]

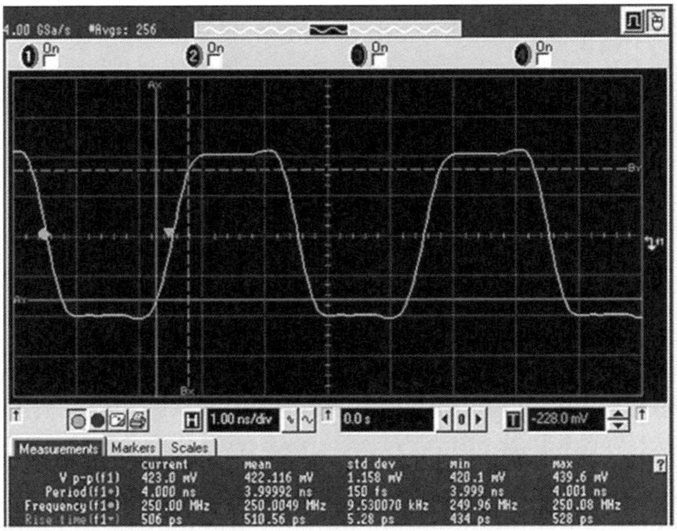

Fig. 52. Measured output waveform for slew rate measurements of the 250 MHz differential amplifier. The 250 MHz input signal was provided to saturate the amplifier. A peak output voltage swing of 420 mV was measured from the output buffer corresponding to an internal input swing of 9.0 V. An average rise/fall time of 510 psec was measured.[45]

420 mV was measured from the output buffer stage. Correcting for the measured attenuation of the output buffer (25.5 dB from S-parameters) the corresponding voltage swing of the amplifier core is 9 V peak to peak (4.5 V single ended) and the average rise/fall time was 510 psec. Based on the signal swing of the amplifier core, this corresponds to a measured slew rate of 1.26×10^4 V-μsec. Similar results were achieved for other differential amplifier design variants and for different wafers highlighting the manufacturability of the approach.

Conclusion

Integration of III-V semiconductor devices with Si technology is a broad and complex field. This chapter has focused on two different approaches to monolithic integration, examining the progress in material structural properties and in the performance of fabricated devices. Material and device figures-of-merit approaching those measured on reference structures grown on lattice-matched InP substrates have been demonstrated. The MBE growth of III-V QWFET structure directly on Si substrates described here demonstrates the potential for future reduction in device characteristic size, matching contemporary Si device architecture such as the incorporation of high-K dielectric materials, and the use of 3D gate topology. The growth process for HBT epilayers in windows on the patterned SOLES substrates enabled the fabrication of clusters of III-V devices in close proximity to Si CMOS circuitry, producing differential amplifiers which served as a proof of concept for the fabrication of high-performance, monolithically-integrated hybrid ICs. Moreover, because the use of engineered substrates in combination with MBE patterned growth is a young technology, any achievements should be evaluated as a technology that is still in the early development stage. The integration technologies presented are not limited to these two specific examples; rather the growth technique can be extended to other III-V alloy growths for monolithic integration with Si circuitry of a variety of material systems, devices, and applications.

Acknowledgement

I would like to thank many people who have contributed so much to the efforts discussed in this work. First, I gratefully acknowledge Intel Corporation for authorization to use their previously published illustrations on direct growth of III-V QWFET on Si. Our work with many talented people there, including Mantu Hudait, Niloy Mukherjee, Suman Datta, Marko Radosavljevic, and Robert Chau, has been very rewarding.

The work on direct monolithic integration of III-V and Si CMOS using SOLES was supported in part by the DARPA COSMOS Program (Contract Number N00014-07-C-0629), and I would like to thank Mark Rosker (DARPA), Harry Dietrich (ONR) and Karl Hobart (NRL) for their support. This work was done in collaboration with my colleagues, Amy Liu, Andreas Einsenbach, Andrew Snyder, Joel Fastenau, Mike Rehrig, and Ying Wu at IQE; Abbas Torabi, Bill Hoke, Jeff LaRoche, Katherine Herrick, and Tom Kazior at Raytheon Integrated Defense Systems; Mayank Bulsara and Eugene Fitzgerald at MIT and Paradigm Research, LLC; Miguel Urteaga, Wonill Ha, Joshua Bergman, M. J. Choe and Bobby Brar at Teledyne Scientific; D. Clark, D. Smith and R.F. Thompson at Raytheon Systems Limited; and Nicolas Daval and Charlotte Drazek at Soitec. I truly appreciate all the stimulating technical discussions we had over the course of this work.

Finally, I am in debt to Amy Liu and Joel Fastenau for in-depth discussions of the presented results, critical comments and assistance with manuscript preparation.

References

1. Chia-Hong, J., Agostinelli, M., Deshpande, H., El-Tanani, M.A., Hafez, W., Jalan, U., Janbay, L., Kang, M., Lakdawala, H., Lu, Y-L., Mudanai, S., Park, J., Rahman, A., Rizk, J., Shin, W.-K., Soumyanath, K., Tashiro, H., Tsai, C., Vandervoorn, P., Yeh, J.-Y. and Bai, P. *"RF CMOS Technology Scaling in High-k/Metal Gate Era for RF SoC (System-on-Chip) Applications"*, Intel Corporation, web document, http://www.intel.com/content/dam/doc/technology-brief/32nm-soc-with-rf-cmos-technology-presentation.pdf.
2. Rodwell, M. (2013), Workshop on Technologies for THz Integrated Systems, IEEE International Microwave Symposium.
3. Urteaga, M., Pierson, R., Rowell, P., Jain, V., Lobisser, E. and Rodwell, M. (2011), *Proc. 69th Annu., DRC* p. 281.
4. Chau, R., Datta, S. and Majumdar, A. (2005), *CSIS Digest*, p. 17.
5. Miller, D. H. L., Micovic, M., Lubyshev, D. I., Weizhong Cai, Wen-Yen Hwang and Kai Zhang. (1999), *J. Vac. Sci. Technol.*, **B 16**, p. 1361.
6. Fang, S. F., Adomi, K., Iyer, S., Morkoc, H. and Zabel, H. (1990), *J. Appl. Phys.*, **68**, R31.
7. Bolkhovityanov, Y. B. and Pchelyakov, O. P. (2008), *Phys. Usp.*, **51**, p. 437.
8. Fischer, R., Chand, N., Kopp, W., Morkoc, H., Erickson, L. P. and Youngman, R. (1985), *Appl. Phys. Lett.*, **47**, p. 397.
9. Olshanetsky, B. Z. and Shklyaev, A. A. (1979), *Surf. Sci.*, **82**, p. 445.
10. Cordier, Y. and Ferre, D. (1999), *J. Crys. Growth*, **201/202**, p. 263.

11. Lubyshev, D., Liu, W. K., Stewart, T. R., Cornfeld, A. B., Fang, X.-M., Xu, X., Specht, P., Kisielowski, C., Goorsky, M. S., Mirecki-Millunchick, J., Whelan, C. S., Hoke, W. E., Marsh, P. F. and Svensson, S. P. (2001), *J. Vac. Sci. Technol.* **B19**, p. 1510.
12. Lubyshev, D., Wu, Y., Fang, X.-M., Yurasits, T., Liu, W.K., Cornfeld, A. B., Mensa, D., Jaganathan, S., Pullela, R., Dahlström, M., Sundararajan, P. K., Mathew, T. and Rodwell, M. (2001), *13th IPRM*, WP-40.
13. Baklenov, O., Lubyshev, D., Wu, Y., Fang, X.-M., Fastenau, J. M., Leung, L., Towner, F. J., Cornfeld, A. B. and Liu, W. K. (2002), *J. Vac. Sci. Technol.*, **B20**, p. 1200.
14. Lubyshev, D. I., Fastenau, J. M., Fang, X.-M., Wu, Y., Doss, C., Snyder, A., Liu, W. K., Lamb, M. S. M., Bals, S. and Song C. (2004), *J. Vac. Sci. Technol.*, **B22**, p. 1565.
15. Ishizaka, A., Nakagawa, K. and Shiraki, Y. (1982), *Collected Papers of 2nd Int. Conf. MBE-CST*, p. 183.
16. Asano, T., Ishiwara, H., Lee, H. C., Tsutsui, K. and Furukawa, S. (1986), *Jpn. J. Appl. Phys.*, **25**, p. L139.
17. Li, W., Anan, T. and Schowalter, L. J. (1994), *Appl. Phys. Lett.*, **65**, p. 595.
18. Eisenbeiser, K., Emrick, R., Droopad, R., Yu, Z., Finder, J., Rockwell, S., Holmes, J., Overgaard, C. and Ooms, W. (2002), *IEEE Electron Device Letters*, **23**, p. 300.
19. Yu, Z., Droopad, R., Jordan, D., Curless, J., Liang, Y., Overgaard, C., Li, A., Talin, T., Eschrich, B., Craigo, K., Eisenbeiser, R., Emrick, J., Finder, X., Hu, Y., Wei, J., Edwards, Jr., H., Convey, D., Moore, K., Marshall, D., Ramdani, J., Tisinger, L., Ooms, W., O'Steen, M., Towner, F. and Hierl, T. (2002), *Int. Conf. GaAs Manufacturing Technology*, p. 13E.
20. Droopad, R., Eisenbeiser, K., Hierl, T., Lubyshev, D., Ooms, W., O'Steen, M., Ramdani, J. and Towner, F. (2002), *14th IPRM*, p. A4-7.
21. Pidduck, A. J., Robbins, D. J., Cullis, A. G., Leong, W. Y. and Pitt A. M. (1992), *Thin Solid Films*, **222**, p. 78.
22. Akahori, K., Wang, G., Okumura, K., Soga, T., Jimbo, T. and Umeno M. (2001), *Solar Energy Materials and Solar Cells*, **66**, p. 593.
23. Hudait, M. K., Datta, S., Dewey, G., Fastenau, J. M., Kavalieros, J., Liu, W. K., Loubychev, D., Pillarisetty, R., Radosavljevic, M. and Chau, R. (2007), *IEDM Tech. Dig.*, p. 23.5.
24. Radosavljevic, M., Dewey, G., Fastenau, J. M., Kavalieros, J., Kotlyar, R., Chu-Kung, B., Liu, W. K., Lubyshev, D., Metz, M., Millard, K., Mukherjee, N., Pan, L., Pillarisetty, R., Rachmady, W., Shah, U. and Chau, R. (2010), *IEDM Tech. Dig.*, p. 126.
25. Lau, K. M., Tang, C. W., Li, H. O. and Zhong, Z. Y. (2008), *IEDM Tech. Dig.*, p. 723.
26. Mukherjee, N., Boardman, J., Chu-Kung, B., Dewey, G., Eisenbach, A., Fastenau, J., Kavalieros, J., Liu, W.K., Lubyshev, D., Metz, M., Millard, K., Radosavljevic, M., Stewart, T., Then, H.W., Tolchinsky, P. and Chau, R. (2011), *IEDM Tech. Dig.*, p. 35.1.
27. Balakrishnan, G., Huang, S., Dawson, L. R., Xin, Y.-C., Conlin, P. and Huffaker, D. L. (2005), *Appl. Phys. Lett.*, **86**, p. 034105.

28. Huang, S. H., Balakrishnan, G., Khoshakhlagh, A., Jallipalli, A., Dawson, L. R. and Huffaker D. L. (2006), *Appl. Phys. Lett.*, **88**, p. 131911.
29. Huang, S. H., Balakrishnan, G., Khoshakhlagh, A., Dawson, L. R. and Huffaker, D. L. (2008), *Appl. Phys. Lett.*, **93**, p. 071102.
30. Radosavljevic, M., Chu-Kung, B., Corcoran, S., Dewey, G., Hudait, M. K., Fastenau, J. M., Kavalieros, J., Liu, W. K., Lubyshev, D., Metz, M., Millard, K., Mukherjee, N., Rachmady, W., Shah, U. and Chau, R. (2009), *IEDM Tech. Dig.*, p. 13.1.
31. Christiaens, W., Loeher, T., Pahl, B., Feil, M., Vandevelde, B. and Vanfleteren, J. (2008), *Circuit World*, **34**, p. 3.
32. Alexe, M., Gösele, U. (2004), *Wafer Bonding: Applications and Technology*, (Springer).
33. Knickerbocker, J. U., Andry, P. S., Buchwalter, L. P., Deutsch, A., Horton, R. R., Jenkins, K. A., Kwark, Y. H., McVicker, G., Patel, C. S., Polastre, R. J., Schuster, C., Sharma, A., Sri-Jayantha, S. M., Surovic, C. W., Tsang, C. K., Webb, B. C., Wright, S. L., McKnight, S. R., Sprogis, E. J. and Dang, B. (2005), *IBM Journal of Research and Development*, **49.4.5**, p. 725.
34. Hjort K., (2004), *J. Cryst. Growth*, **268**, p. 346.
35. Patterson, P. R. and Elliott, K. (2008), *214th Electrochemical Society Meeting*, p. E11-2218.
36. Nam, P., Zhou, J., Chang-Chien, P., Tornquist, K., Scott, D., Oyama, B., Thai, K., Phan, W., Sandhu, R., Gutierrez, A., Wang, J., Barsky, M. and Oki, A. (2008), *214th Electrochemical Society Meeting*, p. E12-2245.
37. Lubyshev, D., Fastenau, J. M., Liu, W. K., Wu, Y., Bulsara, M. T., Fitzgerald, E. A. and Hoke, W. E. (2007), *25th NAMBE Conf.*, Tu5.2.
38. Herrick, K. J., Fitzgerald, E., Liu, A., Brar, B., Urteaga, M., Bulsara, M., Clark, D. and Kazior, T. (2008), *33rd GOMACTech Conf.*, p. 9.2.
39. Herrick, K. J., Urteaga, M., Liu, A. W. K., Loubychev, D. I., Fastenau, J. M., Fitzgerald, E., Bulsara, M., Brar, B., Ha, W., Bergman, J., Kazior, T. and Daval, N. (2008), *MRS Spring Meeting*, p. C2.1.
40. Lubyshev, D., Fastenau, J. M., Wu, Y., Liu, W. K., Urteaga, M., Ha, W., Bergman, J., Brar, B., Bulsara, M. T., Fitzgerald, E. A., Hoke, W. E., LaRoche, J. R., Herrick, K. J. and Kazior, T. E. (2008), *20th IPRM Conf.*, p. MoB2.3.
41. Lubyshev, D., Fastenau, J. M., Liu, W. K., Wu, Y., Bulsara, M. T., Fitzgerald, E. A. and Hoke, W. E. (2008), *J. Vac. Sci. Technol.*, **B26**, p. 1115.
42. Ha, W., Urteaga, M., Bergman, J., Brar, B., Liu, W. K., Lubyshev, D., Fastenau, J. M., Wu, Y., Bulsara, M. T., Fitzgerald, E. A., Hoke, W. E., LaRoche, J. R., Herrick, K. J., Kazior, T. E., Clark, D., Smith, D., Thompson, R. F., Drazek, C. and Daval, N. (2008), *66th DRC Conf.*, p. IV.B-9.
43. Liu, W. K., Lubyshev, D., Fastenau, J. M., Wu, Y., Bulsara, M. T., Fitzgerald, E. A., Urteaga, M., Ha, W., Bergman, J., Brar, B., Hoke, W. E., LaRoche, J. R., Herrick, K. J., Kazior, T. E., Clark, D., Smith, D., Thompson, R. F., Drazek, C. and Daval, N. (2008), *15th Int. Conf. on Molecular Beam Epitaxy*, p. THA1.1.
44. Liu, W. K., Lubyshev, D., Fastenau, J. M., Wu, Y., Bulsara, M. T., Fitzgerald, E. A., Urteaga, M., Ha, W., Bergman, J., Brar. B., Hoke, W. E.,

LaRoche, J. R., Herrick, K. J., Kazior, T. E., Clark, D., Smith, D., Thompson, R. F., Drazek, C. and Daval, N. (2009), *J. Cryst. Growth*, **311**, p. 1979.
45. Kazior, T. E., LaRoche, J. R., Lubyshev, D., Fastenau, J. M., Liu, W. K., Urteaga, M., Ha, W., Bergman, J., Choe, M. J., Bulsara, M. T., Fitzgerald, E. A., Smith, D., Clark, D., Thompson, R., Drazek, C., Daval, N., Benaissa, L. and Augendre, E. (2009), *IPRM Conf.*, p. 100.
46. Hoke, W. E., Kennedy, T. D., LaRoche, J; Torabi, A; Bettencourt, J., Saledas, P., Lee, C. D., Lyman, P. S., Kazior, T. E., Bulsara, M. T; Fitzgerald, E. A., Lubyshev, D. and Liu, W. K. (2010), *J. Vac. Sci. Technol.* **B28**, p. C3H1.
47. Lubyshev, D., Fastenau, J. M., Wu, Y., Liu, W. K., Bulsara, M. T., Fitzgerald, E. A., Urteaga, M., Ha, W., Bergman, J., Choe, M. J., Brar, B., Hoke, W. E., LaRoche, J. R., Kazior, T. E., Smith, D., Clark, D., Thompson, R. F., Drazek, C., Daval, N., Benaissa, L. and Augendre, E. (2010), *IBM Materials Research Community Workshop on III-V Compound Semiconductors.*
48. Kazior, T. E., LaRoche, J. R., Lubyshev, D., Fastenau, J. M., Liu, W. K., Urteaga, M., Bergman, J., Choe, M. J., Lee, K. J., Seong, T., Seo, M., Yen, A., Bulsara, M. T., Fitzgerald, E. A., Smith, D., Clark, D; Thompson, R. F., Drazek, C. and Guiot, E. (2010), *IEEE 2010 Compound Semiconductor Conf.*, (1092), TH1C.
49. Kazior, T E., LaRoche, J. R., Hoke, W., Fitzgerald, E. A., Bulsara, M., Urteaga, M., Bergman, J., Choe, M. J., Lee, K. J., Seo, M., Seong, T., Yen, A., Lubyshev, D., Fastenau, J. M. and Liu, A. W. K. (2011), *36th Annual GOMACTech Conf.*, p. 2.4.
50. Dohrman, C. L., Chilukuri, K., Isaacson, D. M., Lee, M. L. and Fitzgerald, E. A. (2006), *Materials Science and Engineering B*, **135**, p. 235.
51. Chilukuri, K., Mori, M. J., Dohrman, C. L. and Fitzgerald E. A. (2007), *Semicond. Sci. Tech.*, **22**, p. 29.
52. Maleville, C. and Mazuré, C. (2004), *Solid State Electron.*, **48**, p. 1055.
53. Smart-Cut™is a registered trademark of Soitec.
54. Sieg, R. M., Ringel, S. A., Ting, S. M., Fitzgerald, E. A. and Sacks, R. N. (1998), *J. Electron. Mater.*, **27**, p. 900.
55. Ting, S. M. and Fitzgerald, E. A. (2000), *J. Appl. Phys.*, **87**, p. 2618.
56. Fitzgerald, E. A. (1989), *J. Vac. Sci. Technol.* **B7**, p. 782.

Dielectrics on Silicon

Chapter 5

Low Dielectric Constant Materials for Nanoelectronics

*Mikhail R. Baklanov, Kris Vanstreels, Chen Wu,**
Yunlong Li, and Kristof Croes
IMEC, Leuven, Belgium
**Also at Department of Metallurgy and Materials Engineering, Katholieke University Leuven, Belgium*

1. Introduction

Since the 1960s, the density of transistors in Integrated Circuit (IC) chip has been doubling every 1.5 years. This progression is known as Moore's law, after Gordon Moore, one of the early IC pioneers and founders of Intel Corporation [Moore, 1965]. The semiconductor industry primarily focused on improvements related to the speed of individual transistors and enhanced performance through scaling — by squeezing more transistors into a single IC device. By the early 1990s, the distances between IC components were becoming incredibly small, and the relative effect of the interconnect delay became a greater portion of the overall signal propagation delay. The total resistance (R) of the interconnect structure became a significant factor affecting chip performance. At the same time, the capacitance (C) between the wires was increasing proportionally to the decreasing spacing between the wires. Both these factors significantly increased the RC delay of IC circuits. Microelectronics community realized the need to improve the interconnect delay by making changes to the materials used for wires and the materials used to insulate the wires.

The 1994 National Technology Roadmap for Semiconductors (NTRS) — that time the US industry's technology strategy document — stated that

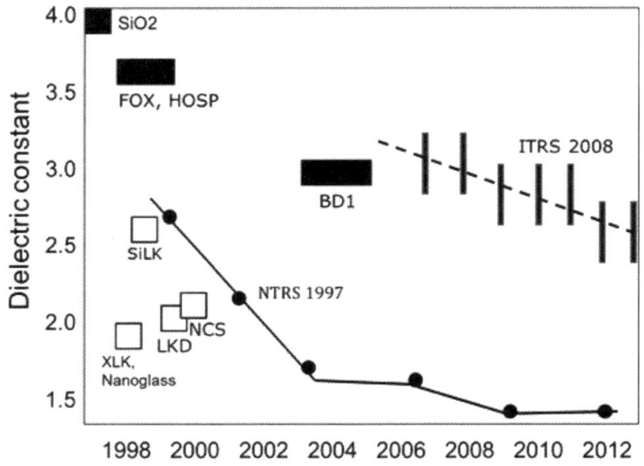

Fig. 1.1. Predictions by NTRS'1997 (solid lines) and ITRS'2008. The delay and corrections are related to difficulty of integration of porous dielectric materials. Only non-porous dielectric materials like FOX (fluorinated oxide), HOSP (dense MSQ) and BD1 (organosilicate glass without artificial porosity) were successfully integrated into IC devices.

materials with a lower dielectric constant would be needed for wires insulating as the feature sizes of IC devices became smaller. The NTRS projected that within 10 years the industry should be able to achieve a standard dielectric constant of less than 1.5 in their production interconnect material. However, the real situation was much more challenging. After several revisions, one of the latest editions of International Technology Roadmap for Semiconductors (ITRS) predicted that low-*k* materials with only $k = 2.5$ will be integrated in 2014 (Figure 1.1). The reasons of such delay are huge challenges that have been met during the integration of porous materials. Porous low-*k* materials are generally soft, mechanically weak, and did not adhere well to dielectric and conductive barrier layers. Further, porous low-*k* materials did not withstand conventional processing (i.e., they degrade during the plasma and chemical processing, crack or delaminate) [Maex *et al.*, 2003], [Hoffman *et al.*, 2005].

By 1997, chip manufacturers started integrating insulating materials with dielectric constants smaller than the value typical for traditional SiO_2 ($k \approx 4.0$). Fluorosilicate glasses (FSG, FOX), which were created by adding fluorine to silicon dioxide, had k values of around 3.6 and required very little change in the production process for semiconductor manufacturers. Thus, FSG was quickly and widely adopted by the industry. However, very little progress had been made towards identifying suitable materials with a

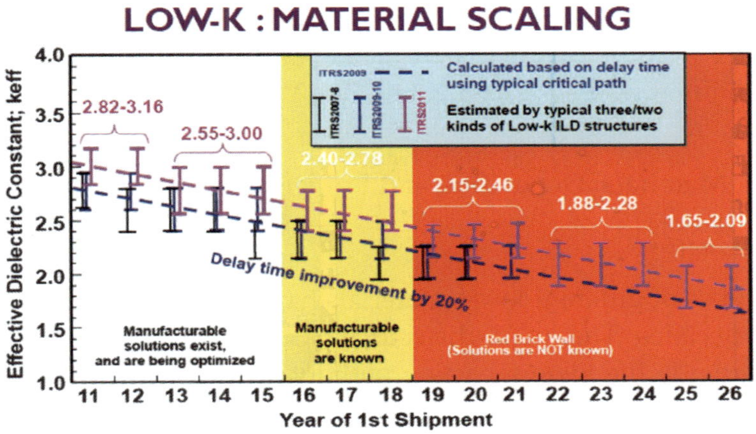

Fig. 1.2. Prediction of low-*k* materials scaling according to the recent edition of ITRS Roadmap.

dielectric constant smaller than 2.8, which would likely require less dense (i.e., more porous) materials.

Progress in the development of low-*k* materials deposition, their properties and integration challenges was recently described and analyzed in several review papers and books [Maex *et al.*, 2003], [Hoffman *et al.*, 2005], [Volksen *et al.*, 2010], [Baklanov *et al.*, 2007], [Baklanov *et al.*, 2012], [Grill *et al.*, 2014]. However, the presently active development of dielectric materials for technology nodes beyond 22 nm requires fast progress in low-*k* implementation. This is the reason why this information needs continuous updates. According to the present knowledge, the low-*k* materials with dielectric constant about 2.4–2.5 still can used for 10 nm technology nodes. The researches and industry are considering k = 2.3 materials for 7 nm technology nodes. It is less clear with integration beyond 7 nm (red brick wall in the Figure 1.2, adapted from the recent edition of ITRS Roadmap). If to keep the present integration schemes, the low-*k* dielectrics should have k = 2.0 and below.

This chapter is an overview of low-*k* materials and present progress in their development and implementation. The first part of the chapter includes overview of different types of low-*k* materials, including their deposition and materials properties, integration challenges (Mikhail R. Baklanov). The second part analyses mechanical properties of ultra low-*k* materials as well as impact of the mechanical properties in the device packaging (Chip-package interaction, CPI) (Kris Vanstreels). The third part describes characterization of electrical properties and reliability with special emphasis on mechanism of leakage current. New test structure (so called planar-capacitors) allowing to investigate intrinsic electrical characteristics and reliability is also described (Chen Wu, Yunlong Li, Kristof Croes).

2. Low-k Materials: Deposition and Properties

2.1. *Introduction*

Several types of different low-*k* materials have been developed during the last two decades. The low-*k* dielectrics based on organic polymers were able to provide the lowest *k* value without requiring the introduction of porosity. Materials with aliphatic C–C, C–H and C–N bonds have lowest dielectric constant but generally become unstable at temperatures greater than 300−400°C and, in some cases, at even lower temperatures. Only materials composed of non-aliphatic C–C, C–O, C–N and C–S bonds, aromatic structures and cross-linked or ladder structures can withstand the temperatures necessary for interconnect technology (450−500°C). Most of these organic low-*k* films with sufficient thermal stability have dielectric constants close to 2.6–2.8. However, the efforts to integrate them into IC circuits have not been successful. In addition to poor mechanical and thermal properties the most important problems were related to relatively high coefficient of thermal expansion (CTE) in comparison with other components of integrated circuits. [Macheta *et al.*, 2003], [Kuchenmeister *et al.*, 1999], [Hasegawa *et al.*, 1999], [Martin *et al*, 2000]. This difference generate catastrophic failure during the packaging.

More successful integration was achieved with Hybrid Organosilicate Glasses (OSG, SiCOH, or carbon doped oxides). The OSG materials can be deposited by both plasma enhanced chemical vapor deposition (PECVD) and spin-on glass (SOG) technology. The PECVD OSG materials have been more popular up to now because of similarity of their properties to traditional silicon dioxide and therefore it was possible to use the equipment and technology developed for deposition of silica based dielectrics. The integration solutions are presently known for the materials with $k \geq 2.5$ when porosity and pore size are relatively small (porosity, $P \leq 30\%$, pore radius, $R \leq 2.0$ nm). Chemical companies also developed several spin-on deposited OSG ultra low-*k* materials with $k < 2.5$: XLK (Dow Corning), Nanoglass (Honeywell), LKD (JSR), NCS (CCIC) etc. [Maex *et al.*, 2003], but their integration is much more challenging because of high porosity and large pore size. The high porosity deteriorates their mechanical properties, makes them more sensitive to degradation during the plasma processing,

CMP and other chemical treatments. According to the recent editions of ITRS Roadmap (Figure 1.2), the manufacturable solutions already exist for low-k materials with $k = 2.55-3.0$, the integration solutions are generally known for 2.40–2.78 materials.

Classically, the dielectric constant of dielectric materials is described by the Clausius–Mossotti equation

$$\frac{k-1}{k+2} = \frac{4\pi}{3} N\alpha \tag{2.1}$$

where $\kappa = \varepsilon/\varepsilon_0$, ε and ε_0 are the dielectric constants of the material and vacuum, N is the number of molecules per unit volume (density) and α is the total polarizability, including electronic (α_e), distortion (α_d) and orientation (α_o) polarizabilities. According to this equation, the dielectric constant of materials can be reduced by decreasing the total polarizability and density. The development of organic low-k materials were based on the idea of reduced polarizability. The materials density can be reduced by introduction of artificial porosity. The OSG low-k materials have skeleton (matrix) similar to silicon dioxide with dielectric constant about 4. Therefore, OSG low-k materials with $k < 3.0$ must be generally porous. The relative dielectric constant of a porous material, k_r, depends on porosity (P) and dielectric constant of the film skeleton (k_s):

$$\frac{k_r - 1}{k_r + 2} = (1 - P) \cdot \frac{(k_s - 1)}{(k_s + 2)} \tag{2.2}$$

Equation (2.2) allows to predict the change of dielectric constant versus porosity (Figure 2.1). Materials with relatively small k_s values (MSQ, organic polymers) provide the same k_r values at lower porosity (Figure 2.1) and this is the reason why different types of matrix materials have been considered for preparation of porous low-k films (Figure 2.2).

As already mentioned, the most presently important ones are hybrid materials: inorganic silica based materials with incorporated organic hydrophobic agents. Pure silica has a tetrahedral elementary unit. To reduce the k value and make them hydrophobic, part of oxygen atoms are replaced with F or alkyl groups CHx. The addition of CHx not only introducing less polar bonds, but also creating free volume. Such organosilicate glasses are normally deposited by PECVD and they are constitutively porous. The carbon concentration in the most of OSG materials is varied between 15 and 20%. The carbon concentration needs careful optimization: it must be

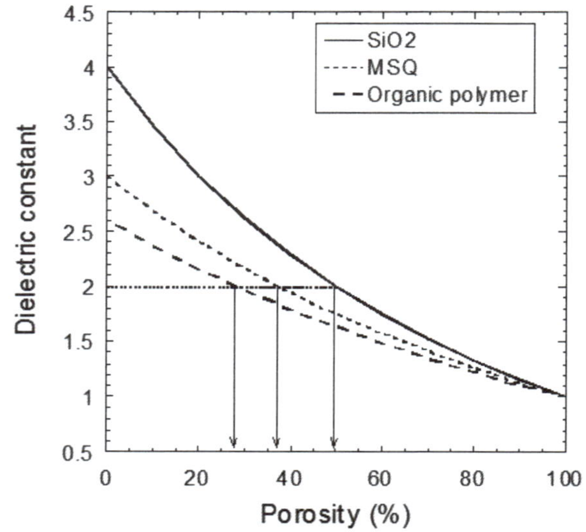

Fig. 2.1. Dependence of dielectric constant on type of matrix materials and porosity. The same k-value can be achieved at lower porosity if the matrix material has lower dielectric constant.

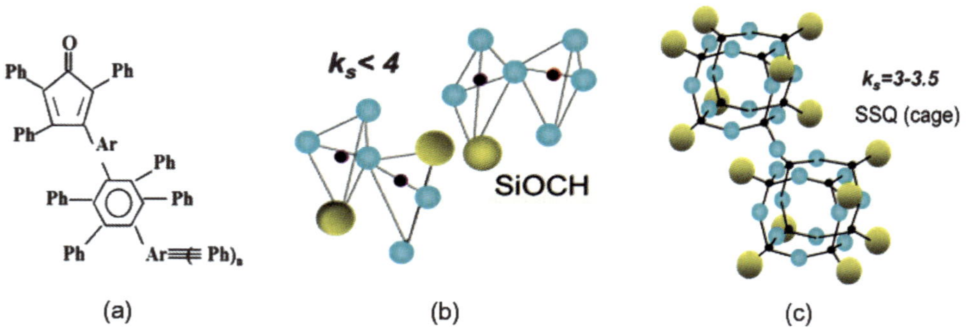

Fig. 2.2. Three typical types of low-*k* matrix materials: a) An organic polymer (SiLK from Dow Chemical). b) Organo-silicate glass (OSG) deposited by Plasma Enhanced Chemical Vapor Deposition (PECVD). Some of oxygen atoms in silica structure are replaced by hydrophobic aliphatic groups (yellow circles) to make the material hydrophobic. c) Alkylsilsesquioxanes (SSQ). - Si atom, - Oxygen atom, - alkyl group, most often CH3.

sufficiently high to provide their hydrophobic properties but small enough to achieve good mechanical properties.

More porous ultralow-*k* materials are deposited by spin-on glass techniques. In typical SSQ (silsesquioxane) type materials, Si and O atoms are arranged in a form of cage or ladder [Baklanov and Maex, 2006]. The cage structure creates free volume, decreasing the material's density and, therefore, it's *k* value. The cages in the polymerized SSQ are connected to each other through oxygen or $-CH_2-$ groups, while other cage corners are terminated by hydrogen (HSQ), methyl (MSQ) or other aliphatic groups. MSQ matrix materials have a lower dielectric constant as compared to HSQ because of the larger size of the CH_3 groups and their lower polarizability. The SSQ cages are metastable and tend to break down to silica tetrahedral, especially during the curing at elevated temperature. The new types of spin-on OSG low-*k* materials are so called Periodic Mesoporous Organosilicates (PMO), which have ordered pores and expected to have more controllable and improved mechanical properties. These materials will be considered below.

The skeleton dielectric constant of all OSG materials (PECVD SiCOH, SOG SSQ and PMO) are defined by polarisability of Si-O bonds and free volume that depends on CHx concentration. Deposition of ultra low-*k* materials needs introduction of artificial porosity in both SOG and PECVD technologies.

2.2. *Deposition by Spin-on Glass (SOG) technology* [Dubois *et al.*, 2007], [Volksen *et al.*, 2010]

In the spin-on deposition, the film coating is performed by dispensing a liquid precursor at the center of the substrate, which is placed on a spinner. Rotation of the substrate creates centrifugal forces that ensure a uniform distribution of material on the surface. The thickness of the coating is a result of balance between centrifugal forces and viscous forces, determined by the viscosity of the solution. Normally, the spinning step is followed by heating or "soft bake" at temperatures typically below 250°C, for the solvents removal. The latter step can also initiate cross-linking of the film. Finally, a sintering at temperatures varying from 350°C to 450°C ("cure", see details in the next paragraph) is required to obtain a stable film. This "cure" step induces the final cross-linking of the polymer chains and results in a mechanically stable film structure.

Numerous methods of introducing subtractive porosity into spin-on deposited materials exist and they can be divided into two main categories. The first category groups all materials where the porosity is introduced exclusively through Sol-Gel processes, while the second group includes the materials where the porosity is formed through the use of sacrificial particles (porogens, templates) that are removed during the film cure.

2.2.1. *Subtractive porosity by Sol-Gel based technologies* [Hench and West, 1990], [Brinker and Scherer, 1990]

There exist two main approaches based on Sol-Gel techniques to the formation of subtractive porosity: the first takes advantage of aging processes and the second relies on a hierarchical organization of the primary particles in the sol (self-assembly). The formation of a more or less rigid skeleton structure before extraction of the liquid from a wet gel is a key point in the formation of high porosity materials. Even if the gel-point is reached after material spinning, a long time is still required before the hydrolysis and condensation reactions are complete. For this purpose, an additional step (aging) before drying the wet gel is introduced. The aim of this step is to accelerate the sol-gel reactions, typically by relying on the pH and the water content in the ambient. Once the network structure is strengthened, extraction of the solvent can take place without collapse of the network backbone. The level of residual porosity is generally tuned through the ratio of solvent to solid content in the sol. In sol-gel science numerous studies have been performed in the synthesis of self-assembled materials. Hierarchical ordering of the aggregates by preferential solvent evaporation during spin coating is reported for a solution of surfactants, swelling agent and soluble silica. By this method, ordered materials with dielectric constant as low as 1.3 have been synthesized. In this case, the final film porosity and pore structure is related to the way in which the primary particles are assembled and ordered.

2.2.2. *Subtractive porosity by macromolecular porogens*

This technique is based on the addition in the dielectric precursor of molecular or supramolecular particles ('porogens') with tailored thermal stability. The stability of these particles is such that they are not affected by the drying step, and they are removed by pyrolysis during the final film sintering or cure (typically in the range from 300 to 400°C). Their volume distribution in the film at the moment of desorption represents the template for the residual pores in the layer. In ideal case, the film's final porous fraction is directly related to the amount of porogen as a function of the total solid part in the precursor solution, and the size of the sacrificial particles is directly related to the final pore size [Baklanov *et al.*, 2002].

There are two ways in which the sacrificial porogens are brought into the precursor solution. One method is dispersion of porogens in the solution. The second is chemically linking sacrificial particles (grafted) to the network polymers. This second method grants an inherent control of the volume distribution of porogens in the dielectric film.

2.2.3. *Periodic mesoporous organosilicates (PMOs)*

Periodic mesoporous organosilicates are ordered mesoporous silica materials of which at least one siloxane bond in the matrix is replaced by an organic bridge. PMOs are usually amorphous, but their pores and pore walls are arranged in a structural order. Typically PMO materials are synthesized by hydrolysis and condensation of bridged organosilanes $(R'O)_3Si-R-Si(OR')_3$ in the presence of a surfactant which acts as structure directing agent [Hoffmann *et al.*, 2006]. After the ageing period, the surfactant can be removed by an extraction procedure or by a thermal treatment under an inert atmosphere, leaving behind a porous PMO material. A typical synthesis scheme is given in Figure 2.3.

The great advantage of PMOs over pure silica ordered mesoporous materials is the presence of uniformly distributed organic functional groups in the walls. This avoids leaching of the functional group or pore narrowing which is often observed in post-functionalized porous silicas. Further, PMOs are hydrothermally and mechanically more stable than mesoporous silicas. These characteristics make them promising candidates for different applications including microelectronics.

By varying synthesis conditions, precursors and surfactants, it is possible to adjust the pore size, porosity, surface area, ordering, wall thickness and morphology of the PMOs [Hunks and Ozin, 2005]. Examples of frequently used surfactants are similar to those for mesoporous silicas. Possible precursors are methylene, ethylene, ethenylene, phenylene, thiophene, etc.

Already since the first publications on PMOs one of the proposed applications was as low-*k* material [Asefa *et al.*,1999]. They are similar

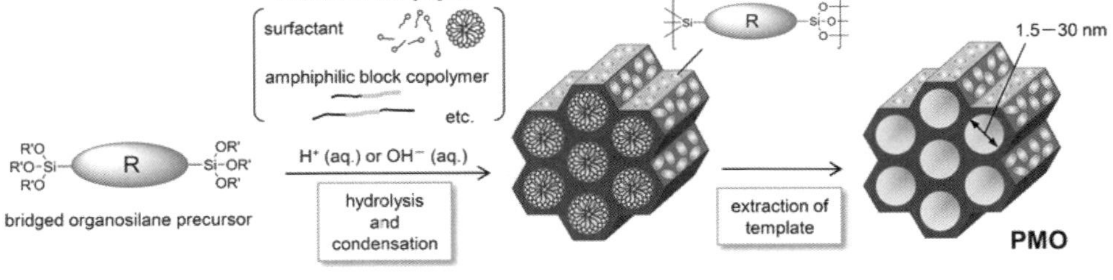

Fig. 2.3. Formation of Periodic Mesoporous Organosilicate (PMO). Reproduced from (N. Mizoshita N. *et al.*, 2011) with permission of The Royal Society of Chemistry.

in chemical composition to the organosilicate based low-*k* materials with the advantage that low polarasible organic groups are integrated in the silica matrix. Especially PMOs with the simplest organic bridges such as ethane and methane are extremely interesting for low-*k* applications because they have the lowest polarisable organic bridge. This means that more network connectivity can be obtained, which should enhances the mechanical stability. Even more bridged organosilane precursors are also incorporated in OSG films resulting in a higher Young's modulus of the film [Landskron *et al.*, 2003].

At this moment, several groups reported the papers related to PMOs and low-*k* applications. They prepared methane, ethane, ethene and ring PMO films using CTAC as template, and investigated their dielectric, mechanical and hydrophobic properties [Hatton *et al.*, 2005], [Goethals *et al.*, 2012(1)], [Goethals *et al.*, 2012 (2)] . They also investigated the influence of the organic bridge on the dielectric constant by mixing the starting PMO precursor with TMOS and came to the same conclusion that the dielectric constant decreases and the mechanical stability increases with organic content. The dielectric constant can be further decreased by annealing the samples above 400°C in a nitrogen atmosphere. The reason for this is an increase in hydrophobic character of the PMO. The more hydrophilic a material is, the more water that will adsorb and this leads to a dramatic increase in dielectric constant. For ethane PMOs, this beneficial effect can be attributed to extra condensation of hydrophilic silanol groups during annealing.

2.3. *Plasma enhanced chemical vapor deposition*

The semiconductor industry has long relied on insulating films of SiO_2 deposited from gas phase and plasma by oxidation of silane (SiH_4) and its derivatives. Therefore, most of the attempts in producing low-*k* materials by different versions of PECVD have been doped versions of SiO_2. The main dopants in the beginning, as already mentioned, were fluorine and carbon. Their introduction is done by replacing standard silane by fluoro- and alkylsilanes like $Si_2H_2F_2$ and $(CH_3)_xSiH_y$ with $(x + y) = 4$. Doping a film with alkyl groups terminates some of the silicon bonds within the oxide lattice and lowers the electronic polarizability of the film. The relatively large molecular volume of the alkyl groups decreases the film density and provides the *k*-value 2.8–3.0.

Various techniques have been employed to produce PECVD SiOCH films with subtractive porosity and the k value smaller than 2.8 [Grill, 2009]. One method utilizes a multiphase deposition. The SiOCH precursor (alkylsilane derivatives and/or organosilicates) is mixed with a thermally unstable hydrocarbon phase (porogen) during deposition. This unstable phase is thermally decomposed and removed from the film during subsequent anneal leaving behind pores in the organosilicate matrix. The resulting porosity depends on the matrix/porogen precursors ratio. Efficiency of porogen removal can be enhanced by UV light and/or electron beam.

This approach has the benefit of allowing manufacturers to use the same PECVD equipment already operating in fab's, modified only by additional gas delivery lines. Thus, adoption of such a process requires a relatively low integration cost and gives the industry a level of confidence in the new process [Grill *et al.*, 2014]. This is the reason why during the last decade PECVD materials have been the most popular and considered as the most possible candidates for integration. However, although the lowest *k* of known PECVD films has been reported as 2.2, only low-*k* materials with $k > 2.4–2.5$ have shown acceptably good electrical characteristics and reliability after integration.

The chemical composition and porosity of OSG low-*k* film is defined my matrix and porogen composition. It is less clear with the exact link between the porogen and the final pore size. In order to test the hypothesis that pore size was related to the molecular volume of the porogen precursor, O'Neill *et al.* have determined pore sizes from both Positron Annihilation Lifetime Spectroscopy (PALS) and Ellipsometric porosimetry (EP). They have compared the results with the molecular volume of the porogen molecule estimated from its molecular weight and liquid density [O'Neill *et al.*, 2006]. The number of porogen molecules per pore ranges from 4 to 10 by this estimation. These results did not show any correlation of pore size with either the molecular volume of the porogen or with its chemical nature (i.e. degree of unsaturation, boiling point, etc.). They concluded that the pores must be formed by a species larger than one porogen precursor molecule (more likely clusters of molecules), which could be created either in the plasma phase, on the film surface, or even in the bulk of the film during the curing stage (Figure 2.4).

One can see that as deposited film is a mixture of low-*k* matrix and agglomerated porogen molecule. If the matrix is already sufficiently cross-linked (for instance, during the deposition), the porogen agglomerates can be removed by thermal curing. The curing temperature is normally about 400–450°C. The pure thermal curing takes quite long times (sometimes, several hours) and for this reason this process can be initiated by electron beam or UV light. One important requirement to curing is that porogen

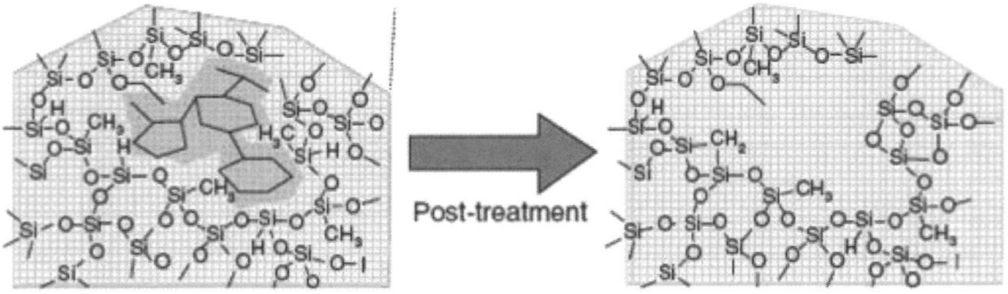

Fig. 2.4. Schematic view of the porogen incorporation.

removal must be sufficiently selective: the reduction of Si-CH_3 bonds concentration should not be significant to keep the hydrophobic properties of low-*k* film. The final porous low-*k* film contains chemical bonds typical for SiO_2 and also some silicon atoms (15–20%) are terminated by hydrophobic CH_3 (Si-CH_3) and H (Si-H) groups. Normally it is preferable to minimize Si-H groups concentration because they have lower chemical stability than Si-CH_3.

PECVD deposition of low-*k* materials has been very popular in industry. Appropriate combination of matrix and porogen together with optimized curing allow to prepare microporous films (with pore diameter smaller than 2 nm) and dielectric constant from 4.0 to 2.4. Presently, some companies are trying to extend PECVD deposition down to $k = 2.0$ materials. However, there are certain applications for which CVD processes are not suited. For example, if gap filling is required, this is often best achieved with spin-on materials. The generation of ULK porous materials with dielectric constants below 2.3 and good mechanical and electrical properties is problematic with CVD, but achievable with spin-on. The deposition of thermally sacrificial layers and features is often more easily accomplished with spin-on materials. If the film properties are enhanced by some organization and long-range order in either hybrid materials or the porous films so derived, this will not be achieved using CVD or PECVD techniques.

2.4. *Alternative low-k materials*

Several other types of materials have also been evaluated as promising low-*k* candidates. Most of these materials are still in the stage of general exploration but some their properties may suggest that they can find integration solutions in future technology nodes. The most known among these materials are, probably, zeolites, boron carbonitrides and metal-organic frameworks.

2.4.1. *Silica zeolites*

Classic zeolites are microporous crystalline materials that consist of open aluminosilicates frameworks derived from $[SiO_4]^{4-}$ and $[AlO_4]^{5-}$ tetrahedra, linked to form cages, channels or cavities of various sizes. The negatively charged frameworks are balanced by positively charged cations of appropriate size located at various positions in the lattice. Because of their well-defined micropores and the presence of acidic sites inside the pores, zeolites have been successfully employed for a diverse range of applications: ion exchange, separations and catalysis (oil refining, petrochemistry, organic synthesis) [Dubois *et al.*, 2007]. Pure silica zeolites have received some attention from semiconductor researches as potential low-*k* dielectrics. Since silicon is tetravalent, pure silica zeolites do not contain framework charges and consequently no cations, that would be detrimental for electronic applications, are present. Due to their dense crystalline structure, better mechanical properties are expected, and since their pore size (<2 nm) is significantly smaller than integrated circuit features.

The first work on the application of zeolites as low-*k* dielectrics involved films prepared by *in-situ* crystallization of Silicalite-1 [Wang *et al.*, 2001], [Yan *et al.* 2002], [Larlus *et al.*, 2004]. In the work by [Wang *et al.*, 2001], thin smooth zeolite silicalite films with a dielectric constant of 2.1 and elastic modulus of 16–18 GPa were obtained. The processing was compatible with the requirements of the semiconductor manufacturers. The significance of zeolite films becomes clear when considering the surpassing mechanical properties compared to other low-*k* dielectrics [Dubois *et al.*, 2007]. This was the starting point for the research on zeolites for low-*k* dielectrics.

A polishing with CMP is often necessary for planarisation of zeolite films. In addition to Silicalite-1, pure-silica zeolite BEA was also deposited by *in-situ* crystallisation and evaluated as low-*k* dielectric [Mitra *et al.*, 2004]. The dielectric constant of as prepared films was 2.3, much lower than as-prepared Silicalite-1 reported to have a dielectric constant of 2.7. Ultra low-*k* values and high porosity is achieved are gained from interparticle voids [Wang *et al.*, 2001]. However, these voids sometimes can be too large generating problems for their integration [Eslava *et al.*, 2008]. The crystalline structure also is one of the main problems limiting their application for damascene integration technology.

2.4.2. *Boron carbonitrides*

Compounds in the boron-carbon-nitrogen are potentially attractive as a base for alternative low-*k* dielectric materials due to their covalent bonding, short bond lengths, and low atomic mass that leads to a unique combination of low dielectric constant but high thermal and mechanical strength [Prakash *et al.*, 2015], [Zhang, 2012]. Some of the already prominently reported properties for these materials include good wear resistance, high hardness, and low coefficient of friction. Amorphous hydrogenated BN (a- BN:H) has been demonstrated as a dielectric with *k*-value of $\sim$ 4.0–5.7. However due to a hygroscopic nature, a-BN:H has exhibited a variety of thin film delamination issues that have limited its implementation. In this respect, carbon can be added to the BN structure to produce boron carbon nitride (BCN) films with potentially reduced moisture sensitivity, but with hardness values still as high as 20 GPa [Morant *et al.*, 2007]. The incorporation of carbon as terminal organic groups can also lead to even lower values of dielectric constant with values as low as 1.9–2.1 reported for some plasma enhanced chemically vapor deposited (PECVD) BCN thin films [Liu *et al.*, 2004], [Sugiuma *et al.*, 2002], [Umeda *et al.*, 2004], [Sulaeva *et al.*, 2014]. PECVD BCN films have also shown the ability to suppress Cu migration and diffusion which is an additional important consideration for low-*k* metal interconnect applications [Umeda *et al.*, 2004].

Beyond PECVD, several other methods for depositing BCN thin films have been reported in the literature including: chemical vapor deposition (CVD) [Watanabe *et al.*, 1995], [Baake *et al.*, 2009] pulsed laser ablation [Perrone *et al.*, 1998], ion beam assisted deposition [Zhou *et al.*, 2006], and DC/RF magnetron sputtering [Ulrich *et al.*, 1999], [Linss *et al.*, 2004].

Polycrystalline boron carbon nitride (BCN) films synthesized by plasma-assisted chemical vapor deposition (PACVD) were reported as materials with dielectric constant as low as 1.9 after annealing at 400°C [Umeda *et al.*, 2004]. Methyl containing BN films were deposited by remote plasma chemical vapor deposition using TMAB gas at 350°C. Extremely good mechanical properties (>26 GPa) at k-value 2.1 were reported [Tokuyama *et al.*, 2008].

However, no serious integration attempts and successes are known up to now. The possible reason is not sufficiently good reproducibility of deposited films and strong modification during the patterning. It was reported that BCN films become hydrophilic after plasma patterning.

2.4.3. *Metal-Organic Frameworks (MOFs)*

Among novel materials, metal-organic frameworks (MOFs) have received much scientific and technological interest. MOFs are three-dimensional structures whose open frameworks contain both inorganic (metal ions) and organic moieties linked together by strong co-ordination bonds [Ferey G., 2008], [Shekhah, O., 2008], [Ma Z. *et al.*, 2011], [Corma A. *et al.*, 2010]. Such frameworks leave a pore network available for the selective sorption of molecules. Their sorption properties have been extensively studied for their application in gas storage, drug delivery, catalysis, and sensors. However, their electrical properties and applications in microelectronics technology remain not sufficiently researched. MOFs should be stiffer and harder than other low-density amorphous inorganic or organic polymers because of their ordered framework and rigid organic linkers. The measured elastic modulus on single crystals of zeolitic imidazolate frameworks (ZIFs) and MOF-5 were found to be from 3 to 10 GPa, which is comparable to the most promising low-κ dielectrics currently under study. Similar values were reported for some other MOF materials (for, instance E=10 GPa was reported for HKUST-1 [Bundschuh *et al.*, 2012]). There is a theoretical calculation for Zn4O(CO2)6-based MOFs, ZIF-8, and other MOFs reporting static κ values below 2.0 [Zagorodnij *et al.*, 2010]. However, these calculations did not take into account the orientational and ionic contributions to the molecular polarizability, which drastically contribute to the dielectric constant at the frequencies at which microelectronic chips operate. Recent experimental evaluation by Eslava and co-authors (ZIF-8 films) [Eslava *et al.*, 2013] and M. Usman and co-authors (Sr-based MOF films) [Usman *et al.*, 2014] suggest that the theoretically estimated dielectric constants of these materials might be underestimated because of the ionic contribution that was omitted in the work of Zagorodniy and co-authors. More complete data are presently under investigation, however, several properties such as crystalline structure, sensitivity to moisture and difficulty of plasma processing allows to assume that MOFs cannot be integrated in the presently used damascene based integration scheme. Only in the case of switching to subtractive approach (metal patterning first) MOFs can be integrated and realize all their advantages.

Probably, all these three types of materials can attract attentions and realize their advantages if industry will be able adopts alternative integration scheme based on metal patterning instead damascene technology.

2.5. *Material properties*

2.5.1. *Instrumentation*

The properties of low-*k* materials after deposition and curing have to be evaluated to make preliminary conclusions about possibility of their integration. Evaluation of low-*k* materials starts with analysis of chemical composition, porosity, pore size and mechanical properties. The most commonly used techniques for the chemical analysis are Fourier-transform infrared spectroscopy (FTIR), X-ray photoelectron spectroscopy (XPS), Time-of-Flight Secondary Ion Mass Spectrometry (TOF SIMS). In some cases additional instrumentations can be used, such as nuclear magnetic resonance (NMR), electron spin resonance (ESR) etc. NMR is efficient for evaluation of efficiency and degree of UV curing because it allows analyzing the number of methyl groups bonded to Si atoms. ESR can be used for analysis of intrinsic and induced defects.

Most of ultralow-*k* materials properties important for integration are related to porosity. Porosity, pore size and their interconnectivity are responsible for the low-*k* properties and their degradation during the technological processing (plasma, chemistries in gas phase and liquids etc.), compatibility with diffusion barriers, mechanical and thermal properties, etc. Porosity and pore size are most commonly evaluated by ellipsometric porosimetry (EP), positron annihilation lifetime spectroscopy (PALS) and small angle X-ray and neutron scattering (SAXS and SANS). Other instrumentations used for evaluation of critical properties of low-*k* materials include Rutherford back scattering (RBS), specular X-ray reflectivity (SXR), surface acoustic wave spectroscopy (SAWS), Brilloin light scattering (BLS), spectroscopic ellipsometry (SE), thermodesorption spectroscopy (TDS) based on mass spectroscopy and thermogravimetry, atomic force microscope (AFM) (Table 1). The low-*k* material defects formed, for instance, during non-optimized porogen/template removal can be evaluated by using UV spectroscopy and electron spin resonance (ESR) spectroscopy. More detailed information related to the materials analysis can be found in the references [Maex *et al.*, 2003], [Plawsky *et al.*, 2007], [Baklanov and Shamiryan, 2011], [Lionti, 2015].

The evaluation of patterned wafers is more challenging. Scanning and transmission electron microscope (SEM and TEM) remain the key instruments to measure etch profile, which is necessary for optimization of etch and cleaning processes (SEM) and for extraction of integrated dielectric constant (TEM). Change of chemical composition and gradients can be evaluated by Energy Filtered TEM (EFTEM), angle resolved XPS and electron energy loss spectroscopy (EELS) often spoken of as being complementary to energy-dispersive X-ray (EDX) spectroscopy. Techniques based on picosecond Laser Ultrasonics have also been demonstrated but they are still not so common as other techniques [Antonelli *et al.*, 2006], [Mechri *et al.*, 2009], [Lomonosov *et al.*, 2012]. Scatterometric porosimetry

Table 1. The most common physical methods used for evaluation of low-*k* materials.

Materials Properties	Method of Evaluation
Chemical composition	FTIR, XPS, TOF SIMS, NMR etc.
Porosity and Pore size	EP, PALS, GISAXS, SANS
Porogen residue evaluation	UV spectroscopy, ESR
Density	RBS, SXR
Mechanical Properties	Nanoidentation, SAWS, BLS
CTE, Stress	SXR, SE, Bending beam
Thermal Stability and Outgassing	TDS
Thermal conductivity	3-Omega test
Adhesion	4 point bending
Roughness	AFM
Cross-section, shape of patterned low-*k* films.	TEM
Composition gradient in patterned structures.	EFTEM, EELS

was recently proposed for pattern sidewalls characterisation [Bouyssou *et al.*, 2010], [Licitra *et al.*, 2011]. The optical characteristcis of damaged porous low-*k* materials under vacuum are close to the optical characteristcis of the undamaged one. In contrast, under moist ambient conditions, water condensation inside the damaged pores of the porous material leads to a dramatic increase of the optical index. It is thus possible to first perform a scatterometric measurement under vacuum to determine the pattern geometry without taking into account the damaged material. In a subsequent experiment conducted under moist ambient conditions, the geometry of the damaged layer can be identified. This technique was validated on complex patterns of porous low-*k* dielectrics integrated at the 45 nm technology node. [Hurand *et al.*, 2011].

Most of these techniques are very common and their description can be found in many published papers and books.

Here we will mainly concentrate on the techniques that are specifically important for general evaluation of ultralow-*k* materials. FTIR spectroscopy is extremely important in general evaluation of low-*k* materials. We will characterize specific absorption peaks typical for OSG low-*k* materials. We also will give information related to porosity evaluation, measurements of dielectric constants. More specific information related to evaluation of mechanical characteristics will be done in the second part of the chapter. The special test structure allowing to evaluate electrical characteristics and intrinsic reliability is included in the third part of the chapter.

2.5.2. *Chemical composition*

The most common method for characterization of chemical composition is Fourier transmission infra-red (FTIR) spectroscopy. This method is fast, simple and allows to detect the presence of most of chemical bonds. The method also allows to compare relative concentration of different groups. A detailed overview of FTIR absorbance peaks in OSG low-*k* films is presented in Table 2. The data presented in the first 3 columns were reported by Grill and Neumayer, [2003]. In their work, OSG dielectrics composed of Si, O, C, and H (SiOCH) have been prepared by plasma enhanced chemical vapor deposition from tetramethylcyclotetrasiloxane (TMCTS) mixed with an organic precursor. FTIR analysis of the films showed that this synthesis route produces a highly cross-linked networked SiOCH film. The dissociation of TMCTS dominates the deposition chemistry as evidenced by the multitude of bonding environments and the formation of linear chains and branches. By mixing TMCTS with an organic porogen, a significant proportion of hydrocarbon fragments is added. Annealing these SiOCH films results in a large release of carbon and hydrogen, coming from the fragmentation and loss of the incorporated organic component. The deconvolution of the Si–O–Si asymmetric stretching band of the annealed films shows the existence of a larger fraction of a cage structure and a correspondingly smaller fraction of a networked (highly cross-linked) structure, when compared to the films deposited from TMCTS only. The next two columns show the chemical composition of PECVD ultra low-*k* films in which the porogen was removed by downstream He/H_2 plasma (DSP) at 280°C before UV assisted thermal curing [Urbanowicz *et al.*, 2010], [Urbanowicz, 2010]. The DSP He/H_2 plasma allows complete porogen removal without the formation of amorphous carbon-like porogen residue. These films are then cured at a temperature of 400–430°C, simultaneously exposed to a monochromatic UV source ($\lambda \approx 172$ nm) to produce ALKA or to a broadband source ($k > 200$ nm) to produce ALKB (ALKA and ALKB are IMEC's internal abbreviations).

Plasma exposure decreases the intensity of the Si–CH_3 peak, makes the material hydrophilic, and peaks related to moisture adsorption do appear (Table 2, No. 29–33). The intensity of these peaks depends on the degree of plasma damage. Infrared vibrations assigned to the O–Si–O skeleton are detected by FTIR from 950 to 1200 cm^{-1}, and they are similar to PECVD SiO_2. Deconvolution analysis from 950 to 1250 cm^{-1} by fitting the data to a sum of three Gaussian functions centered at 1020, 1061, and 1140 cm^{-1} shows that they correspond to Si suboxide, network,

Table 2. FTIR peak assignments for PECVD deposited OSG low-k materials. First three columns show composition of TMCTS based films deposited without porogen (TMCTS) and with organic porogen ($k = 2.8$ and $k = 2.05$). Next 2 columns show the composition of presently popular ultralow-k PECVD films where porogen was removed by downstream H_2/He plasma before UV curing by source A (172 nm) and source B (broadband UV light with $l > 200$ nm). Major contributors are listed first for each vibration. ν = stretching, δ = bending, ρ = rocking, a = antisymmetric, and s = symmetric [Baklanov *et al.*, 2013].

#	TMCTS	k=2.8 (thermal curing)	k=2.05 (thermal curing)	ALK A (k=2.0)	ALK B (k=2.0)	Mode	Comment
1	2968	2969	2968	2968	2968	ν^aC-H_3	sp^3 CH_3
2	2906					ν^sC-H_3	sp^3 CH_3
3		2916	2932	2929	2929	ν^aC-H_2	sp^3 CH_2
4		2880	2875	2873	2873	ν^sC-H_2	sp^3 CH_2
5		2232		2250		ν^sSi-H	H-SiO_3
6		2178		2178		ν^sSi-H	H-SiO_2Si
7	2165					ν^sSi-H	H-SiOSi
8			1740			ν C=O	As deposited
9			1714				
10			1461			δ C-H_2	CH_2 isolated from Si
11	1405	1412	1412			δ^a C-H_3	$SiMe_x$
12		1358	1379			δ C-H_2	Si-CH_2-Si
13	1259	1273	1274	1273	1273	δ^s C-H_3	$SiMe_x$
14		1135	1140	1135	1135	ν^a Si-O-Si	Cage
15						ν C-O	Si-O-C
16					1105	ν^a Si-O-Si	Ladder
17	1063	1063	1065	1063	1063	ν^a Si-O-Si	Network
18		1023	1035	1023	1023	ν^a Si-O-Si	Suboxide
19	890			890		δ H-Si-O	H-SiO_3
20	865					δ H-Si-O	H-SiO_2Si
21		848	843			δ H-Si-O	Network smaller angle
22						ν Si-C, ρ^a CH_3	$SiMe_3$
23		802	800			ν Si-C, ρ^a CH_3	$SiMe_2$
24	754					ν Si-C, ρ CH_3	$SiMe_1$
25		773	779	775	775	ν Si-C, ρ CH_3	$SiMe_1$
26	710	730	720			ν^s Si-O-Si	
27		440	440			δ of H-Si-O	Network and ring opening
28				1600		C=C	Porogen residue
29		3300–3600				H-bonded H_2O (O-H stretching)	
30		3650–3800				Stretching modes of isolated and terminal OH	
31		3690				Hydroxyl terminals	
32		3750				Isolated surface silanols Si–OH, (sharp peak)	
33		3200–3650				H-bonded Si–OH in chain (Si–OH stretching)	

and cage structures, respectively (Table 2). The organic porogen absorption is located between 2750 and 3000 cm^{-1}. Other peaks correspond to the (CH_3)asym ~2960 cm^{-1}, (CH_3)sym ~2878 cm^{-1}, (CH_2)asym ~2918 cm^{-1}, and (CH_2)sym ~2854 cm^{-1} vibration modes [Lubguban *et al.*, 2002]. UV

(172 nm) assisted thermal porogen removal decreases the porogen peaks but simultaneously incorporates SiH bonds. The reduction of Si–CH_3 bonds concentration (1270–1280 cm^{-1}) and formation of SiH bonds are observed if the curing light uses a wavelength shorter than 200 nm. A specific feature of UV curing by light with wavelength shorter than 200 nm can be related to the formation of porogen residues. This residue forms low intensity peaks in the range of 1400–1800 cm^{-1}. Normally, a long UV cure is required to improve mechanical properties. This is a critical step because exposure of the film to light at 172 nm is more efficient for the improvement of mechanical properties, by opening the pathway to cross-linking via Si–O–Si and Si–CH_2–Si bonds. However it reduces the concentration of Si–CH_3 groups in the low-*k* matrix (see absorption peaks at 1260–1290 cm^{-1}), decreases the total CHx concentration (2750–3000 cm^{-1}), and increases the concentration of Si–H bonds (2100–2300 cm^{-1}). The matrix composition of OSG low-*k* films is normally not changing when the light has a wavelength >200 nm, but more time is needed for porogen removal.

Elemental composition of OSG low-*k* materials are quite similar. Normally, the carbon concentration is close to 12–16% but in some cases this range can be extended. Lower carbon concentration allows to improve mechanical properties (smaller concentration of terminating CH_3 groups and more efficient matrix cross linkage) but these materials have lower chemical stability and are more suffering on plasma damage. Higher carbon concentration can be introduced to improve the plasma resistance [Ryan *et al.*, 2008] but it often results the reduction of mechanical properties. Bauer *et al.* [2001] determined atomic composition and atomic film density of OSG low-*k* films with different porosity using RBS for Si, C and O and Forward Recoil Elastic Spectroscopy (FRES) for Hydrogen. The composition determined in this work is quite typical for OSG low-*k* materials and presented in the Table 3.

The terminal CH_3 groups are not uniformly distributed in low-*k* matrix. According to the recent data and conclusions, their concentration is higher on pore wall than in the matrix [Lopaev, 2014].

2.5.3. *Porosity evaluation*

Porosity evaluation is extremely important for low-*k* materials. The films porosity defines most of critical characteristics of low-*k* materials including their compatibility with processing plasmas (etch, strip and barrier deposition), chemistries (cleaning, CMP, surface modification), adhesion

Table 3. Atomic Composition and Atomic Film Density from RBS for Si, C and O and FRES for Hydrogen. [Bauer *et al.*, 2001]

Sample	Porosity (%)	Hydrogen	Carbon	Oxygen	Silicon	Atomic Density (10^{18} at/cm^2)
S1	20	0.28 ± 0.05	0.14 ± 0.05	0.36 ± 0.05	0.22 ± 0.05	1.98 ± 0.05
S2	26	0.31	0.14	0.35	0.20	1.75 ± 0.04
S3	33	0.32	0.14	0.34	0.20	1.59 ± 0.05
S4	37	0.30	0.16	0.34	0.20	0.87 ± 0.02

with barrier, resistance to cracking, thermal conductivity etc. Three major techniques, which are presently popular in microelectronics are ellipsometric porosimetry, positron annihilation spectroscopy (PAS and PALS) and small (grazing) angle X-ray and neutron scattering techniques (SAXS, SANS, GISAXS). These techniques provide valuable information related to the properties of low-*k* materials itself but also can be used for evaluation of the sealing efficiency and barrier performance. All these instrumentations are not absolute and reflect interaction of the used probes with low-*k* materials. For this reason, cross-comparison has been done many times and sometimes show some differences [Baklanov and Mogilnikov, 2002], [Kondoh *et al.*, 2001], [Grill *et al.*, 2003], [Mogilnikov *et al.*, 2004]. Normally, EP and PALS give quite similar results while the scattering techniques often give smaller pore size. More detailed information related to these techniques are given below.

Ellipsometric porosimetry is the most presently common technique. This system is simple and relatively cheap and the obtained results are identical to ones generated by traditional nitrogen porosimetry. Porosity of the films in EP is calculated from change of refractive index (RI) of materials with empty pores and pores filled by an appropriate liquid adsorptive (e.g. toluene). According to Lorentz-Lorenz equation the RI of two-component systems depends on refractive indices of each component (indices 1 and 2) and their fractions:

$$\frac{n_{re}^2 - 1}{n_{re}^2 + 2} = V\frac{n_1^2 - 1}{n_1^2 + 2} + (1 - V)\frac{n_2^2 - 1}{n_2^2 + 2} \tag{2.3}$$

where n_{re} is the measured refractive index, n_1 and n_2 are refractive indices of the material inside the pores and of the film skeleton, respectively.

V is the pore volume. The pore volume is calculated by using Eq. (2.4):

$$V = \left(\frac{n_{rl}^2 - 1}{n_{rl}^2 + 2} - \frac{n_{re}^2 - 1}{n_{re}^2 + 2} \right) \Big/ \left(\frac{n_{ads}^2 - 1}{n_{ads}^2 + 2} \right) \tag{2.4}$$

The porosity calculation needs refractive indices of the film with empty pores n_{re}, pores filled by adsorptive, n_{rl}, and refractive index of the liquid adsorptive, n_{ads}. Presence of closed pores (pores not accessible for the probe molecules) can be estimated by analysis of calculated skeleton refractive index. For instance, the refractive index of dense OSG is close to $RI_{SiO_2} = 1.46$. If the skeleton refractive index calculated by EP software (based on Eq. (2.3)) is equal to 1.4, the volume of micropores non-accessible for toluene (closed) will be equal to $\approx 11\%$.

Calculation of the pore size distribution uses the phenomenon of progressive emptying of a porous system initially filled at $P = P_0$, where P and P_0 are current and saturated toluene pressures, respectively [Gregg and Sing, 1982]. The adsorptive vapor condenses in pores at the vapor pressure P that is lower than the equilibrium pressure of a flat liquid surface P_0. Dependence of the relative pressure (P/P_0) on the pore radius is described by the Kelvin equation:

$$\ln \left(\frac{P}{P_0} \right) = -\frac{\gamma \cdot V_L \cos\theta}{r_K \cdot RT} \tag{2.5}$$

where γ and V_L are surface tension and molar volume of the liquid adsorptive, respectively. The θ is wetting angle of the adsorptive. If the radius of a cylindrical pore is r_p, then $r_p = r_k + t$, where t is the thickness of the layer adsorbed on the pore walls. The t values are obtained from the adsorption of the same adsorptive on a non-porous sample having a chemically similar surface and are defined by the BET equation.

Porosity measurements are performed by using *in-situ* ellipsometry combined with a high-vacuum system [Baklanov *et al.*, 2000]. Initial experimental data for the porosity and PSD calculation are the ellipsometric characteristics Δ and Ψ measured during the adsorption and desorption cycles. EP software calculates the change of refractive indices and thickness of the film during the adsorption and desorption, pore size distribution (PSD) and specific surface area.

Figure 2.5 shows typical experimental data obtained for different types of low-*k* films. One can see that each type of the film has its own characteristic shape of the isotherm. Figure 2.5(a) shows results of porosity evaluation in a CVD low-*k* film with open porosity equal to 30%. The

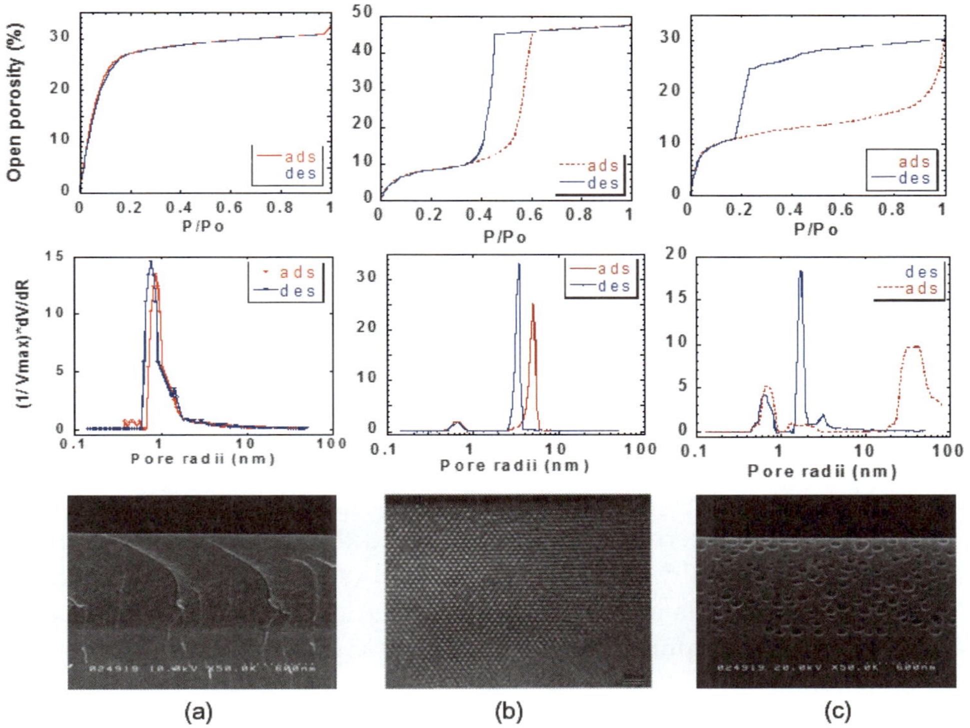

Fig. 2.5. Adsorption-desorption isotherms and pore size distributions in different types of low-*k* films.

adsorption/desorption isotherms do not have hysteresis loop, which is a characteristic feature of microporous films. CVD carbon-doped silica films (SiCOH) are typically microporous and the Dubinin–Radushkevitch method is used for the pore radius calculation [Baklanov and Mogil'nikov, 2002].

Figure 2.5(b) exhibits results obtained for porous films with ordered cylindrical pores (PMO). The pore geometry significantly affects thermodynamic properties of fluids, their adsorption, desorption and diffusion behavior. In the case of semi-infinite cylindrical pores, a cylindrical interface is formed between the adsorbed layer and the vapor during the adsorption, while evaporation (desorption) is related to the formation of a hemispherical meniscus between the condensed fluid and the vapor. Therefore, according to the Kelvin equation, in the case of infinite cylindrical pores the pore radius calculated from adsorption curve is twice the radius calculated from desorption curve and the width of hysteresis loop is defined by Eq. (2.5).

Figure 2.5(c) shows the results of evaluation of a porous film with large quasi-closed cavities. This film contains small micropores (necks) interconnecting the larger cavities. The necks are not visible in the SEM. Evaporation from a pore, which is connected to the bulk phase by narrower pores, is delayed and occurs spontaneously after the emptying of one of the adjacent pores. This is effect of pore blocking controlled desorption, which is commonly called the ink-bottle effect. More detailed description of analysis of such films can be found elsewhere.

In addition to porosity and pore size evaluation, ellipsometric porosimetry has demonstrated its efficiency for some additional applications. The measurement results can be used for evaluation of mechanical properties of thin films as well as integrity of diffusion barriers deposited on top of low-*k* dielectrics [Mogilnikov and Baklanov, 2002], [Shamiryan *et al.*, 2003] . Replacing toluene (standard adsorptive) with water, EP can be used for evaluation of the internal (bulk) hydrophilicity of low-*k* materials. This application is especially important for evaluation of processing damage of low-*k* materials [Baklanov *et al.*, 2006]. Scatterometric porosimetry is also an example of new application of ellipsometric porosimetry for patterned structures [Bouyssou *et al.*, 2010].

Positron Annihilation Spectroscopy (*PALS and PAS*). Positronium (Ps) is an "exotic atom" consisting of an electron and a positron. The orbit of the two particles and the set of energy levels are similar to that of the hydrogen atom. The ground state of Ps, like that of hydrogen, has two possible configurations depending on the relative orientations of the spins of the electron and the positron. The singlet state with antiparallel spins ($S = 0$, $Ms = 0$) is known as para-positronium (p-Ps) and denoted $^1S_0$. It has a mean lifetime of 125 picoseconds and decays preferentially into two gamma quanta with energy of 511 keV each. Detection of these photons allows reconstruction of the vertex of the decay and is used in the positron emission tomography. Para-positronium can decay into even number of photons (2, 4, 6,. . .), but the probability quickly decreases. The triplet state with parallel spins ($S = 1, Ms = -1, 0, 1$) is known as ortho-positronium (o-Ps) and denoted $^3S_0$. The triplet state in vacuum has a mean lifetime of 142 nanoseconds and the leading mode of decay is three gamma quanta.

Positronium formed in porous materials annihilates via a number of channels each producing one or more gamma rays as shown in Figure 2.6. The most important channel for the pore size evaluation is pick-off annihilation o-Ps (c) because it is related to number of collisions with pore

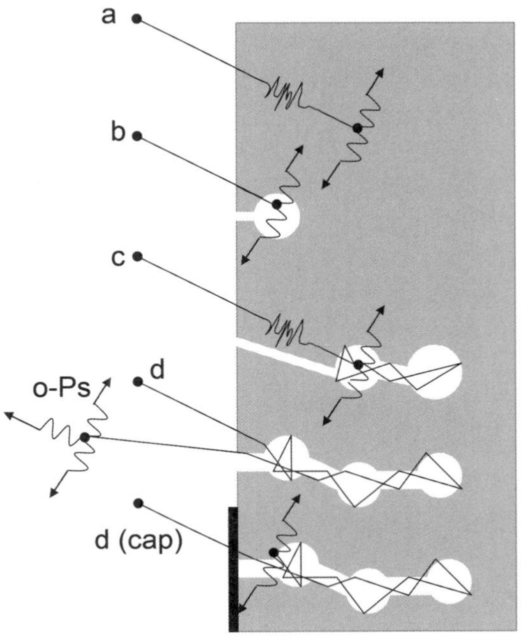

Fig. 2.6. Positronium formation and annihilation in porous media.

wall and therefore contains information related to pore size and shape (geometry).

The most probable annihilation channels produce two or three photons, depending on the relative spin configuration of the electron and positron. The positron implanted in the material is thermalized and has its energy lowered to a few electron volts. Positronium may be formed by the positron capturing a bound molecular electron or recombining with a free electron generated by ionizing collisions. The positronium atoms formed are thus p-Ps and o-Ps (b) and once formed they begin to diffuse and thermalize in the material. The o-Ps formed has intrinsically a much longer lifetime than p-Ps and therefore more readily localize in a pore, thermalize and finally annihilate with molecularly bound electrons through the 'pick-off' process on the pore wall (c). "Pick-off" annihilation is a 2γ process that can decrease the Ps lifetime to values as low as ~1 ns. The o-Ps may also diffuse over long distance if pores are interconnected with the possibility of escaping into the surrounding vacuum and undergoing 3γ annihilation with a natural lifetime of 142 ns (d). The analysis of the fate of the positronium can provide valuable information in the characterization of porous materials. To date, two major techniques have been employed for such analysis, namely the $3\gamma/2\gamma$ ratio technique (PAS) and the positronium lifetime technique (PALS).

When films are irradiated with a focused beam of several keV positrons, they penetrate to a certain depth, which depends on the beam energy. Then they either annihilate or form a positronium. PALS measures the lifetime of positronium, whereas PAS extracts information from the Ps annihilation reaction. Reduction of the natural lifetime of o-Ps in porous materials yields information about the pore size and geometry. Generally, the lifetime depends on number of collisions with pore wall (depends on ratio V/S similar to chemistry of radicals), where V and S are pore volume and internal surface area, respectively). This is the reason why Ps lifetime also depends on pore geometry. In the case of pores with a spherical shape $V/S = 1/3^*R$, where R is pore radius. However, if all pores are open and interconnected, Ps can escape to vacuum and the information related to pore size cannot be obtained (Figure 2.6, d). In this case one needs to deposit a capping layer to measure the pore size (Figure 2.6, d (cap)) [Gidley *et al.*, 2000, Sun *et al.*, 2001].

To obtain information on the pore size, Ps lifetimes are recorded (Figure 2.7), and the lifetime distribution curves are obtained with a fitting program developed for this purpose. The distribution curves can then be transformed into pore size data with a certain assumption about the pore geometry. For the calculation, it is necessary to perform calibration. For

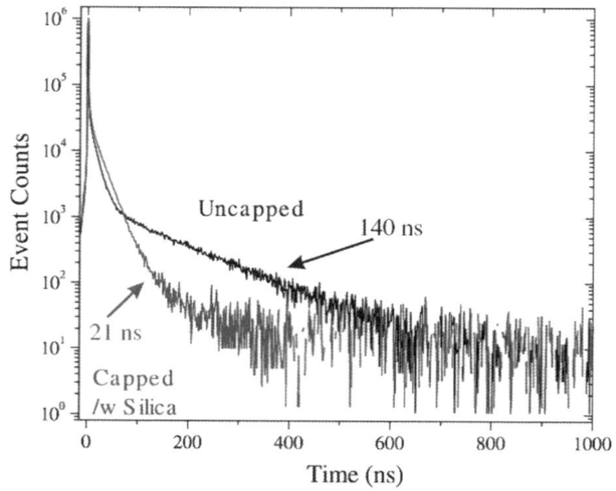

Fig. 2.7. PALS spectra of an uncapped and caped porous MSQ low-*k* film of 22% porosity Reprinted with permission from Jia-Ning Sun, D. W. Gidley *et al.* J. Appl. Phys. 89, 5138 (2001). Copyright 2001, AIP Publishing LLC.

very large pores the calibration is performed using known (standard) high porosity silica powders. However, when the pores have size larger than 30 nm, the size calculation is challenging because the lifetime is becoming to close to the time of self-annihilation [Sun *et al.*, 2002]. In the sub-nm regime, the quantum mechanical Tao-Eldrup model has been empirically used to calibrate Ps lifetime of several nanoseconds with pore size [Gidley *et al.*, 1999, Gidley *et al.*, 2006].

PALS and PAS are very efficient in evaluation the pore sealing and dielectric barriers. Depending on the annihilation reaction in PAS 2γ or 3γ photons are generated. The self-annihilation of o-Ps generates 3γ photons and related to lifetime 142 ns in vacuum, while annihilation with a molecular electron from the pore wall results in the emission of 2γ photons. The probability of the latter depends on the collision frequency with pore walls which affects the o-Ps lifetime. However, even in this case the fractions of 3γ emission exist. That fraction is determined by the branching ratio into 3γ vs. 2γ and the branching ratio is given by the ratio of decay rates, (decay rate into 3γ)/(decay rate into 2γ). Decay rates are the inverse of lifetimes so the 3γ decay rate is the inverse of the Ps lifetime in vacuum, $1/142\,\text{ns}$ or $\sim 0.007\,\text{ns}^{-1}$. For the 10 nm void (assumed spherical diameter) the lifetime is expected to be 95 ns and hence the total decay rate is $0.0105\,\text{ns}^{-1}$ (this is the sum of 3γ and 2γ decay rates). The fraction of all Ps decaying in such a 10 nm void by 3γ should be $(0.007)/0.0105) = 66\%$. The other 34% decays by 2g. The 3g fraction is much lower for a 2 nm diameter void. The Ps lifetime should now be only 23 ns to give a total decay rate of $0.0435\,\text{ns}^{-1}$ so the 3γ fraction is $(0.007)/(0.0435) = 16\%$ [Gidley, 2014]. The rate of "pick-off" annihilation influences the ratio of the 2γ and 3γ annihilation intensity and can therefore be used to derive porosity properties.

Small angle X-ray Scattering (*SAXS and GISAXS*). Small angle X-ray scattering (SAXS) technique (and its modification grazing incidence SAXS [GISAXS]) was developed to probe pore structure in a non-intrusive way. It is built on the following principles. As shown in Figure 2.8, if an X-ray incident at a thin film on a substrate at an angle Θ, part of the beam is reflected at the specular angle Θ and part is refracted into the film. The refracted beam can be scattered inside the film at a defect that has different electron density from the rest of the film (a pore in our case). The scattered beam will exit the film at a different angle $(\Theta + \delta\Theta)$.

The main difference between SAXS and GISAXS is in the incidence angle Θ: SAXS uses normal incidence ($\Theta = 90°$) [Huang *et al.*, 2002]

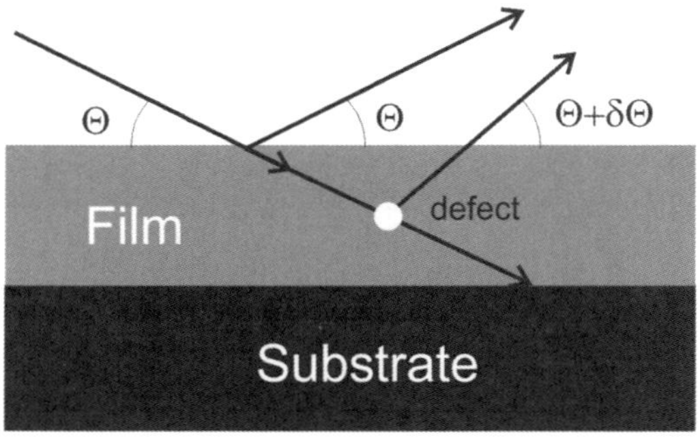

Fig. 2.8. A scheme of X-ray scattering in a film by a defect (a pore).

while GISAXS uses angles close to zero. Application of normal incidence SAXS is limited by small scattering volume, weak scattering of thin porous films, and high absorbance of the silicon necessitates a high intensity X-ray source. At present, a synchrotron source is needed to obtain a large enough flux for transmission SAXS measurements with sufficient signal. Unlike transmission SAXS, the GISAXS geometry, where the scattering signal is collected at the thin-film side, can increase the scattering volume by two orders of magnitude. The X-ray beam does not penetrate into the silicon substrate, but is reflected from the surface. Experimental techniques have been developed and demonstrated the capability of collecting GISAXS data within a time period less than 60 s. GISAXS is a powerful tool to study nanostructured surfaces and thin films, combining the accessible length scales of SAXS and the surface sensitivity of grazing incidence diffraction (GID) [Als-Nielsen *et al.*, 2001]. and diffuse X-ray Reflectivity. The technique was originally introduced by Joanna Levine *et al.*, [1989], but has only recently come to full flourish for the study of nanostructured systems.

In order to make X-ray scattering surface sensitive, a grazing incidence angle α is chosen between about half the critical angle α_c and several critical angles of the film material. The actual choice depends on the system to be studied. For free-standing quantum dots, an incident angle below α_c may be chosen to make the scattering exclusively surface-sensitive. Largest scattering cross sections are achieved when the incident angle is in between the critical angles of the film and the substrate.

The intensity of the scattered light I depends on the form factor F of the defects (pores) [Saski, 2005]:

$$I(q) = |F(q)|^2 \quad (2.6)$$

where $q = \frac{4\pi}{\lambda}\sin\theta$ is the scattering vector, θ is the scattering angle and λ is the X-ray wavelength. Assuming that pores have a spherical shape of diameter D with the average diameter D0 and distribution described by a Γ-function:

$$P(D, D_0, \sigma) = \frac{1}{\Gamma\left(\frac{1}{\sigma^2}\right)}\left(\frac{1}{\sigma^2 D_0}\right)^{\frac{1}{\sigma^2}} D^{-1+\frac{1}{\sigma^2}} \exp\left[-\frac{D}{\sigma^2 D_0}\right] \quad (2.7)$$

where σ is the normalized variance. The scattering intensity then can be written as follows:

$$\left|F(q, D_0, \sigma)\right|^2 = |\Delta\rho|^2 \int_0^\infty \left|\frac{4\pi}{q^3}\left(\sin\left[\frac{qD}{2}\right] - \frac{qD}{2}\cos\left[\frac{qD}{2}\right]\right)\right|^2 P(D, D_0, \sigma)\frac{D_0^3}{D^3}dD \quad (2.8)$$

where $\Delta\rho$ is the density difference between the film and the pore.

The measurements are done in the following way [Omote *et al.*, 2003, Suzuki *et al.*, 2006]: a detector is slightly shifted from the angle Θ to avoid strong specular reflection by about 0.1° and scattering intensity is recorded as function of the incident angle Θ. Then Eq. (2.8) is fitted into the experimental curve using D0 and σ as fitting parameters and the resulting pore size distribution can be plotted using Eq. (2.7). Examples of such fitting and resulting pore size distribution for three different low-*k* films are shown in Figure 2.9.

Diffuse scattering from the film surface can sometimes interfere with the scattering from the pore structure. In that case the contribution of the surface roughness should be estimated. That can be done by measuring rocking scan patterns at several 2Θ positions in addition to the offset scan. Omote *et al.* [2003] reported that the contribution of surface roughness is dominant only at very low incident and exit angles. The main advantage of SAXS is that it can probe all pores regardless their connection to the surface. The main drawback of the method is the fact that the pore size distribution is valid only for spherical pore shape and presumed to be described by a Γ-function. Other shapes of the pores and more complicated distributions would require more complicated mathematical approach, multiple scattering effects have to be taken into account to properly model the data. It is also necessary to mention that the scattering techniques are sensitive to the density gradient. Therefore, sometimes it is difficult to distinguish pores and dense inclusions present in the film.

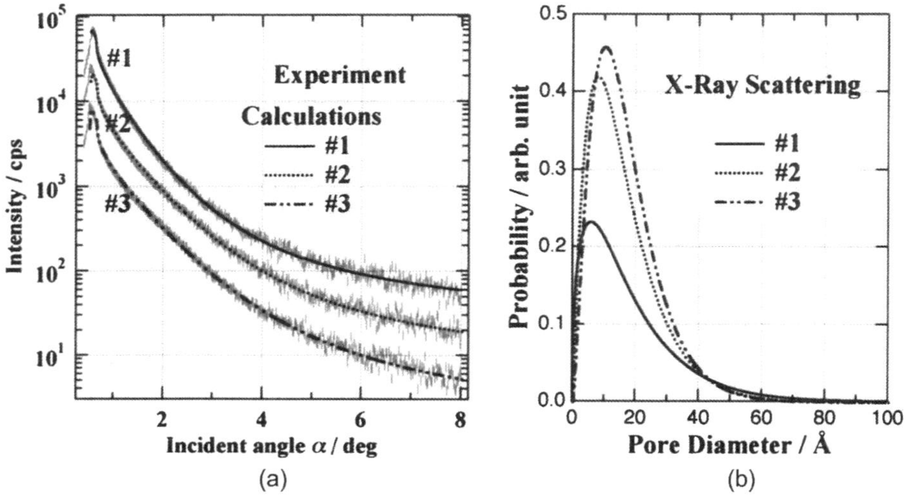

Fig. 2.9. (a) Experimental and calculated X-ray small-angle scattering patterns for the three porous MSQ films. (b) Pore-size distribution of the films obtained by x-ray scattering. Reproduced with permission from Omote K. *et al.*, (2003) Appl. Phys. Lett. 82 p. 544. Copyright 2003, AIP Publishing LLC.

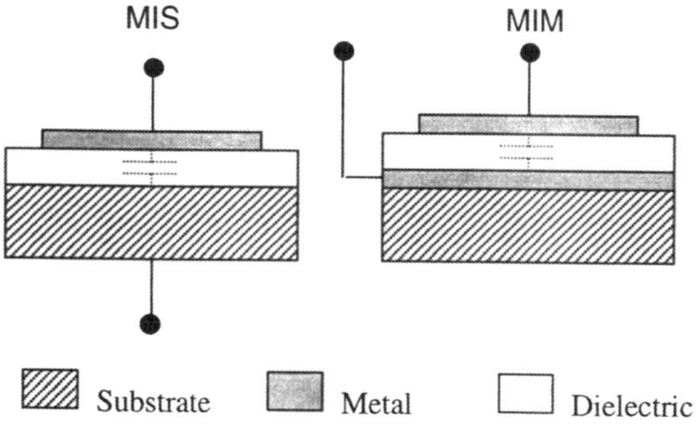

Fig. 2.10. MIS and MIM capacitors for the k-value extraction for blanked low-*k* films.

2.5.4. *Electrical characterization of blanked films*

Electrical performance of low-*k* materials either in thin film or structured in narrow features is the key issue. Exposure to ambient and to process environments can have a large influence on the materials and on its electrical properties. The most basic electrical measurements of the dielectric constant of pristine low-*k* materials is obtained by an area capacitor which consists of either a metal/dielectric/metal (MIM) or a metal/dielectric/Si (MIS) sandwich (Figure 2.10) [Ciofi *et al.*, 2010].

The dielectric constant is obtained directly through the defining equation:

$$k = Ct\varepsilon_0^{-1}A_m^{-1} \tag{2.9}$$

in which C is capacitance, t is the thickness of the dielectric film (most often measured by ellipsometry), A_m the Pt dot area, and ε_o the permittivity of vacuum. It is important to ensure that the Si substrate is in accumulation. This is a valid assumption only if the fringing capacitance is negligible. For square capacitors with a top plate area $>1\ \text{mm}^2$ the error due to fringing is about 0.2%. Since the water adsorption by porous dielectrics may occur readily upon exposure to air (especially in the case of plasma exposed or chemically treated materials), the capacitor has to be sealed carefully by a passivation layer during process or moisture can be desorbed before/during the electrical measurement. Possibility of metal penetration during the dots deposition may also have some influence.

Mercury probe measurements have been used as a quick method to obtain a MIS type capacitor without the need for metal patterning. The interaction of the probe with porous dielectrics sometimes is unclear but the results obtained by this technique in the most of cases have good agreement with the measurements done with metal dots. Area capacitance measurements also provide information about the breakdown properties and leakage currents.

It is more difficult to generate the data related to intrinsic reliability of low-k films. Meanwhile it is very important because the reliability data generated after complete integration very much reflects quality of integration process. This makes difficult the initial search and for this reason recently was developed a new technique, so called "planar capacitors" that is presently very popular tool for selection of new low-k materials. The detailed description of this tool as well as results of evaluation will be reported in the third part of this paper.

The interline capacitors of integrated structures are more realistic structures as compared to the area capacitors, because they represent the capacitance between interconnect lines resulting from technology constraints (minimum dimensions, type of dielectric layers, process conditions). As with the area capacitors, interline structures may also suffer from moisture when not sealed properly, barrier penetration etc. The discussion of this important subject is out of the scope of this paper and we only can refer to appropriate publications [Stucchi and Maex, 2002, Ciofi *et al.*, 2012].

2.6. *Porogen/template removal and matrix densification (curing)*

The most common approach in preparation of ultra low-*k* materials in both Spin-on and PECVD processes are based in utilization of templates or sacrificial porogens, that are co-deposited together with matrix material and have to be removed after deposition and skeleton formation. The efficient removal of sacrificial porogen and formation of sufficiently stiff matrix is the key requirements. The porogen molecules are more volatile and less thermally stable than the film matrix. The porogens are cracked into volatile by-products and have to gas out at elevated temperature. This is the reason why the most traditional curing technologies are based on annealing (thermal, or furnace curing). However, the thermal curing doesn't allow to achieve sufficient throughput (too slow) and therefore, significant efforts have been related to the development of alternative curing technologies (by using electron beams, plasma, UV light etc.). Some of the curing processes are "damaging" low-*k* matrix by breaking Si-CH_3 bonds and making low-*k* materials hydrophilic. Presently, thermal curing ($\approx$ 400°C) assisted with UV light with wavelength longer than 200 nm is considered as the most promising [Prager *et al.*, 2008]. Such curing provides sufficiently high porogen removal rate, provides sufficient densification and cross-linkage of the film skeleton and doesn't change Si-CH_3 concentration in the low-*k* matrix.

Figure 2.11 shows change of dielectric constant and the film shrinkage happening during the UV curing of a PECVD film. After co-deposition of the matrix material and porogen molecules the film is not porous and normally has dielectric constant >3.0, which reflects the porogen/matrix ratio and free volume created by CH_3 groups bonded to Si. The porogen removal leads to reduction of dielectric constant because of porosity formation. The minimum dielectric constant corresponds to the complete porogen removal and to maximum porosity. The low-*k* film shrinks during the porogen removal and after the complete removal. The shrinkage of the film after the complete porogen removal is related to the film densification and increases dielectric constant. The changes happening after the complete porogen removal depend on the wavelength of UV light. The degree of shrinkage is small and no change of the matrix chemical composition if the wave length of UV light is longer than 200 nm, while more energetic photons lead to significant reduction of Si-CH_3 concentration, shrinkage and hydrophilisation.

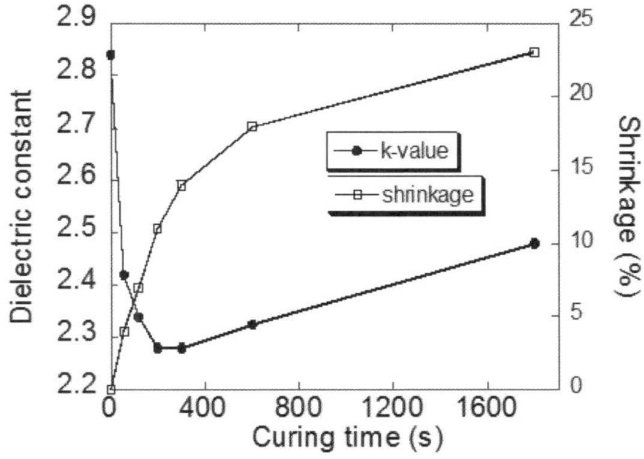

Fig. 2.11. Change of dielectric constant and thickness during UV assisted thermal curing.

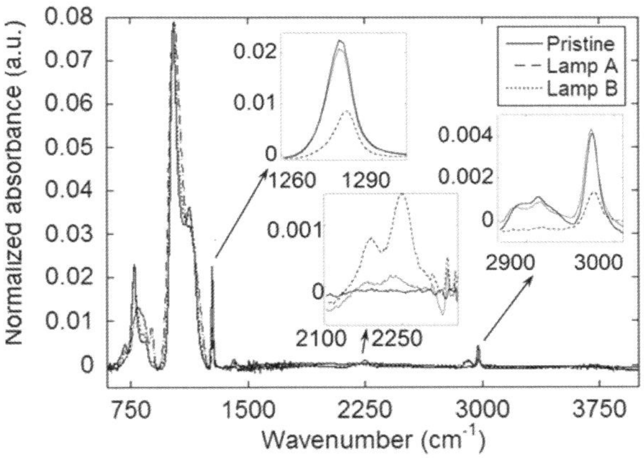

Fig. 2.12. FTIR spectra of a low-*k* film after curing by UV light with wavelength 172 nm (Lamp A) and broadband light with wavelength >200 nm (lamp B). Reprinted with permission from M. R. Baklanov *et al.* J. Appl. Phys., 113, 041101 (2013). Copyright 2013, AIP Publishing LLC.

Figure 2.12 shows the film composition during the curing of OSG low-*k* material when porogen has already been mostly removed. This graph shows that the light with wavelength >200 nm is almost not changing chemical composition of a low-*k* film and FTIR spectra are identical to pristine sample. Exposure of the film to the light of 172 nm significantly reduces concentration of Si-CH_3 groups (the absorption peaks at 1260–1290 cm-1), total CHx concentration (2900–3000 cm-1) and increases concentration of Si-H bonds (2100–2300 cm-1).

Mechanism of the reduction of Si-CH_3 groups concentration and formation of Si-H groups during UV curing was analyzed by Prager *et al.*

[2008] and Marsik *et al.* [2011]. Prager *et al.*, [2008] used quantum-chemical calculations on the model substances octamethyl- and tetramethyl-cyclotetrasiloxane as well as on hexamethyl- and tetramethyldisiloxane resulted in threshold wavelengths for the excitation of the molecule into the first excited singlet state of 190 and 198 nm as well as of 189 and 192 nm, respectively. After excitation and intersystem crossing in an excited triplet state the scission of the Si–CH_3 bond may occur gaining an energy benefit of around 50 kcal mol-1. These findings reveal the presumption that only photons with $\lambda < 200$ nm can generate Si centered radicals which subsequently attract protons from neighboring methyl groups. It is believed that formation of Si-H groups has a negative impact in chemical resistance of low-*k* films. Therefore, application of 172 nm light for UV curing is becoming limited despite of significant benefit in improvement of mechanical properties.

2.6.1. *Porogen residue*

It is generally expected that the porogen molecules are cracked into volatile by-products and have to gas out at elevated temperature. Unfortunately, the photo-induced porogen cracking leads also to the formation of non-volatile carbon-rich reaction products (porogen residue). The issue of incomplete porogen removal based on C=C signature in FTIR and NMR has been, for instance, reported for UV and thermal cures [Favennec2 *et al.*, 2007], [Zenasni *et al.*, 2007]. This residue mainly contains sp2 carbon. FTIR spectrometry is normally not sufficiently sensitive to non polar bonds like sp2 carbon. This is the reason why researches often do not pay much attention to the formation of porogen residue. Meanwhile detection of such groups in pristine low-*k* materials is extremely important because sp2 carbon residues are conductive and therefore they can be a reason of high leakage current and reduced breakdown field in the integrated low-*k* materials [Baklanov *et al.*, 2011], [Van Besien *et al.*, 2012], [Afanas'ev *et al.*, 1996]. These residues can be detected by Raman spectroscopy but application of these measurements for thin low-*k* films deposited on silicon substrate is quite challenging.

Marsik *et al.* [2010] developed a method for detection of porogen residue from UV absorption spectra measured by ultraviolet ellipsometry. It was found that the absorption bands of as deposited porogen in the ultraviolet range has maximum at 6.5 eV. UV curing decreases this band but this peak does not completely disappear even after long curing time. Moreover, new maximum at 4.5 eV (as a shoulder) forms during the curing.

The peak at 4.5 eV suggests formation of porogen residues in the low-*k* material. The incomplete removal is also observed for deposited porogen-only layers treated with the 172 nm UV-cure process. It was concluded that these residues to be mainly sp2 bonded (amorphous or aromatic) carbon because of the appearing absorption of C=C bond vibration in the infrared and the $\pi - \pi^*$ electronic transitions between sp2 carbon orbitals in the ultraviolet. The intensity of this peak is smaller when broadband lamp with $\lambda > 200$ nm was used for UV curing.

Figure 2.13 shows absorption spectra of UV cured PECVD low-*k* films with different porosity. To deposit materials with lower dielectric constant, it is necessary to increase the ratio of the porogen precursor to the matrix precursors. This figure clearly shows that the deposition of highly porous films generates more residue during the UV curing although the type and amount of porogen residue depends on the type of porogen. Electron spin resonance spectroscopy (ESR) study showed that the carbon clusters related to porogen residues are the major sources of paramagnetic defects in low-*k* insulators [Afanasiev *et al.*, 2013], [Stessmans *et al.*, 2013].

Urbanowicz *et al.* [2010] developed process allowing to form porogen residue free low-*k* films using PECVD process. The matrix material and porogen were co-deposited at 300°C. Assuming, that the temperature is sufficient for formation of a stable skeleton and agglomeration of the porogen molecules, the porogen was completely removed before UV curing using He/H_2 downstream plasma. The plasma system has a special grid

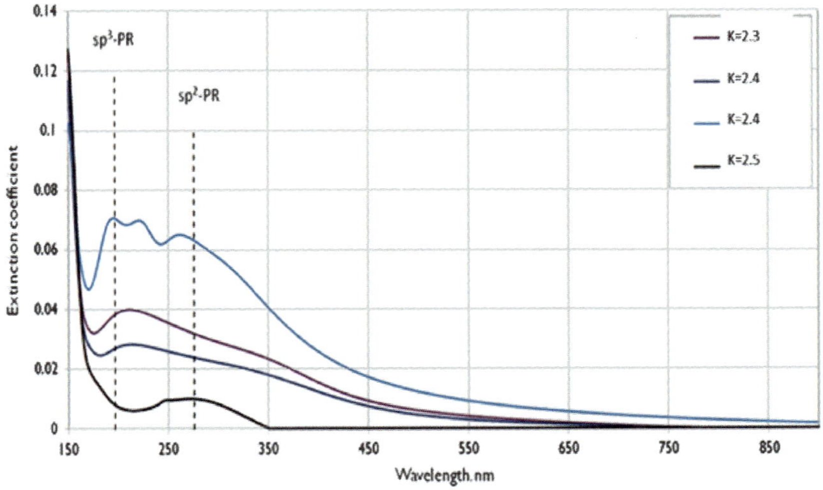

Fig. 2.13. UV absorption spectra of low-*k* films prepared by different technologies.

excluding influence of energetic ions and VUV photons from the plasma to the film. Then the low-*k* films are UV cured with broadband UV light with $\lambda > 200$ nm to finalize the formation of cross-linked and rigid skeleton. The proposed method allows to obtain porogen residue free low-*k* films with variable thicknesses. The obtained films demonstrate Young Modulus of 5–9.5 GPa after curing by lamp A and 3.5–4 GPa after the lamp B. The open porosity of these films were in the range of 41%–46%, the k-value of 1.8–2.4. The presented method show the potential for fabrication of PECVD ultra low-*k* dielectric films for further microelectronic technology nodes.

2.7. *Processing challenges low-K materials*

Integration of porous ultralow-*k* materials is always very challenging. Most of the observed issues are related to their porosity that degrade chemical (etch-selectivity issues and plasma damage) and thermo mechanical and interface properties. The barrier effectiveness also depends on porosity and pore size (Figure 2.14) [Hoofman *et al.*, 2005]. Some of these issues related to the plasma modification, interface issues and electrical reliability will be analyzed in this chapter.

The integrated *k*-value of low-*k* materials is normally larger than the values measured on pristine samples. The reason is degradation of low-*k* materials during the processing. The low-*k* materials are degrading during etching (patterning), cleaning, CMP and the barriers deposition. The degradation during the optimized CMP is relatively small and do not exceed

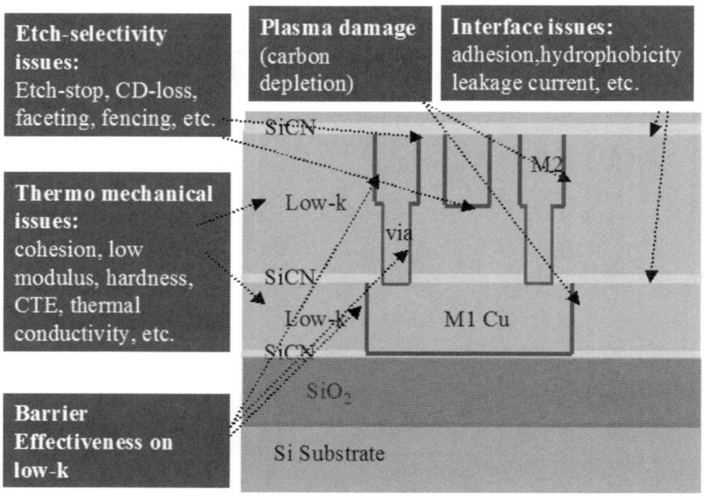

Fig. 2.14. A schematic illustration representing the different integration challenges accompanied with low-*k* introduction in interconnects.

few surface layers. For this reason, the restoration of this damage is relatively easy. Most of presently available cleaning solutions do not introduce much damage. Commonly used diluted HF solutions remove the top part of low-*k* film damaged during the plasma etching (because after carbon loss this layer becomes SiO_2-like). Therefore, HF solutions remove the plasma etch residues (fluorocarbon polymers) by lift–off mechanism. It is predicted that diluted HF solutions may have limited applications for narrow pitch technology because the removal of damaged part will change CD of low-*k* lines. This is the reason why many chemical companies are developing the cleaning chemistries that not etching low-*k* films. These chemistries might be quite efficient in residue removal but in some cases deposition of the cleaning solutions components (surfactants) inside the pores is observed. Their removal need additional efforts like thermal annealing.

The most significant damage is related to plasma processing (plasma damage) and therefore, this subject will be discussed in more detail.

2.7.1. *Plasma damage*

Plasma damage of low-*k* dielectrics is a complex phenomenon involving both physical and chemical effects; the former arises from ion bombardment (energy distribution and flux for each ionic species) while the latter involves chemical reactions between the plasma and low-*k* constituents and/or induced by plasma radiation (photochemistry). The degradation (increase) of dielectric constant directly correlates with the changes in bonding configuration, formation of carbon-depleted layer, film shrinkage, and surface densification. The carbon depleted layer becomes hydrophilic and adsorbs moisture that drastically increases the *k*-value and leakage current, deteriorates reliability. The depth (and degree) of "chemical" plasma damage is mainly determined by the diffusion of active radicals (O, H, F, etc.) into the pores of the low-*k* material and the following removal of organic hydrophobic groups. The penetration depth of active radicals depends on pore size and connectivity, diffusion rates, and their recombination (loss) probability. In the case of fluorine based etch plasma the depth of damage depends not only on the above mentioned factors but also on the etch rate [Iba *et al.*, 2008]. The depth of etch damage increases when the etching speed is lower than the speed of the damage propagation. Therefore it can be reduced by using fast etching recipes. Intensive ion bombardment is also useful because the surface densification reduces the penetration of active radicals into the pores. This is the reason why CCP reactors are more suitable for patterning of low-*k* materials.

During the processing, the plasma constituents (ions, photons, and radicals) contribute and interact in a complex manner to induce damage. A good demonstration of the damage features and mechanisms was reported by Kunnen *et al.*, [2010]. He exposed a OSG low-*k* film in oxygen plasma using an ICP etch chamber. It was found that 3 different conditions (top power only (TPO), bottom power only (BPO), and mixed regime with both top and bottom applied powers (T&BP)) show completely different phenomena. The condition with BPO is similar to what is observed in a CCP reactor: the low-*k* surface is bombarded by energetic ions. The top power only generates a high concentration of radicals, but the intensity of ion bombardment is very low. The results are shown in Figure 2.15 and 2.16.

In the TPO case, the oxygen radicals deeply penetrate into the pores reaching the bottom of the film and resulting in complete carbon depletion. Similar observation was made using EFTEM (Figure 2.16). The dark area corresponds to low-*k* with depleted carbon, and also one can see that the degree of carbon depletion is huge in the case of TPO and very small for BPO. The observed phenomena suggest that ion bombardment in the case BPO densifies the top surface preventing ions and radicals penetration into the low-*k* film. The situation in T&BP condition is more close to BPO, which suggests that the efficiency of surface densification mostly depends on ion bombardment and not much on the radical concentration.

In the case of TPO, the pores remain open and the oxygen radicals are deeply diffusing into pores. Kinetics of plasma damage depends on

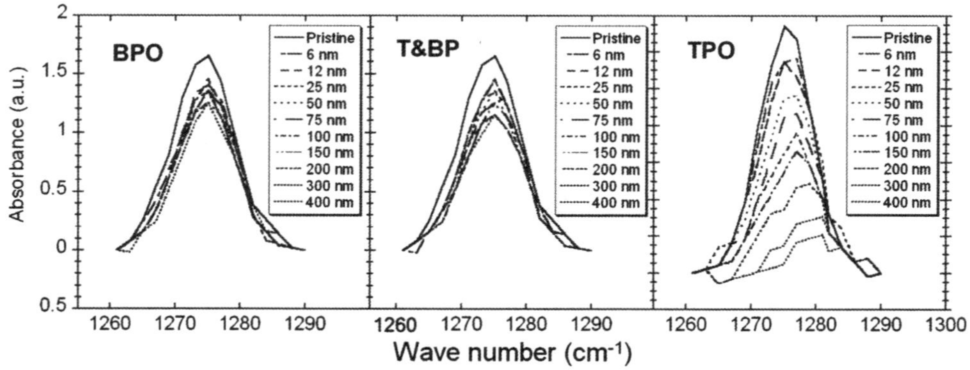

Fig. 2.15. Change in FTIR absorbance at 1275 cm-1 after exposure in TPO, BPO, and T&BP conditions during the time necessary to remove certain amount of resist (legend). In the case of TPO the absorbance completely disappears, while for BPO and T&BP the absorbance partially remains. Reprinted with permission from E. Kunnen *et al.* J. Vac. Sci. Technol. B 28(3), 450 (2010). Copyright 2010, American Vacuum Society.

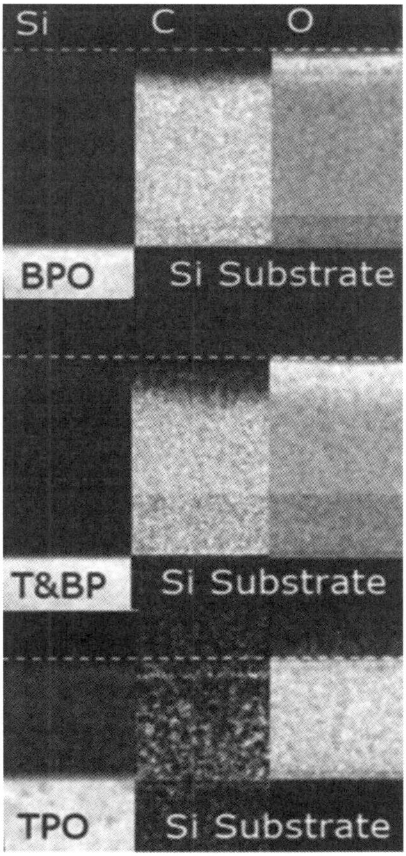

Fig. 2.16. EFTEM elemental profiles (Si, C, and O) of low-k samples exposed in BPO, T&BP, and TPO conditions. The surface densification in BPO and T&BP conditions prevents penetration of oxygen radicals into the pores and reduces the depth of carbon depletion. Reprinted with permission from E. Kunnen *et al.* J. Vac. Sci. Technol. B 28 (3), 450 (2010). Copyright 2010, American Vacuum Society.

reactivity of plasma formed radicals and their diffusion in the pores. For an idealized cylindrical pore and an isothermal, isobaric, and first-order reaction, the material balance equation along a differential length dx of the pore can be written as

$$-\frac{\pi d_p^2}{4} D_A \left(\frac{dC_A}{dx}\right)_x + \frac{\pi d_p^2}{4} D_A \left(\frac{dC_A}{dx}\right)_{x+dx} + \pi d_p dx k_R C_A(1) = 0 \qquad (2.10)$$

The first term of Eq. (2.10) reflects the flux of radicals (A) into the region between x and $x + dx$, the second term is the flux out of this region Figure 2.17). The third term represents the radical consumption inside this

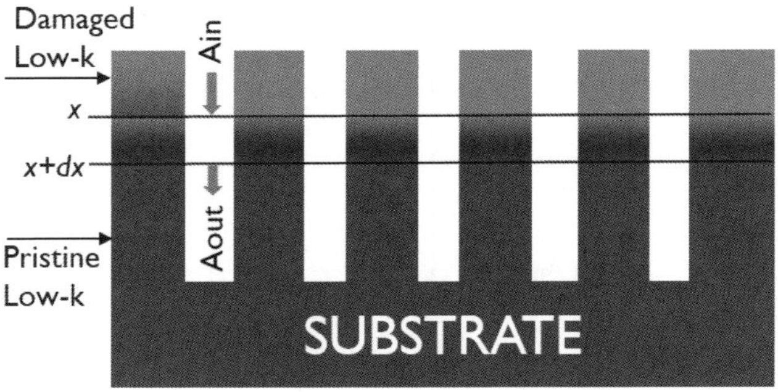

Fig. 2.17. Schematic representation of component A diffusing in a low-k film with idealized cylindrical pores.

region due to the reaction with Si-CH_3 (k_{react}) and surface recombination of oxygen radicals (k_{recomb}): $k_R = k_{react} + k_{recomb}$. The sum $k_R = k_{react} + k_{recomb}$ defines the total loss of radicals A and the depth of their penetration into pores. Analytical solution of Eq. (2.10) shows that the depth of radical penetration depends on ratio of k_R to $D_A(\lambda)$, which is termed as Thiele Modulus (TM)

$$\frac{d^2C_A}{dx^2} - \lambda^2 C_A = 0, \quad \text{where } \lambda = \left(\frac{4k_r}{d_p D_A}\right)^{1/2} \tag{2.11}$$

The modified diffusion-recombination models of plasma damage by oxygen radicals have also been proposed by Kunnen [2011] and Safaverdi [2011].

Braginsky *et al.* [2010] carried out extensive study of low-k damage by oxygen radicals generated in fast flow rf CCP plasma in 90% O2/10% Ar mixture at pressure of 10 Torr. Oxygen atoms loss probabilities were measured by plasma induced fluorescence method as described in Ref. [Rakhimova *et al.*, 2009]. X-ray fluorescence (XRF) and FTIR spectroscopy were focused to reveal the evolution of carbon and oxygen concentration in the films as a function of the exposure time to atomic oxygen flux. Then, the CH_3 groups removal was modeled by 1-D Monte-Carlo simulation using a simplified structure of the regular vertical pore channels providing the same porosity as the material in experimental study. It was shown that in low-k films with pore size approaching to 1–1.5 nm, the classic diffusion approach for calculating O atom concentrations in nanoporous material is becoming irrelevant and more adequate description of experimental results can be obtained by using ideas of random walk theory. The depth of penetration is proportional to pore size (d) and inversely proportional to the sum of recombination and chemical reactions coefficients of the active radicals:

$$L_{pen} \sim d \cdot \left(\frac{1}{k_R}\right) \tag{2.12}$$

The ultra low-*k* materials (the materials with lowest dielectric constant) will inevitably have a higher porosity and a large pore size and, therefore, a high degree of the pore interconnectivity [Baklanov *et al.*, 2013]. This leads to a deeper penetration of the active radicals into the bulk low-*k* material. The depth of damage can be reduced or prevented by reducing d in the Eq. 2.12 (sealing of the uppermost layers of low-*k* films, and/or deposition of a thin layer impenetrable for O atoms on top of low-*k* films), and increasing k_R — chemical modification of the pore surface composition to increase substantially the surface recombination rate of oxygen atoms. The latest can be achieved, for instance, by increasing of carbon concentration in low-*k* films [Ryan *et al.*, 2008].

It is necessary to mention that other diffusion models have also been proposed. Goldman *et al.* [2009] proposed a diffusion-based Deal-Grove type of model, Arkhincheev *et al.* [2011] developed a random walk theory based on so called comb diffusion model when active species are captured in active sites and concluded that the depth of damage is described by logarithmic law as observed in experiments. The reduction of the pore diameter down to values approaching molecular dimensions can result in a dramatic decrease in diffusivity due to non-fickian single file diffusion [Levitt, 1973], [Kollmann, 2003].

The chemical reactions between the oxygen radical and the Si–CH_3 groups located on the pore surface start with detachment of H, because H_2C-H is the weakest bond

$$\equiv \text{Si–CH}_3 + \text{O} -> \equiv \text{Si–CH}_2^* + \text{OH} \tag{2.13}$$

Further reaction with oxygen atoms leads to the complete loss of methyl group and formation of surface active sites (≡Si*) that can chemisorb water molecules forming ≡SiH and ≡SiOH groups

$$\equiv \text{Si–CH}_2^* + \text{O} -> \text{CH}_2\text{O} + \equiv \text{Si}^*$$

$$\equiv \text{Si}^* + \text{H(OH)} -> \equiv \text{SiH}(\equiv\text{SiOH})$$

Reaction of CH_2O molecules leads to formation of CO, CO_2 and H_2O [Braginsky *et al.*, 2010].

The chemical reactions describing interactions of H radicals with Si-CH_3 groups from the low-*k* material have been actively discussed in the literature. The replacement of –CH_3 groups by H atoms with a formation of more polar Si–H groups and the breakage of Si–O–Si groups with a formation of hydrophilic silanol groups were proposed. The estimated enthalpies at 298 K (ΔHr) of the exothermic reactions (2.14) and (2.15) were reported as

$$\equiv \text{Si}-\text{CH}_3 + 2\text{H} -> \equiv \text{Si}-\text{H} + \text{CH}_4 (\Delta Hr = -411\,\text{kJ/mol}) \tag{2.14}$$

$$\equiv \text{Si}-\text{O}-\text{Si} \equiv +2\text{H} -> \equiv \text{Si}-\text{H} + \text{Si}-\text{OH}\,(\Delta Hr = -325\,\text{kJ/mol}) \tag{2.15}$$

The negative enthalpy at 298 K of the reactions (2.14–2.15) allows to assume that these reactions are presumably possible at elevated temperatures. However, detailed analysis of the available data shows that the ability of the H_2-based plasma to break Si–CH_3 bonds strongly depends on experimental conditions [Worsley *et al.*, 2005]. According to our study, H radicals formed in afterglows of H_2 and He/H_2 plasmas do not reduce the concentration of Si–CH_3 groups at the temperature range of 20–280°C [Urbanowicz *et al.*, 2010]. This observation agrees with the literature data, demonstrating no low-*k* damage in afterglow of H_2, He/H_2, and He/Ar plasmas [Worsley *et al.*, 2005], [Han *et al.*, 2005]. Only certain additional activation (for instance, ion radiation and UV light) make these reactions possible, explaining the difficulty to achieve H_2-based damage-free processes in reactive ion etching (RIE) condition. For instance, Matshushita *et al.*, [2003] showed that the damage during the H_2 plasma treatment depends on the type of chamber used. The process stays damage-free in the absence of VUV photons and ion bombardment.

The damage free processing in H radicals can be explained by using the results reported by Rakhimova *et al.* [2009] and Lazzeri *et al.* [2006]. [Rakhimova *et al.*, 2009] used a method based on gas phase analog model and predicted that the reaction of H atoms with a ≡Si–CH_3 group starts with the detachment of a hydrogen atom

$$\equiv \text{Si} - \text{CH}_3 + \text{H} - > \equiv \text{Si–CH}_2^* + \text{H} \tag{2.16}$$

This reaction is similar to (2.13) but if the further reaction of the ≡Si–CH_2^* radical with oxygen atoms leads to its complete destruction, the reaction with hydrogen atoms "restores" the ≡Si–CH_3 group (≡Si–CH_2^* + H −>≡ Si–CH_3). This mechanism can be evidenced by the results presented by Lazzeri *et al.* [2006]. In this work, OSG low-*k* films were exposed to a deuterium plasma. The deuterium replaces hydrogen in the low-*k* matrix, but the total concentration (H + D) remains constant. Therefore, the dangling bonds formed by hydrogen detachment from low-*k* films are saturated by the reaction with deuterium ($\equiv$Si–CH_2^* + D ->$\equiv$Si–CH_2D) and the (H + D) sum remains constant. UV light and ions promotes direct detachment of CH_3 groups from Si and make this reaction irreversible.

Nitrogen and ammonia are often used during the plasma etching/strip and metal barriers deposition (TaNTa, MnN, AlN...). The addition of nitrogen to a hydrogen plasma has a strong effect on low-*k* damage although the N radicals themselves have a low effect on low-*k* materials showing small degree of Si–CH_3 group removal and hydrophilization. However, a mixture of N_2 and H_2 radicals (as well as ammonia plasma) has a detrimental effect on the low-*k* films. This is presumably due to increased amount of possible reaction pathways that leads to the film hydrophilization. One of the possible hydrophilization mechanisms is the replacement of Si–CH_3 bonds by hydrophilic Si–NH_2 bonds, which can then be replaced by Si–OH bonds (the latter as result of hydrolysis with ambient moisture). The formation and existence of Si–NH_2 bonds was proven by FTIR (Shi *et al.*, 2012]. The carbon loss in low-*k* materials can also be related to formation of volatile HCN molecules [Mankelevich, 2013] [Yamamoto *et al.*, 2011(a)], [Yamamoto *et al.*, 2011(b)]. The low-*k* damage during the metal barrier deposition needs more attention. Although it was concluded that low-*k* damage was primarily caused by Ta ions not nitrogen because of the same damage for Ta barrier as for Ta(N) deposition [Zhao *et al.*, 2011], our recent results show the role of nitrogen still needs additional clarification. In the case of metal barrier deposition, some damage can also be related to so called reactive pre-clean step in Ar or He/H_2 plasma.

Interaction of fluorocarbon plasmas with porous SiOCH low-*k* films are the most important processes in low-*k* patterning. Different aspects of these processes leading to plasma damage, surface roughening and electrical degradation were published in many papers. Some of them are analyzed in the review by Baklanov *et al.*, [2013]. Concerning the plasma damage, Iba *et al.*, [2010] studied the etch damage on NCS (nanocrystalline silica [Nakata *et al.*, 2010]) film in a 13.56 MHz capacitively coupled-plasma (CCP) etcher. The design of the experiments allowed to study RIE and pure radical etch separately by using gap structures [Uchida *et al.*, 2008], [Takeda *et al.*, 2011]. It is shown that the etching radicals, which comes from CF_4 plasma, diffuse into NCS film and reduce Si–CH_3 bonds concentration and the Si–CH_3 loss is proportional to the amount of diffused fluorine in NCS film. Some of Si–CH_3 bonds are converted to Si–F bonds then. As a result, the low-*k* performance is degraded and especially the leakage current heavily increases. The degree of sidewall damage was proportional to the value of $CR^{-0.5}$, where C is the damage diffusion coefficient, which is derived from Si–CH_3 decrement ratio from a radical treatment result and R is the etching rate, which is derived from a RIE treatment result under the same plasma condition. The value of $CR^{-0.5}$ depends on the etching condition and must be decreased as much as possible in order to reduce the sidewall damage during RIE.

Posseme *et al.* 2003 studied the etch rates of SiOCH, SiO_2 and SiCH using blanket wafers in a medium density fluorocarbon based plasma (CF_4/Ar/N_2). It was established that the etching of SiOCH and SiCH materials is controlled by a fluorocarbon interaction layer formed on top of the dielectrics under plasma exposure. The etch rate of the dielectrics is strongly modulated by the thickness of the fluorocarbon layer. When the fluorocarbon layer thickness increases, less ion energy is dissipated in the bulk dielectric material, thereby inducing a slowdown of the etch rate. The fluorine content in the fluorocarbon layer has also an impact on the etch rate. A decrease in fluorine content in the fluorocarbon layer induces lower etch rate due to less free fluorine available to reach the dielectric interface. A fluorinated layer (SiCF and SiOCF, respectively) forms at interface between low-*k* material and top fluorocarbon (CFx) layer in the steady state etching regime of SiCH and SiOCH films. This fluorinated layer acts as the fluorine feedstock for the etching reactions. The formation of the fluorocarbon layer is governed by the plasma operating conditions: source power injected in the plasma source, pressure and gas flow in the etch chamber and the nature of the chemistry used: mainly the polymerizing nature of the gas. The addition of a highly polymerizing gases such as C_4F_6 or CH_2F_2 to CF_4/N_2/Ar generates thicker fluorocarbon layer with the F/C ratio of the polymer is unchanged. The Ar dilution in the CF4/Ar gas mixture forms a fluorocarbon layer on SiOCH and SiCH with lower fluorine content. The chemical composition of the dielectric material has also a strong impact on the formation of the fluorocarbon layer. The oxygen concentration in the film induces thinner fluorocarbon layer and higher F/C ratio on SiO_2 than on SiOCH. The carbon and hydrogen concentration favors the formation of thicker fluorocarbon layer and lower F/C ratio on SiCH than on SiOCH.

Another challenge of fluorocarbon plasma application during the etching of OSG materials is the low-*k* surface roughening [Bailly *et al.*, 2010], [Yin *et al.*, 2006, 2007, 2008]. [Lazzeri *et al.*, 2005] proposed that the larger surface area of porous materials leads to an incomplete surface coverage by the fluorocarbon layer, compared to dense dielectrics. Some pores may remain open, leading to plasma species diffusion, deep interaction with the low-*k* material and roughness formation. [Yin *et al.*, 2006, 2007] attributed the etch induced roughness to micromasking by fluorocarbon species, non-uniformly deposited on the surface. Both these conclusions are in agreement with the results of Tatsumi *et al.*, [2005] who proposed that low-*k* materials are roughened when a thin fluorocarbon layer (4 nm) is deposited on the material surface, while no fluorocarbon or thicker fluorocarbon layers do not lead to roughening. Bailly *et al.*, [2010] used atomic force microscopy to investigate the kinetics of SiOCH materials roughening when they are etched in fluorocarbon plasmas. They showed that the root mean square (RMS) roughness and the correlation length do linearly increase with the etched depth, after some initiation period. It was proposed that: (1) during the first few seconds of the etch process, the surface of porous SiOCH materials gets denser, (2) cracks are formed, leading to the formation of deep and narrow pits, (3) plasma radicals diffuse through those pits and the pore network and modify the porous material at the bottom of the pits, and (4) the difference in material density and composition between the surface and the bottom of the pits leads to a difference in etch rate and an amplification of the roughness.

Rakhimova *et al.* [2015] studied interaction of OSG low-*k* films with atomic fluorine. Effects of carbon fluoride, ions and VUV photons were excluded by using SF_6 as a plasma source and special design of experimental chamber. It is shown that fluorination (F chemisorption) of the pore walls is the fastest process of SiOCH modification occurring without activation barrier. The higher porosity and the higher pore connectivity promote the deeper F penetration and fluorination into the material. The slower next stage of low-*k* SiOCH is H atoms abstraction by F atoms from $-CH_3$ groups with formation of -CFxHy surface species. The combined random walk and kinetic model results in evolution of chemical modification inside the SiOCH films depending on F atom dose. The effective etch probability of ULK SiOCH (interaction with Si-O) matrix) materials calculated per one F atom is found to be close to the etch probability of dense SiO_2 at $P < 15\%$ then significantly increases when $P < 30\%$.

VUV light generated in plasma plays an important role. VUV photons can damage OSG low-*k* materials by partial or complete destruction of $Si–CH_3$ bonds with the resulting hydrophilization and forming electrically active dangling bonds [Bittel *et al.*, 2010, Matsunaga *et al.*, 2010]. The depth and degree of plasma damage strongly depends on wave length.

It was shown that the light with wavelength shorted than 200 nm has sufficient energy to break Si-CH_3 bonds [Prager *et al.*, 2008]. Efficiency of the bonds breaking is increasing with further decreasing wavelength but the integral degree of damage at very short wavelength (<150 nm) can be smaller because of high absorption coefficient and limited depth of the light penetration. This is the reason why the most damaging wavelengths correspond to the range close to 150 nm when photons have sufficient energy to break Si-CH_3 bonds and the depth of light penetration is still higher than the typical film thickness [Lee and Graves, 2010], [Lee and Graves, 2011], [Rakhimova T. V. *et al.*, 2013], [Rakhimova T. V. *et al.*, 2014]. The etch plasma often use Ar as a carrier gas. The Ar plasma emits 106 nm VUV photons that are damaging OSG low-*k* materials. Ar replacement to He (58.4 nm) can reduce the integral plasma damage because of higher absorption coefficient. The most damaging plasma is Xe (147 nm) but the light with similar wavelength can also be generated by CF_2 radicals [Jinnai *et al.*, 2010], [Proshina *et al.*, 2015].

VUV light also can degrade the electrical characteristics of low-*k* materials [Ren *et al.*, 2011], [Bittel *et al.*, 2010], [Sinha *et al.*, 2010, 2012]. The VUV-related damage is attracting more and more attention with integration of very low density ultra low-*k* materials. For instance, recently has been shown that UV light might stimulate diffusion of Cu ions into the low-*k* material [Guo *et al.*, 2015]. Although the mechanism of this phenomenon is quite obvious and related to hydrophilization of low-*k* material and further Cu oxidation by adsorbed water with formation of Cu ions, this possibility must be taken into account during the integration. Scaling down of interconnects structures requires to take into account more complicated factors like charging, the secondary electron-electron emission on trench and via surfaces leading to non-uniform etching of the trench bottom, undesirable etching of its sidewall, etch stop, and breakdown of lower level device elements.

2.7.2. *Integration approaches reducing plasma damage*

Optimization of process integration is an alternative and, probably, the most promising way to reduce the plasma damage [Lionti *et al.*, 2015]. Calvert and Gallagher first proposed a "post integration porogen removal" approach [Calvert and Gallagher, 2003]. In this method, the deposited low-*k* dielectric contained MSQ-based (methylsilsesquioxane) matrix and acrylic polymer nanoparticles as porogen material. Instead of removing the porogen before the plasma etching and ashing processes, the porosity was created after

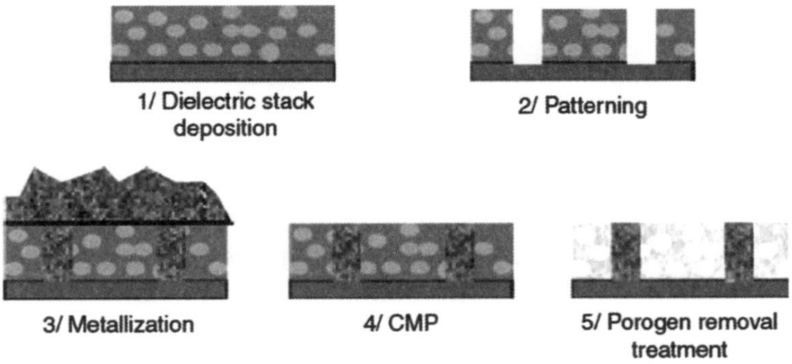

Fig. 2.18. Post processing (post etch and CMP) porogen removal approach.

patterning (or even after copper filling and CMP) by thermal curing. Due to the presence of polymeric nanoparticles (porogen) during the plasma ash process, the penetration of plasma species was avoided, and as a consequence the plasma damage was reduced (Figure 2.18).

This strategy is able to solve major issues of porous dielectrics integration but requires the development of a specific hybrid film with defined characteristics. In particular, the porogen thermal stability should be improved towards a high temperature (typically above 325°C) to allow the use of conventional BEOL integration processes [Jousseaume *et al.*, 2005]. Afterwards, the porogen removal treatment should allow the creation of porosity with minimum film shrinkage at the standard curing temperature (typically 400°C). Favennec *et al.* studied the thermal stability of different hybrid films in order to use this "solid first" integration scheme with PECVD materials (using diethoxymethylsilane (DEMS) as the matrix precursor) [L. Favennec1 *et al.*, 2006]. They showed that the cured films deposited using DEMS and cyclohexene oxide (CHO) mixture still behave as a dense film at temperatures close to 380°C, without significant porogen degradation. A $k \approx 2.3$ material was then successfully integrated without a hardmask on one metal level using conventional BEOL processes [Jousseaume *et al.*, 2006]. This post-integration porogen approach may overcome the main integration issues encountered with porous ULK integration and is extendible to next generation nodes: a lower final k-value only means a larger porogen ratio in the composite material, the processed material being always nonporous. Indeed, several other authors used the same approach to improve the integration yield [Fayolle *et al.*, 2004], [Caluwaerts *et al.*, 2003], [Eon *et al.*, 2007], [Castex *et al.*, 2005]. However, a small shrinkage of low-*k* films was observed during the postprocessing porogen removal, leading to delamination of the barrier layers and a drastic decrease of the reliability of the integrated structures, ruling out this integration strategy.

An alternative solution, Post Processing Plasma Protection (also known as P4 approach and pore stuffing), was recently proposed by Frot *et al.* [2011, 2012]. Their strategy is based on the introduction of protective polymer after the completed curing, i.e., once porogen and porogen residues are fully removed from the matrix. In this case the resulting composite material is less likely to shrink since all irreversible changes related to the porogen removal are done. Application of this strategy leads to a significant reduction of the plasma damage for highly porous materials. The efficiency of shielding the pores with an organic polymer was first demonstrated on ULK 2.0 blanket films, with up to 77% reduction of plasma induced damage. This protective effect depends on both the polymer nature and level of porosity fill. This postintegration sacrificial polymer approach could be a competitive solution for advanced technology nodes when porosity becomes too important and the pore size is close to (or larger than) 2 nm. There are several challenges of this approach and one of them is difficulty to avoid VUV damage [Heyne *et al.*, 2014], [Zhang *et al.*, 2015].

One additional approach which is also based on "pore stuffing" uses cryogenic (low temperature) etch and strip [Iacopi *et al.*, 2011], [Zhang *et al.*, 2013a], [Zhang *et al.*, 2013b]. The underlying principle is that during cryogenic etch, some of the active reactants and reaction products can condense inside the pores and provide "temporary" sealing. The "stuffing" layer is removed by warming the sample back to room temperature. In some cases soft annealing is needed if the reaction products are not sufficiently volatile at room temperature. The process and chemistries can be similar to well-known cryogenic etching that is used for anisotropic deep Si etching (SF_6 plasma). The pore "stuffing" can also be provided by etch by-products. For instance, during the stripping in O_2 plasma, the water molecules (reaction product) will also condense on the pore walls, decrease the pores size, and prevent diffusion of oxygen radicals into the pores.

2.7.3. *Dielectric recovery*

The etch and strip plasmas induce loss of methyl groups from $\equiv$Si–CH_3 groups in OSG low-*k* film. As a result, the low-*k* materials become hydrophilic and adsorb moisture, which increases dielectric constant and degrade electrical performance and reliability. The restoration of the damaged low-*k* materials has been an important subject of research starting from the first attempts to introduce porous low-*k* materials into interconnect technology. Several excellent reviews give extended information about these activities [Shi *et al.*, 2012, Bohm *et al.*, 2013, Lionti *et al.*, 2015]. All dielectric recovery attempts can be divided in three different approaches. The first one is to restore the methyl groups by applying silylation agents such as hexamethyldisilazane (HMDS), trimethylchlorosilane (TMCS), dichlorodimethylsilane (DMDCS), and others. The second option is based on the deposition of hydrophobic agents from hydrocarbon plasma. The third option is thermal treatment allowing to eliminate the adsorbed hydroxyl (OH) groups and the physisorbed water. The thermal treatment are often assisted by UV light.

It is necessary to note that the all restoration technologies are not able to restore initial chemical structure of low-*k* materials. The loss of methyl groups bonded to silicon is often irreversible process because of reconstruction of OSG chemical bonds or steric hindrance effects of silylating agents. For instance, the neighboring silicon dangling bonds formed after -CH_3 groups removal can recombine forming siloxane bridges and the sylilation agents can't break these bonds. Even in the case when the recombination of dangling bonds are not happening and they are terminated by OH groups, the restored structure is quite different. First, the sylilating group always has larger size and therefore, it reduces porosity and increases the *k*-value in comparison with pristine material and also some hydroxyl groups can remain in the film because of their steric hindrance (Figure 2.19).

It was proposed that difunctional silylation agents like DMDCS can react with two neighboring silanols simultaneously but it clear that in the case of high concentration of silanols the probability of complete sylilation is very small [Hayward and Trapnell, 1964]. It has also been assumed that it is possible to replace the OH groups by CH_3 by using methyl radicals generated in CH_4 plasma [Aimadeddine *et al.*, 2006, Bao *et al.*, 2007, Shi *et al.*,

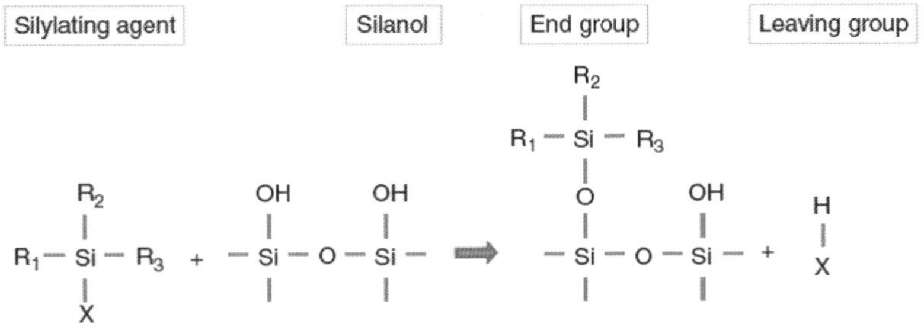

Fig. 2.19. The schematic diagram of the silylation process on a silanol bonds of a damaged low-*k*.

2007]. It was shown by using XPS analysis that the total carbon concentration in low-*k* films can be restored and also significant improvement of electrical characteristics was demonstrated. However, FTIR spectra show that the introduced carbon containing groups are not bonded to Silicon as in pristine materials.

The results from the CH_4 plasma treatment and silylation are mostly confined to the dielectric surface, limiting the recovery of the plasma damage, which is normally throughout the bulk of the film. This has led to the development of the UV process for low-*k* dielectric recovery. Post-deposition UV curing is commonly used to improve the mechanical properties of as-deposited low-*k* dielectrics and to remove porogens. After plasma damage, UV radiation is expected to eliminate the silanol and physisorbed water from the bulk of the low-*k* dielectric, reducing the dielectric constant and leakage current. In a recent study [Shi *et al.*, 2007], k ~2.5 CDO films were first damaged by an oxidative RIE plasma and then UV treated at 350°C for about 300 seconds. After UV treatment, the film thickness was reduced from 83.1 to 78.5 nm, while the refractive index was reduced from 1.333 to 1.323 and the dielectric constant reduced from 3.26 to 2.55. The following reaction scheme was proposed:

$$\equiv \text{Si–OH} + \text{HO–Si} \equiv \;{-}{>}\; \equiv \text{Si–O–Si} \equiv + H_2O$$

$$\equiv \text{Si–OH} + H_3C\text{–Si} \equiv \;{-}{>}\; \equiv \text{Si–O–Si} \equiv + CH_4$$

$$\equiv \text{Si–OH} + \text{H–Si} \equiv \;{-}{>}\; \equiv \text{Si–O–Si} \equiv + H_2$$

$$\equiv \text{Si–OH} + \text{HOC–Si} \equiv \;{-}{>}\; \equiv \text{Si–Si} \equiv + H_2 + CO_2$$

The UV curing recovery can be optimized with the process parameters, including UV wavelength and intensity, substrate temperature, UV curing time, chamber pressure and ambient gas. Although the UV treatment on plasma-damaged OSG low-*k* dielectrics can reduce the bulk –OH bond and moisture, the surface hydrophobicity was not recovered. On the other hand, the silylation process excels in surface recovery. A combination of UV treatment and silylation can lead to an even better recovery, as was demonstrated by Huang *et al.* [2010].

Lionti *et al.* (2015) reviewed recent patents related to restoration activities based on combination of UV curing and silylation [Yim *et al.*, 2012, Chan *et al.*, 2014]. Reactive carbon precursors in the presence of heat and UV radiation were proposed by Applied Materials. Suitable carbon containing reagents include acetylene, ethylene, 1,3-butadiene and isoprene. The UV light was promoting the carbon incorporation. The most efficient UV wavelength is dictated by the nature of the carbon containing precursor, e.g. $\lambda = 200-220$ nm for butadiene and $\lambda = 120-180$ nm for acetylene. From this, it is apparent that highly energetic UV radiation is required to convert the unsaturated carbon precursors into reactive species, which are capable of capping pendant SiOH groups. Radiation wavelength profile is chosen to fall primarily below 300 nm. According to Xie *et al.* [2014] the UV assisted sylilation allowed to achieve dielectric constant repair to within 5% of the original value for a k = 2.2 material which had been plasma damaged to k = 2.5. Corresponding improvements in leakage current were also reported, although the values for repaired samples fall halfway in between those for the pristine and damaged samples. Water contact angle measurements indicated a full recovery from 20 degrees for the damaged sample back to 100 degrees. In addition, TEM-EELS characterization of TaN incursion of the previously damaged and then repaired porous matrix revealed significant pore sealing by this approach. In addition, silane based chemistries in combination with UV light have also been proposed by Novellus [Varadarajan *et al.*, 2013] and Applied Materials. The wavelength is selected to achieve k recovery with little or no dielectric material shrinkage and where the majority of the UV radiation has $\lambda \sim 300$ to 450 nm and less than 10% of the UV radiation with $\lambda < 300$ nm.

2.8. *Conclusions*

Different classes of low-*k* materials have been developed and studied during the last 20 years. The first low-*k* materials were organic polymers that allowed to achieve reduced dielectric constant due to the low polarizability of chemical bonds. However, their integration has not been successful so far, mainly because of too large coefficient of thermal expansion in comparison with other components of interconnect structure.

Porous organosilicate based low-*k* films are the most popular low-*k* materials at the present because of similarity of their properties to traditional SiO_2 and to the best compatibility with existing microelectronic technology. All OSG materials have quite similar chemical composition but different porosity, which increases with scaling of dielectric constant. If the materials with k > 2.5 have relatively low porosity (<15%) and pore size (<2 nm), the new ultra low-*k* films developed for advanced technology nodes have porosity approaching to 50% and pore size >3 nm. The largest problem during the integration of OSG ultra low-*k* materials is related to their degradation (damage) during the exposure to plasma. The degree of degradation increases with pore size and porosity and, therefore, these challenges are becoming even more important for ultra low-*k* materials that are needed for future technology nodes.

The present strategy is the development and exploration of new materials and development of damage free processes. Some new materials are completely different than OSG: metal-organic frameworks (MOFs) and boron carbonitrides (BCN). Development of damage free processes include different versions of pore stuffing, including stuffing by porogen, sacrificial polymers and condensation of the etch chemistries at cryogenic temperature. However, these approaches meet serious challenges and for our opinion the key solution should be based on replacement of presently used damascene approach to subtractive one: first pattern metal and then fill the gaps (via and trenches) by low-*k* materials.

References

Afanasiev V. *et al.* (1996) *Phys. Rev. B*, **54**, p. 10820.

Afanasiev V. *et al.* (2013) *Applied Physics Letters*, **102**, p. 172908.

Aimadeddine M. *et al.* (2006) *Proceedings of the IEEE International Interconnect Technology Conference*, p. 81.

Als-Nielsen J. and McMorrow D. (2001) *Elements of Modern X-ray Physics*, John Wiley & Sons, New York.

Antonelli G. A. *et al.* (2006) *MRS Bull.*, **31**, p. 607.

Asefa T *et al.* (1999) *Nature*, **402**, p. 867.

Baake O. *et al.* (2009) *Analytical and Bioanalytical Chemistry*, **395**, p. 1901.

Bailly F *et al.* (2010) *J. Appl. Phys.*, **108**, p. 014906.

Baklanov *et al.* (2000) *J. Vac. Sci. Technol. B.*, **18**, p. 1385.

Baklanov M. R. *et al.* (2002) *MRS Proceedings of Advanced Metallization Conference in 2001*, p. 271.

Baklanov M. R. and Mogilnikov K. P. (2002) *Microelectr. Eng.*, **64**, p. 335.

Baklanov M. R. and Maex K. (2006) *Phil. Trans. R. Soc. A.*, **364**, p. 201.

Baklanov M. R. *et al.* (2006) *Microelectr. Eng.*, **83**, p. 2287.

Baklanov M. R. *et al.* (2007) *Dielectric Films for Advanced Microelectronics*, Wiley.

Baklanov M. R. and Shamiryan D. (2011) Porous Polymers, Eds. M. S. Silverstein, N. R. Cameron, M. A. Hillmyer (Wiley), Chapter 6, Nondestructive Evaluation of Critical Properties of Thin Porous Films. p. 205.

Baklanov M. R. *et al.* (2011) *Microelectronic Eng.* **88**, p. 990.

Baklanov M. R. *et al.* (2012) *Advanced Interconnects for ULSI Technoilogy*, Wiley.

Baklanov M. R. *et al.* (2013) *Appl. Phys. Rev.*, **113**, p. 041101.

Bao, J. J. *et al.*, (2007) *Proceedings of the IEEE 2007 International Interconnect Technology Conference*, p. 147.

Bauer B. J. *et al.* (2001) *J. Electron. Mater.*, **30**, 4, p. 304.

Bittel B. C. (2010) *Appl. Phys. Lett.* **97**, p. 063506.

Bohm O. *et al.* (2013) *Microelectronic Engineering*, **112**, p. 63.

Braginsky S. (2010) *J. Appl. Phys.* **108**, p. 073303.
Bouyssou R. *et al.* (2010) *J. Vac. Sci. Technol.* B, **28**, p. L31.
Brinker C. J. and Scherer G. W. (1990) *Sol-Gel Science: The Physics and Chemistry of Sol-Gel Processing*, Academic Press: New York 1990.
Bundschuh S. *et al.* (2012) *Appl. Phys. Lett.*, **101**, 101910.
Castex A. *et al.* (2005) *Microelectron. Eng.*, **82**, p. 416.
Caluwaerts R. *et al.* (2003) *Proc. IEEE International Interconnect Technology Conference. (IITC)*, p. 242.
Calvert J. M. and Gallagher M. K., (2003) *Semicond. Int.*, **26**(12), p. 56.
Chan K. *et al.* (2014) *U.S. Patent* 20140004717.
Ciofi I. *et al.* (2010) *Microelectron Eng.*, **87**, p. 2391.
Ciofi I. *et al.* (2012) *IEEE Trans. Electron Devices*, **59**, 6, p. 1607.
Corma, A. *et al.* (2010) *Chem. Rev.*, **110**, p. 4606.
Dubois G. *et al.* (2007), *Dielectric Films for Advanced Microelectronics*, eds. Baklanov M., Green M. and Maex K. Chapter 2 "Spin-on Dielectric Materials" (Wiley), p. 33.
Eon D. *et al.* (2007) *J. Vac. Sci. Technol.*, B **25**, p. 715.
Eslava S. *et al.* (2008) *Adv. Mater.*, **20**, 16 p. 3110.
Eslava S. *et al.* (2013) *Chem. Mater.*, **25**, p. 27.
Favennec L.1 *et al.* (2007) *Proceedings of the Advanced Metallization Conference 2006 — Asian Session*, Tokyo, Japan, p. 345.
Favennec L.2 *et al.* (2007) *J. Appl. Phys.*, **102**, p. 064107.
Fayolle M. *et al.* (2004) *Proceedings of the IEEE International Interconnect Technology Conference*, p. 208.
Frot T. *et al.* (2011) *Adv. Mater.*, **23**(25), p. 2828.
Frot T. *et al.* (2012) *Adv. Funct. Mater.*, **22**(14), p. 3043.
Gidley D. W. *et al.* (1999) *Phys. Rev. B*, **60**, p. 5157.
Gidley D. W. *et al.* (2000) *Mat. Res. Soc. Symp. Proc.*, **612**, p. D4.3.1.
Gidley D. W. *et al.* (2000) *Appl. Phys. Lett.*, **76**, p. 1282.
Gidley D. W. (2006) *Annu. Rev. Mater. Res.*, **36**, p. 49.
Gidley D. W. (2014) Private communication.
Goethals F.(1) *et al.* (2012), *Chem. Commun.*, **48**, p. 2797.
Goethals F.(2) *et al.* (2012) *J. Mater. Chem.*, **22**, p. 8281.
Goldman M. A. *et al.* (2009), *J. Appl. Phys.*, **106**(1), 01311(2009).
Gregg S. J.; Sing K. W. (1982) *Adsorption, Surface Area, and Porosity*. London: Academic Press.
Grill A. and Neumayer D. A., (2003) *J. Appl. Phys.*, **94**(10), p. 6697.
Grill A. *et al.* (2003) *J. Appl. Phys.*, **94**, p. 3427.
Grill A. (2009) *Annu. Rev. Mater. Sci.*, **39**, p. 49.
Grill A. *et al.* (2014) *J. Appl. Phys.*, **1**, p. 011306.
Guo X. *et al.* (2015) *Appl. Phys. Lett.*, **106**, p. 012904.
Han Q. Y. *et al.* (2005) *Solid State Phenom*, **103–104**, p. 341.
Hayward D. O. and Trapnell B. M. W., (1964) *Chemisorption*, 2nd ed., Butterworth, Washington, D. C.
Hasegawa, T. *et al.* (1999) *Proc. IEEE Electron Devices (IDEM)*, p. 623.
Hench L. L. and West J. K., (1990) *Chem. Rev.*, **90**, p. 33.
Heyne M. *et al.* (2014) *J. Vac. Sci. Technol. B.*, **32**(6) p. 062202.

Hoffman R. J. O. M. *et al.* (2005) *Microelectronic Eng.*, **80**, p. 337.
Hoffmann F. *et al.* (2006) *Angewandte Chemie-International Edition*, **45**, p. 3216.
Huang E. *et al.* (2002) *Appl. Phys. Lett.*, **81**, p. 2232.
Huang, H. *et al.* (2010) *Mater. Res. Soc. Symp. Proc. D*, **1079**, N02–10.
Hunks W. J. and Ozin G. A. (2005) *J. Mater. Chem.*, **15**, p. 3716.
Hurand R. *et al.* (2011) *Proceedings of the IEEE International Interconnect Technology Conference*, p. 3.
Iacopi F. *et al.* (2011) *Phys. Chem. Chem. Phys.*, **13**, p. 3634.
Iba Y. *et al.* (2008) *Jpn. J. Appl. Phys.*, **47**, p. 6923.
Iba Y. *et al.* (2010) *Microelectronic Engineering*, **87**, p. 451.
Jinnai B. *et al.* (2010) *J. Appl. Phys.*, **107**, p. 043302.
Jousseaume V. *et al.* (2005) *Proceedings of the IEEE International Interconnect Technology Conference*, p. 60.
Jousseaume V. *et al.* (2006) *Appl. Phys. Lett.*, **88**, p. 182908.
Kollmann Markus (2003) *Phys. Rev. Lett.*, **90**(18), 180602.
Kondoh E. *et al.* (2001) *Jpn. J. Appl. Phys.*, **40**, p. L323.
Kuchenmeister F. *et al.* (1999) *Proc. Advanced Metallization Conference 1998, MRS*, 237.
Kunnen E. *et al.* (2010) *J. Vac. Sci. Technol. B*, **28**(3), p. 450.
Kunnen E. *et al.* (2011) *Microelectronic Engineering*, **88**, p. 631.
Landskron K. *et al.* (2003) *Science*, **302**, p. 266.
Larlus O. *et al.* (2004) *Appl. Surf. Sci.*, **226**, 1–3, p. 155.
Lazzery P. *et al.* (2005) *J. Vac. Sci. Technol.*, B **23**, p. 1491.
Lee J and Graves D B. (2010) *J. Phys. D: Appl. Phys.*, **43**, p. 425201.
Lee J and Graves D B. (2011) *J. Phys. D: Appl. Phys.*, **44**, p. 325203.
Levine J. R. *et al.* (1989) *J. Appl. Cryst.*, **22**, p. 528.
Levitt D. G. (1973) *Phys. Rev. A*, **8**(6), 3050.
Licitra C. *et al.* (2011) *Thin Solid Films*, **519**, p. 2825.
Linss V. *et al.* (2004) *Thin Solid Films*, **467**, p. 76.
Lionti K. *et al.* (2015) *ECS J. Sol. St. Sci. Technol.*, **4**(1), p. N3071.
Liu J. *et al.* (2004) *J. Appl. Phys.*, **96**, p. 6679.
Lomonosov A. M. *et al.* (2012) *ACS Nano*, **6**, 2, p. 1410.
Lopaev *et al.* (2014) unpublished data.
Lubguban, J. Jr. *et al.* (2002), *J. Appl. Phys.*, **92**, p. 1033.
Ma, Z. *et al.* (2011) *Coord. Chem. Rev.*, **255**, p. 1623.
Machetta P. *et al.* (2003) *Mater. Sci. in Semiconductor Processing*, **5**, p. 285.
Maex K. *et al.* (2003) *J. Appl. Phys.*, **93**, 11, 8793–8841.
Marsik P. *et al.* (2010) *Thin Solid Films*, **518**, 15, p. 4266.
Marsik P. *et al.* (2011) *Thin Solid Films*, **519**, 11, p. 3619.
Martin S. J. *et al.* (2000) , *Adv. Mater.*, **12**(23), p. 1769.
Matsunaga N. *et al.* (2010) *Jpn. J. Appl. Phys.*, **49**, p. 04DB06.
Matsushita A. *et al.* (2003) *IEEE International Interconnect Technology Conference*, Burlingame, CA.
Mechri C. *et al.* (2009) *Appl. Phys. Lett.*, **95**, 9, p. 091907.
Mitra A. *et al.* (2004) *Ind. Eng. Chem. Res.*, **43**, p. 2946.
Mizoshita N. *et al.* (2011) *Chem. Soc. Rev.*, **40**, p. 789.

Mogilnikov K. P.; Baklanov M. R. (2002) *Electrochem. Sol. State Lett.*, **5**, p. F29.
Mogilnikov K. P. (2004) *Jap. J. Appl. Phys.*, **43**, p. 247.
Moore G. E. (1965) *Electronics*, 114; (1998) *Proc. IEEE*, **86**(1).
Morant C. *et al.* (2007) *Diam. Relat. Mater.*, **16**, p. 1441.
Nakata Y. *et al.* (2010) *FUJITSU Sci. Tech. J.*, **46**(1) p. 120.
Omote K. *et al.* (2003) *Appl. Phys. Lett.*, **82**, p. 544.
O'Neill M.L., *et al.* (2006) *Materials Research Society Symposium Proceedings*, vol. 914, p. 3.
Perrone A. *et al.* (1998), *Appl. Surf. Sci.*, **133**, 239 (1998).
Plawsky J. L. *et al.* (2007). *Dielectric Films for Advanced Microelectronics*, Eds. M. Baklanov, M. Green, K. Maex. (Wiley) Chapter 4. Mechanical and transport properties of Low-*k* Dielectrics, p 137.
Prakash A. *et al.* (2015) ECS *J. Sol. St. Sci. Technol.*, **4**(1) p. N3122.
Posseme N., *et al.* (2003) *J. Vac. Sci. Technol.* B **21**, 2432.
Prager L. *et al.* (2008) *Microelectronic Eng.*, **85**, p. 2094.
Proshina O. *et al.* (2015) *Plasma Sources Science and Technology*. Submitted.
Rakhimova T V *et al.* (2013) *Appl. Phys. Lett.*, **102**, p. 111902.
Rakhimova T. V. *et al.* (2014) *J. Phys. D: Appl. Phys.*, **47**, p. 025102.
Ren H., *et al.* (2011) *Appl. Phys. Lett.*, **98**, p. 252902.
Ryan E. T. *et al.* (2008) *J. Appl. Phys.*, **104**, p. 094109.
Safaverdi S. *et al.* (2011) *Phys. Rev. B.*, **83**, p. 245320.
Saski A. (2005) *The Rigaku Journal*, **22**, p. 31.
Shamiryan D. *et al.* (2003) *J. Vac. Sci. Technol. B*, **21**, p. 220.
Shekhah, O. *et al. Chem. Soc. Rev.*, 2011, **40**, p. 1081.
Shi, H.1 *et al.* (2007), *Mater. Res. Soc. Proc. B*, **990**, p. B03–12.
Shi, H.2. *et al.* (2007) *Proceedings of the Advanced Metallization Conference* **23**, p. 447.
Shi H. *et al.* (2012) *Advanced Interconnects for ULSI Technology*, eds. M. R. Baklanov, P. S. Ho, E. Zschech, Chapter 3 "Plasma Processing of Low-*k* Dielectrics" (Wiley), p. 79.
Sinha H. *et al.* (2010) *J. Vac. Sci. Technol.*, A **28**(6), p. 1316.
Sinha H. *et al.* (2012) *Appl. Phys. Rev.*, **112**, p. 111101.
Stesmans, A. *et al.* (2013) *Microelectronic Engineering*, **109**, p. 240.
Stucchi M. and Maex K., (2002) *IEEE Trans. Instrum. Meas.* **51**, p. 537.
Sugiyama T. *et al.* (2002) *Appl. Phys. Lett.*, **80**, p. 4214.
Sulaeva V. S. *et al.* (2014) *Thin Solid Films*, **558**, p. 112.
Sun J.-N. *et al.* (2001) *J. Appl. Phys.*, **89**, p. 5138.
Sun J.-N. *et al.* (2002) *Mater. Res. Soc. Symp. Proc.*, 726, Q10.5.1.
Takeda K. *et al.* (2011) *J. Appl. Phys.*, **109**, p. 033303.
Tatsumi T. *et al.* (2005) *J. Vac. Sci. Technol.*, A **23**, 938.
Tokuyama S. *et al.* (2008) *Jap. J. Appl. Phys.*, **47**, 4, p. 2492.
Uchida S. *et al.* (2008) *J. Appl. Phys.*, **103**, p. 073303.
Ulrich S. *et al.* (1999) *Surf. Coat. Tech.*, **116**, p. 742.
Umeda S. *et al.* (2004) *Diamond and Related Materials*, **13**, p. 1135.
Urbanowicz A. *et al.* (2010) *J. Appl. Phys.*, **107**, p. 104122.

Urbanowicz A. (2010), "Study of origin, impact and solutions of processing damage in low dielectric constant materials for advanced interconnect applications," Ph.D. dissertation (IMEC and KU Leuven, 2010).
Varadarajan B. N. *et al.* (2013) *US Patent*. 8,465,991.
Volksen W. *et al.* (2010) *Chem. Rev.*, **110**(1), p. 56.
Wang Z. B. *et al.* (2001) *Adv. Mater.*, **13**, p. 746.
Watanabe M. *et al.* (1995) *J. Appl. Phys.*, **78**, p. 2880.
Worsley M. A. *et al.* (2005) *J. Vac. Sci. Technol.*, B **23**(2), p. 395.
Xie B. *et al.* (2014) *Proc. of IEEE International Interconnect Technology Conference*, p. 335.
Yamamoto1 H. *et al.* (2011) *J. Appl. Phys.*, **110**, p. 123301.
Yamamoto2 H. *et al.* (2011) *J. Appl. Phys.*, **109**, p. 084112.
Yan Y. S. *et al.* (2002) *US Patent* 0,134,995.
Yim K. S. *et al.* (2012) *U.S. Patent* 8,216,861.
Yin Y., *et al.* (2006) *J. Vac. Sci. Technol.*, A **24**, p. 2360.
Yin Y., *et al.* (2007) *J. Vac. Sci. Technol.*, A **25**, p. 802.
Yin Y., *et al.* (2008) *J. Vac. Sci. Technol.*, A **26**, p. 151.
Zagorodniy, K. *et al.* (2010). *Appl. Phys. Lett.*, **97**, p. 251905.
Zenasni A. *et al.* (2007) *J. Appl. Phys.*, **102**, p. 094107.
Zhang X. (2013) *Thin Solid Films*, **544**, p. 2.
Zhang L.1 *et al.* (2013) *ECS Solid State Lett.*, **2**, 2, p. 5.
Zhang L.2 *et al.* (2013) *ECS Sol. St. Sci. & Technol.*, **2**(6), p. N131.
Zhang L. *et al.* (2015) *ECS Solid St. Sci. & Technol.*
Zhao Larry *et al.* (2011) *Microelectronic Engineering*, **88**, p. 3030.
Zhou F. *et al.* (2006) *Thin Solid Films*, **497**, p. 210.

3. Mechanical Stability of Porous Low-k Dielectrics and its Impact on Chip Package Interaction

3.1. *Introduction*

For the past three decades, the semiconductor industry has been improving the performance of microelectronic devices and the functionality of advanced integrated circuits through the continuous scaling of integrated circuits and the maximization of transistor density [Maex, 2003]. Consequently, this encourages the introduction of new materials, processes, chip designs, and packaging strategies into micro-/nano-electronic products. Among these material innovations are insulating materials with a dielectric constant (*k*) less than that of SiO_2, so called low-*k* dielectrics. During the last 15 years, various materials and methods have been developed for fabricating low-*k* dielectric films, among which the plasma enhanced chemical vapor deposition (PECVD) technology of porous organosilicate glasses (OSG) is the most popular due to its better compatibility with the technological needs [Burkey and Gleason, 2003] [Li *et al.*, 2009a] [Baklanov *et al.*, 2007] [Sreenivasan and Gleason., 2009]. The reduction in *k* for low-*k* dielectrics is being pursued through the introduction of controlled levels of porosity. However, finding a good low-*k* material has proven to be much more challenging than first expected due to a number of integration and reliability issues [Maex, 2003]. Besides the need of being compatible with the different lithography, etching, stripping and cleaning processes that are used in state-of-the-art integration schemes, they must also have sufficient mechanical strength to withstand the high shear stresses as well as harsh chemical environments that are involved during the chemical mechanical polishing process without cohesive or adhesive failure occurring [Guyer and Dauskardt, 2003] [Guyer and Dauskardt, 2004a] [Guyer and Dauskardt, 2005]. On top of that, since low-*k* dielectrics exhibit intrinsic tensile stresses and have increased coefficients of thermal expansion compared to SiO_2, thin film cracking and adhesion are serious thermal-mechanical reliability issues for low-*k* dielectric materials (Figure 3.1) [Guyer *et al.*, 2006] [Kim and Dauskardt, 2009]. Thermo-mechanical deformation of the package during package assembly and subsequent reliability tests can induce large local stresses that can initiate and propagate cohesive and/or adhesive cracks in different back-end of line (BEOL) layers [Wang *et al.*, 2005]. Therefore, a careful characterization of the mechanical integrity of potential new low-*k* candidates is required to integrate these new materials and Cu interconnects and to assure reliability during chip packaging and under field conditions.

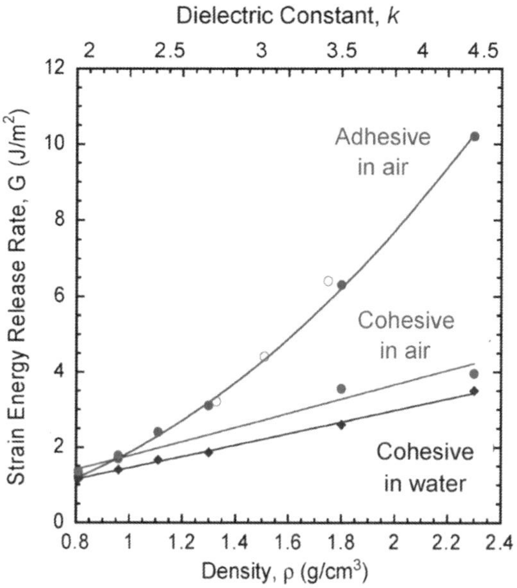

Fig. 3.1. Strain energy release rate as a function of the film density (ρ) and the corresponding dielectric constant (k) of organosilicate thin film glasses. Adhesive and cohesive fracture energies were measured in air and water using respectively 4-point bending and double cantilever beam specimens. Courtesy of Dauskardt. [Kim and Dauskardt, 2009]

This chapter reviews the mechanical stability of porous ultralow-*k* dielectrics with the focus on chip package interaction related issues. First, a brief overview of the most commonly used measurement techniques to characterize the elastic and fracture properties of low-*k* dielectrics will be given. Next, the impact of porosity, pore morphology, network structure, UV radiation and reactive species in the environment on the mechanical stability of low-*k* dielectrics will be discussed and an overview of the most common potential candidates for future generation low-*k* dielectrics will be given. Finally, the impact of the mechanical stability of low-*k* dielectrics on chip package interaction will be discussed.

3.2. *Techniques to quantify the mechanical stability of porous low-k*

In microelectronics, the elastic modulus and fracture properties (cohesive and adhesive fracture toughness) are commonly used parameters to assess the mechanical behavior, fracture behavior and process compatibility of low-*k* dielectrics. When the elastic modulus, cohesive and adhesive fracture toughness cannot meet a minimum standard, the material simply cannot be used for multilevel integration.

3.2.1. *Elastic modulus*

Accurately measuring the elastic modulus of nanoporous low-*k* dielectric films is not trivial [Grill and Patel, 2001] [Urbanowicz *et al.*, 2009] [Zhou *et al.*, 2011] [Bailey *et al.*, 2013] [Guo *et al.*, 2013]. This is because the measured elastic modulus of a porous film is actually an effective modulus, which evolves toward the elastic modulus of the matrix material near zero porosity. An important challenge here is that the actual techniques used for characterizing bulk materials are hardly applicable in the case of nanoporous thin films because of the small volume size of the material under investigation. Therefore, several advanced destructive and nondestructive techniques have been developed to characterize the elastic properties of thin films on substrates, ranging from mechanical tests, optical tests, laser acoustics, and ultrasonics, among others. However, most of these techniques are still not commonly known and require a detailed analysis in order to obtain reliable results, especially in the case of porous films. Complexities of measuring the elastic modulus include substrate influences, material/probe interactions, and data collection and analysis techniques. Moreover, there is a lack of fundamental understanding on how the measured elastic modulus value is influenced by the stiffness of the matrix, porosity, pore structure, and specific technique related effects. To date, nanoindentation is the most commonly used technique in the semiconductor industry to determine elastic modulus of low-*k* dielectric films [Doerner and Nix, 1986] [Pharr *et al.*, 1992] [Oliver and Pharr, 1992]. The main advantage of this technique is that it requires no special sample preparation and tests can be performed quickly and inexpensively. Additionally, nanoindentation combined with frequency-specific dynamic analysis also provides access to viscoelastic measurement of near surface properties of polymeric low-*k* materials. However, several factors may influence the extracted mechanical properties of thin porous low-*k* films. Most notable among these are effects due to the presence of the substrate, which may lead to overestimation of the elastic modulus and hardness values [Chen and Vlassak, 2001] [King, 1997] [Bhattacharya and Nix, 1988], issues with tip calibration [McElhaney *et al.*, 1998], surface roughness, densification of the film as a result of large hydrostatic stresses [Xiang *et al.*, 2004] [Fleck *et al.*, 1992], and size effects as a result of the inhomogeneous strain field [McElhaney *et al.*, 1998]. Besides NI, a number of non-destructive dynamic techniques are available to determine the mechanical properties of thin films on substrates, including surface acoustic wave spectroscopy (SAWS) [Farnell and Adler, 1972], surface Brillouin scattering (SBS) [Carlotti *et al.*, 1996], the Picosecond Ultrasonic Method [Chapelon *et al.*, 2006], among others. However, these techniques typically require knowledge of the density of the film and only provide the elastic behavior of the films. Recently, K. Vanstreels and co-authors demonstrated that the measured elastic modulus of nanometer scale porous films can be severely influenced by specific technique related effects, which in turn also appeared to be sensitive to the specific arrangement and morphology of the pores in the film [Vanstreels *et al.*, 2013]. For this purpose, they compared various destructive and nondestructive experimental techniques, which are based on different physical principles for the calculation of the elastic properties of nanoporous films, including nanoindentation (NI) with sharp and spherical indenter probes, laser-induced acoustic wave spectroscopy (LAwave) and ellipsometric porosimetry (EP). On the basis of this comparative study, they concluded that nanoindentation using a sharp Berkovich probe tends to set an upper limit on the elastic modulus, rather than give an absolute value, while spherical indentation is less sensitive to the local pore structure and may lead to more realistic values. Possible reasons for this overestimation were explained in terms of anisotropic elasticity behavior, densification upon indentation and the characteristic length, which defines the dimension in which the stress–strain analysis is nearly homogeneous. Nevertheless, careful nanoindentation combined with FEM simulations still offers the best approach for a complete mechanical characterization of nanoporous low-*k* films.

3.2.2. *Fracture properties*

From a thermodynamic point of view, crack growth is driven by the release of stored strain energy in the material and the driving force for fracture is defined as the amount of strain energy released per unit area of crack growth. Fracture toughness encompasses the energy required to facilitate crack propagation to rupture, which can be a cohesive or adhesive crack [Dieter, 1976] [Callister, 2003]. Furthermore, the resistance to crack growth is the energy required to break bonds and create free surfaces from the bonded materials, and generate dislocations or other defects near the crack tip. Fracture properties of low-*k* dielectric films are a key component for the reliability assessment of microelectronic devices. Measuring the cohesive and adhesive fracture toughness as a property of the material or interface is therefore a critical procedure for materials characterization for interconnects and packaging [Bakir and Meindl, 2008]. Over the last two decades, advances in fracture mechanics for thin films and layered structures provided a good foundation for the development of experimental techniques for the measurement of both cohesive fracture toughness and interfacial adhesion [Suo, 2003] [Hutchinson and Suo, 2002]. Techniques to quantify the adhesive fracture toughness of thin films fall into different groups of test methods including mechanical flexure tests [Jiang *et al.*, 2000] [Jaeger *et al.*, 2000] [Gille and Wetzig, 1983] [Gille, 1984] [Cox, 1988], scratch tests [Bellido-Gonzalez *et al.*, 1995] [Li and Bhushan, 2001] [Sundararajan and Bhushan, 2001] [Panjan *et al.*, 2003] [Voevodin *et al.*, 1995] [Harry *et al.*, 1999A] [Shum *et al.*, 2004], indentation techniques [Marshall and Evans, 1984] [Kriese and Gerberich, 1999] [Volinsky *et al.*, 1999] [Volinsky *et al.*, 2002] [Lawn *et al.*, 1980] [Lawn and Winslaw, 1975], buckling tests [Cotterell and Chen, 2000] [Chen *et al.*, 2002] [Britvec, 1973], and tensile tests [Drory *et al.*, 1995] [Harry *et al.*, 1998] [Harry *et al.*, 2000] [Harry *et al.*, 1999B] [Hu and Evans, 1989]. Among many different measurement techniques, the mechanical flexure or bending tests, such as the four-point beam bending (FPB) technique [Lacombe, 2006] [Suo and Hutchinson, 1989] [Ma, 1997] [Dauskardt *et al.*, 1998] [Charalambides *et al.*, 1989] [Roham *et al.*, 2004] and double cantilever beam (DCB) [Hutchinson and Suo, 2002] [Fernlund and Spelt, 1994] are the most popular techniques in microelectronics applications that provide accurate quantitative results for the adhesive fracture toughness. This is mainly due to its simple geometry, easy sample preparation, and the relatively easy beam mechanics involved in the fracture analysis. The only limitation of the flexure tests is the complicated and time-consuming experimental methodology and the lack of a measurement automation process. In both techniques, one or more layers of thin-film material are sandwiched between two thick substrates, typically Si (Figure 3.2). Because the films are much thinner than the substrates, the energy release rate for an interfacial crack that is advancing between a substrate and a film or in between two films can be calculated from the far-field loading on the substrates by neglecting the thin films, i.e, the homogeneous solutions given by Hutchinson and Suo [2002]. In case of the DCB test, the energy release rate under a symmetric loading is given by

$$G = \frac{12\left(1-\nu^2\right)P^2a^2}{EB^2H^3}, \tag{3.1}$$

where E and ν are respectively the elastic modulus and Poisson's ratio of the substrate, P is the applied force, a is the crack length, H is the substrate thickness and B is the beam width. With a predetermined crack

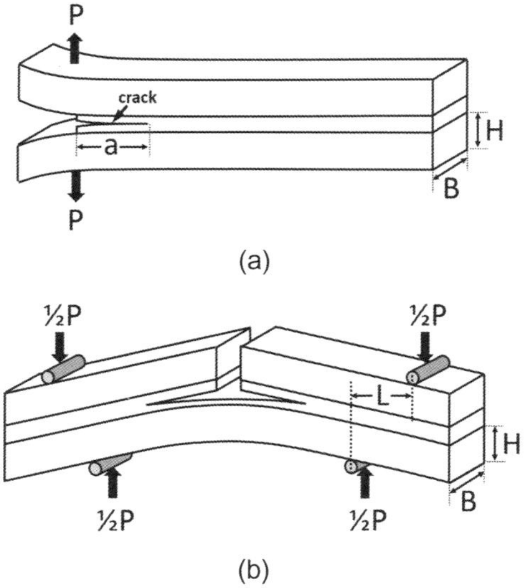

Fig. 3.2. schematic overview of (a) double cantilever beam test and (b) four-point bending test.

length, a critical load *Pc* for the crack to grow can be determined from the load-displacement curve, hence the interface toughness is then calculated as $\Gamma = G(Pc)$. In case of the FPB test, the crack growth along the interface reaches a steady state with the energy release rate independent of the crack length,

$$G = \frac{21\left(1-\nu^2\right)P^2L^2}{16EB^2H^3}, \tag{3.2}$$

where *L* is the distance between the inner and outer loading pins. The mode mixity for the sandwiched samples during DCB or FPB depends on the film thickness and the specific film properties. A common practice has been to specify the mode mixity for sandwiched samples by the far-field phase angle, $\psi\infty$, which is defined by,

$$\psi_\infty = \tan^{-1}\left(\frac{K_{II}^\infty}{K_I^\infty}\right) \tag{3.3}$$

where K_I^∞ and K_{II}^∞ are respectively the opening and shearing modes stress-intensity factors at the crack tip. For the DCB test, $\psi\infty = 0°$, hence a mode I in far field, while for the FPB test, $\psi\infty$ is close to 43°. Other mode mixes can be obtained by using sandwiched samples loaded under end notched flexure conditions (mode II). On the other hand, the nanoindentation techniques offer a much easier experimental procedure to characterize the interfacial properties, including conical indentation [Marshall and Evans, 1984] [Rossington *et al.*, 1984], micro wedge indentation [De Boer *et al.*, 1996A] [De Boer *et al.*, 1996B], wedge indentation [Begley *et al.*, 2000] [Schulze and Nix, 2000] [Vlassak *et al.*, 1997], cross-sectional indentation [Elizalde *et al.*, 2003] [Sanchez *et al.*, 1999] [Wang *et al.*, 2012], super layer nanoindentation [Bagchi *et al.*, 1994] [Bagchi *et al.*, 1996] [Kinbara *et al.*, 1998] [Volinsky *et al.*, 2003], among others. However, interpretation of the indentation data to obtain the interfacial adhesion properties from those tests requires a detailed knowledge of the mechanics of the interface which in many cases can only be obtained with finite element simulations. The cohesive fracture toughness can be easily measured in case of bulk materials and thick films according to ASTM standards [ASTME-399, 1987]. However, for thin film on substrates, there is neither a standard procedure nor commonly accepted methodology to follow. Stress-based test methods examine the stress state near the tip of the crack and include bending tests with pre-crack, scratch in consideration of a critical tensile stress, crack length and spacing, and indentation in consideration of the critical load and corresponding crack length, among others. However, difficulties in these methods lie in the formation of the pre-crack, the determination of the crack length and the critical stress state. These problems are not easily resolved due to the thickness dimension involved. On the other hand, energy-based methods focus on the system's energy state before and after fracture of the film occurs. This category comprises bending tests without a pre-crack in the film, buckling, indentation with chipping, tensile tests, and so on. Among the various available test methods, nanoindentation is the far most popular method to assess the cohesive fracture toughness of low-*k* dielectric films. For certain indenter tip geometry, the fracture toughness can be calculated based on the mechanical properties of the investigated material, the maximum applied indentation load and the observed crack length [Harry *et al.*, 1999A] [Shum *et al.*, 2004] [Kriese and Gerberich, 1999]. However, difficulty lies in the existence of a crack formation threshold, locating the indent and the determination of the crack length. Moreover, in case of thin films on substrates, an appropriate theoretical model should be developed to account for both film thickness and stress and preferably also film porosity. This is because typically the crack generated in these thin films by indentation does not show a half penny shape as is typically seen for bulk materials. Lawn and Winslaw have provided a detailed review of the indentation-induced cracking in bulk brittle materials [Lawn and Winslaw, 1975]. Recently, it was demonstrated that cube corner indentation can be used to determine the through-thickness cohesive toughness of a thin low-*k* film based on its mechanical properties [Yeap *et al.*, 2010], the maximum applied indentation load, the observed crack length and after applying the contact-mechanics based fracture model, as developed by Morris and Cook [2008]. This model is based on radial cracks driven by the indentation wedging mechanism and takes into account the effect of substrate constraint on the indentation crack shape, the contribution of film stresses to the overall fracture response and the influence of the substrate on the indentation stress fields.

3.3. *Mechanical stability of porous low-*k *dielectrics*

3.3.1. *Impact of porosity and pore morphology on mechanical stability*

The overall mechanical properties of porous low-*k* dielectrics are governed by the properties of the solid matrix and the pore morphology. Therefore, the successful integration of these materials requires a thorough understanding of how their mechanical stability varies with porosity and pore microstructure. Although much advancement has been made in this field over the past decades, there is still relatively little fundamental understanding of the impact of porosity and pore morphology on elastic properties and fracture behavior of porous low-*k* dielectrics. Extensive theoretical studies have been directed to modelling porosity effects on the elastic properties of materials, including micromechanics models [Hahsin, 1983] [Aboudi, 1991] [Christensen, 1979], semi-empirical formulae derived from the minimum solid area model [Rice, 1996A] [Rice, 1996B], or computationally solve the equations of elasticity for digital models of microstructure [Garboczi and Day, 1995] [Poutet *et al.*, 1996], among others. Micromechanics models can capture the effect of porosity and pore shape, but are generally limited to structures with only idealised pore geometries [Roscoe, 1952] [Christensen and Lo, 1979] [Benveniste, 1987]. On the other hand, the microstructure that corresponds to the minimum solid area model is not clearly defined, which limits the predictive and interpretative power of this method. As demonstrated by Roberts and co-authors, finite element simulations represent a very promising approach towards a more thorough understanding of the impact of porosity and pore geometry [Roberts and Garboczi, 2000] [Roberts and Garboczi, 2002]. These simulations allow various geometrical factors to be tuned separately to clarify their roles in the structure-property relationship. From the experimental point of view, investigating the impact of porosity and pore morphology is often complicated by the difficulty to separately control the porosity, pore structure and matrix properties during fabrication [Grill and Patel, 2001] [Urbanowicz *et al.*, 2009]. Moreover, there is little fundamental understanding on how the measured elastic modulus value is influenced by the stiffness of the matrix, porosity, pore morphology, and specific technique related effects. Recently, Vanstreels and co-authors demonstrated the intrinsic effect of porosity on the elastic properties and fracture behavior of ultralow-*k* dielectrics by using a special curing process that allows a separate control of porosity and matrix properties (Figure 3.3)

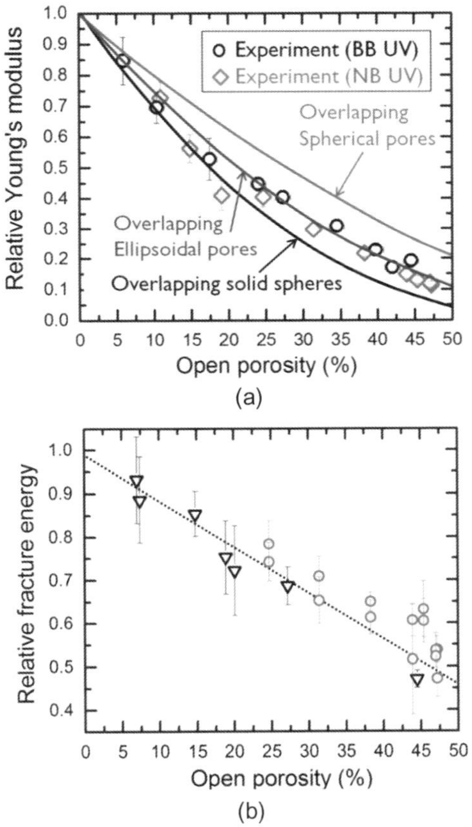

Fig. 3.3. (a) Relative Young's modulus values as a function of porosity, compared with finite element simulations of various pore microstructures. The open dots are experimental results using the same pore template structure but different UV treatment (BB UV: broad band UV lamp, NB UV: narrow band UV lamp); the solid lines are results from simulation models (overlapping spherical pores, overlapping ellipsoidal pores or overlapping solid sphere) published in the literature [Roberts and Garboczi, 2000]; (b) effect of porosity on the relative fracture energy of PECVD SiCOH films [Vanstreels *et al.*, 2012]); The open circular dots are experimental results using 4-point bending [Vanstreels *et al.*, 2012], while the open triangular dots are experimental results using the double cantilever beam technique [Li *et al.*, 2009A]. Reprinted with permission from K. Vanstreels, C. Wu, P. Verdonck and M. R. Baklanov. Appl. Phys. Lett., 101, 123109 (2012). Copyright 2012, AIP Publishing LLC.

[Vanstreels *et al.*, 2012]. The observed change in elastic modulus suggest that the microstructure evolves with increasing porosity and evolves from a structure containing mesopores interconnected with microporous channels at porosities below 25%, to one with heavily interconnected mesopores at higher porosity levels, as confirmed by positronium annihilation lifetime spectroscopy (PALS), ellipsometric pososimetry (EP), and diffusion experiments [Vanstreels *et al.*, 2013]. Moreover, it was found that the fracture energy decreases linearly with increased porosity. This was consistent with a planar through-pore fracture mechanism, in which the solid area that must be fractured changes linearly with film porosity and effects associated with pore shape, size, or distribution are screened out for a random pore microstructure. Li and co-authors have demonstrated an experimental procedure to separate out the effects caused by the matrix at different levels of porosity [Li *et al.*, 2009A]. Such a separation relies on the characterization of the density of various molecular bonds in the low-*k* dielectric film, and the experimentally established correlation between the various bond densities (Si-O-Si, Si-CH2-Si, Si-CH3, among others) determined from the fourier transform infra-red (FTIR) absorption spectra and mechanical and fracture properties for similar but dense low-*k* dielectrics. It was found that for dense OSG dielectrics, the elastic modulus correlates strongly with the densities of networking bonds (Si-O-Si and Si-CH2-Si) of the materials over a broad range of process parameters [Lin *et al.*, 2008]. The elastic modulus of the matrix, Ematrix, of a porous film can be determined from the knowledge of the network bond densities in the matrix. The intrinsic porosity effect on the elastic modulus can be expressed using a dimensionless function as $f(P) = E_{film}(P)/E_{matrix}(P)$, where P denotes film porosity (Figure 3.4). In a similar way, the intrinsic porosity effect on the cohesive fracture energy can be described by the dimensionless function g(p). They found that in case of OSG low-*k* dielectrics, the fracture energy decreases linearly with increasing porosity for porosity values up to 45%, which is in agreement with Vanstreels and co-authors [Vanstreels *et al.*, 2012]. Alternatively, Guyer and co-authors reported a non-linear porosity dependence of fracture energy for methyl-silsesquioxane (MSSQ)-based OSG coatings [Guyer *et al.*, 2006]. This was attributed to several factors including the non-uniformity of the MSSQ density through the coating thickness and contributions from frictional contact between the crack faces [Guyer *et al.*, 2006]. Although increasing the total porosity of low-*k* dielectrics leads to substantial reductions in film stiffness and fracture resistance, there is still little experimental evidence whether or not the stiffness

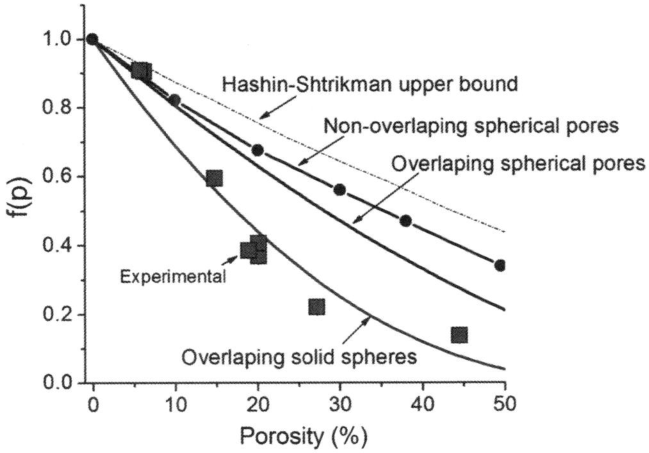

Fig. 3.4. Experimental values of the function, f(P), compared with finite element simulations of various pore microstructures and theoretical upper bound. Han Li, Youbo Lin, Ting Y. Tsui and Joost J. Vlassak. The effect of porogen loading on the stiffness and fracture energy of brittle organosilicates. Journal of Materials Research, 24, 01, 107 (2011). Reproduced with permission from Cambridge University Press.

and fracture properties of porous low-*k* dielectrics can be substantially changed through variations in pore architecture. This is because in realistic porous microstructures, the features of the pore morphology like pore size and shape, distribution and interconnectivity have no obvious definition. Simplified microstructural models are commonly used to experimentally extract those features, however, this may lead to inconclusive trends in the measured modulus-pore size relationship. Moreover, the dependence on pore structure is often overshadowed by the simultaneous scaling of pore size with porosity. Theoretical studies offer a good solution to avoid any interlinkage between pore size, porosity and matrix stiffness and hence studying the intrinsic effect of pore morphology on the elastic properties of materials. Roberts and co-authors used finite element simulations to separately tune various geometrical factors of different arbitrary pore structures [Roberts and Garboczi, 2000] [Roberts and Garboczi, 2002]. They concluded from their simulations that for a given porosity, the overlapping solid sphere model results in a relatively weak structure due to small solid contacts between the solid spheres and the highly interconnected porosity. Overlapping spherical pores provide high (near optimal) stiffness, while overlapping ellipsoidal pores tend to weaken a structure more than spherical pores. In a similar way, W. J. Lee and co-authors modelled and simulated different idealized porous microstructures using a random

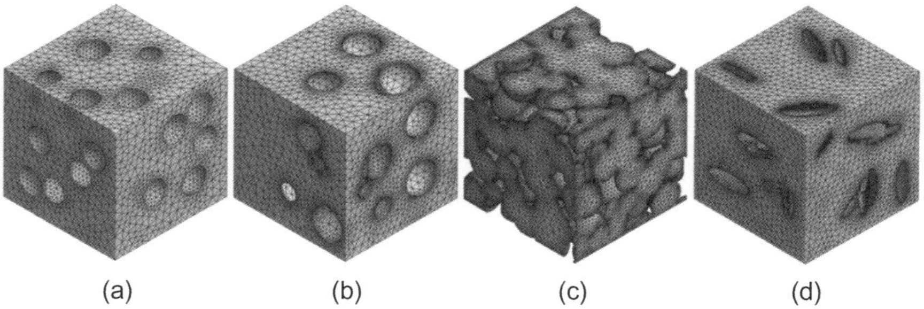

Fig. 3.5. geometries and meshes of RVE (representative volume element) models for porous microstructure, taken from [Lee *et al.*, 2011]; (a) isolated spherical pores; (b) overlapped spherical pores; (c) overlapped spherical solids and (d) overlapped ellipsoidal pores. Reprinted from Lee, W.J.; Cho, Y.J.; Lee, H.S.; Park I.M.; Park, Y.H. Effect of pore morphology on elastic, heat conduction and thermal shock fracture behaviors of porous ceramics, Procedia Engineering, 10, 2459 (2011), Copyright (2011), with permission from Elsevier.

sequential adsorption algorithm and a three dimensional finite element method (Figure 3.5) and found that the elastic properties can indeed be substantially influenced by their pore morphologies [Lee *et al.*, 2011]. Recently, Krishtab and co-authors experimentally showed that the cohesive strength of liquid phase self-assembly (LPSA) low-*k* dielectrics could be improved by tailoring its pore structure [Krishtab *et al.*, 2014A]. The main advantage of LPSA over PECVD low-*k* dielectrics is the ability to suppress the coupling between pore size and porosity, and thereby separating their contribution to the film strengthening. Two major distinctions between the dielectric films were found, i.e. type of bonding within the organosilica matrix and size of pores. A better cohesive strength was found for dielectric films with smaller pores despite a remarkably lower interconnectivity of its organosilica backbone, as compared to films with larger pores, thereby emphasizing the prevailing role of the structural aspects in the tested films.

3.3.2. *Impact of network structure and UV radiation on mechanical stability*

Lowering the k-value of organic-inorganic hybrid low-*k* dielectrics is achieved by introducing organic terminal groups into the silica-like backbone structure, which leads to a decrease in the film density and lowering the bond polarizability, both beneficial for the dielectric performance. However, from a mechanical perspective, these terminal groups disrupt the covalent molecular network and degrade the mechanical integrity of the materials. Experiments have demonstrated that the elastic modulus and fracture toughness of OSG low-*k* materials can be improved by an ultraviolet

(UV)-assisted-thermal curing process [Iacopi *et al.*, 2006] [Urbanowicz *et al.*, 2010] [Urbanowicz *et al.*, 2011]. Such improvements are associated with a more densely cross-linked network structure attained by removing terminal groups and by forming networking bonds. However, not only the UV absorption, but also the direct photo dissociation of the chemical bonds and bond rearrangement by UV light determines the mechanical robustness of these films after UV-curing. In the first approximation, one can say that OSG low-*k* materials contain energetically strong Si-O bonds as well as weaker C-H and Si-CH3 bonds. The Si-O bonds are mostly responsible for the mechanical stability of the films, while the less polar Si-CH3 bonds provide the hydrophobic properties and their lower dielectric constant. Direct Si-CH3 photo dissociation results in more cross-linking opportunities for the film skeleton due to generation of active silyl states, which can react with another silyl state or silanol (Si-OH), which then leads to the creation of Si-O-Si links in the skeleton. Recent work of Urbanowicz and co-authors showed that the presence/absence of porogen molecules (usually cyclic hydrocarbons) in OSG films prior to UV-assisted-thermal curing has a strong impact on the UV-curing mechanism and the obtained mechanical properties after UV-cure [Urbanowicz *et al.*, 2010] [Urbanowicz *et al.*, 2011]. This was explained in terms of changes in absorption induced by porogen removal and an additional cross-linking mechanism due to direct Si-CH3 bond photo dissociation. Improved mechanical properties and ultralow dielectric constant could be obtained by controlled decomposition of the porogen molecules prior to the UV hardening step. M. S. Oliver and co-authors used a combined computational and synthetic approach to establish fundamental scaling laws for the elastic response and fracture properties of hybrid glasses [Oliver *et al.*, 2010]. The specific glasses that were investigated are ethane-bridged oxycarbosilane (Et-OCS) and methylated ethane-bridged oxycarbosilane (Et-OCS(Me)), as illustrated on Figure 3.6. They show that Si-X-Si connectivity provides the best measure of network connectivity because it provides an absolute measure that is independent of the precursor structure. The scaling of bulk modulus with Si-X-Si connectivity could be described by a simple mean-field model that was based on rigidity percolation occurring at Si-X-Si connectivity (Figure 3.6b). The fracture energy of brittle hybrid glasses scales in a similar way as the number of bonds broken during fracture (Figure 3.6c). Therefore, while the elastic properties are very sensitive to changes in connectivity induced during UV curing, the cohesive crack can meander through the network on a path towards regions of low connectivity and low density

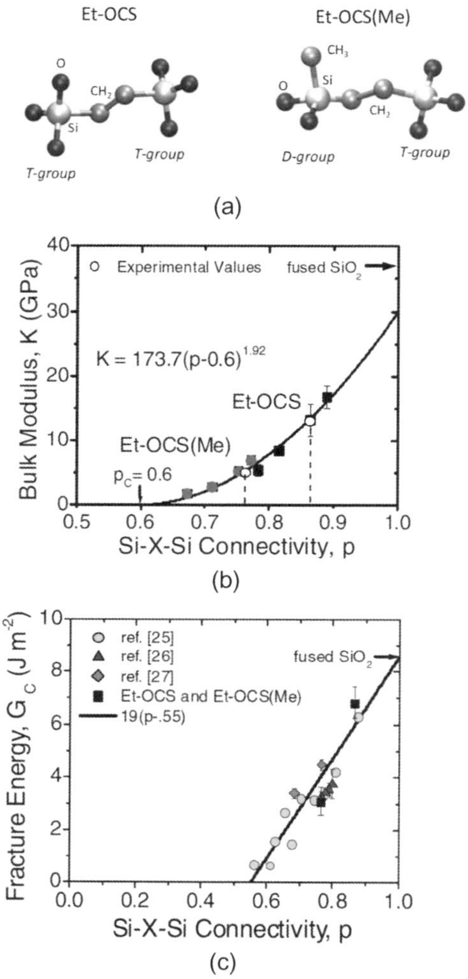

Fig. 3.6. (a) Molecular structure of Et-OCS and Et-OCS(Me) organosilicate glasses showing the basic structural units; (b) Model predictions and experimental measurements of bulk modulus (K) plotted versus Si-X-Si connectivity (p). The solid lines are power-law fits to the model data; (c) Experimentally measured fracture energies (Gc) of various hybrid glasses are plotted against the Si-X-Si connectivity; Figures were taken from [Oliver *et al.*, 2010]. The references cited in the figure are taken from [Oliver *et al.*, 2010] with permission from John Wiley and Sons.

within the glass network and is therefore insensitive to UV cure. This was experimentally confirmed by Gage and co-authors, who showed that UV cure exposure results in a higher degree of film connectivity in the material and significantly enhances the interfacial adhesion strength, but barely influences the fracture toughness of PECVD carbon-doped oxide films [Gage *et al.*, 2008]. H. Li and co-authors [Li *et al.*, 2009A] and Y. Lin and

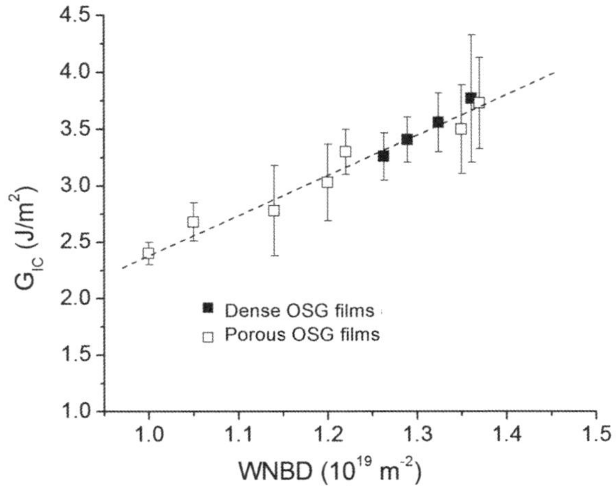

Fig. 3.7. Fracture energy of dense and porous OSG as a function the weighted networking bond density (WNBD). (Figure was taken from Han Li, Youbo Lin, Ting Y. Tsui and Joost J. Vlassak. Journal of Materials Research, 24, 01, 107 (2011), Reproduced with permission from Cambridge University Press.

co-authors [Lin *et al.*, 2008] demonstrated that the fracture energy of OSG materials, measured under mode-I condition using double-cantilever beam technique, scales linearly with the total number density of the two main networking bonds in an OSG material (Figure 3.7). This is because fracture of these materials is a bond breaking process and the energy dissipated during fracture scales with the number of broken networking bonds and with the energy of the bonds. In literature, molecular dynamics (MD) methods have been used to develop atomistic models of low-*k* materials and the mechanical properties as well as the fracture behavior were analyzed [Gao *et al.*, 2009] [Cuong *et al.*, 2011] [Tajima *et al.*, 2006] [Tajima *et al.*, 2007] [Yuan *et al.*, 2007] [Yuan *et al.*, 2008]. H. Li and co-authors used molecular dynamics simulations to investigate the fundamental structure–property relationships of OSG low-*k* with a primary focus on the mass density and elastic properties [Li *et al.*, 2011]. To examine the role of the bridging units and terminal groups, two idealized classes of molecular models were considered that are built out of the backbone structure of amorphous silica. The first class of models (type-I OSG) consists of a Si–O–Si and Si–CH2–Si network without terminal groups. These models are created by replacing a fraction of the bridging oxygen atoms in the silica backbone structure with methylene groups. The second class of models (type-II OSG) have only Si–O–Si crosslinking bonds in the network, but terminal groups (–OH or –CH3) are allowed. These models were created by breaking the Si–O–Si

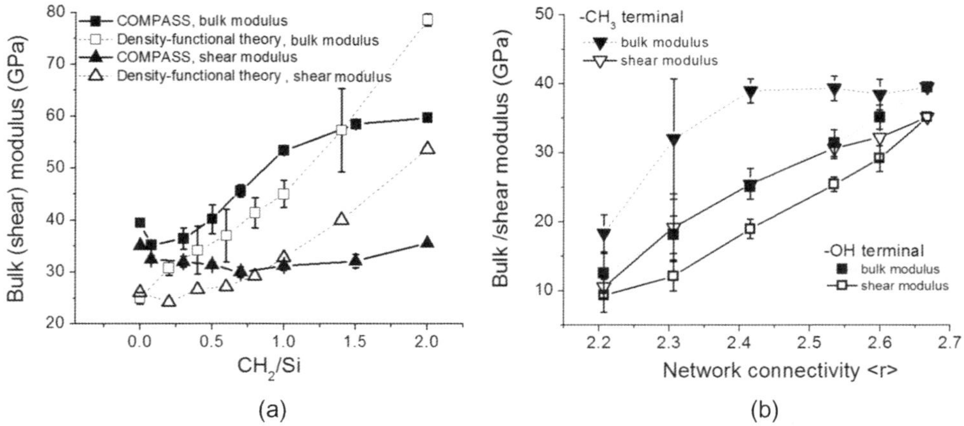

Fig. 3.8. (a) Bulk modulus and shear modulus of type-I OSG as a function of CH2/Si ratio; (b) Bulk and shear modulus of type-II OSG as a function of network connectivity $< r >$ (Figure was reprinted from Han Li, Jan M. Knaup, Efthimios Kaxiras, Joost J. Vlassak, Stiffening of organosilicate glasses by organic cross-linking, Acta Materialia, 59,1, 44 (2011). Copyright (2011), with permission from Elsevier.

bonds and replacing the oxygen atom with two terminal groups (either –OH or –CH3). The results show that the incorporation of –CH2– as bridging unit into the silica network considerably affects the elastic properties of the resultant OSG low-*k*. The greater bending rigidity of Si–C–Si relative to Si–O–Si was found to be responsible for the increased bulk modulus of type-I OSG with increasing –CH2– population (Figure 3.8a). Type-II OSG materials contain a number of terminal groups (–OH, –CH3) interrupting the network structure of amorphous silica. The degree of interruption can be described by the network connectivity number $\langle r \rangle$, defined as the average coordination number per network forming atom [Thorpe, 1983]. The MD simulations show that the bulk modulus and the shear modulus of type-II OSG are reduced significantly with increasing fraction of terminal groups or decreasing $\langle r \rangle$ (Figure 3.8b).

3.3.3. *Impact reactive species in the environment on mechanical stability*

Besides the intrinsic fracture properties of low-*k* dielectrics, time-dependent fracture behavior is another concern for chip fabrication and long term reliability [Guyer and Dauskardt, 2004] [Vlassak *et al.*, 2005]. Moisture-assisted crack growth is a time-dependent mode of failure that occurs at loads well below those required for critical fracture [Lane *et al.*, 2001] [Lane *et al.*, 2000] [Dauskardt *et al.*, 1998] [Cook and Liniger, 1999]. Many materials are known to be susceptible to moisture-assisted cracking including ceramics, glasses, polymers, among others. It is well known that the fracture behavior of OSG low-*k* can be significantly degraded due to the ingress of water and other reactive species [Lin *et al.*, 2007] [Tsui *et al.*, 2006] [Li *et al.*, 2009] [Lin *et al.*, 2009] [Guyer *et al.*, 2007] [Guyer and Dauskardt, 2004]. This is because the strained Si-O bonds in their backbone structures at crack tip can react chemically with water molecules to assist the bond rupture process. More recently, the impact of water diffusion on the fracture behavior of nanoporous dielectric film stacks were studied for various OSG materials [Lin *et al.*, 2007] [Tsui *et al.*, 2006]. Li and co-authors studied various nanoporous OSG dielectrics following the "soak-then-test" approach [Li *et al.*, 2009]. They demonstrated that the adhesion of the OSG/SiCN interface decreases over time (Figure 3.9), while the cohesive fracture energy of the nanoporous OSG layer remains largely constant. This was explained by the tendency of water molecules to diffuse along the OSG/SiCN interface rather than through the bulk of the OSG, and was attributed to the less hydrophobic character of the interface. Moreover, degradation of the cohesive fracture energy was observed when the OSG film stack was exposed to aqueous solutions with organic additives that enhance wetting of the OSG [Lin *et al.*, 2009] [Guyer *et al.*, 2007]. Since various processing steps may cause loss of hydrophobicity to certain extent it should therefore be pointed out that common organic additives such as isopropyl alcohol or acetic acid to water could drastically

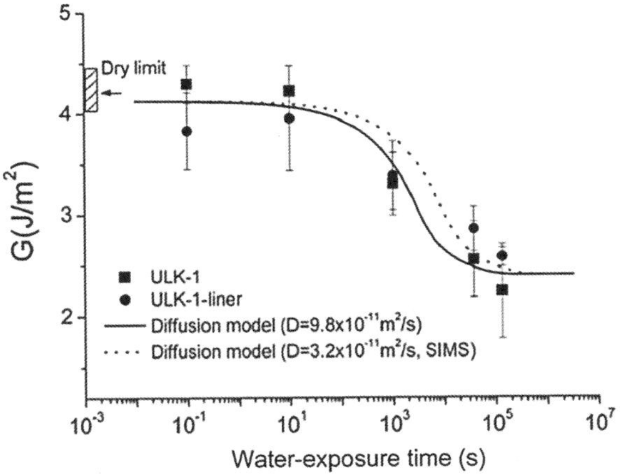

Fig. 3.9. Interfacial adhesion energy of the OSG/SiCN interface as a function of water immersion time for samples ULK-1 (square) and ULK-1-liner (circle) (Figure was taken from [Li *et al.*, 2009B]). Reprinted with permission from Han Li, Ting Y. Tsui, Joost J. Vlassak. Water diffusion and fracture behavior in nanoporous low-*k* dielectric film stacks. *J. Appl. Phys.*, 106, 3, (2009) Copyright 2009, AIP Publishing LLC.

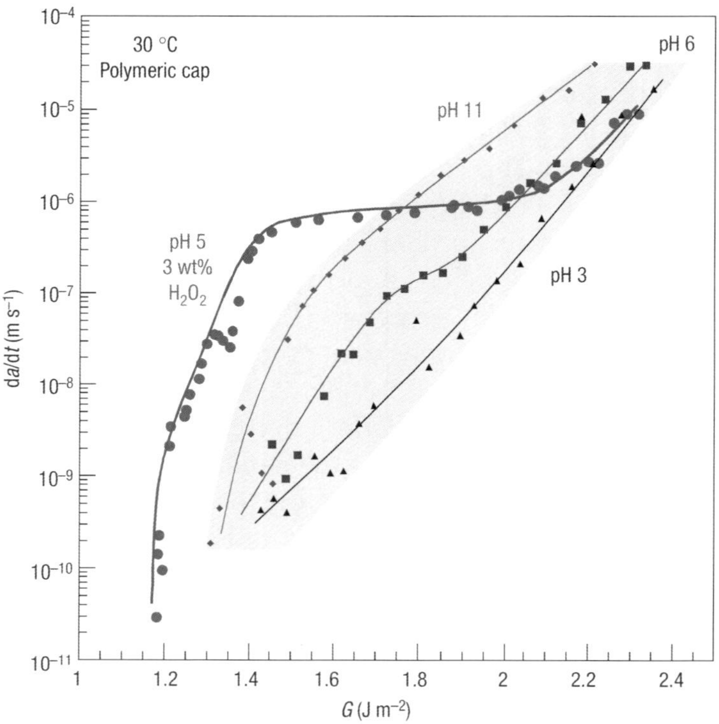

Fig. 3.10. Crack propagation rate versus the applied strain energy release rate (G) measured in hydrogen peroxide solution with different pH. The crack growth behavior clearly differs depending on the solution pH (Figure was taken from [Guyer and Dauskardt, 2004B]). Reprinted by permission from Macmillan Publishers Ltd: (Eric P. Guyer and Reinhold H. Dauskardt. Fracture of nanoporous thin-film glasses. Nature Materials, 3, 1 (2003), copyright (2003).

change the diffusion behaviour. E. P. Guyer and R. H. Dauskardt show that besides moisture, also other environmental species can induce accelerated cracking [Guyer and Dauskardt, 2004] [Guyer and Dauskardt, 2005]. The effect of the solution chemistry on the fracture of nanoporous glasses exposed to aqueous solutions was investigated. Crack growth rates were inhibited by acidic solutions and enhanced by basic solutions (Figure 3.10). They concluded that the successful implementation of nanoporous glasses depends on the selection of appropriate chemistries that inhibit crack-growth reaction kinetics.

3.3.4. *Low-k dielectrics for future technology nodes*

Integration is the deciding factor driving the selection of the most promising materials for a given technology node. During the past decades, PECVD technology has been the method of choice for depositing silicon dioxide (SiO2), fluorine-doped oxides (F-SiO2), carbon-doped oxides (OSG) and porous carbon-doped oxides (p-OSG). Until now, this technology is still the most popular choice because of the good compatibility with existing ultra large scale integration (ULSI) technology and the possibility to deposit relatively good quality films with dielectric constant in the range between 3.5 and 2.4. In order to achieve dielectric constants of 2.4 and beyond, the introduction of additional artificial porosity into the low-*k* dielectrics (p-OSG) becomes crucially important. However, the addition of porosity exacerbated already known integration issues due to the difficulty of combining low-*k* values with other manufacturing requirements such as sufficient mechanical strength, good chemical stability, reduced pore size, and hydrophobicity [Hoofman *et al.*, 2007]. At this moment, the integration of PECVD based OSG films with dielectric constants below 2.4 are still under research. It seems however that there is a limit to values around 1.9–2.0 due to pore collapsing at higher porogen loadings [Urbanowicz *et al.*, 2011] [Verdonck *et al.*, 2011] [Verdonck *et al.*, 2014]. As a result, the list of potentially viable low-*k* candidates has narrowed substantially over the years. For advanced and future technologies nodes below 10 nm, the PECVD technology will no longer be able to satisfy all the integration requirements due to the uncontrollable process of pore formation and the porogen residues that are formed during the UV curing process. As an alternative, spin-on self-assembling technology is considered as more promising for future generations of ultralow-*k* dielectrics due to an easier removal of template and a better control over pore-organization compared to PECVD. Potential improvements of the mechanical stability could be obtained by tailoring the pore structure and/or by skeleton engineering, i.e. when the material is ordered or has high matrix connectivity. Potential low-*k* candidates for future generations include periodic mesoporous organosilicas (PMO), metal-organic frameworks (MOF), among others.

PMO are ordered mesoporous silica materials of which at least one siloxane bond in the matrix is replaced by an organic bridge [Asefa *et al.*, 1999] [Landskon *et al.*, 2003] [Hoffmann *et al.*, 2006] [Hunks and Ozin, 2005] [Hatton *et al.*, 2005] [Mizoshita *et al.*, 2011]. PMO are usually amorphous, but their pores and pore walls are arranged in a structural order. They are similar in chemical composition compared to OSG with the advantage that low polarizable organic groups are integrated in the silica matrix. Especially, PMO with the simplest organic bridges such as ethane and methane are extremely interesting for low-*k* applications because they have the lowest polarizable organic bridge. The great advantage of PMO over pure silica ordered mesoporous materials is the presence of uniformly distributed organic functional groups in the walls and the fact that they are hydrothermally and mechanically more stable. By varying synthesis conditions, precursors and surfactants, it is possible to adjust the pore size, porosity, surface area, ordering, wall thickness and morphology of the PMO. These characteristics make them promising candidates for different applications including microelectronics. Typically PMO materials are synthesized by hydrolysis and condensation of bridged organosilanes in the presence of a surfactant which acts as structure directing agent. After an ageing period, the surfactant can be removed by an extraction procedure or by a thermal treatment under an inert atmosphere, leaving behind a porous PMO material. Compared to PECVD materials, PMO materials enable a better control of the pore and skeleton structure and easier removal of the template. Moreover, the porous structures of these materials are formed without the use of sacrificial porogen and without the risk of residue formation during the UV curing. At this moment, several groups reported the papers related to PMO and low-*k*, in which the dielectric, mechanical and hydrophobic properties were studied [Goethals *et al.*, 2010] [Goethals *et al.*, 2012] [Van Der Voort *et al.*, 2013]. Although there is some variation in the reported k-values of PMO films, the microelectronic industry recognized that PMOs are one of the most promising ultralow-*k* materials to integrate in actual devices. However, still a lot of future research is required, including plasma-damage effects, adhesion, electrical characteristics and reliability.

Metal-organic frameworks have received much scientific and technological interest because they possess a well-defined nanoscale pore size, crystallinity and extremely low dielectric constant, i.e. a combination which is barely accessible for traditional low-*k* dielectrics. MOF are nanoporous, crystalline materials that consist of metal ions that are connected by organic linkers [Shekhah *et al.*, 2011] [Meek *et al.*, 2011] [Ma and Moulton, 2011] [Corma *et al.*, 2010] [Hu and Zhang, 2010] [Davis, 2002]. By varying the metal ion or linker, various sizes and shapes of channels or cavities can be obtained [Eddaooudi *et al.*, 2002]. The interest in these materials first appeared after the theoretical paper by K. Zagorodniy and co-authors, who calculated the dielectric constant of a series of MOF based on $Zn_4O(CO2)_4$ units connected by different organic linkers and found that all the investigated MOF have k-values below 2.0 with good mechanical properties [Zagorodniy *et al.*,

2010]. However, recent experimental evaluation by S. Eslava and co-authors (ZIF-8 films) [Eslava *et al.*, 2013] and M. Usman and co-authors (Sr-based MOF films) [Usman *et al.*, 2014] suggest that the theoretically estimated dielectric constants of these materials might be underestimated because of the ionic contribution that was omitted in the work of Zagorodniy and co-authors. Also, the presence of metal ions in the low-*k* material may also raise questions about the electrical reliability (leakage and breakdown) of these materials. Furthermore, appropriate depositions methods are required to provide high quality thin dense homogeneous films with minimal surface roughness and good bulk crystallinity. From this respect, surface-anchored metal-organic framework (SurMOF) films are one step closer to satisfy the integration conditions. To date, HKUST-1 (also known as Cu-BTC) is by far one of the most investigated SurMOF materials due to the relative simple preparation and the presence of open metal sites which makes it advantageous for application in catalysis and gas storage. Recently, it was demonstrated that these SurMOF materials have sufficiently good mechanical properties [Bundschuh *et al.*, 2012], with Young's modulus close to 10 GPa, which is in very good agreement with theoretical predictions for these materials [Tafipolsky *et al.*, 2010]. However, some concerns are related to the relatively large spread that is found in the mechanical properties across those films, surface roughness and the intrinsic poor hydrophobicity of the HKUST-1 matrix material. Even a small amount of water within the material can eliminate any advantage brought by low density and predicted low electronic polarizability of MOF, as was recently reported by Krishtab [Krishtab *et al.*, 2014B]. Finally, the low chemical stability of MOF materials makes it difficult to integrate them in damascene technology. Nevertheless, these materials may still find an application in subtractive technology that is explored in sub-7 nm technology nodes. Subtractive technology is an alternative Cu patterning process to that of conventional damascene technology, which includes Cu etch and low-*k* replacement activities.

Other potential candidates for future generation low-*k* dielectrics include hyper cross-linked polymer (HCP) networks [Schade *et al.*, 2014], surface-anchored covalent organic frameworks (SurCOF) materials [Tsotsalas *et al.*, 2014], MOF/fluorocarbon composites, Al and Zn based MOF using solvent free synthesis [Stassen *et al.*, 2013] and polyimide-polyoxometalate (PI-POM) nanocomposites [Keshtov *et al.*, 2012], among others.

3.3.5. *Impact of mechanical stability on chip package interaction*

Chip package interaction (CPI) has not been a serious problem for chips with SiO2 as the interlayer dielectric material. Unfortunately, for packaging assembly of Cu/low-*k* chips, CPI has gained a lot of importance in the microelectronics industry due to the increased number of reported reliability problems at integrated circuit (IC) level [Wang *et al.*, 2005] [Iacopi *et al.*, 2004]. During package assembly and subsequent reliability tests or even under operation conditions, the thermo-mechanical deformation of the package can induce large stresses which can be directly coupled into the BEOL to initiate and drive cohesive and/or adhesive crack propagation in multilevel Cu/low-*k* interconnect structures. The nature of CPI is quite complex and its reliability impact on BEOL has recently generated extensive interest [Mercado *et al.*, 2002] [Mercado *et al.*, 2003] [Wang *et al.*, 2003] [Zhao *et al.*, 2004] [Landers *et al.*, 2004] [Uchibori *et al.*, 2006] [Liu *et al.*, 2007] [Liu *et al.*, 2011] [Garner *et al.*, 2005] [Chin *et al.*, 2010] [Auersperg *et al.*, 2010] [Kuechenmeister *et al.*, 2012] [Ryan *et al.*, 2012]. Especially for the advanced technology nodes, structural integrity of Cu/low-*k* interconnect structures is a major reliability concern for the semiconductor industry. The nature of the problem depends on the thermal and chemical treatments used in the different fabrication steps of the chip. Although the assembly or test temperatures of the package are considerably lower than the chip processing temperatures, the thermo-mechanical interaction between the chip and the package can exert additional stresses onto the Cu/low-*k* interconnects. The origin of the problem is the weak thermo-mechanical properties of low-*k* dielectrics and the large thermal stress that is induced by package deformation due to the mismatch in thermal expansion between the silicon die with Cu/low-*k* interconnects and the organic substrate in the package. Therefore, and early assessment of the mechanical stability of low-*k* and its impact on the integrity of BEOL structures is required to assure reliability during chip packaging and under field conditions. A common way to study the different driving forces for CPI related failure modes is by using three dimensional finite element models (FEM). However, for advanced flip-chip packages, the scale differences between package and chip level features could be up to 5 order of magnitude, which makes CPI analysis computationally challenging. Therefore, there is a great need for global-local modeling (also known as the sub-modeling approach), multiple substructures as well as fracture and damage mechanics using cohesive zone elements, visco-elasticity, plasticity and creep of materials. Recently, X. Zhang and co-authors compared the CPI induced delamination for different porous

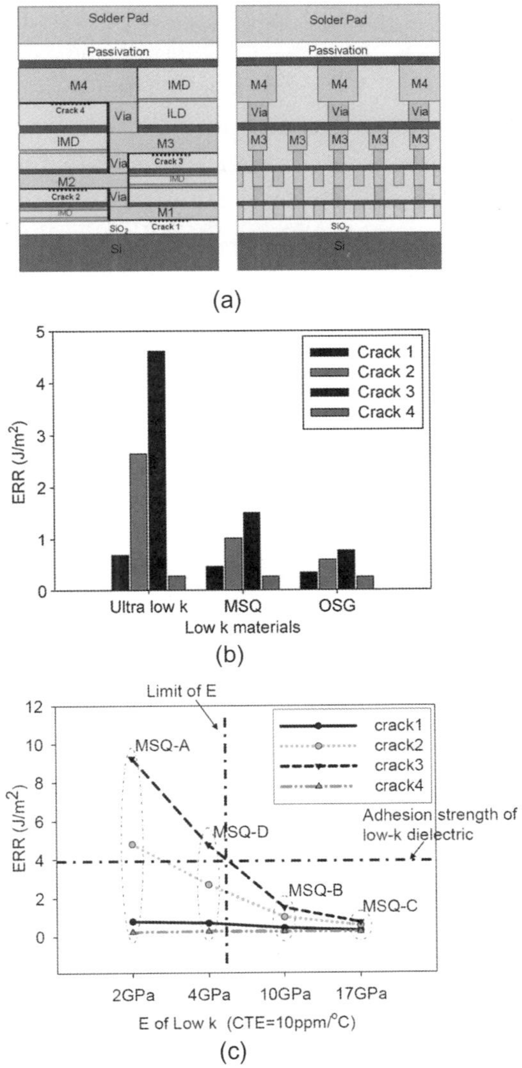

Fig. 3.11. (a) FEM model of 3-D 4-layer interconnect; (b) Comparison of CPI-induced energy release rates in the four-level interconnects with low-*k* and ultralow-*k* ILDs.; (c) Effect of low-*k* material properties on ERR. (Figure taken from [Zhang *et al.*, 2012]).

low-*k* materials with different thermo-mechanical properties by using the multilevel sub-modeling approach [Zhang *et al.*, 2012]. The energy release rates of interfacial cracks in 3-dimensional (3D) four-level interconnect finite element models were calculated by introducing initial cracks at several relevant interfaces as shown on Figure 3.11a. The energy release rates were found to increase significantly from OSG to methyl silsesquioxane (MSQ) and became even higher for ultralow-*k* (ULK) dielectrics (Figure 3.11b). The corresponding elastic modulus of these materials is 17 GPa (OSG), 10 GPa (MSQ) and 4 GPa (ULK), respectively. They showed that by reducing the elastic modulus, the corresponding low-*k* interconnect structure would need increased adhesion energy in order to maintain a mechanical stability equivalent to that of low-*k* dielectrics with higher elastic modulus, hence CPI will be a major concern to interconnect reliability with the implementation of future ultralow-*k* dielectric materials. Furthermore, they concluded that for low-*k* dielectrics, the elastic modulus was the dominant property with regard to mechanical reliability and that in order to achieve reliable low-*k* integration, a dielectric material with elastic modulus larger than 5 GPa is preferred. This is illustrated on Figure 3.11c, where likelihood of low-*k* delamination under CPI increased rapidly as the E of low-*k* dielectrics was reduced. Even more recently, M. Gonzalez and co-authors calculated the energy release rates for predefined cracks in a 3D finite element model of a BEOL containing 7 metal layers [Gonzalez *et al.*, 2013]. The results from these simulations show that as the elastic modulus of the low-*k* decreases more deformation is induced in the vias, thereby increasing the via stress (Figure 3.12b). Moreover, stiffer low-*k* makes the overall structure less deformable, resulting in lower ERR, as shown in Figure 3.12c. Also, the ERR and stress in the via increased nonlinearly with reduced low-*k* stiffness. Finally, they also demonstrate that combining finite element simulations and *in-situ* electrical measurement of stress sensors during a BABSI test are an ideal combination to quantitatively compare the strength of BEOL layers. Despite the capability of the sub-modeling technique to capture the effect of stiffness of low-*k* dielectrics on the rigidity of the BEOL under CPI loading, the main disadvantage is that the results depend on the displacements at the boundary of the sub-model taken from global model simulation result, which can cause misleading results if the global model is not precise enough. Additionally, the mechanical response of newly developed low-*k* materials in BEOL may also raise the question if these materials can still be handled as homogeneous materials. As a solution to this problem, molecular dynamics (MD) techniques increased in popularity to simulate and understand the mechanical behavior of these materials [Chen *et al.*, 2005] [Fan *et al.*, 2006] [Matsumoto *et al.*, 2006] [Dermitzaki *et al.*, 2008]. However, due to the huge amount of computational resources necessary for appropriate MD simulations, direct incorporation of homogenized MD-models inside global-local FEM models

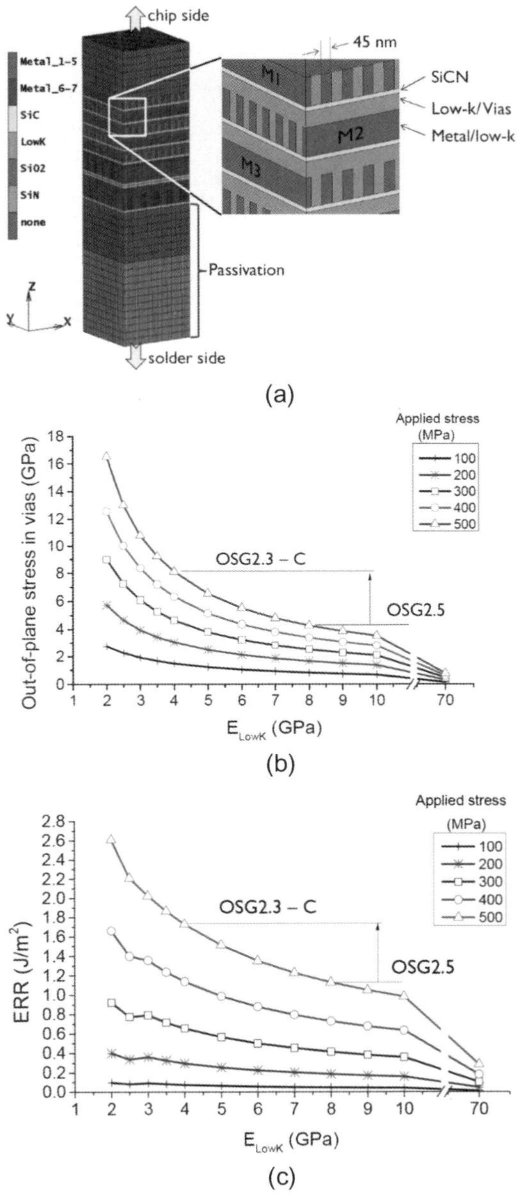

Fig. 3.12. (a) A 3D FEM of a Representative volume element (RVE). The model represents a BEOL containing 7 metal layers and passivation; (b) Influence of Elastic Modulus of low-*k* on the induced stress in the metal vias; (c) Influence of Elastic Modulus of low-*k* on the induced ERR at the crack tip. (Figure taken from [Gonzalez *et al.*, 2013] and re-drawn by permission of M. Gonzalez).

is not straightforward. Experimental extraction of mechanical properties for use in FEM is therefore still preferred at the moment. However, serious issues are connected with the collection of appropriate material properties in the miniaturized range, elastic modulus, initial yield stress, hardening, fracture properties, adhesion, in particular. Although these properties are commonly measured on thin film level, they do not sufficiently represent the failure behavior and structure stability of multilevel interconnects after integration and CPI. Therefore, the semiconductor industry needs reliable, reproducible testing methods and strategies to monitor these parameters, not only for low-*k* dielectric films, but also for complete interconnect stacks. The most commonly used experimental techniques at the moment to qualify the structure stability of multilevel interconnects are listed below. Bump-Assisted BEOL Stability Indentation-Scratch Technique (BABSI) is commonly used as an early screening method to assess the mechanical robustness of BEOL interconnect structures before assembly and packaging processes are performed [Geisler *et al.*, 2011A], [Geisler *et al.*, 2011B] and [Geisler *et al.*, 2013]. This test has the capability to determine the conditions at the onset of fractures in a sensitive, fast and straightforward way. H. Geisler and co-authors demonstrated that *in-situ* BABSI tests can be used to study crack initiation. However, due to the complexity of the deformation induced by the shear displacement of the indenter tip, the local stresses underneath the copper bump cannot easily be quantified. Recently, K. Vanstreels and co-authors demonstrated an advanced experimental BEOL stability test by coupling a BABSI test with *in-situ* monitoring the response of n-FET stress sensors below the Cu bump [Vanstreels *et al.*, 2014]. A good agreement was found between the applied loads to the BEOL stack, the response of stress sensors below the Cu bump and FEM simulations. Recently, Hsing and co-authors used a microprobe metrology system to experimentally measure how interconnect stacks with different dielectric porosities behave under various shear loading conditions and a wide range of temperatures [Hsing *et al.*, 2011]. Other early screening methods include Single Bump Wedge Shear Test [Liu *et al.*, 2011], High load indentation test [Usami *et al.*, 2013], and cross sectional indentation [Ocana *et al.*, 2006], among others. Each of these techniques provides some unique information about the susceptibility of the BEOL levels to package-induced mechanical failures. Combining the outcome of those techniques, combined with multilevel sub-modeling enables a more thorough understanding of relevant CPI interactions.

3.4. *Concluding remarks*

This paper reviews the mechanical and fracture properties of porous ultralow-*k* dielectrics and provides an overview of the most commonly used experimental techniques to characterize the corresponding properties on thin films and multilevel interconnect stacks. It was shown that the mechanical and fracture properties of porous ultralow-*k* dielectrics are intimately linked with porosity, network structure and deposition technologies, while fracture properties of low-*k* are also sensitive to reactive species in the environment. Therefore, the low-*k* survivability upon integration and CPI is not a simple factor of mechanical properties and fracture properties, but it is rather a combination of all these properties, together with a relatively high mechanical or thermo-mechanical load and environmental effects that will determine its faith. For future generations of ultralow-*k* dielectrics, potential improvements of the mechanical stability could be obtained by tailoring the pore structure and/or by skeleton engineering using self-assembly technology. The main potential low-*k* candidates for future generations include PMO and MOF, among others.

References

Aboudi, J. (1991), *Mechanics of Composite Materials: A Unified Micromechanical Approach*, Elsevier, Amsterdam.

Asefa, T.; MacLachlan, M. J.; Coombs N.; Ozin, G. A. (1999), *Nature*, **402**, p. 867.

ASTME-399 (1987), *Standard Test for Plane Strain Fracture Toughness of Metallic Materials American Society for Testing and Materials*, Philadelphia, PA.

Auersperg, J.; Vogel, D.; Lehr, M.U.; Grillberger, M.; Michel, B. (2010), *Proc. EuroSimE Conference* 2010, p. 125.

Bagchi, A.; Lucas, G.; Suo, Z. and Evans, A. (1994), *J. Mater. Res.*, **9**(7), p. 1734.

Bagchi, A.; Evans, A. (1996), *Thin Solid Films*, **286**, p. 203.

Bailey, S.; Mays, E.; Michalak, D. J.; Chebiam, R.; King, S.; Sooryakumar, R. (2013), *J. Phys. D: Appl. Phys.*, **46**, 045308.

Baklanov, M. R.; Green, M. and Maex, K. (2007), *Dielectric Films for Advanced Microelectronics*, Chapter 1, Wiley, West Sussex.

Begley, M. R.; Mumm, D. R.; Evans, A. G. and Hutchinson, J. W. (2000), *Acta Mater.*, **48**, p. 3211.

Bellido-Gonzalez, V.; Stefanopoulos, N.; Deguilhen, F. (1995), *Surf. Coat. Technol.*, 74–75, p. 884.

Benveniste, Y. (1987), *Mechanics of Materials*, **6**(2), p. 147.

Bhattacharya, A. K.; Nix, W. D. (1988), *Int. J. Solids Struct.*, **24**, p. 1287.

Britvec, S. J. (1973), *The Stability of Elastic Systems*, Pergamon Press, New York, 1973.

Bundschuh, S.; Kraft, O.; Arslan, H. K.; Gliemann, H.; Weidler, P. G.; Wöll, C. (2012), *Appl. Phys. Lett.*, **101**, 101910.
Burkey, D. D. and Gleason, K. K. (2003), *J. Appl. Phys.*, **93**, p. 5143.
Callister, Jr., W. D. (2003), *Materials Science and Engineering An Introduction*, 6th ed., New York, Wiley, 2003.
Carlotti, G.; Doucet, L. and Dupeux, M. (1996), *J. Vac. Sci. Technol.*, B **14**, p. 3460.
Chapelon, L. L.; Vitiello, J.; Neira, D.; Torres, J.; Royer, J. C.; Barbier, D.; Naudin, F.; Tas, G.; Mukundhan, P. and Clerico, J. (2006), *Microelectronic Engineering*, **83**, p. 2346.
Charalambides, P. G.; Lund, J.; Evans, A. G. and McMeeking, R. M. (1998), *J. Appl. Mech.*, **56**, p. 77.
Chen, N.; Lusk, M. T.; van Duin, A. C. T.; Goddard III, W. A. (2005), *Physical Review B*, **72**, 085416.
Chen, X.; Vlassak, J. J. (2001), *J. Mater. Res.*, **16**, p. 2974.
Chen, Z.; Cotterell, B.; Wang, W. (2002), *Eng. Fract. Mech.*, **69**, p. 597.
Chin, M.; Marathe, A. (2010), *Proc. IRPS*, 2010, 4C.6.
Christensen, R. M. (1979), *Mechanics of Composite Materials*, Wiley, New York.
Christensen, R. M.; Lo, K. H. (1979), *J. Mech. Phys. Solids* **27**(4), p. 315.
Cook, R.; Liniger, E. (1999), *J. Electrochem. Soc.*, **146**, p. 4439.
Corma, A.; Garcia, H.; Llabres i Ximena, F. X. (2010), *Chem. Rev.*, **110**, p. 4606.
Cotterell, B.; Chen, Z. (2000), *Int. J. Fract.*, **104**, p. 169.
Cox, L. C. (1988), *Surf. Coat. Technol.* **36**, p. 807.
Cuong, N. T.; Otani, M.; Okada, S. (2011), *Phys. Rev. Lett.*, **106**, 106801.
Dauskardt, R. H.; Lane, M.; Ma, Q.; Krishna, N. (1998), *Eng. Fract. Mech.*, **61**, p. 141.
Davis, M. E. (2002), *Nature*, **417**, p. 813.
De Boer, M. P. and Gerberich, W. W. (1996A), *Acta Mater.*, **44**, p. 3169.
De Boer, M. P. and Gerberich, W. W. (1996B), *Acta Mater.*, **44**, p. 3177.
Dermitzaki, E.; Wunderle, B.; Bauer, J.; Walter, H.; Michel, B. (2008), *Proc. 9th Conf. EuroSimE*, 2008 p. 204.
Doerner, M.; Nix, W.D. (1986), *J. Mater. Res.*, **1**, 601.
Drory, M. D.; Dauskardt, R. H.; Kant, A.; Ritchie, R. O. (1995), *J. Appl. Phys.*, **78**(5), p. 3083.
Eddaoudi, M.; Kim, J.; Rosi, N.; Vodak, D.; Wachter, J.; O'Keeffe M.; Yaghi, O. M. (2002), *Science*, **295**, p. 469.
Elizalde, M. R.; Sanchez, J. M.; Martinez-Esnaola, J. M.; Pantuso, D.; Scherban, T.; Sun, B. and Xu, G. (2003), *Acta Mater.*, **51**, p. 4295.
Eslava, S.; Zhang, L.; Esconjauregui, S.; Yang, J.; Vanstreels, K.; Baklanov M. R.; Saiz, E. (2013), *Chem. Mater.*, **25**, p. 27.
Fan, H. B.; Chan, E. K. I.; Wong, C. K. Y.; Yuen, M. M. F. (2006), *Proc. Electronic Components and Technology Conference*, 2006, p. 1425.
Farnell, G. W.; Adler, E. L. (1972) *Physical Acoustics*, Academic Press, New York and London.
Fernlund, G., and Spelt, J. K. (1994), *Composites Science and Technology*, **50**, p. 441.
Fleck, N. A.; Otoyo, H. and Needleman, A. (1992), *Int. J. Solids Struct.*, **29**, p. 1613.
Dieter, G. E. (1976), *Mechanical Metallurgy*, 2nd ed., McGraw-Hill.

Gage, D. M.; Stebbins, J. F.; Peng, L.; Cui, Z.; Al-Bayati, A.; MacWilliams, K. P.; M'Saad, H.; Dauskardt, R. H. (2008), *J. Appl. Phys.*, **104**(4), 043513.

Gao, M. Z.; Zhang, J. Y.; Wang, Y.; Yu, Z. P. (2009), *Jpn. J. Appl. Phys.*, **48**, 04C017.

Garboczi, E. J.; Day, A. R. (1995), *J. Mech. Phys. Solids*, **43**, p. 1349.

Garner, L. *et al.* (2005), *Intel Tech. J.*, **9**(4), p. 297.

Geisler, H.; Hofmann, P.; Brueckner, M.; Lehr, M.U.; Mayer, U.; Lehmann, L.; Roessler, T.; Breuer, D.; Engelmann, H.-J. (2011A), *Proc. Advanced Metallization Conf.*, 2011, abstract 12.

Geisler, H.; Lehr, M.; Kuechenmeister, F.; Grillberger, M. (2011B), German Patent DE 10 2010 002 453 A1 (September 01, 2011), for GLOBALFOUNDRIES.

Geisler, H.; Schuchardt, E.; Brueckner, M.; Hofmann, P.; Machani, K.V.; Kuechenmeister, F.; Breuer, D.; Engelmann, H., *Proc. of International Reliability Physics Symposium (IRPS)*, 2013, 5C.1.1.

Gille, G. (1984), *Thin Solid Films*, **111**(3), p. 201.

Gille, G.; Wetzig, K. (1983), *Thin Solid Films*, **110**(1), p. 37.

Goethals, F.; Ciofi, I.; Madia, O.; Vanstreels, K.; Baklanov, M. R.; Detavernier, C.; Van Der Voort, P.; Van Driessche, I. (2012), *J. Mater. Chem.* **22**, p. 8281.

Goethals, F.; Meeus, B.; Verberckmoes, A.; Van der Voort P.; Van Driessche, I. (2010), *J. Mater. Chem.*, **20**, p. 1709.

Gonzalez, M., Vanstreels, K.; Cherman, V.; Croes, K.; Kljucar, L.; De Wolf, I.; Tőkei, Zs. (2014), *Proc. of International Reliability Physics Symposium (IRPS)*, 2014, 2F.2.1–2F.2.6.

Grill, A.; Patel, V. (2001), *Appl. Phys. Lett.*, **79**, p. 803.

Guo, X.; Jakes, J. E.; Nichols, M. T.; Banna, S.; Nishi, Y.; Shohet, J. L. J. (2013), *J. Appl. Phys.*, 114, 084103.

Guyer, E. P., Patz, M. and Dauskardt, R. H. (2006), *J. Mater. Res.*, **21**(4), p. 882.

Guyer, E. P.; Dauskardt, R. H. (2003), *Proc. International Interconnect Technology Conference*, p. 89.

Guyer, E. P.; Dauskardt, R. H. (2004A), *Proc. International Interconnect Technology Conference*, p. 236.

Guyer, E. P.; Dauskardt, R. H. (2005), *J. Mater. Res.*, **20**(3), p. 680.

Guyer, E. P.; Dauskardt, R. H. (2004B), *Nature Materials*, **3**(1), p. 53.

Guyer, E. P.; Gantz, J.; Dauskardt, R. H. (2007), *J. Mater. Res.*, **22**(3), p. 710.

Harry, E.; Ignat, M.; Pauleau, Y.; Rouzaud, A.; Juliet, P. (2000), *Surf. Coat. Technol.*, **125**, p. 185.

Harry, E.; Ignat, M.; Rouzaud, A.; Juliet, P. (1999B), *Surf. Coat. Technol.*, **111**, p. 177.

Harry, E.; Rouzaud, A.; Ignat, M.; Juliet, P. (1998), *Thin Solid Films*, **332**, p. 195.

Harry, E.; Rouzaud, A.; Juliet, P.; Pauleau, Y.; Ignat, M. (1999A), *Surf. Coat. Technol.*, 116–119, p. 172.

Hashin, Z. (1983), *J. Appl. Mech.*, **50**, p. 481.

Hatton, B. D.; Landskron, K.; Whitnall, W.; Perovic, D. D.; Ozin, G. A. (2005), *Adv. Funct. Mater.*, **15**, p. 823.

Hoffmann, F.; Cornelius, M.; Morell J.; Froba, M. (2006), *Angewandte Chemie-International Edition*, **45**, p. 3216.

Hoofman, R. J. O. M., Nguyen, V. H., Arnal, V., Broekaart, M., Gosset, L. G., Besling, W. F. A., Fayolle, M. and Iacopi, F. (2007), Integration of low-*k* dielectric films in damascene processes, in *Dielectric Films for Advanced Microelectronics* (eds. Baklanov, M., Maex, K. and Green, M.), John Wiley & Sons, Inc., New York.

Hsing, A.; Dauskhardt, R. H.; Iacopi, F.; Lehr, M. U.; H Geisler, H.; Machani, K.; Ryan, V.; Paul, J.; Besser, P. (2011), *Proc. of Advanced Metallization Conf.*, 2011, abstract 22.

Hsing, A.; Geisler, H.; Ryan, V.; Cheng, M.; Machani, K.; Breuer, D.; Lehr, M. U.; Paul, J.; Iacopi, F.; Dauskhardt, R. H. (2012), *Interconnect Technology Conference (IITC)*, 2012.

Hu, M. S.; Evans, A. G. (1989), *Acta Metall.*, **37**(3), p. 917.

Hu, Y. H.; Zhang, L. (2010), *Adv. Mater.*, **22**, E117.

Hunks, W. J.; Ozin, G. A. (2005), *J. Mater. Chem.*, **15**, p. 3716.

Hutchinson, J. W., and Z. Suo (2002), *Advances in Applied Mechanics*, **29**, p. 63.

Iacopi, F.; Brongersma, S.H.; Vandevelde, B.; O'Toole, M.; Degryse, D.; Travaly, Y.; Maex, K. (2004), *Microelectronic Engineering*, **75**, p. 54.

Iacopi, F.; Travaly, Y.; Eyckens, B.; Waldfried, C.; Abell, T.; Guyer, E. P.; Gage, D. M.; Dauskardt, R. H.; Sajavaara, T.; Houthoofd, K.; Grobet, P.; Jacobs, P.; Maex, K. (2006), *J. Appl. Phys.*, **99**, 053511.

Jaeger, G.; Endler, I.; Heilmaier, M.; Bartsch, K.; Leonhardt, A. (2000), *Thin Solid Films*, 377–378, p. 382.

Jiang, Z.; Lu, F. X.; Tang, W. Z.; Wang, S. G.; Tong, Y. M.; Huang, T. B.; Liu, J. M. (2000), *Diamond Relat. Mater.*, **9**, p. 1734.

Keshtov, M.; Said-Galiev, E.; Kochurov V.; Khokhlov, A. (2012), *AIP Conf. Proc.*, **1459**, p. 277.

Kim, T.-S.; Dauskardt, R. H. (2009), *IEEE Transactions on Device and Materials Reliability*, **9**(4), p. 509.

Kinbara, A.; Kusano, E.; Kamiya, T.; Kondo, I.; Takenaka, O. (1998), *Thin Solid Films*, **317**, p. 165.

King, R. B. (1987), *Int. J. Solids Struct.*, **23**, p. 1657.

Kriese, M. D., Gerberich, W. W. (1999), *J. Mater. Res.*, **14**(7), p. 3007.

Krishtab *et al.* (2014B), presented at *2014 MRS Spring Meeting, Symposium CC: New Materials and Processes for Interconnects, Novel Memory and Advanced Display Technologies*, CC3.05, April 21-25, 2014, San Francisco.

Krishtab, M.; Vanstreels, K.; Savage, T.; Matsunaga, K.; De Gendt, S.; Baklanov, M. R. (2014A), presented at the *23rd Materials For Advanced Metallization Conference* (*MAM 2013*), submitted to *Microelectronic Engineering*, in 2014.

Kuechenmeister, F.; Breuer, D.; Geisler, H.; Paul, J.; Shah, C.; Machani, K.V.; Kosgalwies, S.; Agarwal, R.; Gao, S. (2012), *Proc. of Electronics Packaging Technology Conference* (*EPTC*), 2012, p. 430.

Lacombe, R. (2006),"Adhesion Measurement Methods — theory and practice", FL: CRC/Taylor & Francis.

Landers, W.; Edelstein, D.; Clevenger, L. (2004), *Proc. International Interconnect Technology Conference*, 2004, p. 108.

Landskron, K.; Hatton, B. D.; Perovic, D. D.; Ozin, G. A. (2003), *Science*, **302**, p. 266.

Lane, M.; Dauskardt, R. H.; Krishna, N.; Hashim, I. (2000), *J. Mater. Res.*, **15**, p. 203.

Lane, M.; Snodgrass, J. M.; Dauskardt, R. H. (2001), *Microelectron. Reliab.*, **41**, p. 1615.

Lawn, B.; Winslaw, R. (1975), *J. Mater. Sci.*, **10**(6), p. 1049.

Lawn, B. R.; Evans, A. G.; Marshall, D. B. (1980), *J. Am. Ceram. Soc.*, **63**(9-10), p. 574.
Lee, W. J.; Cho, Y. J.; Lee, H. S.; Park I. M.; Park, Y. H. (2011), *Procedia Engineering*, **10**, p. 2459.
Li, H.; Knaup, J. M.; Kaxiras, E.; Vlassak, J. J. (2011), *Acta Mater.*, **59**(1), p. 44.
Li, H.; Lin, Y.; Tsui, T. Y. and Vlassak, J. J. (2009A), *J. Mater. Res.*, **24**, p. 107.
Li, H.; Tsui, T. Y.; Vlassak, J. J. (2009B) *J. Appl. Phys.*, **106**(3), 033503.
Li, X.; Bhushan, B. (1999), *J. Mater. Res.*, **14**(6), p. 2328.
Lin, Y. B.; Tsui, T. Y.; Vlassak, J. J. (2007), *Acta Mater.*, **55**(7), p. 2455.
Lin, Y. B.; Tsui, T. Y.; Vlassak, J. J. (2010), *J. Electrochem. Soc.*, **157**(2), G53.
Lin, Y.; Xiang, Y.; Tsui, T. Y.; Vlassak, J. J. (2008), *Acta Mater.*, **56**(17), p. 4932.
Liu, X. H.; Shaw, T. M.; Lane, M. W.; Liniger, E. G.; Herbst, B. W.; Questad, D. L. (2007), *Proc. International Interconnect Technology Conference.*
Liu, X. H.; Shaw, T. M.; Bonilla, G.; Questad, D.; Liniger, E. G.; Hosadurga, S. (2001), *ASME.*
Liu, X. H.; Shaw, T. M.; Bonilla, G.; Questad, D.; Liniger, E. G.; Hosadurga, S. (2011) *ASME*, 2011.
Ma, Q. (1997), *J. Mater. Res.*, **12**, p. 840.
Ma, Z.; Moulton, B. Coord. (2011), *Chem. Rev.*, **255**, p. 1623.
Maex, K.; Baklanov, M. R.; Shamiryan, D.; Iacopi, F.; Brongersma, S. H.; Yanovitskaya, Z. S. (2003), *J. Appl. Phys.*, **93**(11), p. 8793.
Marshall, D. B.; Evans, A. G. (1984), *J. Appl. Phys.*, **56**, p. 2632.
Matsumoto, R.; Nakagaki, M. (2006), *JSME Int. Journ. Series A*, **49**(4), p. 513.
McElhaney, K. W.; Vlassak, J. J. and Nix, W. D. (1998), *J. Mater. Res.*, **13**, p. 1300.
Meek, S. T.; Greathouse, J. A.; Allendorf, M. D. (2011), *Adv. Mater.*, **23**, p. 249.
Mercado, L.; Goldberg, C.; Kuo, S.-M. (2002), *Proc. International Interconnect Technology Conference*, 2002. p. 119.
Mercado, L.; Kuo, S.-M.; Goldberg, C.; Lee, T.-Y. (2003), *Proc. 53rd Electronic Components and Technology Conference*, 2003. p. 1784.
Mizoshita, N.; Taniab, T.; Inagaki, S. (2011), *Chem. Soc. Rev.*, **40**, p. 789.
Morris, D. J.; Cook, R. F. (2008), *J. Mater. Res.*, **23**(9), p. 2443.
Ocana, I., Molina-Aldareguia, J. M.; Gonzalez, D., Elizalde, M. R., Sanchez, J. M.; Martinez-Esnaola, J. M.; Gil Sevillano, J.; Scherban, T.; Pantuso, D.; Sun, B.; Xu, G.; Miner, B.; He, J.; Maiz, J. (2006), *Acta Materialia*, **54**, p. 3453.
Oliver, M. S.; Dubois, G.; Sherwood, M.; Gage, D. M.; Dauskardt, R. H. (2010), *Adv. Funct. Mater.*, **20**, p. 2884.
Oliver, W. C; Pharr, G. M. (1992), *J. Mater. Res.*, **7**, p. 1564.
Panjan, P.; Cekada, M.; Navinsek, B. (2003), *Surf. Coat. Technol.*, 174–175, p. 55.
Pharr, G. M.; Oliver, W. C.; Brotzen, F. (1992), *J. Mater . Res.*, **7**(3), p. 613.
Poutet, J.; Manzoni, D.; Hage-Chehade, F.; Jacquin, C. G.; Bouteca, M. J.; Thovert, J. F.; Adler, P. M. (1996), *J. Mech. Phys. Solids*, **44**, p. 1587.
Rice, R. W. (1996A), *J. Mater. Sci.*, **31**(1), p. 102.
Rice, R. W. (1996B), *J. Mater. Sci.*, **31**, p. 1509.
Roberts A. P.; Garboczi, E. J. (2002), *Proceedings of the Royal Society of London Series a-Mathematical Physical and Engineering Sciences 458*, 2021, p. 1033.
Roberts, A. P.; Garboczi, E. J. (2000), *J. Ameri. Ceram. Soc.*, **83**(12), p. 3041.
Roham, S.; Hardikar, K.; Woytowitz, P. (2004), *J. Mater. Res.*, **19**, p. 3019.

Roscoe, R. (1952), *British J. Appl. Phys.*, **3**(8), p. 267.
Rossington, C.; Evans, A. G.; Marshall, D. B. and Khuriyakub, B. T. (1984), *J. Appl. Phys.*, p. 2639.
Ryan, V.; Breuer, D.; Geisler, H.; Kioussis, D.; Lehr, M. U.; Paul, J.; Machani, K.; Shah, C.; Kosgalwies, S.; Lehmann, L.; Jaesik Lee; Kuechenmeister, F.; Ryan, E.T.; Karimanal, K. (2012), *Proc. of Reliability Physics Symposium (IRPS) 2012*, 2E.1.1.
Sanchez, J. M.; El-Mansy, S.; Sun, B.; Scherban, T.; Fang, N.; Pantuso, D.; Ford, W.; Elizalde, M. R.; Martinez-Esnaola, J. M.; Martin-Meizoso, A.; Gil-Sevillano, J.; Fuentes M. and Maiz, J. (1999), *Acta Mater.*, **47**, p. 4405.
Schade, A.; Monnereau, L.; Muller, T.; Bräse, S. (2014), *ChemPlusChem*, **79**(8), p. 1176.
Schulze, M. and Nix, W. D. (2000), *Int. J. Sol. Struc.*, **37**, p. 1045.
Shekhah, O.; Liu, J. Fischer, R. A.; Woll, C. (2011), *Chem. Soc. Rev.*, **40**, p. 1081.
Shum, P. W.; Li, K. Y. ; Zhou, Z. F.; Shen, Y. G. (2004), *Surf. Coat. Technol.*, **185**, p. 245.
Song, M.; Chen, L., *Macromolecular Theory and Simulations*, **15**(3), p. 235.
Sreenivasan R. and Gleason, K. K. (2009), *Chemical Vapor Deposition*, **15**, p. 77.
Stassen, I.; Campagnol, N.; Fransaer, J.; Vereecken, P.; De Vos D.; Ameloot, R. (2013), *Cryst. Eng. Comm.*, **15**(45), p. 9308.
Sundararajan, S.; Bhushan, B. (2001), *J. Mater. Res.*, **16**(2), p. 437.
Suo, Z. (2003), *Interfacial and Nanoscale Failure of Comprehensive Structural Integrity*, **8**, p. 265.
Suo, Z.; Hutchinson, J. W. (1989), *Mater. Sci. Eng.*, **A107**, p. 135.
Tafipolsky, M.; Amirjalayer, S.; Schmid, R. (2010), *J. Phys. Chem. C*, **114**(34), p. 14402.
Tajima, N.; Ohno, T.; Hamada, T.; Yoneda, K.; Kobayashi, N.; Hasaka, S.; Inoue, M. (2006), *Appl. Phys. Lett.*, **89**, 061907.
Tajima, N.; Ohno, T.; Hamada, T.; Yoneda, K.; Kondo, S.; Kobayashi, N.; Shinriki, M.; Inaishi, Y.; Miyazawa, K; Sakota, K.; Hasaka, K. S.; Inoue, M. (2007), *Jpn. J. Appl. Phys.*, **46**, 5970.
Thorpe, M. F. (1983), *J. Non-Cryst. Solids*, **57**(3), p. 355.
Tsotsalas, M.; Liu, J.; Tettmann, B.; Grosjean, S.; Shahnas, A.; Wang, Z.; Azucena, C.; Addicoat, M.; Heine, T.; Lahann, J.; Overhage, J.; Bräse, S.; Gliemann, H.; Wöll, C. (2014), *J. Amer. Chem. Soc.*, **136**, p. 8.
Tsui, T. Y.; McKerrow, A. J.; Vlassak, J. J. (2006), *J. Mechanics and Physics of Solids* **54**(5), 887.
Uchibori, C. J.; Zhang, X.; Ho, P. S.; Nakamura, T. (2006), *Proc. International Interconnect Technology Conference*, 2006, p. 196.
Urbanowicz, A.; Vanstreels, K.; Shamiryan, D.; De Gendt, S.; Baklanov, M. R. (2009), *Electrochem. Solid-State Lett.*, **12**(8), H292.
Urbanowicz, A.; Vanstreels, K.; Verdonck, P.; Shamiryan, D.; De Gendt, S.; Baklanov, M. R. (2010), *J. Appl. Phys.*, **107**, 104122.
Urbanowicz, A.; Vanstreels, K.; Verdonck, P.; Van Besien, E.; Trompoukis, C.; Shamiryan, D.; De Gendt, S.; Baklanov, M. R. (2011), *J. Vac. Sci. Technol. B*, **29**(3), 032201.
Usami, T.; Nakamura, T.; Yashima, I. (2013), *Proc. of Interconnect Technology Conference*, (IITC) 2013, p. 1–3.
Usman, M.; Lee, C.-H.; Hung, D.-S.; Lee, S.-F.; Wang, C.-C.; Luo, T.-T.; Zhao, L.; Wu, M.-K.; Lu, K.-L. (2014), *J. Mater. Chem. C*, **2**, p. 3762.

Van Der Voort, P.; Esquivel, D.; De Canck, E.; Goethals, F.; Van Driessche, I.; Romero-Salguero, F. J. (2013), *Chem. Soc. Rev.*, **42**, p. 3913.

Vanstreels, K.; Cherman, V. ; Gonzalez, M.; De Wolf, I.; Van der Plas, G.; De Vos, J.; Bömmels J.; Tőkei, Zs (2014), presented at *MAM 2014 Conference, 2–5* March 2014, Chemnitz; accepted for publication in *Microelectronic Engineering*, 2014.

Vanstreels, K.; Wu, C.; Gonzalez, M.; Schneider, D.; Gidley, D.; Verdonck P. and Baklanov, M. R. (2013), *Langmuir*, **29**, p. 12025.

Vanstreels, K.; Wu, C.; Verdonck, P.; Baklanov, M. R. (2012), *Appl. Phys. Lett.*, **101**, 123109.

Verdonck, P.; Van Besien, E.; Vanstreels, K.; Trompoukis, C.; Urbanowicz, A.; De Roest, D.; Baklanov, M. R. (2011) *Jpn. J. Appl. Phys.*, **50**, 05EB05.

Verdonck, P.; Wang, C.; Le, Q. T.; Souriau, L.; Vanstreels, K.; Krishtab, M.; Baklanov, M. R. (2014), *Microelectronic Engineering*, **120**, p. 225.

Vlassak, J. J.; Lin, Y.; Tsui, T. Y. (2005), *Materials Science and Engineering a-Structural Materials Properties Microstructure and Processing*, **391**(1-2), p. 159.

Vlassak, J. J.; Drory, M. D. and Nix, W. D. (1997), *J. Mater. Res.*, **12**, p. 1900.

Voevodin, A. A.; Rebholz, C.; Schneider, J. M.; Stevenson, P.; Matthews, A. (1995), *Surf. Coat. Technol.*, **73**, p. 185.

Volinsky, A .A.; Moody, N. R.; Gerberich, W. W. (2002), *Acta Mater.*, **50**(3), p. 441.

Volinsky, A. A.; Tymiak, N. I.; Kriese, M. D.; Gerberich, W. W.; Hutchinson, J. W. (1999), *Mater. Res. Soc. Symp. Proc.*, **539**, p. 277.

Volinsky, A. A.; Vella, J. B.; Gerberich, W. W. (2003), *Thin Solid Films*, **429**, p. 201.

Wang, G. T.; Ho, P. S.; Groothuis, S. (2005), *Microelectronics Reliability*, **45**(7–8), p 1079.

Wang, G. T.; Merrill, C.; Zhao, J. H.; Groothuis S.; Ho, P. S. (2003), *IEEE Trans Device and Materials Reliability*, **3**, p. 119.

Wang, X.; Wang, C.; Atkinson, A. (2012), *Acta Mater.*, **60**(17), 6152.

Xiang, Y.; Tsui, T. Y.; Vlassak, J. J. and McKerrow, A. J. (2004), *Proc. IEEE Int. Interconnect Tech. Conf.*, p. 133.

Yeap, K. B.; Hangen, U.; Ritz, Y.; Kim, T.-S.; Dauskardt R.; Zschech, E. (2010), *AIP Conf. Proc.*, 1300, p. 159.

Yuan, C. A.; Sluis, O.; Zhang, G. Q.; Ernst, L. J.; Driel, W. D.; Flower, A. E.; van Silfhout, R. B. R. (2008), *Appl. Phys. Lett.*, **92**, 061909.

Yuan, C. A.; Sluis, O.; Zhang, G. Q.; Ernst, L. J.; Driel, W. D.; van Silfhout, R. B. R. (2007), *Microelectron. Reliab.*, **47**, p. 1483.

Zagorodniy, K., Seifert, G.; Hermann, H. (2010), *Appl. Phys. Lett.*, **97**, 251905.

Zhang, X; Wang, Y.; Im, S. H.; Ho, P. S. (2012), *IEEE Transactions on Device and Materials Reliability*, **12**(2), p. 462.

Bakir, M. S. and Meindl, J. D (2008), *Integrated Interconnect Technologies for 3D Nanoelectronic Systems*, Chapter 2: Chip-package interaction and reliablity impact on Cu/Low-*k* interconnects, by Zhang, X.; Im, S. H.; Huang, R. and Ho, P. S, p. 23–59; Artech House, Inc. Norwood, MA, USA.

Zhao, J. H.; Wilkerson, B.; Uehling, T. (2004), *AIP Conf. Proc.*, **714**, p. 52–61.

Zhou, W.; Bailey, S.; Sooryakumar, R.; King, S.; Xu, G.; Mays, E.; Ege, C.; Bielefeld, J. (2011), *J. Appl. Phys.*, **110**, 043520.

4. Electrical Reliability Challenges of Advanced Low-k Dielectrics

4.1. *Introduction*

The development and integration of reliable low-*k* materials is a key challenge for next generation interconnect technologies. Driven by the requirement of performance increase, highly porous low-*k* dielectrics are incorporated in deeply scaled back-end-of-line (BEOL) interconnects. In general, there are three important concerns during the study of low-*k* dielectric electrical properties and reliability: (a) the k-value of the dielectric after integration should maintain relatively unchanged, (b) the leakage current of the dielectric between metal lines should be low, and (c) the time dependent dielectric breakdown (TDDB) failure time of the integrated BEOL structure at operating conditions should meet its specifications [Baklanov *et al.*, 2007].

Porous SiOCH low-*k* dielectrics deposited by plasma enhanced chemical vapor deposition (PECVD) techniques are the dominant choice for $k > 2.2$ film depositions and are also potential candidates for ultralow-*k* ($k < 2.2$) dielectric applications [ITRS, 2013]. In the past years, significant efforts have been spent to understand their material properties and integration challenges. The bond structures of low-*k* dielectrics is complex due to the non-equilibrium nature of the PECVD technique [Baklanov *et al.*, 2007]. Such amorphous structures are further complicated by the use of porogen, where the porogen based approach is widely used in PECVD processes to produce micro pores in dielectric films. Pure organic molecules are co-deposited with matrix materials as sacrificial porogen. During a curing step, most of this organic material is thermally decomposed, where the original occupied sites are converted into micro pores. A higher porosity in the low-*k* film can be achieved by increasing the ratio of porogen versus matrix during deposition, where high porosities of around 50% have been reported [Vanstreels *et al.*, 2013B]. Above this value, film shrinkage caused by low mechanical strength starts to limit further porosity increase. The low-*k* film quality highly depends on the matrix and porogen deposition and the curing conditions [Baklanov *et al.*, 2012]. Different matrix and porogen compositions result in various network structures in the final film [Grill *et al.*, 2014]. Non-optimized curing steps also introduce hazardous contaminants [Baklanov *et al.*, 2012]. During the integration of dual damascene copper interconnects, low-*k* materials are subjected to processes such as patterning, barrier metal deposition and chemical mechanical polishing (CMP) that modify the material structure and introduce damage. The increase of structure defect densities, the change of film's hydrophobicity, metal diffusion, stress accumulation and delamination seriously lower dielectric reliability performance.

In this review, we focus on the results where the nature of process induced defects are studied, since they are important factors to reveal material properties. These defects can be directly probed through physical characterization techniques. We also summarize learnings about leakage current and breakdown behaviors of low-*k* dielectrics, as leakage current and breakdown are sensitive to disordered structures, where low leakage current and long failure time represent a solid matrix arrangement. The dominant conduction mechanism provides information about the nature of electrically active defects. Insights in these mechanisms are useful during the optimization of film development, integration processes and reliability assessment. In addition, low leakage current levels ensure good reliability as more and more literature data point out that BEOL electrical degradation is dominated by charge injection and that energetic charges facilitate both intrinsic chemical bond breakage and extrinsic metal ion diffusion [Lloyd *et al.*, 2005A] [Wu *et al.*, 2014] [Chen and Shinosky, 2014]. This paper consists of three parts. We will first explain the basic knowledge of bandgap structures and conduction mechanisms in low-*k* dielectrics. Deposition related issues, including the influences of porogen residue, intrinsic defects and porosity will be discussed in the second part. The third part will review process induced material modifications and reliability challenges during BEOL integration.

4.2. *Bandgap and conduction mechanisms in low-k dielectrics*

A bandgap is defined as the difference between the energy of the lowest conduction band and that of the highest valence band in a crystalline solid [Sze, 1981]. As a result, for porous low-*k* dielectrics, there is no real bandgap. But as King *et al.* [2013] argued, the concept of a bandgap is extendable to low-*k* dielectrics and has been successfully used to explain a number of empirical measurement results. To characterize bandgap in porous SiOCH type low-*k* dielectrics, various techniques are being used. The electron energy loss spectroscopy measurements from Cheynet *et al.* [2009] and Singh *et al.* [2014] determine the low-*k* dielectric bandgap to be 8.5 eV and 10 eV, respectively. This bandgap range of low-*k* dielectrics is similar to the one of amorphous SiO_2, which is between 8.0 and 8.8 eV. The reflection electron energy loss spectroscopy results from King *et al.* [2013], the ellipsometric data from Marsik *et al.* [2011], the vacuum ultraviolet

(VUV) spectroscopy measurements from Zheng *et al.* [2014] and the X-ray photoelectron spectroscopy analysis from Nichols *et al.* [2015] support that the bandgap of most porous SiOCH type low-k dielectrics ($k = 2.0 - 3.3$) are in the range between 7.5 to 10 eV. In addition, the barrier height at both the low-k/metal and the low-k/Si interfaces can be measured with internal photoemission experiments. Shamuilia *et al.* [2006] and Atkin *et al.* [2008] suggest that this value is around 4 eV for both interfaces. This value further proves that the bandgap of most of the SiOCH type low-k dielectrics is similar to that of SiO2, indicating that the carbon content in low-k films is not incorporated in the matrix network but primarily exists as terminal methyl groups. If carbon would be present in the network bonds by forming Si-C-Si, the bandgap value would drop dramatically. The study from King *et al.* [2011] indicates that the barrier height at the copper/low-k dielectric interface is dependent on the amount of network carbon in the film and is in the range from 1 to 4 eV.

Three conduction mechanisms have been commonly reported for low-k dielectrics. Schottky emission (SE) and Poole-Frenkel (PF) emission describe field-enhanced thermal excitation of electrons entering the conduction band from the low-k interface and the trap states with coulomb potentials respectively. Fowler-Nordheim (FN) tunneling conduction is caused by electrons tunneling from the metal Fermi energy or trapping sites in the material itself into the low-k dielectric conduction band. The equations describing SE, PF emission and FN tunneling are listed below [Sze, 1981] [Ohring, 1998],

$$\text{Schottky emission} \quad J_{SE} = A^* \cdot T^2 \cdot \exp\left[\frac{-q(\varphi_{SE} - \sqrt{qE/4\pi\varepsilon_0\varepsilon_r})}{kT}\right] \tag{4.1}$$

$$\text{Poole-Frenkel (PF) emission} \quad J_{PF} \sim E \cdot \exp\left[\frac{-q(\varphi_{PF} - \sqrt{qE/\pi\varepsilon_0\varepsilon_r})}{kT}\right] \tag{4.2}$$

$$\text{Fowler-Nodheim (FN) tunneling} \quad J_{FN} \sim E^2 \cdot \exp\left[\frac{-8\pi\sqrt{2m^*}(q\varphi_{FN})^{3/2}}{3qhE}\right] \tag{4.3}$$

where J is current density, A^* is Richardson constant, T is temperature, q is the elementary charge, φ is barrier height, E is electric field, ε_r is dielectric constant, ε_0 is permittivity of free space, m^* is effective electron mass and h is Planck's constant. An efficient way to determine the dominant conduction mechanism is to study the leakage current behavior as a function of field at different temperatures. As suggested by equation 4.1 and 4.2, the slope extracted from the relation between $\ln(J)$ or $\ln(J/E)$ versus $E^{0.5}$ should be similar to the theoretical value as suggested by equation (4.4), if either SE or PF emission dominates.

$$\beta_{PF} = 2\beta_{SE} = \sqrt{\frac{q^3}{\pi\varepsilon_0\varepsilon_r}} \tag{4.4}$$

FN tunneling current exhibits a strong field dependence and is independent of temperature. The plot of $\ln(J/E^2)$ and $1/E$ should also yield a straight line. In addition, in each conduction mechanism, trap depth or barrier height φ can be extracted by employing simple fitting procedures.

Due to the high barrier height being around 4 eV at the low-*k* dielectric interface, the interface controlled SE only happens when there are strong modifications at the interface or when there are changes of emission paths due to the existence of other low barrier height interfaces, such as the etch stop layer/metal interface (less than 2.0 eV) [Chen and Shinosky, 2014]. Instead, PF emission is more likely the dominant conduction mechanism in low-*k* dielectrics especially at low fields [Lloyd *et al.*, 2005A] [Wu *et al.*, 2014]. FN tunneling conduction can occur at high field ranges [Gischia *et al.*, 2010] [Wu *et al.*, 2013]. It is shown that leakage currents can be used as a direct measure of dielectric structure quality so that the influences of deposition conditions and integration processes can be examined. Several examples will be given in the following sections to further emphasize the importance of leakage currents and we will show its application to probe process induced modifications and damages of dielectric properties.

4.3. *Dielectric properties after deposition*

(a) Low-*k* dielectrics with porogen residues

Porogen based PECVD deposition methods are widely used to fabricate porous SiOCH low-*k* dielectric films. To generate porosity, hydrocarbon porogen is used as a sacrificial material in the co-deposited hybrid (matrix/porogen) film, where the porogen is removed during post-deposition ultraviolet (UV) treatments leaving the pores behind. If non-optimized UV curing conditions are applied, the formation of nonvolatile graphitized-carbon residues is likely to happen. Porogen residues are the major defects in low-*k* dielectrics. The existence of sp2 hybridized carbon like porogen residues is well studied using UV spectroscopic ellipsometry. As reported by Baklanov *et al.* [2011], the amount of porogen residue increases with increasing curing time as more original porogen is transformed into porogen residues. The increased amount of porogen residues enhances the leakage current level since these graphitic like structures are conductive. In the same study, the self-assembled low-*k* material prepared without porogen and UV curing shows lower leakage currents compared to low-*k* dielectrics prepared with porogen and curing. In the study from Van Besien *et al.* [2012], the high leakage currents and low breakdown fields in higher porosity low-*k* films are also attributed to the presence of porogen residues: to prepare higher porosity films, more porogen is added during the deposition, and consequently more porogen residues remain inside the pores.

The presence of porogen residues in low-*k* dielectrics is also reported by other authors. For different wavelengths used during UV cure, the efficiency of porogen removal is also different, where UV penetration depths play an important role [Afanas'ev *et al.*, 2011]. In general, depositing thinner low-*k* dielectrics is more favorable as thin films are cured by the UV treatment over the whole depth of the film. The electron-spin resonance (ESR) results from Afanas'ev *et al.* [2011] show that signals associated with a carbon dangling bond in a cluster are observed for 500 nm thick low-*k* dielectrics but are absent for 60 nm thick low-*k* dielectrics, where King *et al.* [2013] pointed out that the defect band between 2–6 eV in the bandgap of the low-*k* dielectric is attributed to these carbon dangling bonds from porogen residues. Also, the UV and VUV irradiated results from Lauer *et al.* [2010] show that the defect centers originating from porogen residues are able to capture and exchange electrons. In addition, Ren *et al.* [2011] reported using ESR measurements that increasing porosity result in an increased defect density in low-*k* films due to the presence of porogen residues. These defects were later interpreted by Pormorski *et al.* [2014] as carbon dangling bond centers again using ESR measurements, which confirms the findings of Afanas'ev *et al.* Their studies further indicate that this type of defect plays a dominating role in the electrical transport at low fields. In conclusion, porogen based PECVD processes should incorporate an improved curing step to avoid porogen residues as they have a negative impact on electrical properties of low-*k* dielectrics. High defect numbers and leakage currents result in an insufficient reliability performance.

(b) Low-*k* dielectrics free of porogen residues

Lately, a scheme to fabricate porogen residue free low-*k* dielectrics is proposed by Urbanowicz *et al.* [2010B]. After co-deposition of porogen and matrix, an H2 after-glow treatment is applied prior to UV curing. Similar to the photomask removal mechanism, hydrogen atoms promote the transformation of high mass porogen chains to volatile short chain molecules. This step enables complete removal of porogen without scission of Si-CH3. Such porogen residue free low-*k* dielectrics exhibit enhanced mechanical properties due to more cross linked oxygen atoms in the network. Recent electrical characterization results from Wu *et al.* [2013] suggest that J-E relationships in this type of low-*k* dielectric are insensitive to temperature in the high electric field range. A barrier height of 4 eV of low-*k*/metal interface is calculated based on the theory of FN tunneling. This value suggests that the backbone of this particular low-*k* dielectric is similar to SiO_2.

The influence of porosity has been studied on this porogen residue free low-*k* dielectric. The trend of leakage current versus porosity is different from the one characterized from low-*k* dielectrics containing porogen residues, where higher leakage currents are observed in higher porosity films. Vanstreels *et al.* [2013A] reported the leakage and breakdown behavior of porogen residue free films with 30%–50% porosity, where it was shown that the leakage current is independent of porosity. However, increasing porosity results in lower breakdown fields, which is explained by a higher amount of cage structures in the film. We extended this film porosity range to 10%–45%. A dedicated test vehicle (low-*k* planar capacitor or p-cap) was used to study the pristine properties of low-*k*/barrier systems [Zhao *et al.*, 2009A] [Zhao *et al.*, 2009B], of which the cross-section is shown in Figure 4.1. Since no plasma exposures are involved after low-*k* deposition and the

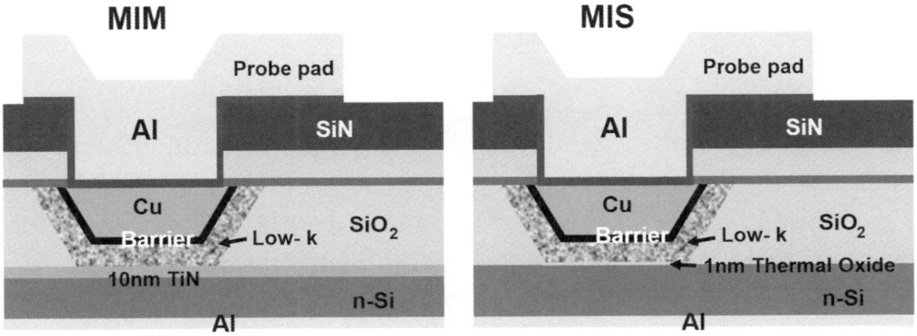

Fig. 4.1. Cross sections of MIM and MIS planar capacitors.

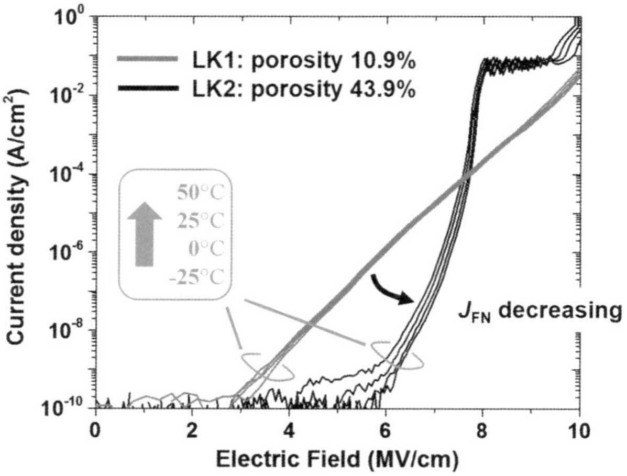

Fig. 4.2. Leakage current density versus applied electric fields of LK1 (10.9% porosity) and LK2 (43.9% porosity) in the temperature range from −25°C to 50°C (Figure was taken from [Wu *et al.*, 2015]. © The Author(s) 2014. Published by ECS.).

low-*k* material is not directly exposed to CMP chemicals, the material is kept pristine. Due to optimized rounded profiles, edge failures do not occur and nice area scaling was demonstrated [Gischia *et al.*, 2010]. Figure 4.2 shows the results of leakage current density versus applied electric fields of LK1 (10.9% porosity) and LK2 (43.9% porosity) measured in the temperature range from −25°C to 50°C. FN tunneling current of higher porosity film is significantly lower, indicating a better matrix quality which has a lower defect density. FTIR results in Figure 4.3 confirm that high porosity films have a better matrix quality as more network bonds are formed and the bonds related to the sub-oxide peak are reduced. Since the sub-oxide peak is associated with oxygen deficiency related defects [Urbanowicz *et al.*, 2010A], it is argued that it can explain the low leakage current in the high porosity films deposited with the above mentioned process scheme.

(c) Intrinsic defects

The selection of matrix precursors is an important factor influencing the properties of deposited films. As we suggested, films deposited by matrix only precursors show the highest leakage currents due to the formation of oxygen deficiency related defects. The ESR results from Pomorski *et al.* [2014] further explain the defect natures caused by matrix only precursors, where a dense film (deposited without porogen and curing) and a porous film (deposited with porogen and curing), deposited with the same matrix

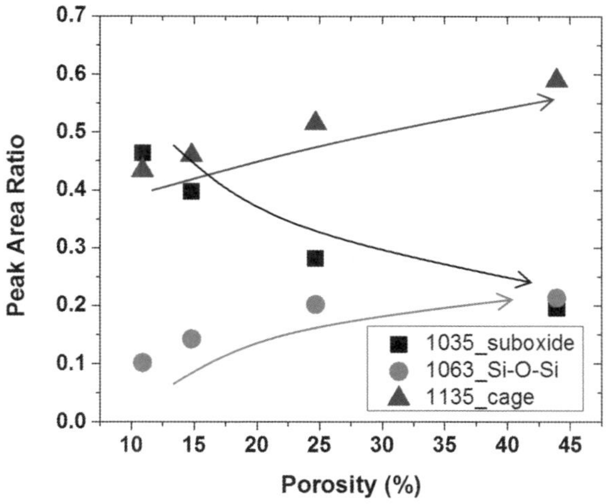

Fig. 4.3. Peak area ratio of the deconvoluted oxide peaks vs porosity of the low-*k* films (Figure was taken from [Wu *et al.*, 2015]. © The Author(s) 2014. Published by ECS.).

precursors, are compared. The dense as well as the porous film contain dangling carbon bonds (in terminal organic groups attached to silicon) and E′ centers (silicon dangling bonds attached to silicon atoms bonded to three oxygen atoms). The presence of these two types of defects are attributed to the use of alkoxysilane as the matrix precursor and are not related to porogen or curing. The authors also emphasized that if specific matrix precursors containing unsaturated terminal organic groups are used, porogen residue related defects are introduced in the film. In addition, defects like CH3 radicals and other types of silicon dangling bond are reported in the same paper.

(d) Field enhancement and strength reduction by porosity

Besides various types of defects in as-deposited low-*k* dielectrics, porosity itself is also reported to influence leakage conduction. Kayaba and Kikkawa [2008] calculated the magnitude of electrical fields in porous low-*k* dielectrics and showed that the maximum field strength always appears at the air/skeleton interfaces. Hong and Milor [2006] used finite element and Monte Carlo simulations to study the electrical field and conduction mechanisms inside porous low-*k* films. Different shapes and interconnectivities of pores change the local field distribution. These local enhanced fields increase with porosity and therefore, poorer insulating properties are expected in higher porosity films. Similar results were reported by Lee *et al.* [2009] suggesting that pore related field enhancement facilitate transportation of charged species and bond breakage, where at

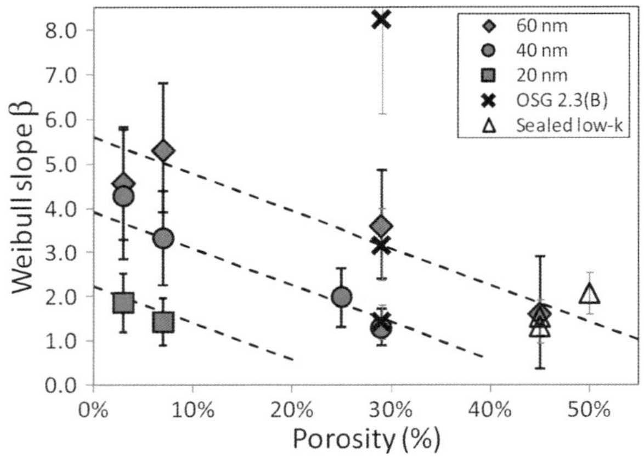

Fig. 4.4. Weibull slopes' values for thicknesses of 20, 40 and 60 nm as a function of low-*k* porosity. The dashed lines show the fits for the three thicknesses. Only the data with filled markers were used in the fit. (Figure was taken from [Barbarin *et al.*, 2013]. Reproduced by permission of K. Croes and M. R. Baklanov.)

operating conditions, this field enhancement is relatively small compared to high fields.

Ogawa *et al.* [2003] used the percolation model to understand the effect of porosity in low-*k* dielectrics. In the percolation model, dielectric breakdown happens when a single column of defective cells is generated between electrodes. Based on the calculations, the authors showed that the pores reduce the time required to generate a percolation path, reduce the breakdown strength and lower the Weibull shape parameter. A smaller Weibull slope limits the failure time prediction at low percentiles [Wu and Vollertsen, 2002]. Barbarin *et al.* [2013] reported similar data based on the electrical measurement results of various low-*k* films. Figure 4.4 and 4.5 show the impact of porosity and thickness on the Weibull slope and the maximum allowed electrical field to meet a 10-year lifetime. It is clear that both increasing porosity and decreasing thickness weaken the electrical robustness of low-*k* films. Their simulations employing an extended percolation model well fitted the measurement data.

4.4. *Reliability after integration*

Integration of low-*k* dielectrics involves many processes that can seriously degrade material properties and reliability. Change of film hydrophobicity and generation of electrically active defects during plasma exposure have been widely reported. Dielectric reliability is also weakened by adsorption of moisture and metal diffusion as well as CMP process induced damages

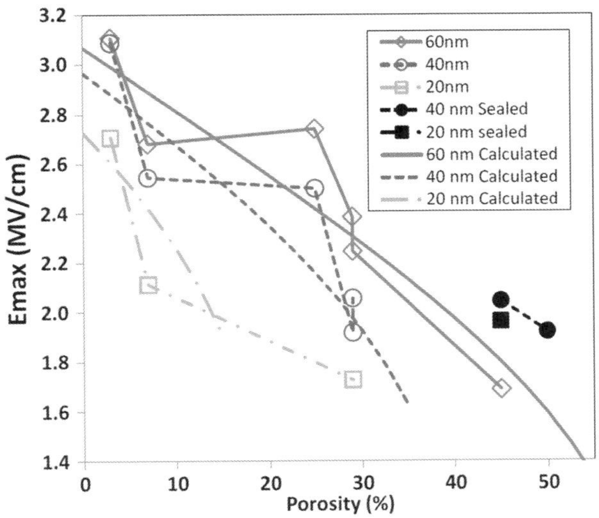

Fig. 4.5. Emax as a function of low-*k*'s porosity for 20, 40 and 60 nm films. The filled markers indicate the low-*k*'s sealed with 12-nm OSG 3.0 layer. The smooth lines without markers are calculated using the percolation model (Figure was taken from [Barbarin *et al.*, 2013]. Reproduced by permission of K. Croes and M. R. Baklanov.)

and contaminants. Stress accumulation and dielectric delamination could also potentially facilitate breakdown. This section provides a closer look at these aspects.

(a) Plasma damage

Plasma contains vacuum ultraviolet (VUV) and UV photons, energetic ions, and highly reactive radicals. Exposure to plasma introduces physical damage and chemical modifications in low-*k* dielectrics. During such processes, methyl groups are extracted from the network leading to densification and *k*-value increase. Since most of low-*k* dielectrics rely on methyl groups for keeping hydrophobicity, severe plasma damage makes them hydrophilic which facilitates moisture uptake during the following integration processes [Baklanov *et al.*, 2013]. In order to understand the complex damage mechanisms in low-*k* dielectrics, the effects of photons, ions, and radicals must be separately clarified.

Jinnai *et al.* [2008] suggested that photon irradiation plays an important role in the damage process and enhances the breakage of Si-C bonds. In their study, Ar and SF6 plasma irradiation was first compared with Ar and SF6 neutral beam irradiation. The plasma irradiations in both Ar and SF6 cases cause drastic extraction of methyl groups due to the presence of photons. The effect of photons was further confirmed with experiments

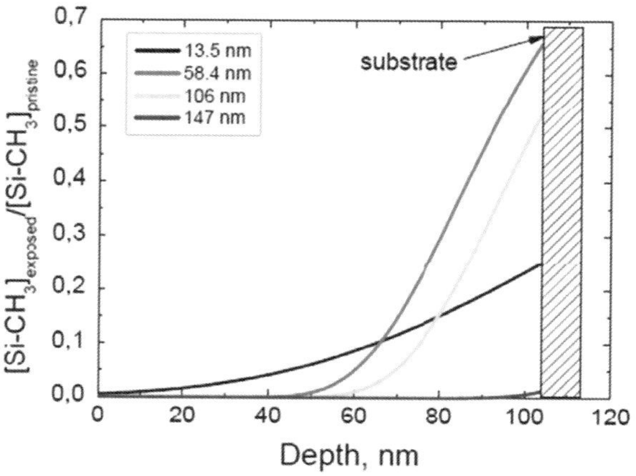

Fig. 4.6. The depth profiles of model Si-CH3 bonds depletion in ALK-B film exposed to 13.5, 58.4, 106 and 147 nm emissions for 7200s (Figure was taken from [Rakhimova *et al.*, 2014]. © IOP Publishing. Reproduced by permission of IOP Publishing. All rights reserved.)

combining an SF6 neutral beam and a 172 nm excimer lamp. In both radical irradiation and neutral beam/radical irradiation cases without photons, no damage was observed while the irradiations with photons created damage. The effect of VUV and extreme ultraviolet (EUV) photons on low-*k* film modification was systematically studied by Rakhimova *et al.* [2014]. The wavelengths studied were 13.5 nm (EUV) and 58.4 nm, 106 nm, 147 nm and 193 nm (VUV). Figure 4.6 shows that the most severe damage over the whole film thickness occurs after exposure with 147 nm light since this emission has the smaller photoabsorption cross section and enables deep penetration into low-*k* films. In Figure 4.7, the effective quantum yield was calculated based on the VUV exposure results of the low-*k* dielectrics with porosity ranging from 20% to 55%. The effective quantum yield in the range between 0 and 1 quantifies the efficiency of CH3 escaping from the porous film with a possible reattachment of –CH3 back to dangling Si-bonds. This figure shows that high interconnectivity in low-*k* dielectrics also facilitates photon penetration, therefore leading to a stronger degradation.

As reported by Nichols *et al.* [2012], dielectric failure times and charges to breakdown decrease for VUV exposed low-*k* dielectrics. The same research group [Nichols *et al.*, 2013] also investigated the influence of photon irradiation on leakage currents. Comparison of leakage currents in low-*k* films before and after UV and VUV exposure show that VUV exposures could have a similar effect to UV cures, where leakage current are reduced due to certain photoactive compounds being decomposed and removed

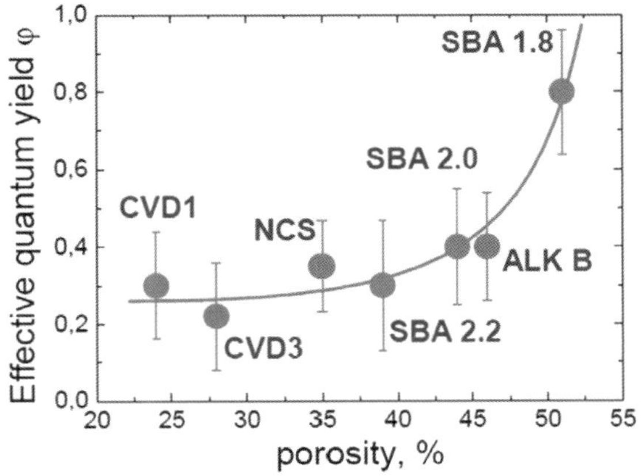

Fig. 4.7. The average effective quantum yield for breaking Si-CH3 bonds at VUV photon absorption depending on low-*k* dielectrics porosity (Figure was taken from [Rakhimova *et al.*, 2014]. © IOP Publishing. Reproduced by permission of IOP Publishing. All rights reserved.)

during the exposure. However, the ESR results from Ren *et al.* [2011] show that VUV photon irradiation generates additional defects and twisted bonds in the structure. Increasing defect densities and leakage currents after UV light irradiation are also discussed in other literature reports [Bittel *et al.*, 2010]. Sinha *et al.* [2012] suggested that photon irradiation produces trapped charges inside the low-*k* material, where such charge accumulation leads to potential reliability problems.

The nature of the defects generated during ion bombardments was studied by Afanas'ev *et al.* [2013] by using three discharging gases: H2, He, and Ar. Knock-off of atoms from the low-*k* material network are believed to happen during the ion sputtering process resulting in the formation of Si vacancies like EX centers or dangling carbon bonds, where the carbon related defects contribute to a higher leakage. Furthermore, the results of King *et al.* [2013] show the generation of surface oxygen vacancies, probably due to the removal of terminal organic groups after Ar+ sputtering, where two related sub-gap surface states are observed at 5.0 eV and 7.2 eV. Nichols *et al.* [2013] studied the effect of the ion energy in the plasma. The wafer bias in the plasma reactor was adjusted to increase the ion energy. An increase of the ion energy resulted in increased leakage currents and reduced breakdown fields, where the formation of an oxide like layer due to loss of carbon near the film surface was put forward as an explanation. It was further hypothesized that defects are more easily generated inside this oxide-like layer. Sinha *et al.* [2012] suggested that ion bombardments lead to ions sticking on the dielectric surface. Similar to their conclusions on photon irradiation, this charge accumulation is considered to have a negative impact on reliability.

The concern of radicals is mainly about the change of the film's hydrophobicity since radicals could also remove methyl groups from the low-*k* matrix. Recent results from Zhang *et al.* [2013] show that carbon depletion can be significantly reduced at cryogenic temperature due to etch byproduct condensation in pores, which blocks active radical diffusion.

(b) Moisture

Process induced dielectric damage makes low-*k* films hydrophilic, where moisture uptake could happen in subsequent process steps. Due to the high dielectric constant of water (~80), a small amount of water adsorption significantly increases the effective dielectric constant. Literature reports [Liniger *et al.*, 2012] show that the adsorbed moisture degrades reliability performance of low-*k* films. Attempts were done to characterize the conduction mechanism of moisturized low-*k* dielectrics, but the exact mechanism remains unknown. The work of Lloyd *et al.* [2005B] shows that "wet" capacitors could have SE or PF emission type of conduction but Michelon *et al.* [2006] suggested a non-linear ln(J)-E0.5 relationship. A hump in the J-E curve is suggested to be related to the presence of moisture [Miyazaki *et al.*, 2008]. An annealing step is supposed to remove moisture and improve the film reliability. However, Li *et al.* [2008] suggested that the removal of physisorbed water by baking at 190°C could only slightly improve dielectric reliability. In contrary, the lifetime is significantly improved after annealing at 400°C due to the removal of chemisorbed moisture. The removal of chemisorbed moisture also helps to reduce the thermal activation energy of time dependent dielectric breakdown. However, recent results from Cheng *et al.* [2014] point out that annealing at 400°C is still not sufficient. As a result, preventing the generation of Si-OH group in the low-*k* dielectric is still the best way to reduce the chemisorbed moisture instead of relying on post integration anneals.

(c) Copper and barrier metal

The influence of copper diffusion in the low-*k* dielectric has been explained in two ways. First, copper could generate shallow energy levels in the band gap of low-*k* dielectrics [Miyazaki *et al.*, 2009]. These states act as defect centers facilitating PF type conduction. Second, the penetration of copper atoms or ions also contribute to field enhancement locally inside the dielectric or at the electrode of electron injection [Lee *et al.*, 2009] [Suzumura *et al.*, 2006]. Since copper surface migration is much easier than bulk migration [Hwang *et al.*, 2005] and the copper diffusion process is strongly activated by temperature [Li *et al.*, 2007], the copper penetration is enhanced in higher porosity low-*k* films and at increased temperatures. The deposition of barrier metal prior to copper plating prevents possible copper migration. Recent studies, however, show that barrier metals like tantalum could also be involved in low-*k* dielectric degradation in a way similar to copper, where the effective dielectric thickness is reduced due to the barrier penetration after deposition [Ou *et al.*, 2008] [Tan *et al.*, 2009]. Zhao *et al.* [2011] quantified the metal barrier induced damage based on its impact on the low-*k* capacitance. The damage caused by PVD (physical vapor deposition) based Ta(N) barrier depositions increases as the low-*k* porosity increases. Atomic layer deposition (ALD) is promising to replace PVD for a conformal thin barrier deposition in high aspect ratio trenches applicable to more advanced BEOL structures. However, ALD precursors easily penetrate into the low-*k* film through interconnected pores [Furuya *et al.*, 2005] [Verdonck *et al.*, 2013] [Travaly *et al.*, 2005], making pore sealing an essential step before the actual ALD deposition. Armini *et al.* [2014] reported that self-assembled monolayers (SAM) deposited on porous low-*k* dielectrics are potential candidates to seal pores, where it was shown that these SAM layers completely prevented metal penetration during ALD barrier deposition.

(d) Chemical mechanical polishing

Chemical mechanical polishing is a process using mechanical polishing, assisted by chemical reaction to produce a topography with improved planarity [Wolf, 2012]. The V-shape of low-*k* trenches in advanced BEOL structures leads to field enhancement along the CMP interface which is a potential critical path for low-*k* dielectric breakdown. Therefore, the quality of this interface after CMP is crucial for BEOL dielectric reliability. Poor reliability performance is obtained when using unsuitable polishing slurries and post-CMP cleaning solutions [Yamada *et al.*, 2008] [Ishikawa *et al.*, 2006] [Hsieh *et al.*, 2013]. As suggested by Izumitani *et al.* [2010], porosity increase influences the film resistance to CMP damage. CMP interface quality degradation involves three concerns: dangling bonds generation, metal contaminants and moisture presence. As discussed in previous sections, these three factors inevitably increase leakage current and shorten dielectric lifetime. Yamada *et al.* [2008] reported that the presence of bond defects introduced by CMP increases PF type leakage currents. Dissolved copper atoms and moisture at the surfaces facilitate electrochemical reaction induced copper ion migration and accelerate dielectric breakdown. Heylen *et al.* [2008] showed a direct correlation between post-CMP surface metallic residues and dielectric reliability. Other issues like surface roughness and polishing scratches also degrade BEOL reliability performance [Lin *et al.*, 2012] [Liu *et al.*, 2009]. Ways to recover bond defects, to minimize copper residues and to restore film hydrophobicity during CMP or after CMP are proposed in several literature reports [Takeda *et al.*, 2001] [Sinapi *et al.*, 2007]. It is evident that developing optimized CMP processes in BEOL integration is crucial for advanced interconnect technologies.

(e) Mechanical stress and delamination

During chip bonding and packaging processes, large thermal and mechanical stresses occur in BEOL structures. Yang *et al.* [2013A] [2013B] studied the influence of mechanical stress on J-E relations of low-*k* dielectrics, where it is shown that tensile stress leads to decreased energy barrier heights and increased PF conduction. On the contrary, compressive stresses increases energy barriers and reduces leakage currents. Tan *et al.* [2008] investigated delamination induced dielectric breakdown in copper/low-*k* interconnects. Instead of mechanical stress, they attributed weak adhesion of interface between SiC(N) and SiOCH to be the cause of delamination initiation since weak adhesion facilitates electrons breaking interfacial bonds [Hwang *et al.*, 2006]. PF emission is determined as the dominant conduction mechanism at the delaminated interface. Delamination is further expanded by Joule heating and eventually leads to breakdown.

4.5. *Summary*

We reviewed recent research work on electrical properties and reliability of SiOCH type of low-*k* dielectrics. During dielectric deposition, the use of porogen based PECVD techniques introduces potential reliability issues. Appropriate selection of matrix and porogen precursors and improvements of post-deposition treatments are key to produce robust dielectrics without porogen residues or other intrinsic defects. A delicate balance between low bulk dielectric constant and sufficient intrinsic reliability performance is required when scaling porosity, where increasing porosity in the dielectric material results in local field enhancement. Also, porosity weakens the electrical robustness of the film leading to reduced failure times and breakdown fields. During damascene integration, low-*k* dielectrics are subjected to processes that create material damage, where plasma exposure is one of the critical processes. During such processes, methyl groups are extracted from the network which leads densification and *k*-value increase. Plasma damage could make the dielectric hydrophilic which facilitates moisture uptake, where moisture adsorption degrades reliability performance and cannot be solved by a post annealing step. A stable barrier/metal system is needed to ensure reliable electrical performance. Copper and barrier metal penetration inside the dielectric during barrier deposition or electrical stress are fatal as they facilitate electrical conduction and enhance local fields. Also, optimal CMP processes are necessary since the potential introduction of dangling bonds, metal contaminants and moisture significantly shorten the failure times. In order to have a successful implementation of advanced ultralow-*k* films in future BEOL interconnects, more research efforts are needed with respect to material screening and integration optimization. As shortly discussed in this paper, fabrication of porogen free low-*k* dielectrics with robust matrix arrangement is a first step to ensure sufficient reliability. However, high porosity and large pore sizes in conventional PECVD low-*k* dielectrics make them sensitive to integration induced damages. Therefore, optimization and innovation of integration approaches have to be achieved to overcome the reliability challenges, such as plasma damage, moisture adsorption and CMP damage etc..

References

Afanas'ev, V. V.; Keunen, K.; Stesmans, A.; Jivanescu, M.; Tőkei, Zs.; Baklanov, M. R. and Beyer, G. (2011), *Microelectronic Engineering*, **88**(7) p. 1503.

Afanas'ev, V. V.; Nguyen, A.; Houssa, M.; Stesmans, A.; Tőkei, Zs. and Baklanov, M. R. (2013), *Appl. Phys. Lett.*, **102**(17) p. 172908.

Armini, S.; Prado, J.; Krishtab, M.; Swerts J.; Verdonck, P.; Meersschaut, J.; Conard, T.; Blauw, M.; Struyf, H. and Baklanov, M. R. (2014), *Microelectronic Engineering*, 120 p. 240.

Atkin, J.; Song, D.; Shaw, T.; Cartier, E.; Laibowitz, R. and Heinz, T. (2008), *J. Appl. Phys.*, **103**(9) p. 094104.

Baklanov, M. R.; De Marneffe, J.; Shamiryan, D.; Urbanowicz, A.; Shi, H.; Rakhimova, T.; Huang, H. and Ho, P. (2013), *J. Appl. Phys.*, **113**(4) p. 041101.

Baklanov, M. R.; Green, M. and Maex, K. (2007) *Dielectric Films for Advanced Microelectronics*, Wiley, England.

Baklanov, M. R.; Ho, P. and Zschech, E. (2012) *Advanced Interconnects for ULSI Technology*, Wiley, England.

Baklanov, M. R.; Zhao, L.; Van Besien, E. and Pantouvaki, M. (2011), *Microelectronic Engineering*, **88**(6) p. 990.

Barbarin, Y.; Croes, K.; Roussel, P.; Li, Y.; Verdonck, P.; Baklanov, M. R. and Tőkei, Zs. (2013) *Proc. IEEE Int. Reliability Physics Symposium*, IRPS, p. 2F.3.1.

Bittel, B.; Lenahan, P. and King, S. (2010), *Appl. Phys. Lett.*, **97**(6) p. 063506.

Chen, F. and Shinosky, M. (2014), *Microelectronics Reliability*, **54**(3) p. 529.

Cheng, Y.; Leon, K.; Huang, J.; Chang, W.; Chang, Y. and Leu, J. (2014), *Microelectronic Engineering*, 114 p. 12.

Cheynet, M.; Volpi, F.; Pokrant, S.; Pantel, R.; Aimadedinne, M. and Arnal, V. (2009), *G.I.T Imaging & Microscopy*, **11**(1) p. 44.

Furuya, A.; Ohtsuka, N.; Misawa, K.; Shimada, M. and Ogawa, S. (2005), *J. Appl. Phys.*, **98**(9) p. 094902.

Gischia, G.; Croes, K.; Groeseneken, G.; Tőkei, Zs.; Afanas'ev, V. V. and Zhao, L. (2010) *Proc. IEEE Int. Reliability Physics Symposium*, IRPS, p. 5A.2.1.

Grill, A.; Gates, S. M.; Ryan, T. E.; Nguyen, S. V. and Priyadarshini, D. (2014), *Appl. Phys. Rev.*, **1**(1) p. 011306.

Heylen, N.; Li, Y.; Kellens, K.; Travaly, Y.; Vereecke, G.; Volders, H.; Tőkei, Zs.; Versluijs, J.; Rip, J.; Van Besien, E. (2008) *Proc. Advanced Metallization Conference*, p. 415.

Hong, C. and Milor, L. (2006) *Proc. IEEE Int. Interconnect Technology Conference*, IITC, p. 140.

Hsieh, Y.; Lin, W.; Lin, Y.; Hsu, H.; Chen, C.; Tsao W.; Hsu C.; Huang R.; Lin C.; Su Y.; Liu K.; Huang C. and Wu J. (2013) *Proc. IEEE Int. Reliability Physics Symposium*, IRPS, p. BD.3.1.

Hwang, N.; Tan, T.; Cheng, C.; Du, A.; Gan, C. and Pey, K. (2006), *IEEE Electron Device Letters*, **27**(4) p. 234.

Hwang, S.; Lee, H.; Ro, H. W.; Yoon, D. Y. and Joo, Y. (2005), *A. Phys. Lett.*, **87**(11) p. 111915.

Ishikawa, A.; Shishida, Y.; Yamanishi, T.; Hata, N.; Nakayama, T.; Fujii, N.; Tanaka, H.; Matsuo, H.; Kinoshita, K. and Kikkawa, T. (2006), *J. Electrochemi. Soc.*, **153**(7) p. G692.

ITRS 2013, http://www.itrs.net/Links/2013ITRS/2013Chapters/2013Interconnect.pdf

Izumitani, J.; Kodama, D.; Kido, S.; Chibahara, H.; Oka, Y.; Goto, K.; Suzumura, N.; Fujisawa, M. and Miyatake, H. (2010), *Japanese J. Appl. Phys.*, **49**(5S2) p. 05FC02.

Jinnai, B.; Nozawa, T. and Samukawa, S. (2008), *J. Vac. Sci. Technol. B: Microelectronics and Nanometer Structures*, **26**(6) p. 1926.

Kayaba, Y. and Kikkawa, T. (2008), *Japanese J. Appl. Phys.*, **47**(7R) p. 5314.

King, S.; French, M.; Jaehnig, M.; Kuhn, M. and French, B. (2011), *Appl. Phys. Lett.*, **99**(20) p. 202903.

King, S.; French, B. and Mays, E. (2013), *J. Appl. Phys.*, **113**(4) p. 044109.

Lauer, J.; Sinha, H.; Nichols, M.; Antonelli, G.; Nishi, Y. and Shohet, J. (2010), *J. Electrochemi. Soc.*, **157**(8) p. G177.

Lee, S.; Oates, A. and Chang, K. (2009) *Proc. IEEE Int. Reliability Physics Symposium*, IRPS, p. 481.

Li, Y.; Ciofi, I.; Carbonell, L.; Heylen, N.; Van, Aelst J.; Baklanov, M. R.; Groeseneken, G.; Maex, K. and Tőkei, Zs. (2008), *J. Appl. Phys.*, **104**(3) p. 034113.

Li, Y.; Groeseneken, G.; Maex, K. and Tőkei, Zs. (2007), *IEEE Transactions on Device and Materials Reliability*, **7**(2) p. 252.

Lin, W.; Lin, J.; Tsai, T.; Hsu, C.; Lin, J.; Hwang, C. and Wu, J. (2012), *Microelectronic Engineering*, 92 p. 115.

Liniger, E.; Shaw, T.; Cohen, S.; Leung, P.; Gates, S.; Bonilla, G.; Canaperi, D. and Rao, S. (2012), *Microelectronic Engineering*, 92 p. 130.

Liu, W.; Lim, Y.; Zhang, F.; Zhang, W.; Chen, C.; Zhang, B.; Tan, J.; Sohn, D. and Hsia, L. (2009) *Proc. IEEE Int. Reliability Physics Symposium*, IRPS, p. 613.

Lloyd, J. R.; Liniger, E. and Shaw, T. (2005A), *J. Appl. Phys.*, **98**(8) p. 084109.

Lloyd, J. R.; Shaw, T. and Liniger, E. (2005B) *Proc. IEEE Int. Integrated Reliability Workshop*, IIRW, p. 39.

Marsik, P.; Urbanowicz, A.; Verdonck, P.; De Roest, D.; Sprey, H. and Baklanov, M. R. (2011), *Thin Solid Films*, **519**(11) p. 3619.

Michelon, J. and Hoofman, R. (2006), *IEEE Transactions on Device and Materials Reliability*, **6**(2) p. 169.

Miyazaki, H.; Kodama, D. and Suzumura, N. (2008) *Proc. IEEE Int. Reliability Physics Symposium*, (IRPS), p. 150.

Miyazaki, H.; Kodama, D. and Suzumura, N. (2009), *J. Appl. Phys.*, **106**(10) p. 104103.

Nichols, M.; Li, W.; Pei, D.; Antonelli, G.; Lin, Q.; Banna, S.; Nishi, Y. and Shohet, J. (2014), *J. Appl. Phys.*, **115**(9) p. 094105.

Nichols, M.; Mavrakakis, K.; Lin, Q. and Shohet, J. (2013), *J. Appl. Phys.*, **114**(10) p. 104107.

Nichols, M.; Sinha, H.; Wiltbank, C.; Antonelli, G.; Nishi, Y. and Shohet, J. (2012), *Appl. Phys. Lett.*, **100**(11) p. 112905.

Ogawa, E.; Kim, J.; Hasse, G.; Mogul, H. and McPherson, J. (2003) *Proc. IEEE Int. Reliability Physics Symposium*, IRPS, p. 166.

Ohring, M. (1998) *Reliability and Failure of Electronic Materials and Devices*, Academic Press, USA.

Ou, Y.; Wang, P.; He, M.; Lu, T.; Leung, P. and Spooner, T. (2008), *J. Electrochem. Soc.*, **155**(12) p. G283.

Pomorski, T.; Bittel, B.; Lenahan, P.; Mays, E.; Ege, C.; Bielefeld, J.; Michalak, D. and King, S. (2014), *J. Appl. Phys.*, **115**(23) p. 234508.

Rakhimova, T.; Rakhimov, A.; Mankelevich, Y.; Lopaev, D.; Kovalev, A.; Vasil'eva, A.; Zyryanov, S.; Kurchikov, K.; Proshina, O.; Voloshin, D.; Novikova, N.; Krishtab, M. and Baklanov, M. R. (2014), *J. Phys. D: Appl. Phys.*, **47**(2) p. 025102.

Ren, H.; Nichols, M.; Jiang, G.; Antonelli, G.; Nishi, Y. and Shohet, J. (2011), *Appl. Phys. Lett.*, **98**(10) p. 102903.

Shamuilia, S.; Afanas'ev, V. V.; Somers, P.; Stesmans, A.; Li, Y.; Tőkei, Zs.; Groeseneken G. and Maex K. (2006), *Appl. Phys. Lett.*, **89**(20) p. 202909.

Sinapi, F.; Heylen, N.; Travaly, Y.; Vereecke, G.; Baklanov, M. R.; Kesters, E.; Van Hoeymissen, J.; Hernandèz, J.; Beyer, G. and Fischer, P. (2007), *Microelectronic Engineering*, **84**(11) p. 2620.

Singh, P.; Knaup, J.; Zimmermann, S.; Schulze, S.; Schulz, S.; Frauenheim, T. and Hietschold, M. (2014), *Microporous and Mesoporous Materials*, 187 p. 23.

Sinha, H.; Ren, H.; Nichols, M.; Lauer, J.; Tomoyasu, M.; Russell, N.; Jiang, G.; Antonelli, G.; Fuller, N.; Engelmann, S.; Lin, Q.; Ryan, V.; Nishi, Y. and Shohet, J. (2012), *Journal of Applied Physics*, **112**(11) p. 111101.

Suzumura, N.; Yamamoto, S.; Kodama, D.; Makabe, K.; Komori, J.; Murakami, E.; Maegawa, S. and Kubota, K. (2006) *Proc. IEEE Int. Reliability Physics Symposium*, IRPS, p. 484.

Sze, S. M. (1981) *Physics of Semiconductor Devices, 2nd Ed.*, Wiley, USA.

Takeda, K.; Hinode, K.; Noguchi, J. and Yamaguchi, H. (2001), *Japanese J. Appl. Phys.*, **40**(1-4B) p. 2658.

Tan, T.; Gan, C.; Du, A. and Cheng, C. (2009), *J. Appl. Phys.*, **106**(4) p. 043517.

Tan, T.; Gan, C.; Du, A.; Tan, Y. and Ng, C. (2008), *J. Mater. Res.*, **23** (6) p. 1802.

TravaIy, Y.; Schuhmacher, J.; Baklanov, M. R.; Giangrandi, S.; Richard, O.; Brijs, B.; Van Hove, M.; Maex, K.; Abell, T.; Somers, K.; Hendrickx, M.; Vanquickenborne, L.G.; Ceulemans, A. and Jonas, A. (2005), *J. Appl. Phys.*, **98**(8) p. 083515.

Urbanowicz, A.; Shamiryan, D.; Zaka, A.; Verdonck, P.; De Gendt, S. and Baklanov, M. R. (2010A), *J. Electrochem. Soc.*, **157**(5) p. H565.

Urbanowicz, A.; Vanstreels, K.; Verdonck, P.; Shamiryan, D.; De Gendt, S. and Baklanov, M. R. (2010B), *J. Appl. Phys.*, **107**(10) p. 104122.

Van Besien, E.; Pantouvaki, M.; Zhao, L.; De Roest, D.; Baklanov, M. R.; Tőkei, Zs. and Beyer, G. (2012), *Microelectronic Engineering*, 92 p. 59.

Vanstreels, K.; Ciofi, I.; Barbarin, Y. and Baklanov, M. R. (2013A), *J. Vac. Sci. Technol. B: Microelectronics and Nanometer Structures*, **31**(5) p. 050604.

Vanstreels, K.; Wu, C.; Gonzalez, M.; Schneider, D.; Gidley, D.; Verdonck, P. and Baklanov, M. R. (2013B), *Langmuir*, **29**(38) p. 12025.

Verdonck, P.; Delabie, A.; Swerts, J.; Farrell, L.; Baklanov, M. R.; Tielens, H.; Van Besien, E.; Witters, T.; Nyns, L. and Van Elshocht, S. (2013), *Microelectronic Engineering*, 106 p. 81.

Wolf, S. (2002) *Silicon Processing for the VLSI Era IV*, Lattice Press, USA.

Wu, C.; Li, Y.; Barbarin, Y.; Ciofi, I.; Croes, K.; Bömmels, J.; De Wolf, I. and Tőkei, Zs. (2013), *Appl. Phys. Lett.*, **103**(3) p. 032904.

Wu, C.; Li, Y.; Barbarin, Y.; Ciofi, I.; Tang, B.; Kauerauf, T.; Croes, K.; Bömmels, J.; De Wolf, I. and Tőkei, Zs. (2014) *Proc. IEEE Int. Reliability Physics Symposium* (IRPS), p. 3A.2.

Wu, C.; Li, Y.; Baklanov, M. R. and Croes, K. (2015), ECS *J. Solid State Sci. Technol.*, **4**(1) p. N3065.

Wu, E. Y. and Vollertsen, R. P. (2002), *IEEE Transactions on Electron Devices*, **49**(12) p. 2131.

Yamada, Y.; Konishi, N.; Noguchi, J.; Jimbo, T.; Kurokawa, S. and Doi, T. (2008), *Japanese J. Appl. Phys.*, **47**(6R) p. 4469.

Yang, Y.; Young, T.; Chang, T.; Hsu, J.; Tsai, T.; Jian, F. and Chang, K. (2013A), *IEEE Electron Device Letters*, **34**(8) p. 1056.

Yang, Y.; Young, T.; Chang, T.; Shen, F.; Hsu, J.; Tsai, T.; Chang, K. and Chen, H. (2013B), *Appl. Phys. Lett.*, **102**(19) p. 192912.

Zhang, L.; Ljazouli, R.; Lefaucheux, P.; Tillocher, T.; Dussart, R.; Mankelevich, Y. A.; De Marneffe, J. -F.; De Gendt, S. and Baklanov, M. R. (2013), *ECS J. Solid State Sci. Technol.*, **2**(6) p. N131.

Zhao, L.; Volders, H.; Baklanov, M. R.; Tőkei, Zs.; Pantouvaki, M.; Wilson, C.; Van Besien, E.; Beyer, G. and Claeys, C. (2011), *Microelectronic Engineering*, **88**(9) p. 3030.

Zhao, L.; Tőkei, Zs.; Gischia, G.; Pantouvaki, M.; Croes, K. and Beyer, G. (2009A) *Proc. IEEE Int. Reliability Physics Symposium*, IRPS, p. 848.

Zhao, L.; Tőkei, Zs.; Gischia, G.; Volders, H. and Beyer, G. (2009B) *Proc. IEEE Int. Interconnect Technology Conference*, IITC, p. 206.

Zheng, H.; King, S.; Ryan, V.; Nishi, Y. and Shohet, J. (2014), *Appl. Phys. Lett.*, **104**(6) p. 062904.

Chapter 6

Silicon Nitride on Si: Electronic Structure for Flash Memory Devices

V. A. Gritsenko
A. V. Rzhanov Institute of Semiconductor Physics
Siberian Branch Russian Academy of Sciences
13, Lavrentieva Ave. 630090, Novosibirsk, Russia

Amorphous silicon oxide SiO_2 and amorphous silicon nitride Si_3N_4 are the two key dielectrics in silicon electronics. Amorphous Si_3N_4 exhibits a memory effect due to its capability of trapping electrons and/or holes injected into this material; the accumulated charge can be stored in nitride during a very long time (typically, ten years at 85°C). Amorphous silicon nitride presents a model material for investigating the localization effect in amorphous materials. This effect in Si_3N_4 is used for developing a new generation of nonvolatile flash-memory devices. In spite of the numerous many-year studies, the nature of defects and traps responsible for localization of electrons and holes in silicon nitride still remains a debated matter. In the present review, we analyze electron paramagnetic resonance experiments, charge transport and luminescence studies, and also results of quantum-chemical calculations devoted to the study of traps localizing electrons and holes in Si_3N_4.

1. Introduction

The capability to exhibit localization of electrons and/or holes is a characteristic feature of amorphous semiconductors and dielectrics [Anderson (1975), Streeti *et al.* (1975), Ziman *et al.* (1979), Mott (1979), Sze (1981)]. The electron localization phenomenon in amorphous selenium gives a basis for the photocoping method. Localization of charge carriers in amorphous semiconductors and dielectrics is also manifested in experiments on charging and in electron and hole transport studies. Experimentally, in both cases an exponential (Arrhenius-type) dependence on reciprocal temperature is observed. The thermal activation energy of conductivity is normally less than half the bandgap energy. Such a behavior is typical of a broad range of materials. Among such materials are silicon nitride Si_3N_4 [Gritsenko (1986a), Roizin *et al.* (2007)], non-stoichiometric silicon nitride SiN_x [Nasyrov *et al.* (2011)], germanium nitride Ge_3N_4 [Gritsenko (1976)], boron nitride BN [10], and also oxides of semiconductors and metals such as silicon-rich silicon oxide SiO_x [11], aluminum oxide Al_2O_3 [Novikov *et al.* (2009)], tantalum oxide Ta_2O_5, zirconium oxide ZrO_2 [Houssa *et al.* (2000)], hafnium oxide HfO_2 [Jegert *et al.* (2010)], and titanium oxide TiO_2 [Kim *et al.* (1996)].

The localization, or memory, effect in silicon nitride was discovered in 1967 [Paol *et al.* (1967)]. Experimentally, the memory effect in silicon nitride was manifested as accumulation of either negative (electron localization) or positive (hole localization) charges in the material with injection of charge carriers out of silicon occurring in strong ($\approx 5 \times 10^6$ V/cm) electric fields (Fig. 1) [Nasyrov *et al.* (2009)]. The thermal energy of electron and hole traps in silicon nitride is about 1.5 eV. Deep traps in silicon nitride guarantee a very long data (charge) storage times in flash memory devices. Localized electrons and holes are retained in silicon nitride at 85°C during a a period of ten years. Presently, the memory effect in silicon nitride is widely used in developing a new generation of nonvolatile flash memory devices for storing data at giga- and terabit scales [Roizini *et al.* (2007), Eitan *et al.* (2000), Ginovker *et al.* (1974), Gritsenko *et al.* (2003), Lee *et al.* (2005), Lisiansky *et al.* (2006)].

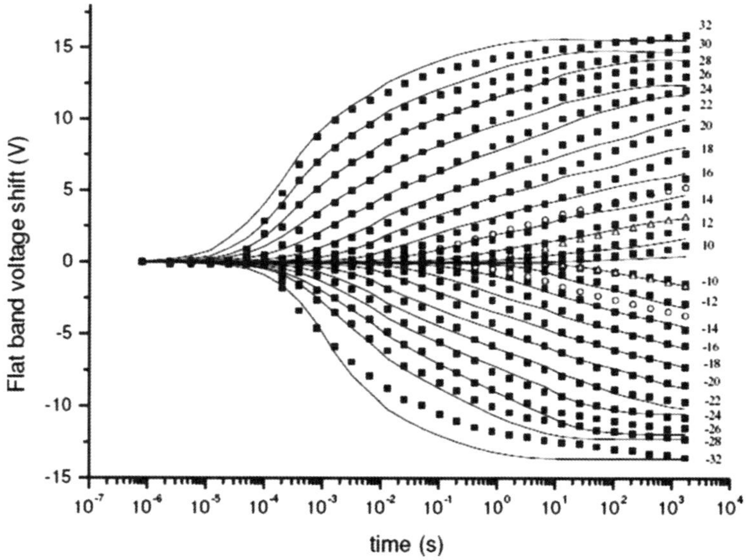

Fig. 1. Accumulation of electrons and holes in the metal-nitride-oxide-semiconductor (MNOS) structure.

The widespread use of the localization effect in silicon nitride has stimulated numerous experimental and theoretical studies in this field. Presently, amorphous silicon nitride is considered as a model material to investigate into the localization phenomenon in amorphous materials. A good deal of studies has been devoted to the study of localization in silicon nitride by means of electron paramagnetic resonance studies [Krick *et al.* (1988a), Watanabe *et al.* (1990), Warren *et al.* (1990), Kanicki *et al.* (1990), Nadolinnyui *et al.* (1991), Gritsenko *et al.* (1999), Pacchioni *et al.* (1999), Aozasa *et al.* (2007)], charge transport studies [Nasyrov *et al.* (2009), Gritsenko *et al.* (1997a), Bachhofer *et al.* (2001), Nasyrov *et al.* (2004), Hong *et al.* (2005), Lee *et al.* (2005), Kim *et al.* (2006), Tzeng *et al.* (2006), Fuks *et al.* (2006), Chen *et al.* (2007), Vishnyakov *et al.* (2009), Seo (2008)], luminescence studies [Pundur *et al.* (1986), Gritsenko *et al.* (1986b), Mo *et al.* (1993), Giorgisr *et al.* (1999), Seo (2000), Park *et al.* (2001), Gritsenko *et al.* (2004), Cen *et al.* (2009), Anutgan *et al.* (2011)], photoelectron spectroscopy [Gritsenko (1986a) Ley *et al.* (1984), Karcher *et al.* (1984), Guraya *et al.* (1990), Hasegawa *et al.* (1991), Hasegawa *et al.* (1992), Gritsenko *et al.* (1998b), Gritsenko *et al.* (2000), Gritsenko *et al.* (2002)], and quantum-chemical calculations [Robertson (1981), Martin-Moreno *et al.* (1987), San-Fabian *et al.* (1989), Robertson (1991), Gritsenko (1993), Gritsenko *et al.* (1997b), Pacchioni *et al.* (2000), Petersen *et al.* (2006), Sorokin *et al.* (2008), Giacomazzi1 *et al.* (2009), Nekrashevich *et al.* (2009), Valentin *et al.* (2011), Grillo *et al.* (2011), Gritsenko (2013)]. Our purpose is to give an overview of modern concepts regarding the nature of localized states responsible for the localization phenomenon in amorphous silicon nitride.

2. Application of Silicon Nitride in Flash-Memory Devices

Currently, flash memory based on TaN-alumina-nitride-oxide-semiconductor (TANOS) structures is under extensive study and launching in mass production. The TANOS memory cell presents a field-effect transistor with a multi-layer dielectric and variable threshold voltage (Fig. 2). The storage medium in such memory cells is formed by amorphous silicon nitride Si_3N_4 with a high (10^{19} cm^{-3}) density of deep ($\approx$1.5 eV) traps for electrons and holes. In TANOS structure, the silicon nitride layer is separated from the silicon substrate by a tunnel SiO_2 layer 1.8 to 5.0 nm thick. For avoiding parasitic injection of electrons and holes into silicon nitride, the silicon nitride is isolated from the gate by a blocking oxide layer. Traditionally, the blocking layer in polysilicon-oxide-nitride-oxide-silicon (PONOS) structures is silicon oxide.

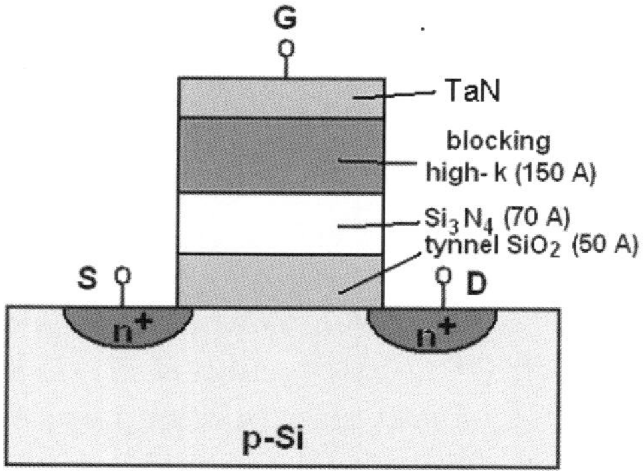

Fig. 2. The TaN-Al_2O_3-Si_3N_4-Si (TANOS) flash memory cell.

Application of a negative potential (voltage pulse of 10^{-3}-sec duration) to the gate triggers the injection of holes from silicon substrate into silicon nitride through the tunnel oxide layer; the injected holes then get captured by deep traps. After removal of the negative gate voltage, the holes accumulated in silicon nitride induce a conductive inversion channel in the field-effect transistor. The transistor passes over to on-state corresponding to logical "0". Unlike in ordinary triggers, due to a very long time of hole retention at traps the logical "0" can be stored in memory without consumption of energy during a ten-year period at 85°C. In this way, information is memorized in a flash memory cell. A nonvolatile multi-element memory involves many flash memory cells. Erasing of data, or reprogramming, is achieved by applying a positive voltage pulse to the gate. In the latter case, electrons are injected out of silicon across the tunnel oxide layer to get captured by deep traps in silicon nitride. After removal of the gate voltage, the nitride layer carries a negative charge that induces an accumulation region in silicon enriched with holes. The conducting layer vanishes, and the transistor comes in off-state. The latter state refers to logical "1". This is the way the traditional PONOS flash memory cell based on silicon nitride operates.

In [Roizini *et al.* (2007), Gritsenko *et al.* (2003)], replacement of the blocking SiO_2 layer in PONOS structure with a high-k dielectric such as, for instance, Al_2O_3, HfO_2, or ZrO_2, was proposed. Band diagrams illustrating the operation of a PONOS memory cell with a ZrO_2 layer used as the blocking layer in data erasure mode are shown in Fig. 3. Since the dielectric constant of ZrO_2 ($\varepsilon = 25$) is much greater than the same constant of SiO_2 ($\varepsilon = 3.9$), at one and the same amplitude of programming/erasure pulses the electric field in the high-k blocking dielectric turns to be lower than the

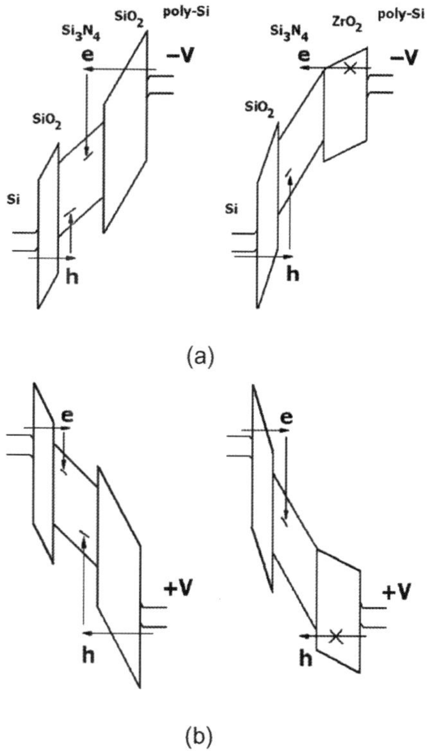

Fig. 3. The schematic comparison write/erase energy diagrams of SONOS structure with different top dielectrics: SiO_2 and ZrO_2. The arrows show injection of electrons or holes through top and bottom dielectrics when the different polarity of applied voltage is used.

field in the blocking SiO_2. Hence, the voltage drop across the high-k blocking dielectric turns out to be smaller than the voltage drop across the blocking SiO_2 layer. The reduction of the voltage drop across the blocking layer due to the use of high-k dielectrics results in that a larger voltage drops across the tunnel silicon oxide layer and, hence, the field in the tunnel oxide increases in intensity. The increase of electric-field strength leads to an enhanced tunnel injection of electrons and holes out of the silicon substrate and, hence, to faster operation of the memory cell in programming and erasure modes at one and the same thickness of tunnel oxide layer. In PONOS structures with blocking SiO_2 layer, a 1.8-nm thick thermally grown tunnel SiO_2 layer is used. The replacement of the blocking SiO_2 layer with a high-k dielectric allows one, with keeping the speed of operation equally high, to increase the thickness of tunnel SiO_2 layer from 1.8 to 3.5 nm [Lee *et al.* (2005), Lisiansky *et al.* (2006), Lee *et al.* (2005)]. The enhanced thickness of tunnel oxide layer improves the reliability of flash memory devices in data storage mode.

Another advantage offered by the use of high-k blocking dielectrics consists in suppression of parasitic injection of electrons and holes from the gate due to the decrease of field intensity in the high-k blocking dielectric. Thus, the use of high-k dielectrics enables achievement of a faster speed of operation, improved reliability, enhanced product yield, and, hence, a lower cost of flash memory. Another promising dielectric for use as blocking dielectric in PONOS structures is aluminum oxide Al_2O_3.

An advanced flash memory cell using silicon nitride as the storage medium is an n-channel field-effect transistor with TaN-Al_2O_3-Si_3N_4-Si cross-sectional structure (see Fig. 2). Here, the 3.5-nm thick silicon oxide layer thermally grown on silicon is a tunnel dielectric through which tunnel injection of electrons and holes out of silicon into silicon nitride occurs. The 5.0-nm thick amorphous silicon nitride layer acts as a storage medium in which injected electron and holes get localized. As a dielectric to block the parasitic injection of electrons and holes out of gate into silicon nitride, a 15.0-nm thick amorphous aluminum oxide layer is normally used. Charge accumulation in such a TANOS flash memory cell occurs on application of either positive or negative potential to the conducting gate (TaN).

The band diagram of a TANOS flash memory cell in different states is shown in Fig. 4. The band diagram in the initial state, i.e. without a charge accumulated in silicon nitride, is shown in Fig. 4(a). The values of the energy barriers at the interfaces between the various materials are given as taken from [Perevalov *et al.* (2010)]. The band diagram of the structure under a positive potential applied to the gate (accumulation of electrons in silicon nitride, data writing mode) is shown in Fig. 4(b). The dielectric permittivity of tunnel-transparent silicon oxide SiO_2 is $\varepsilon = 3.9$, the permittivity of storage-medium silicon nitride Si_3N_4 is $\varepsilon = 7.0$, and the permittivity of blocking Al_2O_3 is $\varepsilon = 10.0$. In Fig. 4(b), and in subsequent figures, the electric field in the dielectric layers of TANOS structure is shown with regard for different values of ε in the various dielectrics. The horizontal arrows in Fig. 4(b) and in subsequent figures show the directions of electron and hole currents flowing in the TANOS structure. Here, the parasitic injection of holes out of TaN gate into aluminum oxide is blocked due to the high height of the barrier for holes and due to the low field strength in Al_2O_3. A short-circuited TANOS structure containing electrons accumulated in the silicon nitride layer is shown in Fig. 4(c). The field due to electrons localized in silicon nitride is screened in silicon by free holes. There is no conducting channel in the transistor. The latter state of TANOS structure refers to logical "1". The band diagram of the TANOS structure under a

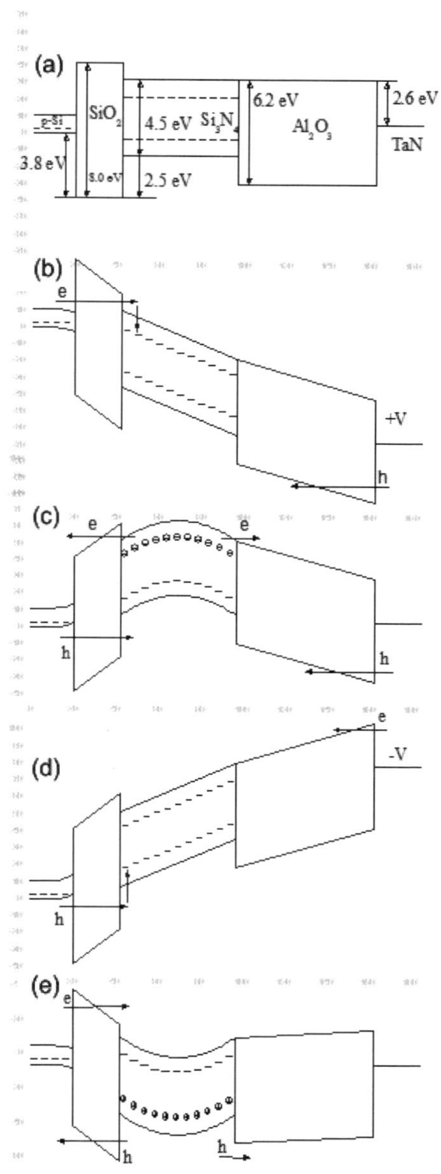

Fig. 4. The band diagram of a TaN-Al_2O_3-Si_3N_4-SiO_2-Si (TANOS) flash memory cell with silicon nitride used as the storage medium: (a) zero bias voltage, (b) positive potential at the metal, injection of electrons into silicon nitride, (c) short-circuited state with accumulated electrons, data storage regime, logical “0”, (d) negative potential at the metal, injection of holes into silicon nitride, (e) short-circuited state with accumulated holes, data storage regime, logical “1”.

negative potential at the gate is shown in Fig. 4(d). Here, tunnel injection of holes out of silicon into silicon nitride across the tunnel oxide occurs. The injected holes get accumulated at deep traps in silicon nitride. In short-circuited TANOS structure the holes localized in silicon nitride induce an inversion layer in silicon Fig. 4(e). In the latter case, a conducting layer forms in the TANOS transistor; this state refers to logical "0". The development of TANOS structure and silicon-nitride-based nonvolatile memory based on this structure stimulated the study of localization of electrons and holes in amorphous silicon nitride. Presently, development of 3D flash memory chips based on poly-silicon-Al_2O_3-Si_3N_4-SiO_2-Si (PANOS) structures in which, for increasing the data storage capacity, many stacked poly-silicon layers are used, is under way [http://www.theregister.co.uk].

3. Atomic and Electronic Structure of Silicon Nitride

3.1. *Electronic structure of stoichiometric silicon*

Silicon nitride is a tetrahedral compound in which each silicon atom is coordinated with four nitrogen atoms, and each nitrogen atom is coordinated with three silicon atoms. In crystalline state, there exist two hexagonal phases of silicon nitride, α-Si_3N_4 and β-Si_3N_4 [Gritsenko *et al.* (1984)], and also a cubic c-Si_3N_4 phase [Mo *et al.* (1999)]. In silicon devices, amorphous silicon nitride is used. The electronic structure of crystalline Si_3N_4 was studied experimentally using X-ray spectroscopy [Gritsenko (1993), Gritsenko *et al.* (1997b), Gritsenko *et al.* (1997c)]. Theoretically, the electronic structure of Si_3N_4 was the subject matter of calculations reported in [Robertson (1981), Martin-Moreno *et al.* (1987), San-Fabian *et al.* (1989), Robertson (1991), Gritsenko (1993), Gritsenko *et al.* (1997), Sorokin *et al.* (2008), Giacomazzi1 *et al.* (2009), Grillo *et al.* (2011)]. The Si-N chemical bond in Si_3N_4 is formed due to hybridization of Si 3s, p, d orbitals of silicon with N 2p orbitals of nitrogen. The valence band of Si_3N_4 consists of two subbands. The wide upper subband is formed by Si 3s, p, d orbitals of silicon and N 2p binding orbitals of nitrogen. Near the top of the valence band, there is an antibinding N $2p_\pi$ orbital of nitrogen. The lower subband is formed by N 2s orbitals of nitrogen with an admixture of Si 3s, 3p orbitals of silicon.

There is much controversy concerning the value of the bandgap energy in Si_3N_4. According to different workers, the bandgap energy of Si_3N_4 falls into the interval from 4.5 eV [Gritsenko (1986a), Gritsenko (1993)] to 5.3 eV [Gritsenko (1982)]. According to optical absorption data, at the fundamental absorption edge the bandgap width of amorphous Si_3N_4 amounts to 4.5 eV

[Philip (1973), Sorokin *et al.* (2009)]. A similar value was obtained while examining the photoemission of electrons from aluminum into Si_3N_4, and also from data gained in studying the photoelectron spectroscopy of the valence band of silicon nitride on silicon [Sorokin *et al.* (2009)].

3.2. *Atomic and electronic structure of silicon-rich silicon nitride*

In extreme cases, two alternative models are used to describe the structure of non-stoichiometric tetrahedral amorphous layers: the random mixture (RM) model and the random bonding (RB) model [Gritsenko (1986a), Gritsenko *et al.* (1998b), Gritsenko *et al.* (2000), Gritsenko *et al.* (2002)]. The RM model implies that SiN_x consists of two phases, amorphous silicon (Si) and Si_3N_4 [Voskoboinikov *et al.* (1976), Novikov *et al.* (2009)]. The RB model represents the structure of SiN_x as an atomic network formed by five types of tetrahedra, SiN_vSi_{4-v} (v = 0, 1, 2, 3, 4) [Gritsenko (2008)]. Both models fail to give a quantitative explanation to the structure of SiN_x. In [Gritsenko *et al.* (2004)], an intermediate model to explain the atomic structure of SiN_x was proposed. The intermediate model implies the presence of individual phases of Si ($SiSi_4$ tetrahedron) and Si_3N_4 (SiN_4 tetrahedron), and, also, the presence of subnitrides consisting of SiN_3Si, SiN_2Si_2, and $SiNSi_3$ tetrahedrons. In this model, the chemical composition of SiN_x is assumed to exhibit local fluctuations in space.

The intermediate model for amorphous SiN_x is schematically illustrated by Fig. 5(a) and (b). Figure 5(b) shows the 2D spatial distribution of silicon phase (black), stoichiometric nitride (Si_3N_4) phase (white), and subnitrides (grey). Figure 5(a) shows the section along the line A-A in the energy space. The solid line in Fig. 5(a) is the energy level from which the electron energy is reckoned (vacuum energy). Symbols E_c and E_v denote the conduction-band bottom and the top of valence band. A decrease of bandgap energy points to the presence of a subnitride phase in silicon nitride. The lowest width of the forbidden band (E_g = 1.6 эВ) corresponds to amorphous silicon phase.

The electronic structure of silicon-rich silicon nitride with an atomic structure obeying the RB model was theoretically examined in [Martin-Moreno *et al.* (1987)]. According to X-ray absorption edge data $SiL_{2,3}$, enrichment of SiN_x with excess silicon is accompanied by a shift of the conduction-band bottom of SiN_x into the forbidden band [Ley *et al.* (1984)]. According to photoelectron spectroscopy data, on increasing the excess silicon concentration the top of the SiN_x valence band also moves into the bandgap [Karcher *et al.* (1984)]. Enrichment of SiN_x with silicon leads to a decrease of the optical bandgap energy (Fig. 6) [Martin-Moreno *et al.* (1987)].

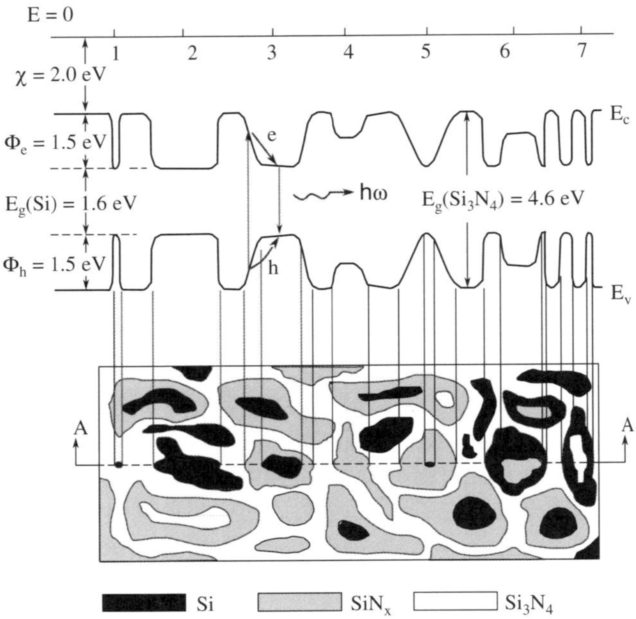

Fig. 5. The intermediate model of the atomic structure of amorphous SiN_x. Bottom: 2D picture illustrating the atomic structure of SiN_x. The black color shows the silicon phase, the white color shows the stoichiometric silicon nitride Si_3N_4, and the grey color, subnitrides. Top: The band diagram of SiN_x in the section A-A. Symbols E_c and E_v denote the conduction band bottom and the valence band ceiling.

This phenomenon is related with the fact that the introduction of nitrogen into SiN_x is accompanied by a shift of binding and anti-binding orbitals in Si-Si bonds towards a greater value of bandgap energy.

4. Electronic Structure of Defects in Silicon Nitride According to Experimental and Quantum-Chemical Simulation Data

4.1. *Triply coordinated silicon atom with unpaired electron*

In silicon nitride irradiated with ultra-violet (UV) radiation, a 14-G wide EPR signal with gyromagnetic ratio $g \approx 2.003$ was detected (Fig. 7) [Lenahan *et al.* (1990)]. The hyperfine splitting of the EPR signal due to the presence of ^{29}Si isotope in powdered silicon nitride was examined (natural abundance of this isotope in silicon is 4.7%). Prior to performing measurements, samples were irradiated with UV quanta whose energy was approximately 5.0 eV. The spectrometer sensitivity in measurement of satellite lines was one hundred times greater than the sensitivity in measuring the central line. The time values in Fig. 7 indicate the duration of UV irradiation. The nature

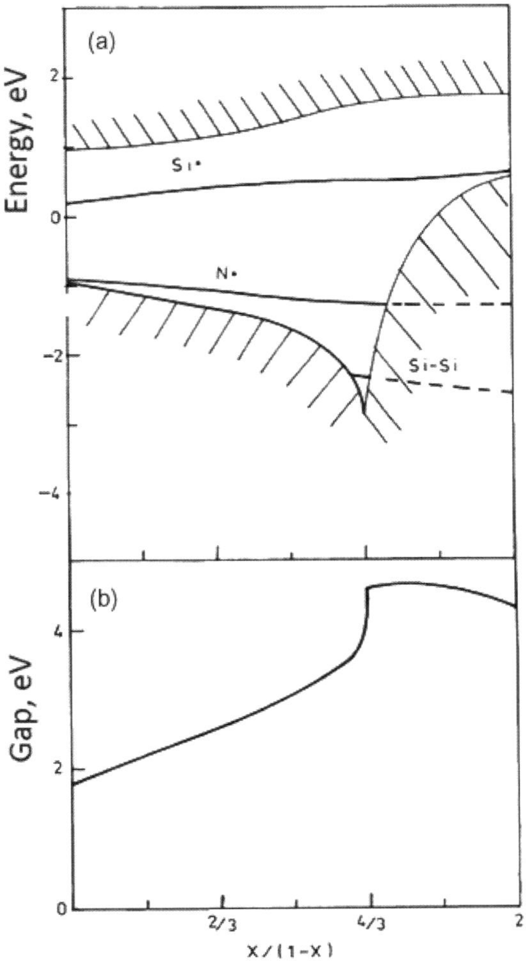

Fig. 6. (a) Band edges and gap state energies of a-$Si_{1-x}N_x$ as a function on N content. Solid lines indicate true localized states introduced by both Si and N dangling bonds and Si-Si wrong bonds. Dashed lines indicate resonances. The position of the "lone-pair" band maximum is also shown in the figure. (b) Optical gap energy as a function on N content.

of defects responsible for the EPR signal was identified in [Warren *et al.* (1990)]. Before UV irradiation, a weak signal with $g \approx 2.003$ was observed in the samples. The UV irradiation induced satellite lines spaced from the central line by 171 G. The intensity of the signal increased on increasing the duration of the exposure to UV irradiation. An analysis showed that the EPR signal was due to an unpaired electron localized at a silicon atom with 20.5% Si 3s plus 55% Si 3p hybridization. In silicon nitride, a silicon atom with unpaired electron triply coordinated with nitrogen, $=N_3Si\cdot$, is called the K-center [Krick *et al.* (1988a)]. A triply coordinated silicon atom

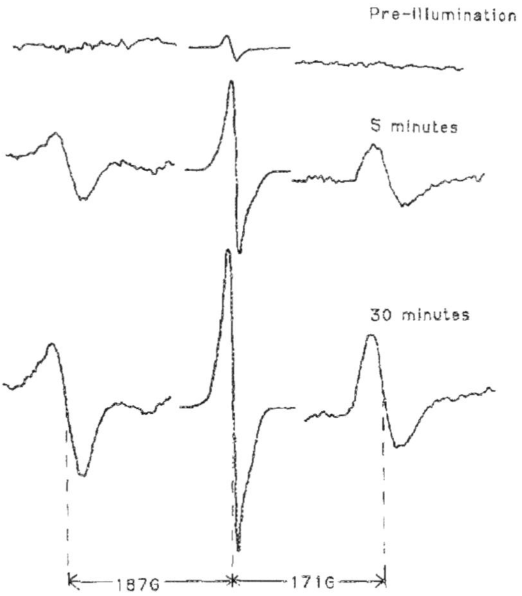

Fig. 7. Center ($g = 2.003$) and satellite lines before and after the indicated periods of ultraviolet ($E < 5\,\text{eV}$) illumination, The spectrometer gain for the satellites is a factor of 100 higher than for the center line. The satellite line amplitudes consistently scale at about $\approx$2% each of the center line.

with unpaired electron $N_3Si\cdot$ in silicon nitride presents an analogue to the well-known triply coordinated silicon in silicon oxide, $O_3Si\cdot$, also called the E′-center [Conley *et al.* (1994)]. The gyromagnetic ratio of the EPR line due to the K-center is $g = 2.003$, the width of the line at half maximum being 13 Gauss.

In [Krick *et al.* (1988a)] it was concluded that K-centers in silicon nitride are responsible for electron and hole localization. According to the concept of [Krick *et al.* (1988a)], the capture of an electron or a hole makes the neutral paramagnetic centers charged diamagnetic defects according to the following reactions:

$$\equiv\text{Si}\cdot + \text{e} \rightarrow \equiv\text{Si}: \quad (1)$$

$$\equiv\text{Si}\cdot + \text{h} \rightarrow \equiv\text{Si} \quad (2)$$

Here, the symbol (-) denotes a normal chemical bond formed by two electrons, and the symbol (·) denotes an unpaired electron.

Thereupon, in [Krick *et al.* (1988a)] it was concluded that it is K-centers in silicon nitride that act as traps for electrons and holes. It should be noted here that the idea that K-centers in silicon nitride act as traps for electrons and holes has been widely discussed in literature [Pacchioni *et al.* (1999), Aozasa *et al.* (2007), Hasegawa *et al.* (1991), Hasegawa *et al.* (1992), Robertson (1981), Pacchioni *et al.* (2000), Kamigaki *et al.* (1990), Yang *et al.* (2000), Hasegawa *et al.* (1996), Robertson (1994), Krick *et al.* (1988b), Libsh *et al.* (1987), Lenahan *et al.* (1989), Shannon *et al.* (1999), Kanicki *et al.* (1995), Gritsenko *et al.* (2003), Aozasa (1999), Fujita *et al.* (1985)].

The capability of K centers to capture electrons and holes was examined by performing *ab initio* quantum-chemical calculations in [Pacchioni *et al.* (1999), Pacchioni *et al.* (2000), Gritsenko *et al.* (2003)]. According to [Pacchioni *et al.* (1999)], the neutral K-center in silicon nitride acts as a trap for electrons. Depending on a particular model, the energy gain due to electron capture falls in the interval 0.8 to 3.2 eV. According to [Pacchioni *et al.* (1999)], the K-center can localize an electron. In [Pacchioni *et al.* (1999), Gritsenko *et al.* (2003)] it was established that trapping of holes at K-centers in silicon nitride is energetically unfavorable.

4.2. *Silicon-silicon bond*

A hypothesis that, in silicon nitride, localization of electrons and holes occurs due to their trapping at Si-Si bonds was first put forward at the Institute of Semiconductor Physics, Russian Academy of Sciences, Siberian Division, Novosibirsk [Gritsenko (1986a), Pundur *et al.* (1986), Gritsenko *et al.* (1986b), Brytov *et al.* (1984), Gritsenko *et al.* (1987)].

The silicon-silicon bond, $N_3Si\text{-}SiN_3$, in crystalline silicon nitride results from the nitrogen vacancy following the saturation of one of the dangling bonds of the neighbor silicon atom, $\equiv$Si·, with hydrogen due to the formation of a $\equiv$Si-H bond [Gritsenko *et al.* (2003a)].

Quantum-chemical calculations performed by the recursion method show that, like the hydrogen molecule, the neutral diamagnetic silicon-silicon bond $N_3Si\text{-}SiN_3$ in silicon nitride has a binding orbital and an anti-binding orbital [Petersen *et al.* (2006), Nekrashevich *et al.* (2009), Robertson (1994)]. At the binding orbital of the Si-Si bond, two electrons with opposite spins are localized. The energy level of the binding σ-orbital in the Si-Si bond is located near the bottom of the valence band. The unoccupied energy level of the anti-binding σ^*-orbital is located near the bottom of the silicon nitride conduction band (Fig. 8).

Gritsenko *et al.* [Gritsenko *et al.* (1998a)] employed the semi-empirical MINDO/3 method to model the distribution of charge and spin density over the Si-Si bond as dependent on the separation between silicon atoms in different charge states. Figure 9 shows the 1D distribution of the electron

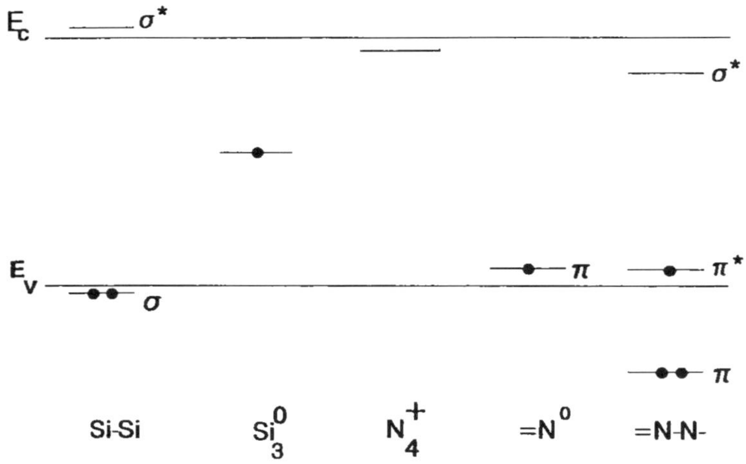

Fig. 8. Energies of defect states in silicon nitride.

density at the silicon atoms of the Si-Si bond in neutral state versus the separation between the silicon atoms [Gritsenko *et al.* (1998a)]. An increase in the silicon-to-silicon separation is accompanied by a reduction of the charge at silicon atom and with an increase of spin density. The 3D distribution of electron density at the neutral Si-Si bond is shown in Fig. 10. In the state with trapped hole, an asymmetric distribution of charge and spin densities is observed (Fig. 11). Similarly, the state with captured electron also features an asymmetric distribution of charge density (Fig. 12). At one of the atom, two electrons are localized while at the other atom, one electron. The spin density is distributed asymmetrically, too. At the atom with two electrons the spin density vanishes, whereas at the atom with localized electron this density assumes a finite value. The model of the Si-Si bond as a trap for electrons and holes in silicon nitride is illustrated in Fig. 13. Quantum-chemical modeling shows that the Si-Si bond in silicon nitride acts as a trap both for electrons and holes.

4.3. *Nitrogen vacancy in silicon nitride*

The nitrogen vacancy in silicon nitride results on withdrawal of one nitrogen atom out of the silicon nitride lattice [Nekrashevich *et al.* (2009)]. Following this removal, three silicon atoms with unpaired electron, ≡Si·, triply coordinated with nitrogen, arise. Subsequent atomic relaxation leads to the formation of a diamagnetic ≡Si-Si≡ bond and a paramagnetic silicon atom with unpaired electron, ≡Si· [Valentin *et al.* (2011)]. In [Nekrashevich *et al.* (2009)] it was shown that the nitrogen vacancy in silicon nitride acts as

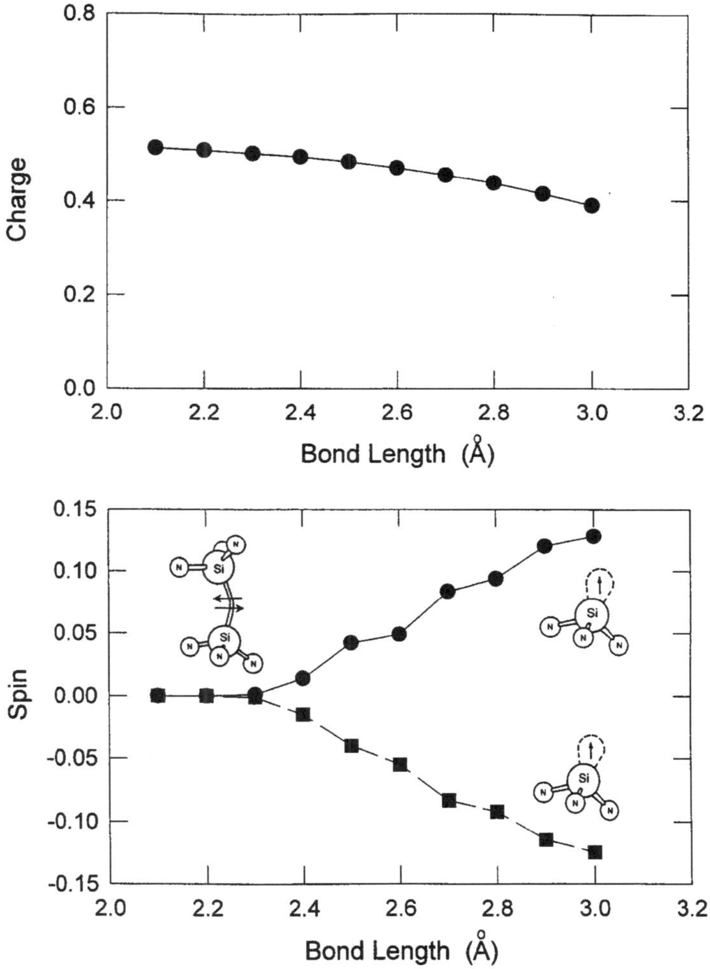

Fig. 9. Charge and spin distribution of silicon atom for the Si-Si bond in Si_3N_4 as a function Si-Si distance.

a deep trap for electrons and holes. Yet, the idea that the traps for electrons and holes in silicon nitride originate just from nitrogen vacancies contradicts to experiments. In the initial, non-polarized state silicon nitride shows no EPR signal. The nitrogen vacancy in neutral state is a paramagnetic center.

4.4. *Silicon clusters in silicon nitride*

N.-M. Park, S.-H. Choi, and S.-J. Park [Park *et al.* (2002)] studied the localization of electrons and holes in SiN_x with amorphous silicon nanoclusters. It was found that such nanoclusters in silicon nitride act as traps

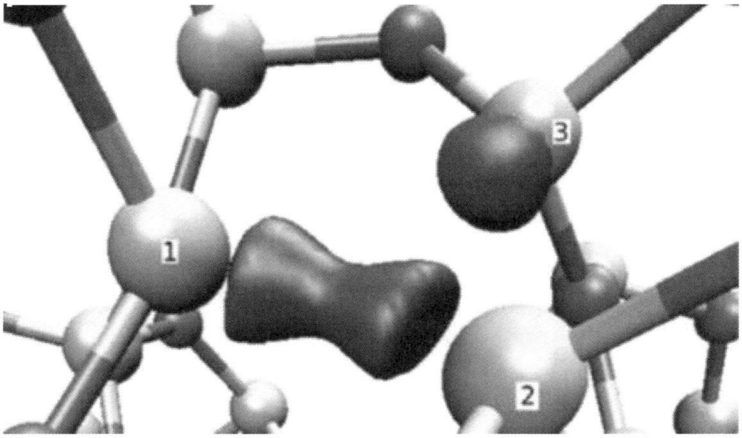

Fig. 10. Bonding orbital between two silicon atoms (1,2), unpaired electron on the atom No. 3.

for electrons and holes. This conclusion was confirmed theoretically by quantum-chemical calculations [Valentin *et al.* (2011)].

5. Electron Paramagnetic Resonance in Silicon Nitride

5.1. *Electron paramagnetic resonance in neutral silicon nitride*

EPR experiments provide data on the atomic and electronic structure of defects in various materials. Electron paramagnetic resonance in near-stoichiometric, neutral silicon nitride was examined in [Warren *et al.* (1993), Shames *et al.* (2001)].

It was found that, in neutral silicon nitride, after synthesis there are no paramagnetic defects whose concentration would be comparable with the concentration of traps (Fig. 14) [Shames *et al.* (2001)]. The latter implies that in non-occupied neutral state the traps in silicon nitride present diamagnetic centers. Simultaneously, after UV irradiation in silicon nitride synthesized from dichlorsilane paramagnetic centers with gyromagnetic ratio $g = 2.003$ arise (Fig. 14). A similar result was also obtained for silicon nitride synthesized from dichlorsilane in [Gritsenko *et al.* (1998)] (Fig. 15).

5.2. *Electron paramagnetic resonance in silicon nitride with localized charge carriers*

Electron paramagnetic resonance in silicon nitride with localized electrons was examined using silicon samples with deposited $SiN_{1.6}$ layers [Warren *et al.* (1993)]. With the help of a corona discharge, electrons were injected

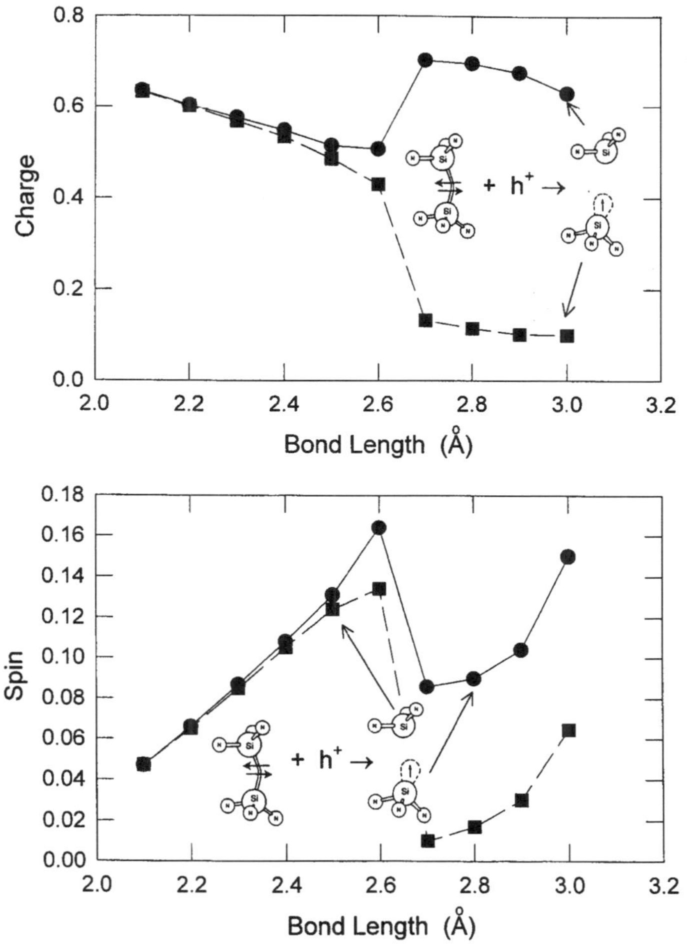

Fig. 11. Charge and spin variation of two silicon atoms of the Si-Si bond with a captured hole in Si_3N_4.

into the silicon nitride. Trapping of electrons in the $SiN_{1.6}$ layer resulted in a shift of capacitance-voltage characteristics towards positive gate potentials. Electrons in silicon nitride get captured at deep traps; the accumulated negative charge resulted in the mentioned shift of measured capacitance-voltage curves [Warren *et al.* (1993)]. However, in an ESR experiment performed on silicon nitride with localized electrons no paramagnetic centers with a concentration comparable with the concentration of negatively charges traps were detected (Fig. 16) [Warren *et al.* (1993)].

Special experiments aimed at observation of an EPR signal from localized holes in silicon nitride were reported in [Gritsenko *et al.* (1998)]. For the experiment, special silicon-oxide (20 nm)-nitride (210 nm)-oxide (50 nm)

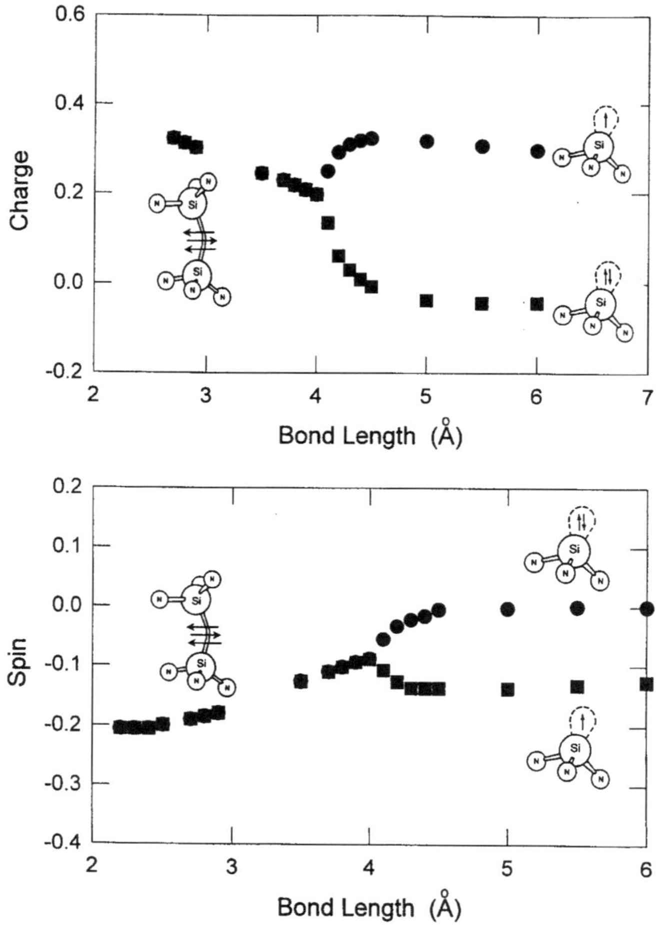

Fig. 12. Charge and spin variation of two silicon atoms of the Si-Si bond with a captured electron in Si_3N_4.

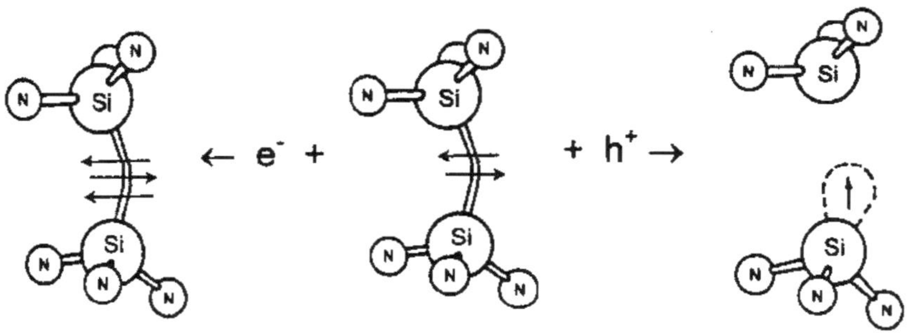

Fig. 13. Atomic model of amphoteric neutral diamagnetic Si-Si defect with electron and hole capturing in Si_3N_4.

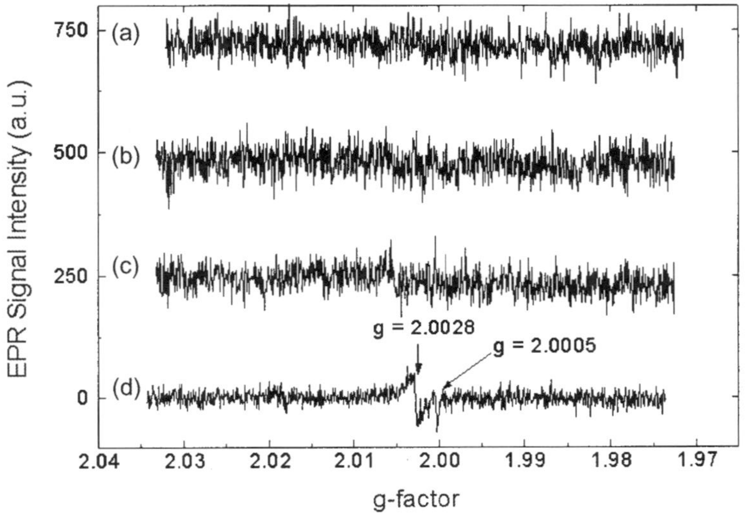

Fig. 14. Room temperature EPR spectra of (a) non-polarized Si_3N_4; (b) Si_3N_4, containing injected $5*10^{12}$ electrons cm^{-2}; (c) Si_3N_4, containing injected $5*10^{12}$ holes cm^{-2}; (d) weak Pitch test sample containing about 10^{13} spins.

(SONO) structures were fabricated. The silicon nitride was deposited onto an oxidized silicon surface from a mixture of dichlorsilane SiH_2Cl_2 and ammonia NH_3 at 850°C. The oxide layers on the silicon and nitride surfaces were used to prevent escape of electrons and holes out of the silicon nitride layer. Figure 17 shows the band diagram of an SONO structure under flat-band conditions (without an applied gate voltage and an accumulated charge). Injection of electrons and holes out of silicon into silicon nitride was organized using corona discharge plasma. A strong electric field in the dielectric ($\approx 10^7$ V/cm) was produced with the help of charged ions impacting onto the dielectric surface out of plasma [Warren *et al.* (1993)]. The injection of electrons and holes out of silicon through the oxide layer was a tunnel injection. Since the concentration of traps in SiO_2 was low, a prevailing fraction of injected charge carriers underwent trapping in Si_3N_4. After polarization, or charge injection, capacitance-voltage measurements were performed to evaluate the sheet density of the charge accumulated in silicon nitride. Then, the silicon wafer was cut into rectangles sized $7 \times 4\,mm^2$, and then it was given grinding such that to reduce its thickness from 400 to 10–15 μm. The obtained samples were stacked together (each stack comprised about one hundred samples), wrapped in a fluoroplastic film, and EPR measurements were performed. For EPR measurements, SONO samples with initial (non-polarized) nitride and SONO samples

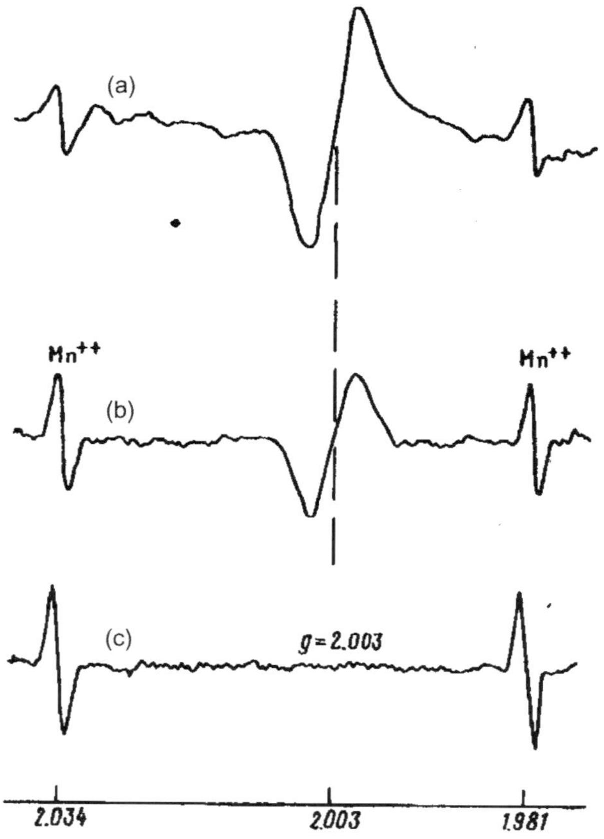

Fig. 15. ESR spectra: (a) Silicon-enriched nitride $SiN_{1,2}$; (b) stoichiometric Si_3N4 after UV irradiation for 2 h with an unfiltered radiation from a 400 W hydrogen lamp; (c) Si_3N_4 before UV irradiation.

with nitride-accumulated holes were prepared. The adhered strategy has allowed the authors of [Warren *et al.* (1993)] to increase, by two orders of magnitude, the total number of Si_3N_4–localized spins in the resonator of their EPR spectrometer in comparison with a situation, being often the case, in which just one sample was used for measurements.

EPR spectra measured on SONO structures with the total number of localized holes $6 \cdot 10^{14}$ are shown in Fig. 18 [Gritsenko *et al.* (1998)]. Here, two signals are observed (see Fig. 18(a)). A broad signal with g=2.0055 (halfwidth $\Delta G=5$ G) is observed in amorphous silicon and on the surface of crystalline silicon. This signal is due to silicon atoms with unpaired electron $\equiv Si_3Si\cdot$ triply coordinated with other silicon atoms. The EPR signal due to this center is also exhibited by defects in grinded silicon; it was observed in non-polarized samples without an accumulated charge (Fig. 18(b)).

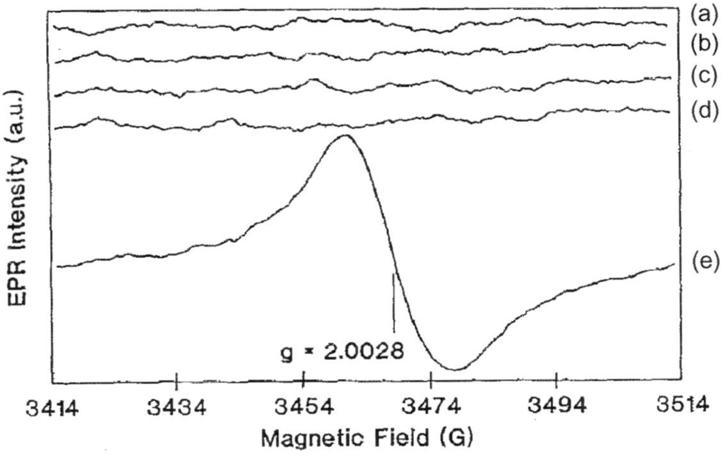

Fig. 16. ESR spectra of unilluminated a-SiN_x:H: Trace (a) virgin, (b)–(d) injected with electrons for 50, 80, and 200 s of positive corona time, respectively, and (f) UV illuminated following electron injection.

A narrow signal ($\Delta G = 1.4\,G$) with $g = 2.00055$ (Fig. 18(c)) is due to the well-known E′-center in SiO_2 [Gritsenko *et al.* (2011)]. The E′-center forms on capture of a hole by a neutral ≡Si-Si≡ defect according to the reaction ≡Si-Si≡ -e →≡Si^+· Si≡ proceeding at the Si/SiO_2 interface. On capture of a hole at the defect ≡Si-Si≡ in Si_3N_4, a signal due to triply coordinated silicon atom with unpaired electron N_3Si·, which forms by analogy with the E′-center according to the reaction =N_3Si-SiN_3 = -e →=N_3Si^+· Si N_3 =, can be expected. Indeed, the silicon atom with unpaired electron with $g = 2.003$ (K-center) was observed in silicon-rich silicon nitride ($SiN_{1.2}$) and in stoichiometric nitride after UV irradiation (Fig. 15).

Thus, in examined samples E′-centers appeared in SiO_2 after injection of holes into the SONO structure, and trapping of holes was accompanied by the formation of positively charged paramagnetic centers. Simultaneously, accumulation of holes in silicon nitride did not lead to the formation of paramagnetic centers. This means that after capture of holes the traps in silicon nitride remain diamagnetic.

5.3. *Anti-Ferromagnetic ordering of electrons due to resonant exchange interaction*

The authors of [Shames *et al.* (2001), Gritsenko *et al.* (1998), Gritsenko (1996a)] discussed the model (possibility) of anti-ferromagnetic pairing of electrons localized at neutral deep traps in a dielectric. They have considered a dielectric with a high concentration of randomly scattered deep

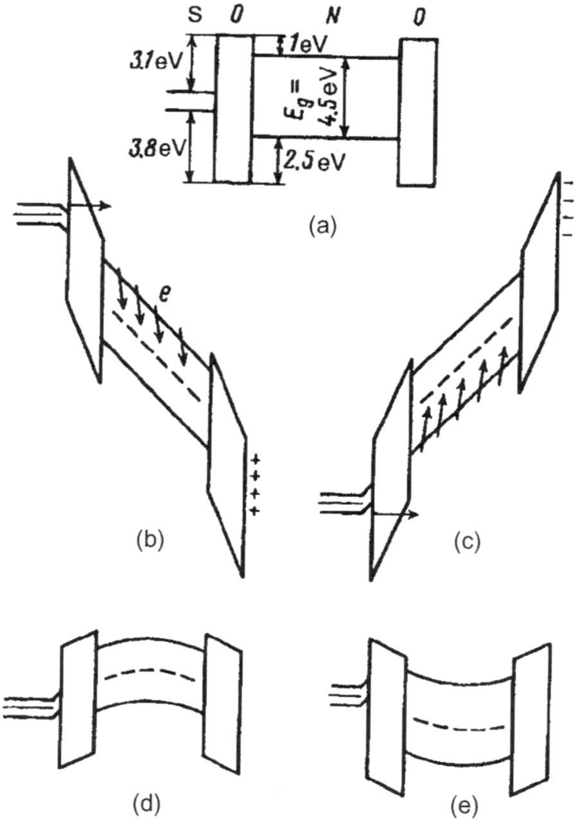

Fig. 17. Energy diagram of silicon–oxide–nitride–oxide (SONO) structures: (a) flat-band regime; (b) polarization in a positive corona discharge, electron injection from Si; (c) polarization, hole injection from Si; (d) structure with localized electrons; (e) structure with localized holes.

neutral traps. Due to the Coulomb repulsion of trap-localized electrons, the concentration of traps occupied with electrons n_t can be much lower than the total concentration of traps N_t. In the proposed model, resonant tunneling of localized electrons through empty traps results in their anti-ferromagnetic pairing. This model therefore explains the experimentally observed absence of an electron paramagnetic resonance (EPR) signal from the electrons and holes localized at deep traps in amorphous silicon nitride.

Consider magnetic properties of electrons in a periodic superlattice formed by traps (Fig. 19(a) and (b)). The exchange integral J for electrons 1 and 2 located at sites A and B has the form

$$J = \int\int \psi_A(1)\psi_B(2)\frac{e^2}{r_1 - r_2}\psi_B(1)\psi_A(2)dr_1dr_2 \quad (3)$$

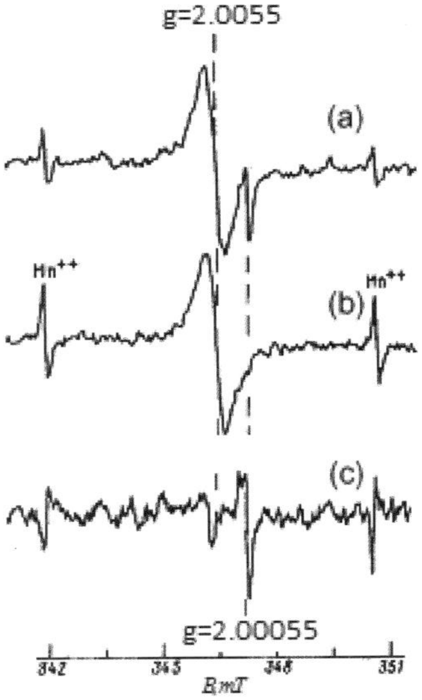

Fig. 18. ESR spectra of SONO structures: (a) with localized holes; (b) initial unpolarized structure; (c) difference spectrum obtained by subsracted the spectra (a) and (b).

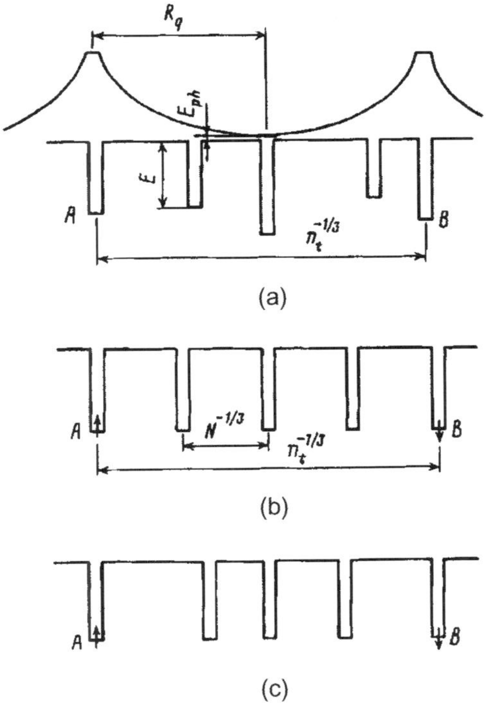

Fig. 19. Potential energy diagram illustrating the formation of a quasiperiodic electron lattice on account of Coulomb repulsion (a) and antiferromagnetic ordering in a periodic lattice on account of the resonance exchange interaction (b). Nonperiodic lattice 4 (c).

Here, the radii r_1 and r_2 are reckoned respectively from sites A and B. For calculating the exchange integral, it is required to find the eigenfunctions for an electron in the lattice. For illustration of the proposed mechanism, we choose not to solve the exact problem about the stationary distribution of spin density in the superlattice. Let us evaluate instead the overlap integral in WKB approximation. The probability P to find electron 1 of well À at a distance R is

$$P = \left(-\frac{2}{\hbar}R\sqrt{2m * E}\right). \quad (4)$$

In (4), m* is the effective electron mass. In the present model (see Fig. 19) the quantity E refers to the energy of the ground state of electron in the well. The energy of the direct exchange interaction of the electrons at sites A and B is

$$J_{AB} = q^2 P n_t^{1/3} / 4\pi\varepsilon\varepsilon_0 \quad (5)$$

Anti-ferromagnetic ordering will occur if the exchange interaction energy is greater than the Zeeman splitting energy of the level due to localized electron:

$$\left(q^2 n_t^{1/3} / 4\pi\varepsilon\varepsilon_0\right) \exp\left(-2n_t^{1/3} / h\sqrt{2m * E}\right) > g\mu H. \quad (6)$$

In (6), μ is the Bohr magneton and H is the magnetic field intensity. A typical value of the Zeeman splitting for electrons localized in a solid is $g\mu H \approx 6 \times 10^{-5}$ eV. For $n_t = 10^{18}\,cm^{-3}$, $\varepsilon = 10$, and $E = 1$ eV for the tunneling probability and for the exchange interaction energy we then obtain: $P = 10^{-27}$ and $E = 10^{-29}$ eV (for $m^* = 0.4m_e$).

Thus, the energy of the direct exchange interaction of electrons at traps is too low for initiating a spin flipping process. The presence of a superlattice in the model under consideration leads to delocalization of spin density. Let us evaluate the magnitude of the exchange interaction between an electron at site A and a neighboring empty well by substituting $R = N^{-1/3} = 10^{-7}$ cm into (21); then, for the tunneling probability we obtain $P = 10^{-3}$. The exchange interaction energy is about 10^{-4} eV. The latter energy is greater than the Zeeman splitting energy 6×10^{-5} eV. This estimate shows that the exchange interaction between trap-localized electrons in the presence of a high density of empty traps can result in a magnetic ordering of spin systems.

Above, we have considered a periodic superlattice. Exchange interaction can also occur in the case in which traps are distributed non-periodically in space yet have identical energies E (Fig. 19(c)). Here, the probability of electron tunneling from well A into well B is not the product of the probabilities of electron tunneling through individual barriers. If the barriers in between neighboring wells are strongly scattered in terms of their width, then the total probability of tunneling, in a first approximation, is defined by the tunnel transparency of the barrier separating the most distant traps (to those traps, the widest barrier corresponds). Note that for the exchange interaction between a pair of filled traps the electron can choose a trap chain with a minimal barrier width (Fig. 19(c)). In the latter case, a percolation problem arises. It seems that, in reality, pairing of electrons can indeed occur.

According to the proposed model, for observation of Wigner crystallization a dielectric with a high concentration of traps in a metal-insulator-semiconductor (MIS) structure can be used. In wide-bandgap amorphous dielectrics, the concentration of extrinsic and intrinsic charge carriers is negligible. That is why, unlike in semiconductors, the charge accumulated in the dielectric layer of MIS structure is screened by mobile carriers in the semiconductor and in the metal.

A specific feature of dielectrics with traps in MIS structures is the space charge generally present in the dielectric layer. Figure 20 shows the band diagram of an MIS structure under flat-band conditions (a), in the regime of tunnel injection of charge carriers (electrons) from the semiconductor (b), and in the state with an electron charge accumulated in the dielectric layer (c). The accumulation of electrons/holes in the dielectric leads to the formation of a space charge $qn_t(z)$ and to the emergence of an electric field normal to the semiconductor-dielectric interface.

Wigner crystallization in the dielectric of an MIS structure can also occur in 3D case. On the other hand, anti-ferromagnetic spin pairing can occur in 2D case, in the plane parallel to the semiconductor-dielectric interface. In the direction normal to the interface the trap levels are displaced due to the electric field potential by energy $\Delta\varphi = qn_n^{-1/3} \int n_t(z)dz/\varepsilon\varepsilon_0$. For typical concentration of traps $n_t = 5 \times 10^{18}\,\text{cm}^{-3}$, the displacement of neighboring traps over energy in the space-charge-induced field is $\Delta\phi \approx 0.5\,\text{eV}$. This displacement hampers the tunnel exchange in the direction normal to the semiconductor-dielectric interface. In the direction parallel to the interface the electric field is zero due to the screening of the charges localized at traps by the charges induced in the semiconductor and in the metal.

The absence of EPR signal in a dielectric with localized charge carriers serves a criterion for the occurrence of a pronounced exchange interaction.

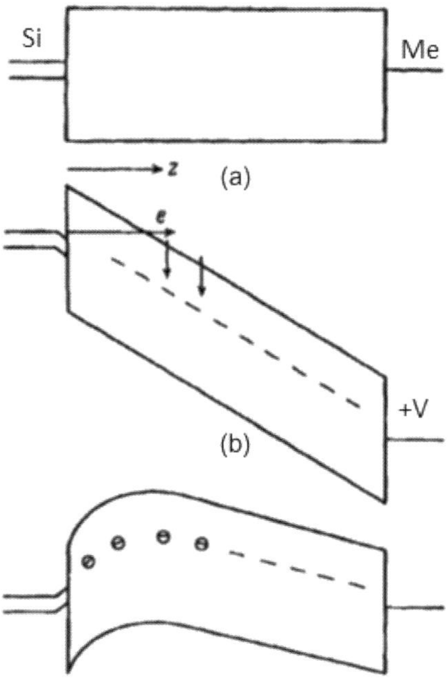

Fig. 20. Energy diagram of a MIS structure: (a) flat band regime; (b) injection of electron from semiconductor and trapping in traps in insulator; (c) shorted state.

Today, several exchange interaction mechanisms are known. The direct Heisenberg exchange mechanism explains the ferromagnetic properties of solids. The Anderson super-exchange mechanism between d-electrons through the p-states of oxygen explains the ferromagnetic properties of oxides. The indirect Rudemann-Kittel mechanism accounts for the interaction between localized spins of lattice ions via free electrons in the metal.

Thus, the resonant exchange mechanism explains the anti-ferromagnetic ordering of electrons and holes and the absence of EPR in dielectrics with a high density of traps.

6. Wigner Crystallization of Electrons in Silicon Nitride

6.1. *Wigner crystallization of localized electrons: Concept and simple models*

The Coulomb repulsion of free electrons leads to their spatial ordering, or Wigner crystallization. Like in the original Wigner model, the crystallization of localized charge carriers occurs due to their Coulomb repulsion. The Coulomb repulsion of electrons also underlies the Coulomb blockade effect.

The Coulomb blockade consists in that two electrons never occupy one and the same point in space because of their Coulomb repulsion. A system of free interacting electrons can undergo an arrangement into a regular Wigner lattice on the condition that the potential energy of electrons due to their Coulomb repulsion exceeds their kinetic energy [Ando *et al.* (1982)].

Possible Wigner crystallization of electrons on liquid-helium surface and in an inversion channel of a semiconductor was analyzed [Ando *et al.* (1982)]. In the present section, we consider a fundamentally different phenomenon, the ordering of electrons at deep traps in a dielectric due to the Coulomb repulsion of the electrons. If the concentration of traps N in a dielectric is high, so that the average separation between them, $N^{-1/3}$, is smaller than the effective capture radius $(\sigma/\pi)^{1/2}$ (here, σ is the capture cross-section), then electrons will be trapped only at part of traps $\mathrm{n_t}$, thus forming a quasi-periodic lattice. The period of the lattice will have a scale of order the characteristic radius of Coulomb repulsion. If, in addition, traps have identical energies W, then pairing of spins at neighboring traps, or anti-ferromagnetic ordering due to resonant exchange interaction, may occur.

Consider a wide-bandgap dielectric containing no intrinsic or extrinsic charge carriers. In the dielectric, there are neutral empty deep traps available in a high concentration $N \sim 10^{21}\ \mathrm{cm}^{-3}$. Consider a chain of traps in which the traps are spaced apart by a mean distance $a = N^{-1/3}$ (Fig. 19(a)). Generally, the traps may have different delocalization energies W. The separation between neighbor traps can also be different. Consider the capture of electrons injected out of a contact to the traps present in such a dielectric. As a trap A captures an electron, another (probe) charge residing at a distance R will be acted upon by the repulsive potential $\mathrm{e}^2/4\pi\varepsilon_p\varepsilon_o R$, where $\varepsilon_p^{-1} = \varepsilon_\infty^{-1} - \varepsilon^{-1}$ is the polaron dielectric permittivity, and ε_∞ and ε are respectively the high- and low-frequency dielectric permittivities. B silicon nitride, we have: $\varepsilon_\infty = 4.0$, $\varepsilon = 7.0$, and $\varepsilon_\mathrm{p} = 10$. A second electron with kinetic energy $W_{ph} \approx kT$ can approach the charged trap A to a distance $2R_q$ such that

$$\frac{q^2}{4\pi\varepsilon_p\varepsilon_o R_q} = W_{ph}. \tag{7}$$

Relation (14) defines the concentration of populated traps $\mathrm{n_t} = (2R_q)^{-3}$.

$$n_t = \frac{(4\varepsilon_p\varepsilon_o W_{ph})^3}{q^6} \tag{8}$$

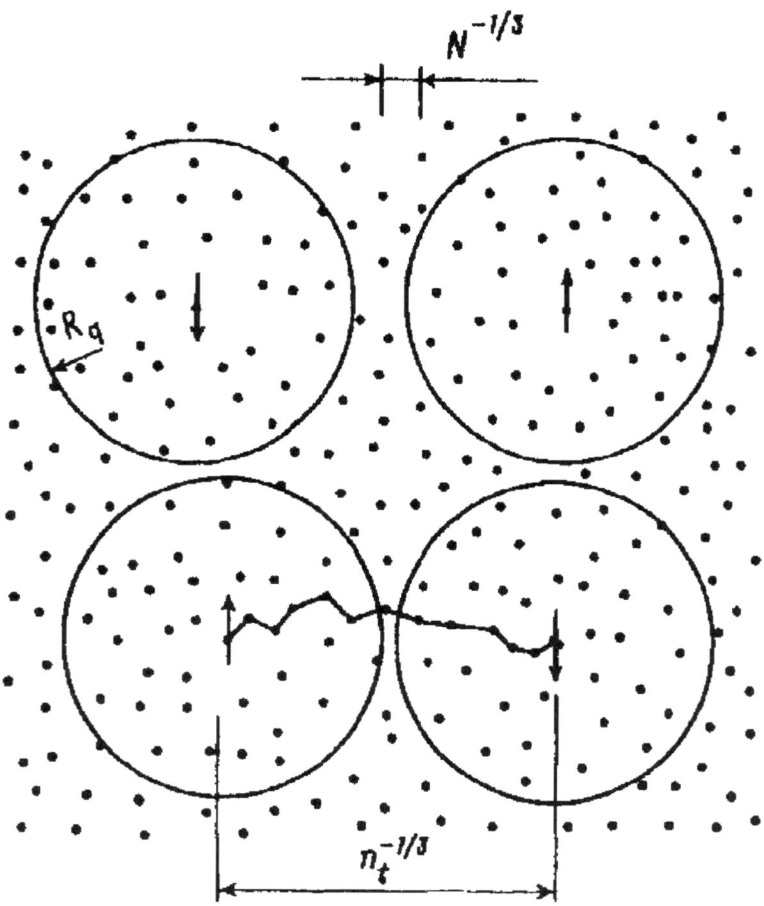

Fig. 21. Two-dimensional scheme of the formation of a quasiperiodic lattice on account of the Coulomb repulsion of electrons localized in traps and antiferromagnetic pairing of electrons on account of a resonance exchange interaction. A tunneling exchange trajectory is indicated for the bottom pair of electrons.

A schematic illustrating the formation of a 2D square lattice with period $2R_q$ is shown in Fig. 21. In the present model, such a lattice is formed by circumferences of radius R_q. The gap between the circumferences varies in proportion to $N^{-1/2}$. The latter value defines the characteristic scale of the spread of lattice periods.

Melting of a Wigner crystal formed by free electrons is known to occur at $\Gamma = r_o/a_B^* \gg 1$ (here, r_o is the mean separation between the electrons, and $a_B^* = 4\varepsilon_p\varepsilon_o\hbar/mq^2$ is the effective Bohr radius). Our model crystal will undergo melting when the Coulomb interaction energy grows in excess of the binding energy of trapped electron W, $e^2n_t^{1/3}/4\pi\varepsilon_p\varepsilon_o \gg W$. Another obvious criterion is $W \gg kT$. It implies that the trap ionization probability is low.

A universal parameter in our model is the Coulomb repulsion radius R_q. Its magnitude in our model is greater than the inter-trap separation $R_q \gg N^{-1/3}$. The capture cross-section σ is defined not only by the individual properties of traps, or by the energy loss mechanism, but also by the squared radius R_q:

$$\sigma = \pi R_q^2 = \frac{\pi q^4}{(4\pi\varepsilon_p\varepsilon_o W_{\text{ph}})^2} \tag{9}$$

Thus, in this model the concentration of filled traps and the capture cross-section are related by

$$n_t^{-1/3}/2 = (\sigma/\pi)^{1/2} = R_q \tag{10}$$

In our model, the quantity R_q has the meaning of the free path with respect to trapping, $R_q \approx (n_t\sigma)^{-1}$. The model dielectric under consideration possesses the property of being capable to localize all injected electrons till trap ionization in the field due to the accumulated space charge or in the external electric field begins.

On substitution of $W_{\text{ph}} = kT$ into (9) and adoption of $\varepsilon_p = 10$ for Si_3N_4, for the characteristic Coulomb radius we obtain an estimate $R_q \approx 5.5$ nm (see Fig. 22). Within error, the calculated Coulomb radius R_q is roughly equal to the mean separation between filled traps as determined from the experiment. This equality proves that the separation between filled traps is defined by the Coulomb repulsion of localized charge carriers. That is why in

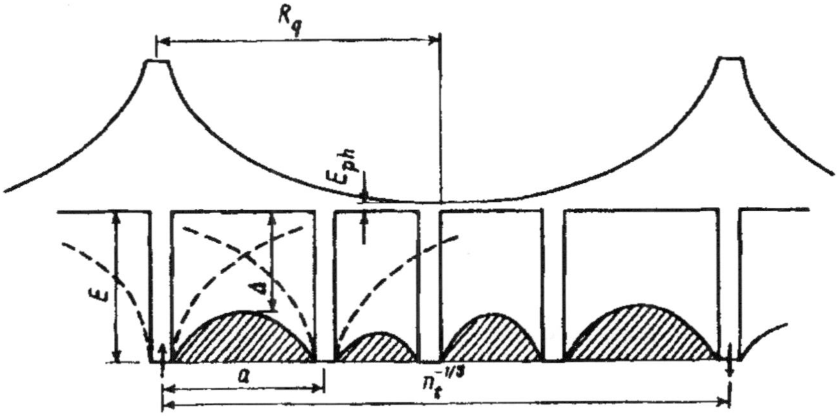

Fig. 22. Potential energy diagram illustrating spin pairing in Si_3N_4 on account of an exchange interaction with tunneling via unfilled states.

2D case in the plane parallel to the silicon-dielectric interface the electrons in silicon nitride form an ordered densely packed lattice of spheres of radius R_q. A 2D picture illustrating the distribution of localized charge carriers forming a square lattice of spheres of radius R_q is shown in Fig. 23. Thus, localized electrons and holes in silicon nitride can form a quasi-periodic lattice, or 2D Wigner crystal (or, more exactly, glass).

Relation (17) holds in a dielectric containing a high concentration of traps if the concentration of traps occupied with charge carriers n_t is low ($n_t \ll N$) in comparison with the total concentration of neutral traps N. The mean capture radii for electron and hole traps in silicon nitride are identical, equal to $R^e = R^h = (\sigma/\pi)^{1/2} = 4.5 \pm 2\,\text{nm}$ [Maslovskii *et al.* (1985)].

In silicon nitride, within inaccuracies the following relation holds for electron traps:

$$(n_t^e)^{-1/3}/2 = (\sigma^e/\pi)^{1/2} = R_q^e \tag{11}$$

Maslovsky *et al.* [Maslovskii *et al.* (1985), Aganin *et al.* (1988)] obtained relation (11) for electrons in more accurate experiments on the kinetics of accumulation of electrons injected from the top poly-silicon electrode in silicon nitride based MIS structures. In that study, samples in which silicon nitride was given various treatments were examined; due to those treatments, the concentration of filled electron traps in the sample n_t^e showed variation. A decrease of n_t^e was accompanied by an increase of σ^e, and for all samples in the experiment relation (11) proved to be valid.

Thus, both the concentration of filled traps and the capture cross-section in silicon nitride are defined by the magnitude of the characteristic Coulomb repulsion radius. The latter finding provides support to the hypothesis that trap-localized electrons in silicon nitride samples form a quasi-periodic 2D lattice. Note that expression (9) involves no parameters dependent on the sign of charge carriers. That is why the magnitude of R_q is identical for electrons and holes, $R^e = R^h = (\sigma/\pi)^{1/2}$.

6.2. *Wigner crystallization of localized electrons: Numerical simulations*

In the previous section, it was assumed that the electrons localized in silicon nitride form a quasi-periodic square lattice. Amorphous silicon nitride can be considered a model dielectric for investigation of effects due to the Coulomb blockade of electrons and holes localized at traps. According to charge transport data gained for Si_3N_4 the concentration of neutral traps in this material amounts to $N_t \sim 10^{19} - 10^{21}\ \text{cm}^{-3}$ [Gritsenko *et al.* (2003b),

Nasyrov *et al.* (2004)], whereas the concentration of filled traps is much lower, $n_t \sim (2-6) \times 10^{18}$ cm^{-3} [Hamptoon (1979)]. The low concentration of filled traps in comparison with the concentration of neutral traps ($n_t << N_t$) points to realization of Coulomb-blockade effects in silicon nitride.

In the present section, we describe results of a study in which numerical simulations were used to study the Wigner crystallization effects in a dielectric with deep traps on the basis of a 2D model. We consider a model 2D dielectric with a surface density of randomly scattered neutral traps N_s. The amorphous non-ordered structure of the dielectric material is modeled on the assumption of a random spatial distribution of traps. This dielectric is being populated with electrons available in current surface density n_s.

The calculated data are presented in Fig. 23. These data show that, if the density of traps $\mathrm{N_s}$ is comparable with the density of the electrons introduced into the system n_s, then the dependence $n_s(r)$ exhibits no specific features, and it reflects a random distributions of traps (see Fig. 23, the case of $n_s/N_s = 0.5$). For the density ratio $n_s/N_s = 0.01$, a certain structure can be traced such that the maxima in the curve $\mathrm{n_s(r)}$ are coincident with the trap-to-trap separations for charged traps typical of a six-fold coordinated lattice (Fig. 23(b)). Thus, in a plane randomly filled with neutral traps the charged traps form a 2D crystal or, more exactly, a glass with hexagonal lattice.

Thus, we showed that at certain values of n_s/N_s in a plane randomly filled with neutral traps an ordered hexagonal lattice forms from charged traps. Earlier, similar results were obtained for 2D free-electron gas [Ando *et al.* (1982)].

In the previous section, the Coulomb blockade phenomenon in silicon nitride was explained on the assumption that localized charge carriers form a square lattice. Results described above refine the structure of the Wigner lattice — in a dielectric with a high density of neutral traps, localized electrons form a hexagonal (not square!) lattice due to their Coulomb repulsion. Thus, in the present section we showed that electrons localized in a dielectric can form a non-equilibrium 2D lattice, or Wigner glass.

7. Properties of Silicon-Rich Silicon Nitride

Silicon nitride used as a storage medium can be obtained by pyrolytic decomposition of silicon- (silane SiH_4, silicon tetrachloride $SiCl_4$, dichlorsilane SiH_2Cl_2) and nitrogen-containing (NH_3) gases. The process is normally held at temperatures in the temperature range 700 to 900°C.

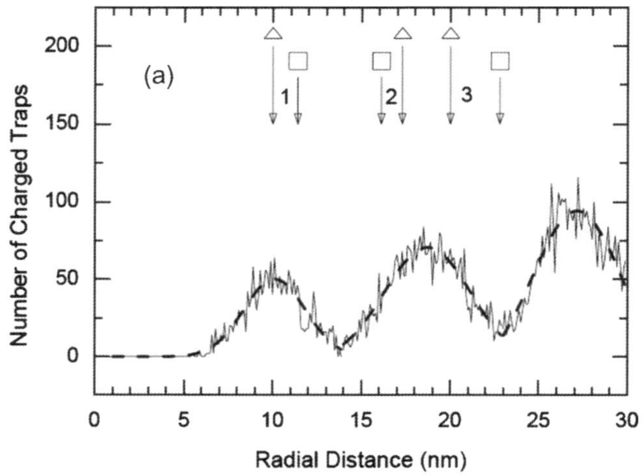

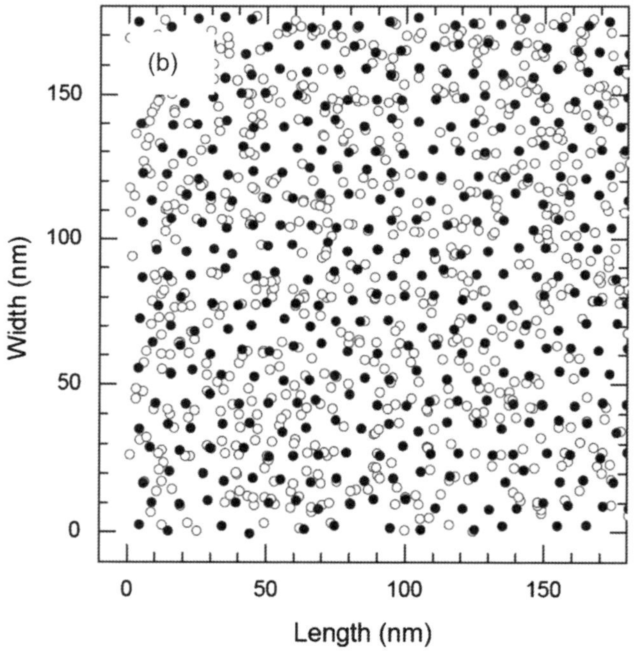

Fig. 23. (a) Calculated radial distribution of charged traps for $ns/Ns = 0.01$ with Gaussian approximation (dashed curve). Three peaks were found and their locations are close to the first, second, and third coordination spheres (marked as 1, 2, and 3, respectively) in a hexagonal (denoted by triangle or a square crystal lattice). (b) Calculated spatial distribution of neutral traps (open circles) and charged traps (filled circles) of the corresponding value of n_S/N_S.

An important parameter defining the chemical composition of synthesized nitrides is the pressure ratio of silicon-containing to nitrogen-containing gas: SiH_4/NH_3, $SiCl_4/NH_3$, SiH_2Cl_2/NH_3. Stoichiometric silicon nitride Si_3N_4 involves just Si-N bonds. As a rule, as the storage medium in metal-nitride-oxide-semiconductor (MNOS), silicon-nitride-oxide-semiconductor (SNOS), metal-oxide-nitride-oxide-semiconductor (MONOS), silicon-oxide-nitride-oxide-semiconductor (SONOS), and TaN-Al_2O_3-Si_3N_3-SiO_2-Si (TANOS) structures, nearly stoichiometric amorphous silicon nitride $SiN_{x\approx 4/3}$ is used. In general case, silicon nitride is a non-stoichiometric material. Non-stoichiometric silicon nitride is denoted with the chemical formula SiN_x:H. Such nitride involves Si-N, Si-Si, Si-H N-H, and Si-O bonds. Si-H and N-H hydrogen bonds arise due to incomplete decomposition of silicon- and nitrogen-containing gases used in the synthesis process of silicon nitride. Oxygen Si-O bonds arise due to oxidation of silicon-containing gases with residual oxygen and water vapor.

The concentration of excess silicon increases on increasing the relative partial pressure of the silicon-containing gas in the reactor (in the given case, on increasing the silane-ammonia ratio (Fig. 24) [Bailey *et al.* (1982)]. Simultaneously, the concentration of electrons trapped in silicon nitride increases on increasing the concentration of excess silicon (Fig. 25) [Bailey *et al.* (1982)]. Similar data were independently obtained in [Fujital *et al.* (1985)]. Figure 26 shows the hysteresis in silicon nitride as dependent on the NH_3/SiH_4 ratio. In a first approximation, the flat-band potential varies

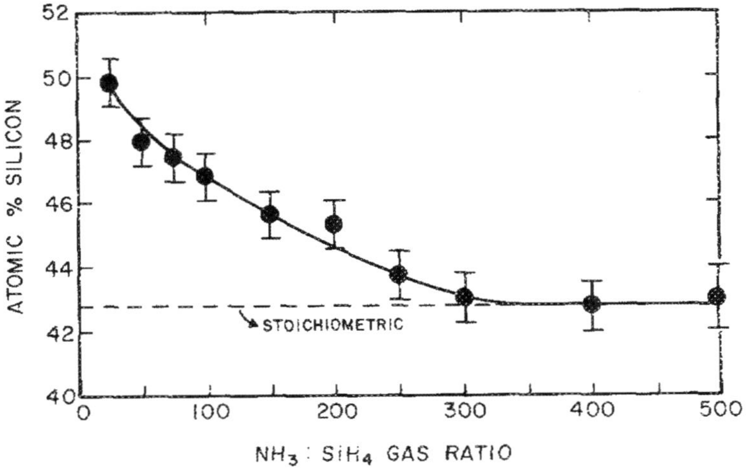

Fig. 24. Silicon concentration in silicon nitride versus the SiH_4/NH_3 ratio in the synthesis process.

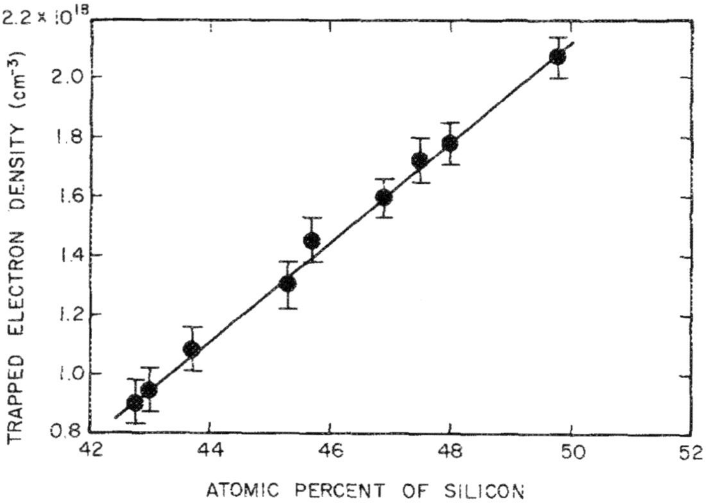

Fig. 25. Concentration of electrons trapped in silicon nitride versus the excess silicon concentration.

in proportion to the concentration of electrons and holes trapped in silicon nitride. Thus, the data of Fig. 26 points to the fact that the concentration of occupied electron and hole traps in silicon nitride increases on increasing the concentration of excess silicon.

Figure 27 shows the spin concentration in silicon nitride versus the silane-to-ammonia ratio SiH_4/NH_3 in the synthesis process [Fujital *et al.* (1985)]. The spin concentration increases from about $2 \times 10^{15}\,cm^{-3}$ at $SiH_4/NH_3 \approx 200$ to $10^{18}\,cm^{-3}$ at $SiH_4/NH_3 \approx 10$. Note that, as the storage medium in memory devices, silicon nitride obtained at $SiH_4/NH_3 \approx 100$ is normally used. In such nitride the concentration of spins, or paramagnetic defects, is typically about $5 \times 10^{16}\,cm^{-3}$. Simultaneously, the concentration of occupied electron and hole traps in silicon nitride is $n_t^e = n_t^h \approx 5 \times 10^{18}\,cm^{-3}$ [Hamptoon (1979)]. Thus, in silicon nitride the concentration of paramagnetic centers is several orders lower than the concentration of occupied traps. The latter means that the electron and hole traps in filled state in silicon nitride are diamagnetic centers. This fact is explained by anti-ferromagnetic pairing of electrons and holes in silicon nitride and, in particular, in silicon-rich silicon nitride SiN_x.

The memory properties of a silicon nitride synthesized from a mixture of dichlorsilane SiH_2Cl_2 and ammonia NH_3 were examined in [Kamigaki *et al.* (1990)]. In the experiment, it was established that, in silicon nitride, there exist two paramagnetic defects, one defect being the nitrogen atom

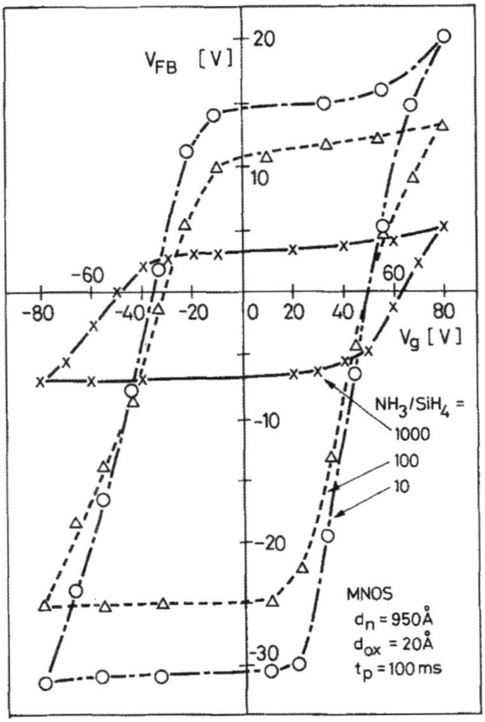

Fig. 26. Hysteresis curve of a metal-nitride-oxide-semiconductor structure grown at different values of the SiH_4/NH_3 ratio during synthesis. Polarization pulse duration width 100 msec.

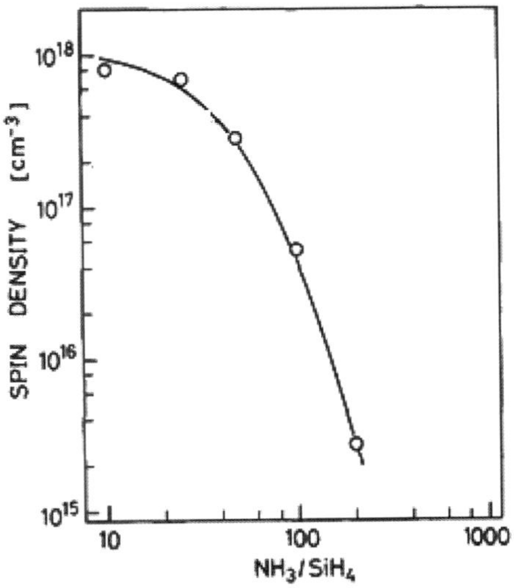

Fig. 27. Spin concentration of EPR signal versus the SiH_4/NH_3 ratio in the synthesis process.

with unpaired electron $\equiv N_3Si\cdot$ triply coordinated with silicon atoms (gyromagnetic ratio 2.002) and the silicon atom with unpaired electron $\equiv Si_3Si\cdot$ triply coordinated with other silicon atoms (gyromagnetic ratio $g = 2.005$) (Fig. 28) [Kamigaki *et al.* (1990)]. In [Kamigaki *et al.* (1990)], it was established that the concentration of electron and hole traps increases with increasing the dichlorsilane-to-ammonia (SiH_2Cl_2/NH_3) ratio implemented during the synthesis of silicon nitride. Normally, the total concentration of paramagnetic centers $\equiv N_3Si\cdot$ and $\equiv Si_3Si\cdot$ in silicon nitride synthesized from dichlorsilane does not exceed 2×10^{17} cm^{-3}. The latter value is one order of magnitude lower than the concentration of filled traps in silicon nitride. Thus, the concentration of paramagnetic centers in silicon nitride synthesized from dichlorsilane is one order of magnitude lower than the concentration of electron and hole traps. From here, it follows that the silicon atoms with unpaired electrons $=N_3Si\cdot$ triply coordinated with three nitrogen atoms (K-centers) cannot be considered as traps responsible for the memory effect in silicon nitride. This conclusion is valid at least for silicon nitride synthesized from silane.

Recently, new EPR data for dichlorsilane-synthesized nitride were reported [Aozasa *et al.* (2007)]. The concentration of traps was found to increase in value on enrichment of nitride with excess silicon. In nitrides grown at $SiH_2Cl_2/NH_3 = 0.01$, 0.1, 1.0, and 2.5 the concentration of K-centers was much lower than the trap concentration. The above findings imply that neither $=N_3Si\cdot$ nor $\equiv Si_3Si\cdot$ defects are responsible for the memory effect in silicon nitride. The latter conclusion is valid at least for dichlorsilane-synthesized silicon nitride. Such nitride is used in silicon nitride based flash memory technology.

8. Charge Transport in Silicon Nitride

Mechanisms of electrical conduction in Si_3N_4 have been under study since 1967 [Sze (1968)]. It was established that the conductivity of Si_3N_4 at room and higher temperatures exponentially increases with the growth of electric field intensity and temperature. Today, this observation is qualitatively interpreted within the Frenkel model. An exponential dependence on field intensity in the Frenkel effect arises due to the reduction of the Coulomb barrier of a charged trap in an electric field (Fig. 29) [Frenkel (1938)]. At high and low temperatures, respectively, thermoionic emission of electrons and tunneling ionization dominate, whereas at intermediate temperatures we have thermally stimulated ionization. Current-voltage characteristics of

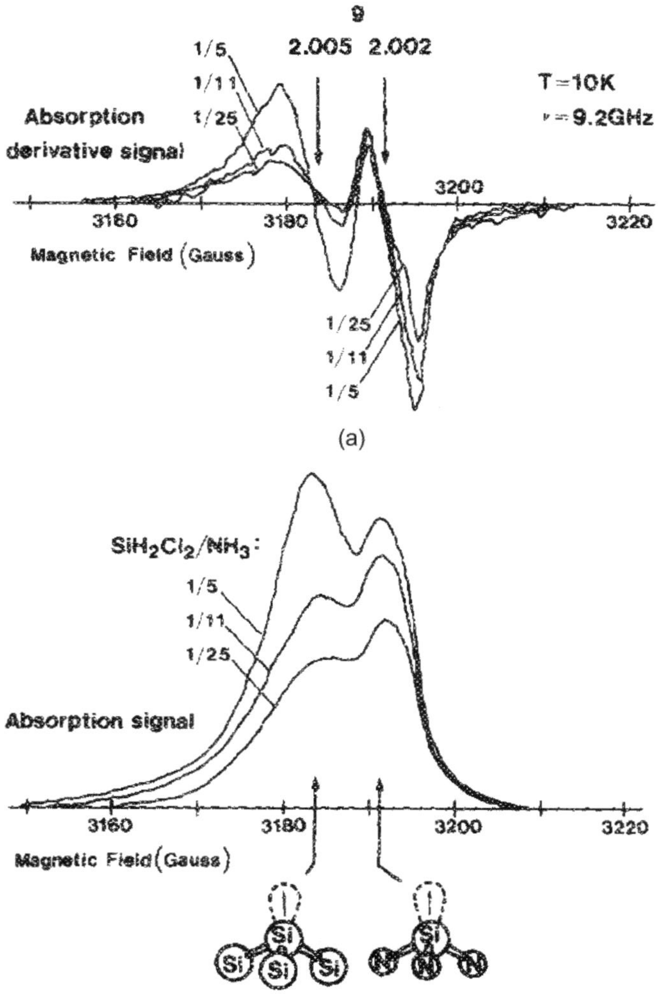

(a)

(b)

Fig. 28. Electron paramagnetic resonance in silicon nitride at different values of the dichlorsilane-to-ammonia (SiH_2Cl_2/NH_3) ratio measured at 10 K: (a) first derivative, (b) integral signal. Two signals are observed. The signal with $g = 2.005$ is due to silicon atoms with unpaired electron $\cdot SiSi_3$ triply coordinated with other silicon atoms, and the signal with $g = 2.002$ is due to silicon atoms with unpaired electron $\cdot SiN_3$ triply coordinated with nitrogen atoms.

Si_3N_4 measured at room and higher temperatures appear as straight lines in the coordinates lgJ-$\sqrt{F}$, where $F = V/d$ is the mean electric field intensity, V is the applied voltage, and d is the thickness of the dielectric layer. It is widely recognized today that the conduction in silicon nitride obeys the Frenkel model.

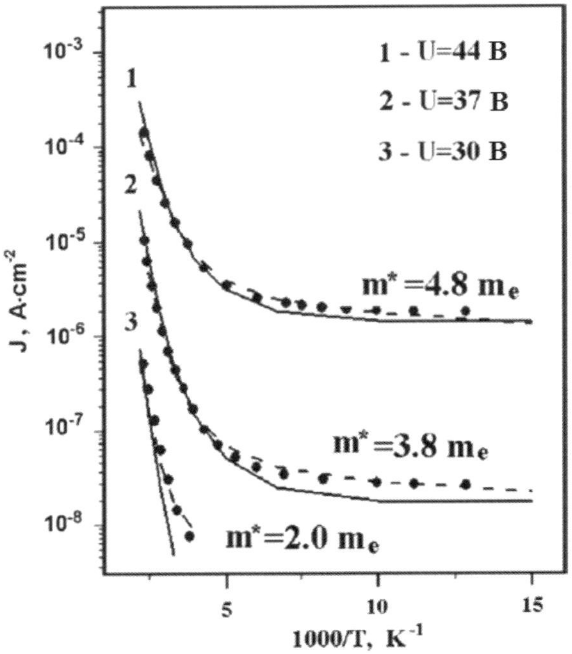

Fig. 29. The current–temperature characteristics of MNOS under different positive voltages (with respect to Al). MNOS geometry: tunnel oxide thickness dox = 1.8 nm, nitride thickness dN = 53 nm. Dots — experimental data, dashed curves — theoretical data predicted by the Frenkel model (with the effective masses indicated at each bias voltage), solid curves — theoretical data predicted by the multi-phonon model. Parameter values: $m_e* = m_h* = 0.5m_o$, $W_{opt} = 2.8$ eV, $W_T = 1.4$ eV, $W_{ph} = 60$ meV, $\sigma_t^e = \sigma_t^h = \sigma_r = 5 \cdot 10^{-13}$ cm^2.

At high temperatures, the trap ionization probability is

$$P_{PF} = \nu \exp\left(-\frac{W_T - \beta\sqrt{F}}{kT}\right); \qquad \beta = \sqrt{\frac{e^3}{\pi\varepsilon_\infty\varepsilon_o}}. \tag{12}$$

In (12), W_t is the trap energy, β is the Frenkel constant, $\varepsilon_\infty = 4.0$ is the high-frequency dielectric permittivity of Si_3N_4, and ν is the pre-exponential factor. At high temperatures, trap ionization occurs due to thermally stimulated tunneling. In the latter case, the ionization probability is given by [Vishnyakov *et al.* (2009)]

$$P_{TAT} = \frac{\nu}{kT}\int_0^{W_T-\beta\sqrt{F}} dE \exp\left(-\frac{\varepsilon}{kT} - \frac{2}{\hbar}\int_{x_1}^{x_2} dx\sqrt{2m^*(eV(x)-E)}\right)$$

$$V(x) = W_T - \frac{e}{4\pi\varepsilon_\infty\varepsilon_0 x} - Fx. \tag{13}$$

In (13), ε is the excitation energy, m^* is the tunneling effective mass, and x_1 and x_2 are the turning points

$$x_{1,2} = \frac{1}{2}\frac{W_T - \varepsilon}{eF}\left(1 \mp \left(1 - \frac{eF}{\pi\varepsilon_\infty\varepsilon_0 (W_t - E)^2}\right)^{1/2}\right). \tag{14}$$

Presently, it is firmly established that the conduction in Si_3N_4 is bipolar [Eitan *et al.* (2000), Gritsenko *et al.* (2003), Gritsenko *et al.* (1997a), Vishnyakov *et al.* (2009), Gritsenko *et al.* (1987), Svensson (1977), Suzuki *et al.* (1979), Schroder *et al.* (1979), Suzuki *et al.* (1982), Watanabe *et al.* (1982)], so that electrons and holes can be injected into silicon nitride respectively from the negatively biased and positively biased electrode. In the near-contact regions of the silicon nitride layer, recombination of electrons and holes occurs (Fig. 30).

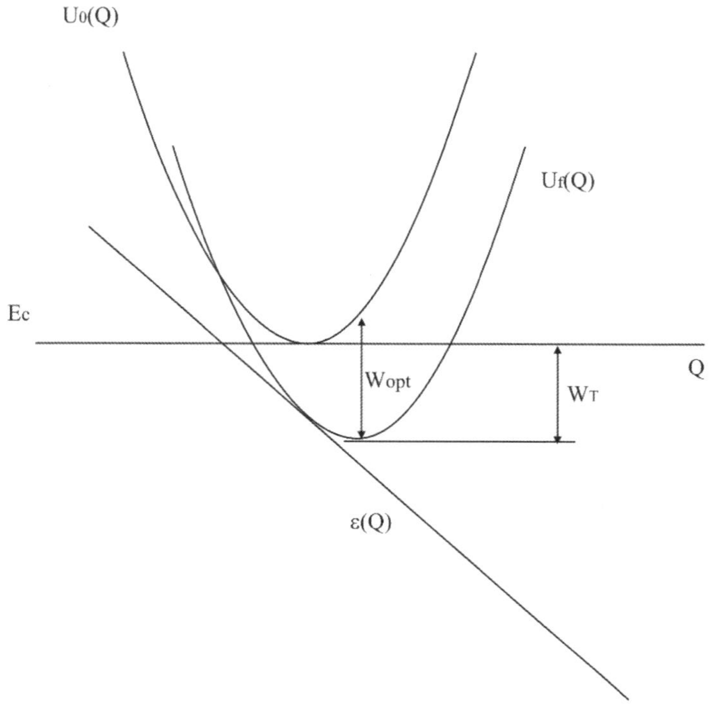

Fig. 30. (a) Configuration diagram of a trap in the theory of multi-phonon ionization. Potentials U_f and U_0 respectively correspond to a trap with a captured electron and an ionized trap.

In exact quantum theory, the ionization probability is [Makram-Ebeid *et al.* (1982)]

$$P = \sum \exp\left\{\frac{nW_{ph}}{2kT} - \frac{W_{opt} - W_T}{W_{ph}} \coth \frac{W_{ph}}{2kT}\right\}$$

$$\times I_n \left(\frac{W_{opt} - W_T}{W_{ph} \sinh\left(\frac{W_{ph}}{2kT}\right)}\right) P_i(W_T + nW_{ph}), \tag{15}$$

$$P_i(W) = \frac{eF}{2\sqrt{2m * W}} \exp\left\{-\frac{4}{3}\frac{\sqrt{2m}}{\hbar eF} W^{\frac{3}{2}}\right\} \tag{16}$$

Here, I_n is the modified Bessel function, and $P_i(W)$ is the probability of electron tunneling through a triangular barrier of height W.

Attempts to perform a quantitative comparison of current-temperature characteristics of Si_3N_4 samples measured at various voltages with the Frenkel model were reported in [Nasyrov *et al.* (2009), Nasyrov *et al.* (2004), Vishnyakov *et al.* (2009)]. In calculating the current-voltage characteristics by Poisson equation, the space charge due to electrons and holes localized in the Si_3N_4 layer was taken into account. It was found that in a wide range of field intensities and temperatures the conductivity of Si_3N_4 admits a quantitative description within the Frenkel model. Yet, a quantitative agreement can only be achieved on the assumption of an abnormally small value of the pre-exponential factor ($\upsilon = 10^8 - 10^9\,\text{sec}^{-1}$). In the original publication by Frenkel, the pre-exponential factor was evaluated as amounting to $\approx 10^{14}\,\text{sec}^{-1}$ [Frenkel (1938)]. Another contradiction with the Frenkel model consists in that at temperatures below room temperature, when the conductivity, limited by the tunneling ionization process, no longer depends on temperature, an agreement with the experiment can only be achieved on an assumption of an abnormally high value of the effective mass of charge carriers $m^* = (2 - 5)m_o$ [Eitan *et al.* (2000), Nasyrov *et al.* (2004)]. The experimental value of the effective mass of electron in Si_3N_4 is close to $m^* = 0.5m_o$ [Gritsenko *et al.* (1997a)]. Moreover, treatment of experimental data within the Frenkel model has showed that the effective mass of charge carriers in Si_3N_4 proved to be dependent on voltage; for instance, for $V = 44, 37$, and $30\,V$ effective mass values $m^* = 4.8m_o, 3.8m_o$, and $2.0m_o$ were obtained (Fig. 29) [Vishnyakov *et al.* (2009)].

In [Eitan *et al.* (2000), Nasyrov *et al.* (2004), Vishnyakov *et al.* (2009)] it was shown that in a broad range of field intensities and temperatures the conduction in Si_3N_4 can be adequately understood on the assumption of multi-phonon ionization of neutral traps. In the latter model, ionization is treated in terms of configuration coordinates. A trap is characterized with two parameters, one parameter being the thermal ionization energy W_t and the other, the optical ionization energy W_{opt}. The electron undergoes emission from a localized state into the energy continuum when the ascending trap level reaches the continuum (Fig. 30).

The energies of electron and hole traps as evaluated from experiments on charge transport in silicon nitride and from quantum-chemical simulation data are given in Fig. 31. According to experimental data, within experimental error the energies of electron and hole traps are identical, equal to $W_t^e = W_t^h = 1.4 \pm 0.1$ eV. An exception here are trap energies $0.8 - 0.9$ eV deduced from experimental data on thermally stimulated depolarization of silicon nitride. However, in those experiments the trap energies could obviously be underestimated because the experimental data were interpreted without taking the electric-field-induced barrier lowering into account.

Let us summarize the main experimental data concerning the characteristics of traps in silicon nitride.

1. Silicon nitride is neutral in non-polarized state, featuring amphoteric nature of traps, or their capability to localize both electrons and holes.
2. The concentrations of charged filled electron traps (n_t^e) and hole traps (n_t^h) are identical within experimental error, $n_t^e = n_t^h = (2 - 6) \times 10^{18}\,\text{cm}^{-3}$.
3. In the multi-phonon ionization model, the concentrations of neutral electron (N_t^e) and hole traps (N_t^h) are identical, equal to $N_t^e = N_t^h = (5 - 9) \times 10^{19}\,\text{cm}^{-3}$.
4. The concentrations of electron and hole traps show an increase on increasing the concentration of excess silicon in silicon nitride.
5. In the Frenkel model, the trap energy for electrons W^e is equal to the trap energy for holes W^h: $W^e = W^h = 1.4 \pm 0.1$ eV.
6. In the multi-phonon ionization model, the thermal energy of electron traps is equal to the thermal energy of hole traps: $W_t^e = W_t^h = 1.4$ eV.
7. In the multi-phonon ionization model, the optical energy of electron traps is equal to the optical energy of hole traps, being simultaneously two times greater than the thermal trap energy: $W_{Opt}^e = W_{Opt}^h = 2W_t^{e,h} = 2.8$ eV.
8. The capture cross-section for electrons σ_t^e is equal to the capture cross-section for holes σ_t^h, both cross-sections having gigantic values $\sigma_t^e = \sigma_t^h = 5 \times 10^{-13}\,\text{cm}^{-2}$.

9. The cross-sections for electron-hole recombination are also gigantic, $\sigma_R \approx 5 \times 10^{-13}\,cm^2$.
10. Silicon nitride in non-polarized state is a diamagnetic material exhibiting no EPR signal.
11. Silicon nitride with localized electrons and holes is also a diamagnetic material exhibiting no EPR signal.

9. Models for Traps in Silicon Nitride

9.1. *Mott polaron model for trapping in silicon nitride*

Silicon nitride is a polar material with ion-covalent bonds featuring a high bond ionicity. In localization of electrons/holes in this material, the polaron effect, or electron-phonon interaction, can be expected to play an important part. The polaron effect consists in that the electron/hole energy decreases in value due to electron- and ion-induced polarization of the lattice. Localization of an electron (formation of a polaron) occurs. Let us evaluate the intrinsic polaron energy in silicon nitride. The general expression for the electron-phonon coupling constant is

$$\alpha = \frac{q^2}{\varepsilon_p}\left(\frac{m^*}{2\hbar^3\omega_L}\right)^{1/2}; \quad \varepsilon_p = \varepsilon_\infty^{-1} - \varepsilon^{-1} \tag{17}$$

In (17), ω_L is the longitudinal optical phonon frequency, ω_∞ is the high-frequency (optical) dielectric permittivity, ε the low-frequency dielectric permittivity, and ε_p is the polaron dielectric permittivity. The frequency ω_L can be found from the Lyddane-Sachs-Teller relation

$$\omega_L = \sqrt{\frac{\varepsilon}{\varepsilon_\infty}}\omega_T, \tag{18}$$

and the frequency of transverse optical phonons in silicon nitride ω_T can be evaluated from the lattice IR absorption maximum ($\lambda = 11.5_{MKM}$). Substituting into (17) numerical values for the constants in silicon nitride, $\omega_\infty = 4.0$ and $\varepsilon = 7.0$, we obtain the value of polaron dielectric permittivity $\varepsilon_p = 10$ and, for the effective mass $m^* = 0.5\,m_e$ and $\hbar\omega_L = 0.14$ эB, the value of the electron-phonon coupling constant $\alpha = 0.7$. The obtained value of α in silicon nitride is close to unity. The latter means that, strictly speaking, neither the weak coupling approximation nor the strong coupling approximation works; here, we have an intermediate case. Let us evaluate the intrinsic polaron energy in silicon nitride in two extreme cases, namely in the weak and strong coupling approximations. In the former approximation, the polaron energy is given by

$$E_p = \alpha \hbar \omega_L \tag{19}$$

For silicon nitride, in the weak coupling approximation for the polaron energy we obtain a value $E_p = 0.1$ эВ.

In the strong coupling approximation the same energy is given by

$$E_p = 0.1\alpha^2 \hbar \omega_L \tag{20}$$

For silicon nitride, formula (20) yields $E_p = 0.007$ эВ.

In conclusion, both in the weak and in the strong coupling approximation the intrinsic polaron energy in silicon nitride is within a value of order 0.1 eV. This value is one order of magnitude smaller than the experimentally observed delocalization energy of electrons and holes in silicon nitride ($W_t^e = W_t^h \approx 1.4\,\text{eV}$).

Another objection against the intrinsic polaron model of electron and hole localization in silicon nitride is the absence of the memory effect, or localization, in amorphous silicon oxide thermally grown on silicon. The electronic structures of silicon nitride and silicon oxide are similar [Gritsenko (1993)]. In both materials the valence band is formed via hybridization of Si sp^3 orbitals of silicon and N 2p orbitals of nitrogen in nitride and O 2p orbitals of oxygen in oxide, respectively. Near the valence-band ceiling, in both cases there are narrow zones due to $2p_\pi$ antibinding orbitals. The effective masses of electrons and holes in silicon nitride and oxide are close in value. The electron-phonon coupling constants in nitride and oxide are also comparable. Thus, due to the formation of intrinsic self-polarons, both in silicon oxide and in silicon nitride a localization effect can be expected. However, the many-year experience in using metal-insulator-semiconductor (MIS) structures based on oxide layers thermally grown on silicon contradicts to the latter assumption. Trapping (localization) of electrons and holes in the volume of silicon oxide is substantially (by several orders) is less pronounced in comparison with silicon nitride. To summarize the discussion, we have to admit the fact that the localization of electrons and holes in silicon nitride cannot be explained by the formation of self-polarons that occurs due to polarization of a defectless polar lattice.

9.2. *D-centers and the model of triply coordinated silicon atom*

The model of D-centers is based on the Anderson's idea that, on drawing together two paramagnetic defects with unpaired electrons D^0, the electrons get redistributed so that both electrons become localized at one of the defects [Anderson (1975)]:

$$2D^o \rightarrow D^- + D^+ \tag{21}$$

To such a transition, a negative Habbard correlation energy $U < 0$ corresponds. Anderson advanced a hypothesis that the gain in energy due to deformation of lattice (electron-phonon interaction) can exceed the energy of the Coulomb repulsion of two electrons localized at one defect. In the model of D-centers, electrons and holes get captured at charged defects D^- and D^+:

$$\begin{aligned} D^+ + e &\rightarrow D^o \\ D^- + h &\rightarrow D^o \end{aligned} \tag{22}$$

The model of D-centers explains the neutrality of silicon nitride, the gigantic capture cross-section for electrons and holes, the diamagnetic properties of silicon nitride in non-polarized state, and the equality between the concentrations of electron and hole traps $N_t^e = N_t^h$.

A particular case of the model of D-centers is the model of VAP (Valence Alternation Pair)-defects [Street *et al.* (1975), Kastner *et al.* (1976)]. This model assumes that transition (22) proceeds with re-coordination of atoms. Kirk analyzed the model of VAP — defects as applied to Si_3N_4 on the assumption that nitrogen undergoes re-coordination in the course of the transition

$$2N_3^o \rightarrow N_4^+ + N_2^-. \tag{23}$$

Here, N^o is normally (triply) coordinated neutral nitrogen atom, N_4^+ is positively charged nitrogen atom four-fold coordinated with silicon atoms, and N_2^- is negatively charged nitrogen atom doubly coordinated with silicon atoms [Kirk (1979)].

According to the model of VAP — defects, for the formation of such defects single N $2p_\pi$ antibinding electrons are required. Theoretical calculations of the electronic structure of Si_3N_4 predict the existence of an isolated pair of N $2p_\pi$ electrons in silicon nitride [Lee *et al.* (2005)]. Yet, direct quantum-mechanical calculations by Robertson of the Hubbard energy for pairing of electrons at nitrogen atoms in Si_3N_4 have showed the transition (23) to be energetically unfavorable [Robertson (1994)]. The model of VAP — defects fails to explain the increase in trap concentration observed on the enrichment of silicon nitride with excess silicon. Besides, the model of VAP — defects fails to explain the gigantic value of recombination cross-section. After trapping of an electron (hole) at a charged center, the center becomes neutral. The latter, however, is inconsistent with the gigantic recombination cross-section observed in the experiment.

According to the model of D-centers, triply coordinated silicon atoms with unpaired electron $\equiv$Si$\cdot$ (K centers) can undergo pairing by the scheme

$$\equiv \mathrm{Si}\cdot + \cdot\mathrm{Si}\equiv \;\rightarrow\; \equiv\mathrm{Si}: + \mathrm{S}\equiv \tag{24}$$

Transition (24) occurs if the correlation energy is negative. The latter implies that pairing of electrons proceeding with the formation of a pair of diamagnetic defects $\equiv$Si: and S$\equiv$ is energetically favorable.

If the correlation energy is positive, then the pairing of electrons is unfavorable, energetically favorable being the existence of a pair of paramagnetic defects $\equiv$Si$\cdot$. In initial state, such a center is neutral and paramagnetic; on an electron capture, this center acquires charge and passes over to diamagnetic state.

In the case of positive correlation energy, the traps for electrons and holes are neutral paramagnetic silicon atoms with unpaired electron $\equiv$Si$\cdot$. The model of triply coordinated silicon atoms with unpaired electrons explains the amphoteric nature of centers, the absence of paramagnetic resonance in nitride with trapped electrons (holes), and the gigantic recombination cross-section (indeed, after the capture of an electron, the defect becomes charged). The latter model, however, fails to explain the gigantic capture cross-section for electrons and holes, because in the initial state the defect is neutral. A fundamental contradiction with the experiment is that the model of triply coordinated silicon atom with unpaired electron does not explain the absence of EPR signal in non-polarized (without trapped electrons and holes) silicon nitride.

10. Concluding Remark: Analogy with Si-Si Bond in SiO_2

The model of neutral diamagnetic Si-Si defect is a universal model offering a reasonable explanation to the localization phenomenon of electrons and holes in amorphous silicon nitride. Yet, the gigantic value of electron and hole capture cross-sections for neutral traps still lacks a satisfactory explanation. It should be noted that in silicon oxide SiO_2 the neutral Si-Si bond acts as a hole trap with hole capture cross-section $3.6 \times 10^{-13}\,\mathrm{cm}^{-2}$, the latter value being close to the similar cross-section in silicon nitride [Gritsenko *et al.* (2011)]. The absence of EPR signal due to localized charge carriers in Si_3N_4 can be attributed to anti-ferromagnetic pairing of charge carriers localized at traps in silicon nitride.

11. Evolution, Problems and Perspectives of Flash Memory Based on Electrons and Holes Localization in Silicon Nitride

Memory effect in silicon nitride was discovered in 1967. The first memory device was Metal(Al)-Nitride(100 nm)-Oxide(1.8 nm)-Silicon (MNOS) structures. Later the Al gate in MNOS was replaced by polysilicon in polySilicon-Nitride(50 nm)-Oxide(1.8 nm)-Silicon (SNOS) structures. To reduce the parasitic electron injection from the gate, a blocking silicon oxide was placed between polysilicon and nitride in polySilicon-Oxide(5 nm)-Nitride(5 nm)-Oxide(1.8 nm)-Silicon (SONOS) structures. SONOS structures are programmed by electrons and holes tunnel injection from the silicon. In case of two bits flash memory devices, hot electrons and holes injections are used for programming SONOS structures. In this case, to increase the retention time the tunnel oxide thickness in SONOS structures was increased up to 5 nm. To block parasitic carrier injection from the gate the blocking oxide in SONOS was replaced by high-κ dielectric, namely by alumina. Later, tantalum nitride (TaN) replaced the polysilicon to increase the electron barrier between the gate and blocking oxide. It was a time of TaN-Alumina(15 nm)-Nitride(5 nm)-Oxide (3.5 nm)-Silicon TANOS flash devices.

In TANOS transistor with the design rule of 22 nm in the nitride layer under the gate only 20 electrons are localized. The future decreasing of design rule in TANOS flash memory producing is limited by fluctuations of f small amount of electrons. a small number of localized electrons in TANOS flash memory is limited by electron Wigner crystallization due to coulomb electron repulsion.

To overcome this limitation and increase the information density, recently three-dimensional 3D TANOS flash memory was developed. In 2014 in mass production 128 gigabits 3D TANOS memory with 32 memory layers was implemented. According to projections in 2017 year 3D flash will occupy 70% of market.

Acknowledgments

This work has been supported by the Russian Science Foundation, grant # 14-19-00192.

References

Aganin, A. P. Maslovskii, V. M. and Nagin, A. P. (1988), *Mikroelektronika*, **17**(4), 348.
Anderson, P. W. (1975) *Phys. Rev. Lett.*, **35**, 953.

Ando, T. Fowler, A. and Stern, F. (1982) *Rev. Mod. Phys.*, **54**(2), 3.
Anutgan, M. Atilgan, T. and Katircioglu, I. B. (2011), *J. Luminescence*, **131**, 1305.
Aozasa, H. Fujiwara, II. Nakamura, A. and Komatsu, Y. (1999), *Jpn. J. Appl. Phys.*, **38**, 1441.
Aozasa, H. Fujiwara, I. and Kamigaki, Y. (2007), *Japanese J. Appl. Phys.*, **46**(9A), 5762.
Bachhofer, H., Reisinger, H., Bertagnolli, E. and Philipsborn, H. (2001) *J. Appl. Phys.*, **89**(5), 2791.
Bailey, R. S. and Kapoor, V. J. (1982), *J. Vac. Sci. Technol.*, **20**, 484.
Brytov, I. A., Gritsenko, V. A. and Romashenko, Yu. N. (1984) Electronic Structure of Amorphous Silicon Nitride, preprint 6–84, (Institute of Semiconductor Physics of Siberian Branch of Russian Academy of Sciences, Novosibirsk, Russia) p. 41.
Cen, Z. H., Chen, T. P., Ding, L., Liu, Y., Wong, J. I., Yang, M., Liu, Z., Goh, W. P., Zhu, F. R. and Fung, S. (2009), *Appl. Phys. Lett.*, **94**, 041102.
Chen, S.-C., Chang, T.-C., Liu, P.-T., Wu, Y.-C., Chin, J.-Y., Yeh, P.-H., Feng, L.-W., Sze, S. M., Chang, C.-Y. and Lien, C.-H. (2007), *Appl. Phys. Lett.*, **91**, 193103.
Chen, D., Viner, J. M., Taylor, P. C. and Kanicki, J. (1994), *Phys. Rev. B*, **49**(19) 13420.
Conley, J. F., Lenahan, P. M., Evans, H. L., Lowry, R. K. and Morthorst, T. J. (1994), *Appl. Phys. Lett.*, **65**(18), 2281.
Eitan, B., Pavan, P., Bloom, I., Aloni, E., Frommer, A. and Finzi, D. (2000), *IEEE Electron Device Letters*, **21**(11), 543.
Frenkel, J. (1938), *Phys. Rev.*, **54**, 647.
Fuks, D., Kiv, A., Roizin, Y., Gutman, M., Bibi, R. A. and Maximova, T. (2006), *IEEE Trans. Electron Devices*, **53**(53), 304.
Fujita, S. and Sasaki, A. (1985), *J. Electrochem. Soc.*, **132**, 398.
Giacomazzi1, L., and Umari, P. (2009), *Phys. Rev. B*, **80**, 144201.
Ginovker, A. S., Gritsenko, V. A. and Sinitsa, S. P. (1974), *Phys. Stat. Sol. B*, **26**(2), 489.
Giorgis, F., Pirri, C. F., Vinegoni, C. and Pavesi, L. (1999), *Phys. Rev. B*, **60**(16), 11572.
Gritsenko, V. A. (2012), *Physics-Uspekhi*, **182**(5), 531.
Gritsenko, V. A. and Wong, H. (2011), *Criti. Rev. Solid State Mater. Sci.*, 36, 129.
Gritsenko, V. A. (2008), Physics-Uspekhi, **51**(7), 699.
Gritsenko, V. A., Gritsenko, D. V., Novikov, Yu. N., Kwok, P. B. M. and Bello, I. (2004), *JETP*, **98**(4), 760.
Gritsenko, V. A., Novikov, Yu. N., Shaposhnikov, A. V., Wong, H. and Zhidomirov, G. M. (2003a), *Physics of the Solid State*, **45**(11), 2031.
Gritsenko, V. A., Nasyrov, K. A., Novikov, Yu. N., Aseev, A. L., Yoon, S. Y., Lee, J.-W., Lee, E.-H. and Kim, C. W. (2003b), *Solid-State Electronics*, **47**(10), 1651.
Gritsenko, V. A., Kwok, R. W. M., Wong, H. and Xu, J. B. (2002), *J. Non-Crystalline Solids*, **297**, 96.
Gritsenko, V. A., Shavalgin, Yu. G., Pundur, P. A., Wong, H. and Kwok, W. M. (2000), *Philos. Mag.*, **80**(10), 1857.
Gritsenko, V. A., Zhyravlev, K. S., Milov, A. D. and Wong, H. (1999), *Thin Solid Films*, 353 p. 20.
Gritsenko, V. A. Novikov, Yu. N. Morokov, Yu. N. and Wong, H. (1998a), *Microelectro. Reliab.*, **38**, 1457.

Gritsenko, V. A., Xu, J. B., Wilson, I. H., Kwok, R. M. and Ng, Y. H. (1998b), *Phys. Rev. Lett*, **81**(5), 1054.
Gritsenko, V. A., Meerson, E. E. and Morokov, Yu. N. (1997a), *Phys. Rev. B*, **57**(4) R2081.
Gritsenko, V. A., Morokov, Yu. N. and Novikov, Yu. N. (1997b), *Applied Surface Science*, **113**(114), 417.
Gritsenko, V. A., Morokov, Yu. N. and Novikov, Yu. N. (1997c), *Solid State Phys., (Engl. Transl.)*, **39**(8) 1191.
Gritsenko, V. A. (1996a), *Lett. JETP*, **64**(7), 525.
Gritsenko, V. A. and Milov, A. D. (1996b), *Lett. JETP*, **64**(7), 531.
Gritsenko, V. A. (1993) Structure and Electronic Structure of Amorphous Dielectrics in Silicon MIS Structures (Science, Novosibirsk, Russia).
Gritsenko, V. A., Meerson, E. E., Travkov, I. V. and Goltvjanskii, Yu. V. (1987), *Microelectronics (Sov)*, **16**(1), 42.
Gritsenko, V. A. (1986a) Electronic Structure and Optical Properties of Silicon Nitride, in *Silicon Nitride in Electronics* (Elsevier New York), 236.
Gritsenko, V. A. and Pundur, P. A. (1986b), *Sov. Phys. Solid State*, **28**(10), 1829.
Gritsenko, V. A., Britov, I. A., Kostikov, Yu. P., Obolenskii, E. A. and Romaschenko, Yu. N. (1984), *Sov. Phys. Solid State*, **26**(6), 1022.
Gritsenko, V. A. (1982) Optical Properties of Silicon Nitride, In *Silicon Nitride in Electronics* (Science, Novosibirsk, USSR), 112.
Gritsenko, V. A. (1976), *Phys. Stat. Sol. A*, **28**(1), 387.
Grillo, M.-E. and Elliott, S. D. (2011), *Phys. Rev. B*, 83 085208.
Guraya, M. M., Ascolani, H., Zampieri, G., Cisneros, J. I., Dias da Silva, J. H. and Cantao, M. P. (1990), *Phys. Rev. B*, **42**(9), 5677.
Hamptoon, F. L. and Cricchi, J. R. (1979), *Appl. Phys. Lett.*, **35**, 802.
Hasegawa, S., Ikeda, M. and Kurata, Y. (1996), *J. Appl. Phys*, **80**(5), 2896.
Hasegawa, S., He, L., Inokuma, T. and Kurata, Y. (1992), *Phys. Rev. B*, **46**, 12478.
Hasegawa, S., Matsuda, M. and Kurata, Y. (1991), *Appl. Phys. Lett.*, **58**(7), 18.
Hill, R. M. (1971), *Phlios. Mag.* **23**(181), 59.
Houssa, M., Tuominen, M., Naili, M., Afanas'ev, V., Stesmans, A., Haukka, S. and Heyns, M. M. (2000), *J. Appl. Phys.*, **87**(12), 8615.
Hong, S. H., Jang, J. H., Park, T. J., Jeong, D. S., Kim, M. and Hwang, C. S. (2005), *Appl. Phys. Lett.*, 87, 152106. http://www.theregister.co.uk/2013/08/19/samsung_launches_damp_3d_ ssd_squib/
Jegert, G., Kersch, A., Weinreich, W., Schroder, U. and Lugli, P. (2010), *Appl. Phys. Lett.*, **96**, 062113.
Kamigaki, Y., Minamo, S. and Kato, H. (1990), *J. Appl. Phys.*, **68**(5), 2211.
Kanicki, J. and Warren, W. L. (1995), *J. Non-Cryst. Solids*, **104**, 1055.
Kanicki, J., Sankaran, M., Gelatos, A., Crowder, M. S. and Tober, E. D. (1990), *Appl. Phys. Lett.*, **57**(7), 698.
Karcher, R., Ley, L. and Johnson, R. L. (1984), *Phys. Rev. B*, **30**, 1896.
Kastner, M., Adler, D. and Mott, N. F. (1976), *Phys. Rev. Lett.*, **37**, 1504.
Kim, T. H., Park, H., Lee, J. D., Shin, H. C. and Park, B.-G. (2006), *Appl. Phys. Lett.*, **89**, 063508.

Kim, H. S., Gilmer, D. C., Campbell, S. A. and Polla, D. L. (1996), *Appl. Phys. Lett.*, **69**(25) 3860.

Kirk, Jr. C. T. (1979), *J. Appl. Phys.*, **50**, 4190.

Krick, D. T., Lenahan, P. M. and Kanicki, J. (1988a), *Phys. Rev. B*, **38**(12), 8226.

Krick, D. T., Lenahan, P. M. and Kanicki, J. (1988b), *J. Appl. Phys.*, **64**(7), 3558.

Lee, C.-H., Hur, S.-H., Shin, Y.-C., Choi, J.-H., Park, D.-G. and Kim, K. (2005), *Appl. Phys. Lett.*, **86**, 152908.

Lee, C.-H., Park, K.-C. and Kim, K. (2005), *Appl. Phys. Lett.*, **87**, 073510.

Lenahan, P. M. and Curry, S. E. (1990), *Appl. Phys. Lett.*, **56**(2), 157.

Lenahan, P. M., Krick, D. T. and Kanicki, J. (1989), *Appl. Surf. Sci.*, **39**, 392.

Ley, L., Karcher, R. and Johnson, R. L. (1984), *Phys. Rev. Lett.*, **53**, 710.

Libsh, F., Roy, A. and White, M. H. (1987), *Appl. Surf. Sci.*, **30**, 160.

Lisiansky, M., Heiman, A., Kovler, M., Fenigstein, A., Roizin, Y., Levin, I., Gladkikh, A., Oksman, M., Edrei, R., Hoffman, A., Shneider, Y. and Claasen, T. (2006), *Appl. Phys. Lett.*, **89**, 153506.

Makram-Ebeid, S. and Lannoo M. (1982), *Phys. Rev. B*, **28**, 6406.

Martin-Moreno, L., Martinez, E., Verges, J. A. and Yndurain, Y. (1987), *Phys. Rev. B*, **35**(18), 9683.

Maslovskii, V. M., Nagin, A. P. and Chernyshev, Yu. P. (1985), *Mikroelektronika*, **14**(3), 230.

Mo, C., Zhang, L., Xie, C. and Wang, T. (1993), *J. Appl. Phys.*, **73**(10), 5188.

Mo, S. D., Ouyang, L., Ching, W. Y., Tanaka, I., Koyama, Y. and Riedel, R. (1999), *Phys. Rev. Lett.* **83**, 5046.

Mott, N. F. and Davis, E. A. (1979) *Electron Processes in Non-Crystalline Materials*, Clarendon Press, Oxford.

Nasyrov, K. A., and Gritsenko, V. A. (2011), *J. Appl. Phys.*, **109**, 097705.

Nasyrov, K. A., Shaimeev, S. S., Gritsenko, V. A. and Han, J. H. (2009), *J. Appl. Phys.* **105**, 123709.

Nasyrov, K. A., Gritsenko, V. A., Novikov, Yu. N., Lee, E.-H., Yoon, S. Y. and Kim, C. W. (2004), *J. Appl. Phys.*, **96**(8), 4293.

Nadolinnyui, V. A., Bolotin, V. P., Vasilev, V. V. and Mikhailovskii, I. P. (1991), *Physica Status Solidi (a)*, **127**, K 135.

Nekrashevich, S. S., Vasilev, V. V., Shaposhnikov, A. V. and Gritsenko, V. A. (2009), *Microelectro. Eng.*, **86**(7), 1866.

Novikov, Y. N., Gritsenko, V. A. and Nasyrov, K. A. (2009), *Appl. Phys. Lett.*, **94**, 222904.

Pacchioni, G. and Erbetta, D. (2000), *Phys. Rev. B*, **61**(22), 15005.

Pacchioni, G. and Erbetta, D. (1999), *Phys. Rev. B.*, **60**(19), 12617.

Pao1, H. and Connel, M. O. (1967), *Appl. Phys. Lett.*, **12**, 260.

Park, N.-M., Choi, S.-H. and Park, S.-J. (2002), **81**, 1092.

Park, N.-M., Choi, C.-J., Seong, T.-Y. and Park, S.-J. (2001), *Phys. Rev. Lett.*, **86**(7) 1355.

Perevalov, T. V. and Gritsenko, V. A. (2010), *Physics Uspekhi*, **53**(6), 561.

Petersen, M. and Roizin, Y. (2006), *Appl. Phys. Lett.*, **89**, 053511.

Philip, H. R. (1973), *J. Electrochem Sos.*, **120**(2), 295.

Pundur, P. A., Shavalgin, Yu. G. and Gritsenko, V. A. (1986), *Phys. Stat. Sol. A*, **94**, K107.
Roizin, Y. and Gritsenko, V. A. (2007) Chapter in book *Dielectric Films for Advanced Microelectronics*, (eds. by M. R. Baklanov, M. Greeen, K. Maex and Wiley & Sons).
Robertson J. (1994), *Phil. Mag. B*, **69**(2), 307.
Robertson J. (1991), *Phil. Mag. B*, **63**, 47.
Robertson J. (1981), *Phil. Mag. B*, **44**, 215.
San-Fabian, E., Lousis, E., Martin-Moreno, L. and Verges, J. A. (1989), *Phys. Rev. B*, **39**(3), 1844.
Schroder, D. K. and White, M. H. (1979), *IEEE Trans. Electron Devices*, **ED-26**(6), 899.
Seo, Y. J., Kim, K. C., Kima, T. G., Sung, Y. M., Cho, H. Y., Joo, M. S. and Pyi, S. H. (2008), *Appl. Phys. Lett.*, **92**, 132104.
Seol, K. S., Watanabe, T., Fujimaki, M., Kato, H. and Ohki, Y. (2000), *Phys. Rev. B.*, **62**(3), 1532.
Shames, A. I., Gritsenko, V. A., Samoilova, R. I., Tzvetkov, Yu. D., Braginsky, L. S. and Roger, M. (2001), *Solid State Communi.*, **118**, 129.
Shannon, J. M. and Morgan, B. A. (1999), *J. Appl. Phys.*, **86**(3), 1548.
Sorokin, A. N., Karpushin, A. A., Gritsenko, V. A. and Wong, H. (2009), *J. Appl. Phys.*, **105**, 073706.
Sorokin, A. N., Karpushin, A. A., Gritsenko, V. A. and Wong, H. (2008), *J. Non-Cryst. Solids*, **354**, 1531.
Street, R. A. and Mott, N. F. (1975), *Phys. Rev. Lett.*, **35**, 1293.
Suzuki, E. and Hayashi, Y. (1982), *J. Appl. Phys.*, **53**, 8880.
Suzuki, E., Hayashi, Y. and Yanay, H. (1979), *J. Appl. Phys.*, **50**(11) 7001.
Svensson, C. M. (1977), *J. Appl. Phys.*, **48**(1), 329.
Sze, S. M. (1981), *Physics of Semiconductor Devices*, Wiley, New York, 362–430.
Sze, S. M. (1967), *J. Appl. Phys.*, **38**(7), 2951.
Tzeng, S.-D. and Gwo, S. (2006), *J. Appl. Phys.*, **100**, 023711.
Valentin, C. D., Palma, G. and Pacchioni, G. (2011), *J. Phys. Chem. C*, **115**, 561.
Vishnyakov, A. V., Novikov, Yu. N., Gritsenko, V. A. and Nasyrov, K. A. (2009), *Solid-State Electronics*, **53**(3), 251.
Voskoboinikov, V. V., Gritsenko, V. A., Efimov, V. M., Lesnikovskaja, V. E. and Edelman, F. L. (1976), *Phys. Stat. Sol. A*, **24**, 85.
Warren, W. L., Seager, C. H., Robertson, J., Kanicki, J. and Poindexter, E. H. (1996), *J. Electrochem. Sos.*, **143**(11), 3685.
Warren, W. L., Kanicki, J., Robertson, J., Poindexter, E. H. and McWhorter, P. J. (1993), *J. Appl. Phys.*, **74**, 4034.
Warren, W. L., Lenahan, P. M. and Curry, S. E. (1990), *Phys. Rev. Lett.*, **65**(2), 207.
Warren, W. L. and Lenahan, P. M. (1990), *Phys. Rev. B*, **42**(3), 1773.
Watanabe, K. and Wakayama, S. (1982), *J. Appl. Phys.*, **53**, 568.
Yang, Y. and White, M. H. (2000), *Solid-State Electronics*, **44**, 949.
Ziman, J. M. (1979) *Models of Disorder* (Cambridge University Press, Cambridge).

Chapter 7

High-k Oxides on Si: MOSFET Gate Dielectrics

Takashi Ando, Unoh Kwon†, Siddarth Krishnan†, Martin M. Frank*, and Vijay Narayanan**

**IBM T. J. Watson Research Center, Yorktown Heights, NY 10598, USA*

†IBM Semiconductor Research and Development Center, Hopewell Junction, NY 12533, USA

High-k Metal Gate (HKMG) technology has enabled the continued scaling of complementary metal-oxide-semiconductor (CMOS) devices at the 45 nm logic technology node and beyond. This type of technology employs high-permittivity ('high-k') gate dielectrics, typically based on hafnium (e.g., HfO_2), to insulate the metal-oxide-semiconductor field-effect transistor (MOSFET) gate electrode from the silicon or silicon germanium (SiGe) transistor channel, thus supplanting the traditional silicon oxide or silicon oxynitride gate insulator. In addition, metals instead of polycrystalline silicon (poly-Si) are used as gate electrode materials. These changes permit raising gate stack capacitance beyond the limits of the poly-Si/SiO(N) system, which improves electrostatic control of band bending in the transistor channel and thus enables continued reductions in gate length. In this chapter, we review key material and process innovations in HKMG technology for equivalent oxide thickness (EOT) scaling and metal effective work function (EWF) control. Then, we review the current status and challenges in the application of HKMG technology to Si-based multi-gate structures. Finally, we give a brief outlook to high-k oxides on III-V compound semiconductor channels, highlighting similarities and differences in oxide-semiconductor interface chemistry between Si- and III-V-based stacks.

1. Introduction

Continued device scaling for future technology nodes requires reduction in EOT of gate dielectrics. Extendibility of the conventional poly-Si/SiO(N) gate structure is challenged due to exponential increase in gate leakage currents [Buchanan, 1999]. As alternatives to SiO(N) gate dielectrics, much work has been done on the research of high-k materials. The capacitance of gate dielectric is expressed as follows.

$$C_{ox} = \frac{k\varepsilon_0 A}{t} \tag{1}$$

where C_{ox} is the capacitance of gate dielectric, k is the dielectric constant of the material, ε_0 is the permittivity of free space (=8.8531023 fF/μm), A is the area of the MOS structure, and t is the thickness of the gate dielectric. C_{ox} needs to be increased as the device size is scaled in order to maintain good electrostatics. This has traditionally been achieved by decreasing t in (1), however, this approach results in exponential increase in gate leakage current when the dimension enters the regime where direct tunneling of carriers through the oxide band gap occurs ($t < 2$ nm). As a result, replacement of SiO(N) gate dielectric with high-k material became mandatory for continued scaling at the 45-32nm technology nodes. Scalability of high-k materials is often evaluated with a metrics called equivalent oxide thickness (EOT), which represents the theoretical thickness of SiO_2 that would be required to achieve the same capacitance density as described in the following equation.

$$EOT = t_{high-k}\frac{k_{SiO2}}{k_{high-k}} = \frac{k_{SiO2}\varepsilon_0 A}{C_{ox}} \tag{2}$$

where t_{high-k} is the thickness of the high-k film, k_{SiO2} is the dielectric constant of SiO_2, and k_{high-k} is the dielectric constant of the high-k film. Capacitance-equivalent thickness (CET) is defined by using total capacitance of the MOS stack instead of C_{ox} in (2). CET is used when the capacitance from the semiconductor substrate via quantum mechanical effects is difficult to estimate. CET in inversion state of MOSFET is termed CET_{inv} or T_{inv}.

Key considerations for the choice of high-k materials are 1) k-value and band gap (E_g), 2) thermodynamic stability on Si, and 3) thermal stability with the electrode material.

It is known that E_g values have a roughly inverse dependence on k-values ($E_g \sim k^{-0.65}$) [Robertson, 2008]. Therefore, materials with too high k-values typically result in excessive direct tunneling currents. HfO_2 exhibits a k-value of approximately 20 with a sufficiently high E_g value of 5.5–6.0 eV, which makes it an attractive candidate from this perspective.

As for the stability on Si, Wilk *et al.* uses ternary phase diagrams for M-Si-O systems (M is a constituting element of the high-k material) to predict suitable high-k materials [Wilk *et al.*, 2001]. If there are no stable ternary compounds in the diagram, the high-k materials may separate into M_xO_y and M_xSi_y phases after thermal processing. Hafnium and Zirconium are known to form stable silicates, resulting in thermally stable high-k films on Si. Therefore, hafnium based high-k materials, such as HfO_2 or $HfSi_xO_y$, have become materials of choice. These films are typically deposited by chemical vapor deposition (CVD) or by a variant of CVD known as atomic layer deposition (ALD). The ALD technique is based on alternating exposures of the surface to precursors with self-saturating surface adsorption properties, e.g., a metal oxide can be grown from a metal containing molecule M and water vapor H_2O according to the sequence M–H_2O–M–H_2O–. . . . Typical Hf precursors include hafnium tetrachloride ($HfCl_4$) and tetrakis(ethylmethylamido)hafnium (TEMAH, $Hf(NC_2H_5CH_3)_4$).

Early research for hafnium based high-k materials were performed in conjunction with poly-Si electrodes for compatibility with the conventional process. This approach, however, resulted in undesired EWF shifts for p-type MOSFETs due to the reaction at the poly-Si/high-k interface [Hobbs *et al.*, 2003]. Therefore, thermally stable metals, such as TiN, were employed to suppress the reaction with the gate dielectric for the implementation of high-k films into CMOS technologies. More details in the selection of metal electrode materials will be reviewed in the later section.

The HKMG devices in the market are categorized as gate-first, gate-last, and their derivatives, depending on the device integration flows. We review pros and cons for each approach and discuss requirements for materials and process in the first section.

Continued gate length (L_g) scaling beyond the 22-nm-node requires EOT ≤ 0.9 nm. Current status and challenges of aggressive EOT scaling of high-k gate dielectrics via higher-k (>20) materials and interfacial layer scavenging techniques are reviewed in the second section.

Another important aspect for CMOS integration is the threshold voltage (V_t) control. A high performance logic large-scale integrated circuit (LSI) requires the metal EWFs to be within 0.3 eV of the conduction band edge (E_c) and valence band edge (E_v) of Si for n- and p-MOSFET, respectively. Material innovations to enable such EWF control are reviewed in the third section.

As conventional scaling on bulk Si or partially depleted SOI (PDSOI) is becoming increasingly challenging, thin body devices are being considered as long-term alternatives. Architectures include extremely thin, fully depleted SOI (ETSOI/FDSOI) FETs as well as multi-gate devices such as FinFETs. Key HKMG technologies to enable such 3-D structures are discussed in the fourth section.

In view of predictions that performance scaling of Si channel devices may end, there has recently been an increased interest in non-Si based high mobility channel materials. Most of the attention has focused on Ge and on III-V compound semiconductors such as GaAs and InGaAs. Key challenges in employing HKMG stacks on these channel materials are discussed in the fifth section.

2. Integration of HKMG Device

2.1. *Gate-first and Gate-last processes*

The HKMG devices in the market are categorized as gate-first (GF), gate-last (GL, sometimes referred to as Replacement Metal Gate or RMG), and their derivatives, depending on the device integration flows. We review pros and cons for each approach and discuss requirements for materials and process in the first section.

For integration of high-k gate dielectrics into a CMOS process, a conventional integration scheme is generally favored. In GF process, formation of high-k metal gate stack is completed prior to dopant activation to maximize the compatibility with the conventional CMOS process. In state-of-the-art CMOS integration schemes, shallow trench isolation (STI) is employed for device isolation as shown in Figure 1-1-1. After the isolation process, a pre-high-k treatment is performed. Either a thermal treatment or a wet chemical process is used as surface preparation. Then a high-k film is deposited, followed by a metal film deposition as a work function setting layer. Then, the work function setting metal films are capped with either a poly-Si film or a low resistivity metal as shown in Figure 1-1-2. The former stacked structure is termed metal inserted poly-Si stack (MIPS) [Jung *et al.*, 2005]. For the latter approach, W is typically used for its thermal stability and low resistivity [Ootsuka *et al.*, 2008]. MIPS approach is widely accepted because of the compatibility with the conventional CMOS integration schemes. For gate patterning, photo resist is used in combination with reactive ion etching (RIE). The device structure after gate etching is shown in Figure 1-1-3.

Prior to source/drain ion implantation, a spacer is formed. The activation of the junction and gate dopants is achieved by an anneal at ~1000°C. The ideal device structure after spacer and junction formation is illustrated in Figure 1-1-4. In order to minimize the junction and gate resistance,

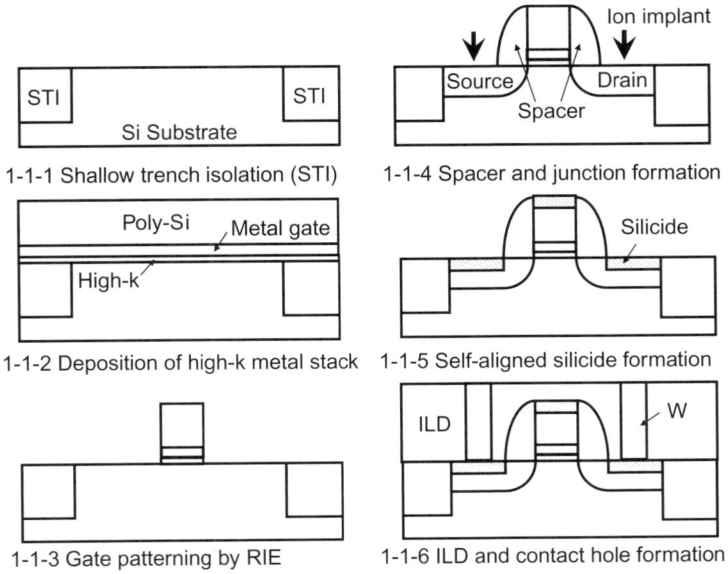

Fig. 1.1. Process flow of gate-first (GF) integration.

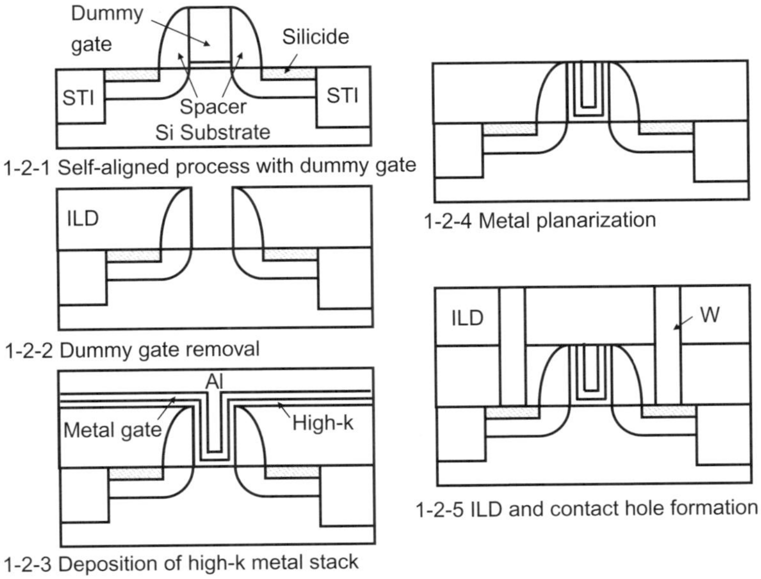

Fig. 1.2. Process flow of high-k-last gate-last (HKLGL) integration.

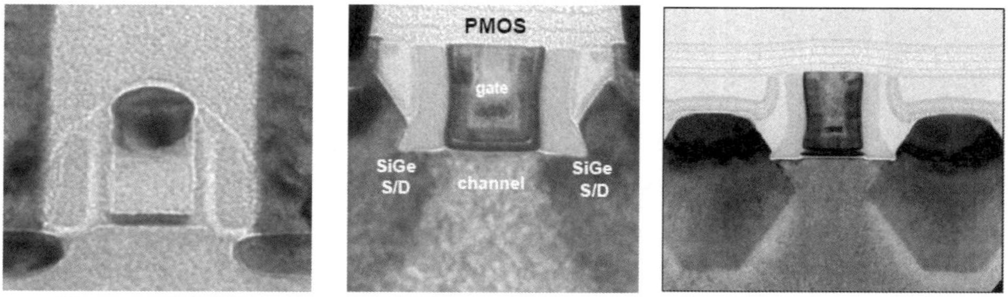

[Greene et al., 2009] (a) | [Packan et al., 2009] (b) | [Mistry et al., 2007] (c)

Fig. 2. Cross-sectional transmission electron microscopy (TEM) images of MGHK stack fabricated with (a) gate-first (GF) integration (after [Green *et al.*, 2009]); (b) high-k-last gate-last (HKLGL) integration (after [Packan *et al.*, 2009]); and (c) high-k-first gate-last (HKFGL) integration (after [Mistry *et al.*, 2007]).

a self-aligned silicide process is used, where Ni is deposited on the wafer first and then a thermal process step is applied to react the metal and the Si forming N-silicide. The device structure after silicide formation is shown in Figure 1-1-5. For device contact, an inter-layer dielectric (ILD) is deposited followed by a planarization step using chemical mechanical polishing (CMP). The contact hole process includes lithography, etch, clean and W deposition to obtain the structure described in Figure 1-1-6. The metallization is achieved using Cu wiring in state-of-the-art CMOS devices. A cross-sectional transmission electron microscopy (TEM) image of HKMG stack fabricated with GF process is shown in Figure 2(a) (after [Green *et al.*, 2009]).

The major advantage of GF process is the high compatibility with the conventional CMOS integration, but on the other hand, a high thermal budget poses more stringent requirements for the high-k material. In the early days of high-k research, the anomalous V_t behavior was attributed to the reaction at the Hf-based oxide/poly-Si interface. Recent studies, however, have revealed that this effect is present even in conjunction with thermally stable metal electrodes [Akasaka *et al.*, 2006]. Therefore, V_t control for high-k/metal gate stack still remains as one of the critical obstacles for implementation of GF integration. The technique to control V_t under the constraints of the high thermal budget of GF process will be discussed more in the later section.

Another way to obtain sufficiently low V_t is employing dual metal gates which have vacuum work functions corresponding to E_C and E_V of Si and limiting the thermal budget to below 500°C. This is achieved by forming junctions prior to the HKMG stack formation [Yamaguchi *et al.*, 2006]. This approach is termed GL integration. The early process steps in GL process follow the conventional CMOS integration using SiO_2/poly-Si gate stacks. The SiO_2/poly-Si gate stacks serve the purpose as dummy gate structures. The source/drain junctions are formed by ion implantation using the dummy gate structure and a dopant activation anneal at ~1000°C. Then the standard self-aligned silicide process is employed to form a low resistivity contact layer on the source/drain area as shown in Figure 1-2-1. After ILD deposition, the poly-Si gate is exposed by CMP process, followed by poly-Si removal using either RIE process or wet etching process. Then the dummy SiO_2 layer is removed to form a thinner interfacial layer as shown in Figure 1-2-2. As seen in Figure 1-2-2, the high-k layer must be grown uniformly at the bottom of the dummy gate. Therefore atomic layer deposition (ALD), which takes advantage of surface adsorption rather than gas phase reaction of the precursor, is the most suitable technique to form high-k layer for GL process. The high-k deposition is typically followed by a post-deposition anneal (PDA). Then work function setting metal layers are deposited on the high-k layer. For fabrication of CMOS devices, a metal layer for the n-type (p-type) MOSFETs needs to be removed from the p-type (n-type) MOSFET area selectively to the underlying high-k layer, followed by a metal layer for the p-type (n-type) MOSFETs. The suitable materials for work function setting metals will be discussed in the later section. Next, the rest of the trench is filled with a low resistivity metal such as Al or W as illustrated in Figure 1-2-3. Then, the metal layers are removed by CMP to expose the ILD and to leave the metal layers only in the trench as shown in Figure 1-2-4. Thus, the devices are isolated with each other in a self-aligned manner. After the metal CMP, the second ILD is deposited, followed by the conventional contact hole and metallization process as discussed in GF integration to obtain the final structure as shown in Figure 1-2-5. This process flow is hereinafter called high-k-last gate-last (HKLGL) to differentiate from the derivative of gate-last process that we describe next. A cross-sectional TEM image of HKMG stack fabricated with HKLGL process is shown in Figure 2(b) (after [Packan *et al.*, 2009]).

High-k dielectrics typically require an anneal at a high temperature (>600°C) to attain a high carrier mobility and a good reliability [Narayanan *et al.*, 2006a]. Such treatment is difficult to incorporate after a deposition of high-k film in a typical HKLGL process flow since source drain junctions and silicide contacts are already formed at that stage. Therefore, a hybrid version of GF and GL was used in some cases to facilitate introduction of

HKMG technology. Let us categorize this integration flow as high-k-first gate-last (HKFGL) in order to differentiate it from the previously described HKLGL. In HKFGL, a high-k layer is prepared with a dummy gate prior to a junction activation anneal just like GF process. Then, only the dummy gate is replaced with a metal gate later such that the high-k layer receives a high temperature anneal whereas a thermal budget on the metal gate is limited. A cross-sectional TEM image of HKMG stack fabricated with HKFGL process is shown in Figure 2(c) (after [Mistry *et al.*, 2007]).

2.2. *Industry trend for HKMG integration*

The semiconductor industry has not yet converged into one way for manufacturing. Table 1 summarizes the industry trend for MGHK manufacturing process. The first product with HKMG stacks was manufactured by Intel with HKFGL process at the 45-nm node [Mistry *et al.*, 2007]. HKFGL process was used by TSMC later at the 28-nm node when they first introduced HKMG technology [Yang *et al.*, 2011], [James, 2014]. Thus, HKFGL process

Table 1. Summary of industry trend for HKMG integration schemes.

		HKMG Integration		
		HKFGL	HKLGL	GF
Technology Node (nm)	45	Intel [Mistry et al., 2007]		
	32		Intel [Natarajan et al., 2008] [Packan et al., 2009]	IBM HP [Greene et al., 2009] Panasonic [1] IBM Alliance LP [Chen et al., 2008]
	28	TSMC [Yang et al., 2011] [James, 2014]		IBM Alliance LP [Arnaud et al., 2009] STMicroelectronics/CEA-Leti [Planes et al., 2012]
	22		Intel [Auth et al., 2012]	IBM HP [Narasimha et al., 2012]
	20		Samsung [Cho et al., 2011] IBM Alliance LP [Shang et al., 2012]	STMicroelectronics [Cesana, 2012]

[1] Panasonic press release
http://panasonic.co.jp/corp/news/official.data/data.dir/en100915-2/en100915-2.html

has been used as an entry technology to achieve high quality high-k films at high temperatures and relatively easy WF control by avoiding high thermal budget on the metal electrodes. However, HKFGL process is a complex technology requiring essences of both GF and GL processes. Therefore, HKLGL was introduced at the 32-nm node by engineering the process to attain high quality high-k films without relying on the thermal budget from the junction activation anneal [Packan *et al.*, 2009]. Another benefit of GL process is strain enhancement in the channel of MOSFETs. It has been reported that removal of dummy gate structure under external stress elements enhances the strain in the channel resulting in higher performance [Wang *et al.*, 2007], [Auth *et al.*, 2008]. Since the modern CMOS technology relies on performance boost from strain engineering, GL process offers another attractive knob which has not been exploited in the previous CMOS generations. The biggest challenge of GL process is the metal fill into the trench and the subsequent metal CMP process. These steps add a lot of complexity to the CMOS integration flow. Also, the L_g scaling with GL process is limited by the gap fill capability, and thus requires continued innovation for future technology nodes.

On the other hand, the first GF products were introduced at the 32-nm node by IBM [Green *et al.*, 2009], its alliance partners [Chen *et al.*, 2008], and Panasonic [press release, 2010]. One of the most attractive characteristics of GF process is its simplicity. Therefore, GF process was used for foundry business at the 32-nm and the 28-nm nodes to take advantage of its high compatibility with the conventional CMOS technology [Arnaud *et al.*, 2009]. In addition, GF process has become a foundation for fully depleted SOI (FDSOI) technologies at the 28-nm node and beyond [Planes *et al.*, 2012], [Cesana, 2012]. Figure 3 summarizes the trend of inversion oxide thickness (T_{inv}) or EOT as a function of technology node for various MGHK integration schemes. For bulk and partially depleted SOI (PDSOI) technologies, GF process has shown smaller T_{inv} values compared to GL process at a given technology node. This is another advantage of GF process and aggressive T_{inv} and L_g scaling were pursued for planer GF technologies [Narasimha *et al.*, 2012]. The requirements for T_{inv} scaling were relaxed for fully depleted device structures such as FinFET and FDSOI and the T_{inv} trends for GF and GL processes are converging for these new device architectures. The biggest challenge of GF process is EWF control under the constraints of high thermal budget and typically additional process/material knobs are required to overcome this issue.

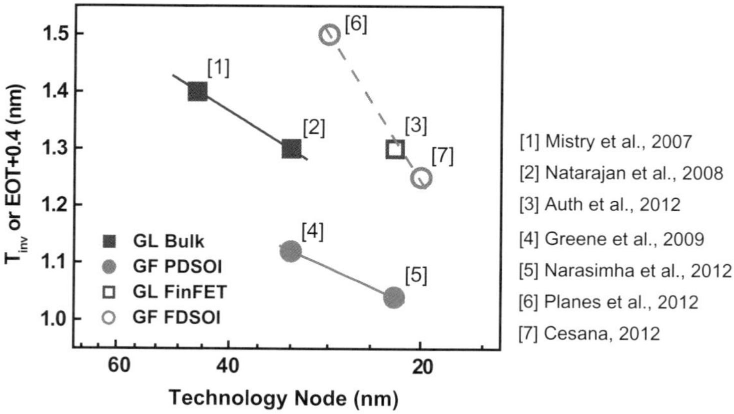

Fig. 3. T_{inv} or EOT + 0.4 nm as a function of technology node comparing various device architectures with gate-first (GF) and gate-last (GL) integrations.

As seen in Table 1, at least low power technologies are converging to HKLGL process at the 22/20-nm node [Auth *et al.*, 2012], [Cho *et al.*, 2011], [Shang *et al.*, 2012] and the trend is expected to continue toward the insertion node of FinFET by most companies. Thus, some level of process commonality has been achieved for HKMG technology.

3. EOT Scaling with HKMG

3.1. *Review of EOT scaling approaches*

After a decade-long search for the appropriate high-k materials, the semiconductor industry has converged on Hf-based oxides for the first generation CMOS products featuring HKMG technology [Mistry *et al.*, 2007], [Green *et al.*, 2009]. The EOT for the first generation HKMG device is approximately 1.0 nm. Continued L_g scaling requires sub-nm EOT to suppress short-channel effects. Fully depleted device structures, such as FinFET or FDSOI, improve short-channel control and thus relax the requirements for EOT scaling. However, the insertion point of such device architectures is expected to be the 22 nm and beyond and sub-nm EOT may be still required at those advanced technology nodes.

A typical HKMG stack structure contains a silicon oxide based interfacial layer (IL), a high-k dielectric, followed by a metal gate electrode. This system is equivalent to two capacitors connected in series. Thus, the total EOT of the HKMG stack can be expressed as follows.

$$EOT = EOT_{IL} + EOT_{HK} \tag{3}$$

where EOT_{IL} and EOT_{HK} are contributions from the IL and high-k layer, respectively. An apparent way to scale EOT_{HK} is to reduce the physical thickness of the high-k layer, however, there is little room in this direction. It has been reported that a thick HfO_2 layer causes significant degradations in carrier mobility and charge trapping both with GF [Kirsch *et al.*, 2006] and GL [Ando *et al.*, 2010] processes. Therefore, the first generation HKMG devices already employ the thinnest possible high-k layer. This leaves us with three possible EOT scaling approaches: (1) Introduce a new high-k material with k-value greater than that of HfO_2 (approximately 20), (2) Increase a k-value of IL, (3) Reduce a physical thickness of IL.

Table 2 shows EOT scaling approaches reported in literature. The scaling trends for dynamic random access memory (DRAM) capacitors and gate dielectrics are summarized together for comparison. A clearly different approach is taken for each application due to different requirements for high-k dielectrics. DRAM capacitors are embedded in the back end of line (BEOL) of CMOS integration for most applications. Therefore, requirements for thermal stability are relaxed compared to those for gate dielectrics. This allows use of thermally instable elements with high polarizability, such as Sr, Ti, and Ba, to enhance k-values [Hwang *et al.*, 1995], [Pawlak *et al.*, 2011], [Kim *et al.*, 2008]. Another difference is that requirements for reliability of DRAM capacitors are not as stringent as those for gate dielectrics. As mentioned previously, high-k layers are typically targeted as thin as possible to qualify reliability specs of MOSFETs. This limitation is not applicable to DRAM capacitor dielectrics and thicker high-k layers can be used. It is easier to control crystallinity of high-k for physically thicker films. Therefore, stabilization of higher-k phase (tetragonal or cubic) of HfO_2 or ZrO_2 via crystallinity control has been widely practiced for DRAM application [Monaghan *et al.*, 2009], [Mise *et al.*, 2009]. There have been reports on k-value increase by higher-k phase stabilization for thinner HfO_2 films via doping of elements such as yttrium [Kita *et al.*, 2005], and silicon [Tomida *et al.*, 2006]. However, controllability of crystallinity in an ultra thin HfO_2 thickness regime (<2 nm) has yet to be demonstrated. Other groups have suggested La-based higher-k materials such as La-Lu-O [Edge *et al.*, 2011] or La-Al-O [Suzuki *et al.*, 2005], [Arimura *et al.*, 2011] in a thin film thickness regime. These La-based materials have shown k-values ranging from 20 to 30 in amorphous form, thus making application to gate dielectrics easier.

A less disruptive way to increase the average k-value of the gate stack is to boost the k-value of the IL in conjunction with Hf-based oxides.

Table 2. Summary of EOT scaling knobs for DRAM capacitor and gate dielectrics.

	HK Scalling			IL Scaling		
Application	Material	EOT (nm)	K-value	Material	EOT (nm)	K-value
DRAM Dielectric	(Ba, Sr)TiO_3 [1]	0.24	325			
	TiO/Sr Tio [2]	0.40	85			
	Al-doped TiO_2 [3]	0.48	47			
	Cubic ZrO_2 [4]	0.90	31			
	Cubic Al-doped HfO_2 [5]	0.70	40			
	Cubic Y-doped HfO_2 [6]	NA	27			
	Cubic Si-doped HfO_2 [7]	NA	27			
	LaLuO [8]	0.58	23			
	LaAIO [9,10]	0.31–0.74	23–25			
Gate Dielectric				SrO [11,12]	0.50–0.60	13
				LaSiO [13]	0.58	12
				TmSiO [14]	0.65	12
				SiO_2 scavenging [15]	0.42	3.9

[1] Huang *et al.*, 1995 [2] Pawlak *et al.*, 2011 [3] Kim *et al.*, 2008 [4] Monaghan *et al.*, 2009 [5] Mise *et al.*, 2009 [6] Kita *et al.*, 2005 [7] Tomida *et al.*, 2006 [8] Edge *et al.*, 2011 [9] Suzuki *et al.*, 2005 [10] Arimura *et al.*, 2011 [11] Marchiori *et al.*, 2011 [12] Frank *et al.*, 2011 [13] Ando *et al.*, 2009 [14] Lita *et al.*, 2013 [15] Ando, 2012.

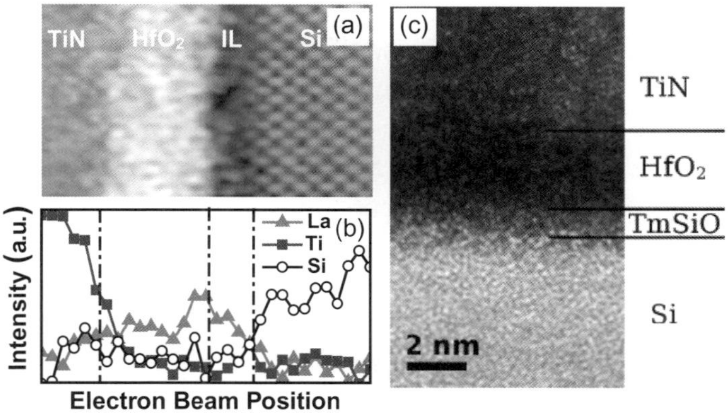

[Ando et al., 2009] **[Litta et al., 2013]**

Fig. 4. (a) Z-contrast image; (b) corresponding Electron Energy-Loss Spectroscopy (EELS) profiles (La, Ti, and Si) of the SiO2/HfO2/La2O3/TiN/poly-Si stack after a 1,000 °C anneal (after [Ando *et al.*, 2009]); and (c) cross-sectional transmission electron microscopy (TEM) image of the $TmSiO/HfO_2/TiN$ stack (after [Litta *et al.*, 2013]).

A La_2O_3 cap on a Hf-based high-k layer has been widely used to adjust EWF of n-type field-effect-transistors (nFET) [Narayanan *et al.*, 2006b]. Some portion of the La elements diffuse through the high-k layer and form La-silicate IL after high temperature anneals [Copel *et al.*, 2009]. This is an effective way to increase the k-value of SiO_2-based IL. Figures 4(a) and (b) show the Z-contrast image and Electron Energy-Loss Spectroscopy (EELS) profiles, respectively, for the $SiO_2/HfO_2/La_2O_3/TiN/poly$-Si stack after a 1000°C anneal (after [Ando *et al.*, 2009]). As seen in Figure 4(b), formation of La-silicate in the IL can be precisely controlled by carefully optimizing the La_2O_3 layer thickness and the thermal budget, resulting in EOT of 0.58 nm. As a potential new alternative, it was recently shown that a thin epitaxial strontium oxide (SrO) interfacial layer prior to HfO_2 deposition results in EOT of 0.50 nm [Marchiori *et al.*, 2011] and 0.60 nm [Frank *et al.*, 2011] for GL and GF integrations, respectively. All of these lanthanide (e.g., La, Lu) or alkaline earth (e.g., Sr) metals are known to form electric dipoles at the SiO_2/high-k interface which shifts the EWF toward E_c of Si. Formation of such dipoles makes application of these materials to p-type field-effect-transistors (pFET) extremely difficult. Development of CMOS-compatible higher-k materials is one of the biggest challenges to overcome. Unlike other lanthanide-based higher-k materials, TmSiO IL showed EOT scaling down to 0.65 nm without much EWF shift. A cross-sectional TEM image for $TmSiO/HfO_2/TiN$ stack is shown in Figure 4(c) (after [Lita *et al.*, 2013]). This unique nature of TmSiO IL may allow CMOS integration.

Considering the maturity of Hf-based high-k gate dielectrics, scaling of SiO_2-based IL in conjunction with Hf-based oxides via scavenging reaction may be more practical in meeting the requirements for the 22 nm node and beyond. We have categorized various IL scavenging techniques into direct-scavenging and remote-scavenging based on the locations of the scavenging elements in the gate stacks [Ando, 2012]. Direct-scavenging schemes incorporate the scavenging elements within the high-k layers, whereas the remote-scavenging schemes isolate the scavenging elements from the high-k layers. Figure 5 shows cross-sectional TEM images of the high-k metal gate stacks for remote-scavenging with (a) medium scavenging metal dose, (b) high scavenging metal dose (after [Ando *et al.*, 2009]), and (c) direct scavenging with TaN-alloy (after [Choi *et al.*, 2009]), respectively. As shown in Figures 5(a) and (b), the final IL thickness is controllable by changing the scavenging metal dose.

3.2. *Materials choice for IL scavenging*

Choice of Scavenging element is one of the most important factors to control IL scavenging reaction. We found that the Gibbs free energy change at 1,000 K (ΔG^0_{1000}) of the following reaction serves as a guiding principle for

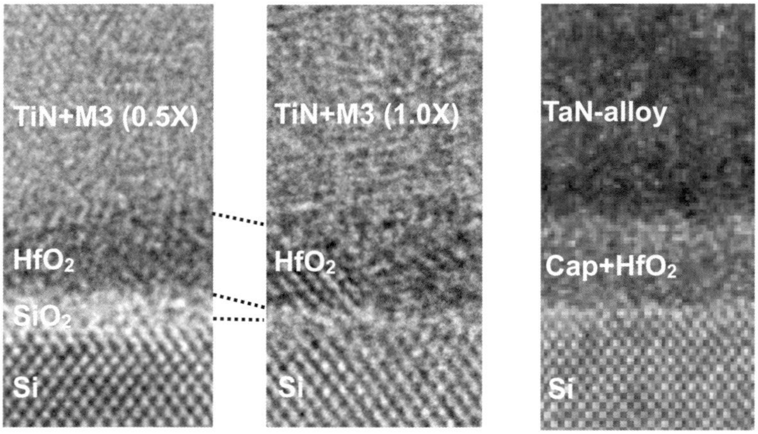

(a) Remote (b) Remote (c) Direct

Fig. 5. Cross-sectional transmission electron microscopy (TEM) images of HKMG stacks for remote-scavenging with (a) medium scavenging metal dose; (b) high scavenging metal dose (after [Ando *et al.*, 2009]); and (c) direct scavenging with TaN-alloy (after [Choi *et al.*, 2009]).

the choice of scavenging element [Ando *et al.*, 2009].

$$Si + \frac{2}{y}M_xO_y \rightarrow \frac{2x}{y}M + SiO_2 \tag{4}$$

where M is the scavenging element in the gate stack. When M with a positive ΔG^0_{1000} value is doped in the gate stack, there is an energy gain for forming a metal oxide (M_xO_y) rather than maintaining SiO_2, which is the driving force for the IL scavenging reaction. In the actual scavenging reaction, oxygen transport from SiO_2 to M requires some energy. However, the experimental data suggest that oxygen diffusion through HfO_2 or ZrO_2 is quite rapid at elevated temperatures and this is not the rate-limiting step for the IL scavenging reaction [Kim *et al.*, 2004], [Ando *et al.*, 2010]. The EOT trend for MIPS [Jung *et al.*, 2005] with SiO_2/HfO_2 dual-layer gate dielectrics in the literature is summarized in Figure 6 as a function of ΔG^0_{1000} per oxygen atom ($\Delta G^0_{1000}/O$) for the scavenging element (after [Ando, 2012]). Here, the scavenging element is defined as the metal with the most positive $\Delta G^0_{1000}/O$ value (after [Hubbard *et al.*, 1996]), existing in either gate dielectrics or work function-setting metal layers. As seen in Figure 6, EOT and $\Delta G^0_{1000}/O$ values show a very strong correlation. Most thermally stable metals studied in the early phase of MGHK development (e.g. W and TaN) have negative $\Delta G^0_{1000}/O$ values, resulting in IL regrowth during high temperature activation anneals. On the other hand, TiN has a slightly positive $\Delta G^0_{1000}/O$ value, providing slight EOT scaling compared to Poly-Si in most cases. The near zero $\Delta G^0_{1000}/O$ value may be the main

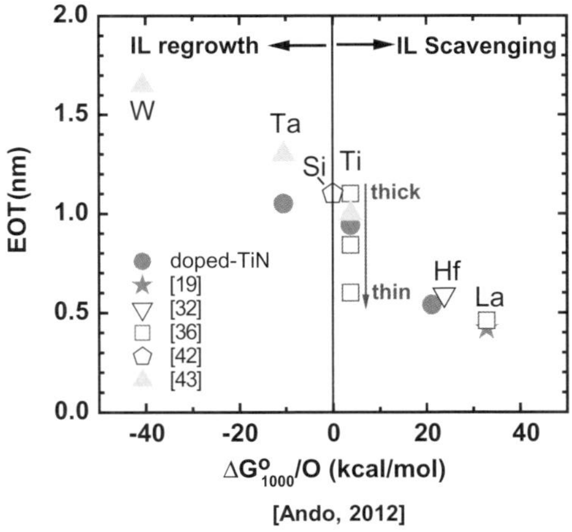

Fig. 6. EOT of SiO2/HfO2 metal-inserted poly-Si stack (MIPS) structure as a function of ΔG^0_{1000} per oxygen atom for scavenging element (after [Ando, 2012]).

reason why TiN has been widely accepted as a baseline metal electrode throughout the industry. The neutral nature of TiN typically results in neither IL regrowth nor scavenging. In order to facilitate the scavenging reaction, we proposed doping of scavenging metals with high $\Delta G^0_{1000}/O$ values into a thermally stable TiN electrode. This technique enables highly controllable IL scavenging for both GF (maximum process temperature: 1000°C) [Ando *et al.*, 2009] and GL (maximum process temperature: 600°C) [Ando *et al.*, 2013] integrations.

3.3. *Device implications of IL scaling*

Cartier *et al.* investigated impacts of remote IL scavenging on reliability [Cartier *et al.*, 2011]. SiO_2/HfO_2 dual-layer stacks with varying IL thicknesses were prepared in both GF and GL processes in this work. The change in the device lifetimes, including positive bias temperature instability (PBTI), negative bias temperature instability (NBTI), and time dependent dielectric breakdown (TDDB), are estimated and they found that lifetime trends for IL scaling are similar for GF and GL processes. This indicates that these trends arise from the fundamental materials properties of the SiO_2/HfO_2 dual-layer stacks and do not depend much on the fabrication method. The BTI lifetimes are predicted to decrease by 50–100× for every 0.1 nm of IL scaling. Drastic lifetime reductions also occur for TDDB. Therefore, calculated IL scaling, considering reliability requirements, is needed for future CMOS nodes. More comprehensive review on reliability issues for HKMG stacks was provided by IBM group [Linder *et al.*, 2014].

The implication of IL scaling for carrier mobility is another important factor for device performance. Although a great deal of progress has been made across the community, electron mobility values for HKMG stacks typically fall below the trend for conventional SiO(N)/Poly-Si stacks. In the case of ideal thermal oxide gate dielectrics, the electron mobility is limited by Coulomb scattering from the substrate impurities (low E_{eff}), phonon scattering from the Si substrate (middle E_{eff}), and surface roughness scattering from the Si/SiO_2 interface (high E_{eff}) [Takagi *et al.*, 1994]. It has been reported that introduction of high-k dielectrics brings about additional degradation in each E_{eff} regime. In the low E_{eff} regime, fixed charges in the high-k layer or interface dipoles localized at the IL/high-k interface cause remote Coulomb scattering (RCS) [Yamaguchi *et al.*, 2002], [Saito *et al.*, 2003], [Ota *et al.*, 2007], [Ando *et al.*, 2009]. In the middle E_{eff} regime, Fischetti *et al.* predicted that soft optical phonons in high-k layers can couple with carrier electrons, resulting in remote phonon scattering (RPS) [Fischetti *et al.*, 2001]. In the high E_{eff} regime, either the fluctuation of permittivity of high-k layers or the surface roughness at the IL/high-k interface can cause remote surface roughness scattering (r-SRS) [Li *et al.*, 1987], [Ando *et al.*, 2009]. All of these additional scattering mechanisms are predicted to show strong dependences on the IL thickness.

Figure 7 shows high field electron mobility as a function of EOT from literature data including various IL scaling approaches with HfO_2. It should be noted that all data fall on the same line with a slope of approximately 20 cm^2/Vs per 0.1 nm scaling, showing the universality of IL thickness dependent mobility in the ultra-thin EOT regime. Thus, EOT scaling with some compromise of carrier mobility is a realistic scaling approach. In this regard, Tatsumura *et al.* predicted a mobility-EOT relationship providing the same drive current (I_{on}) at a given L_{min} (L_g defined by off-state leakage current) by considering both mobility degradation and short-channel effects suppression from EOT scaling. The comparison between the experimental data in the literature and the extracted contour lines from [Tatsumura *et al.*, 2008] is shown in Figure 7. The mobility-EOT slope for IL scaling ($\sim$20 cm^2/Vs per 0.1 nm) is shallower than the breakeven relationship for $L_{min} \leq 30$ nm ($\sim$40 cm^2/Vs per 0.1 nm). This indicates that it is indeed possible to improve the short-channel device performance by employing IL scaling in conjunction with aggressive L_g scaling in future nodes albeit with some electron mobility degradation.

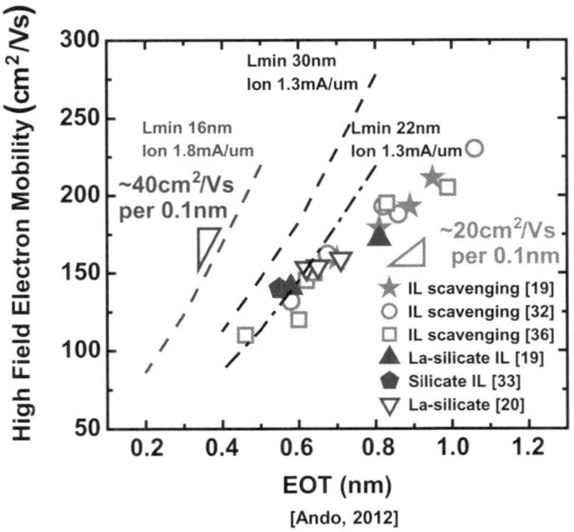

Fig. 7. High field electron mobility as a function of EOT from literature data including various IL scaling techniques (after [Ando, 2012]). Simulated contour lines providing the same I_{on} at L_{min} 16, 22, and 30 nm are shown for comparison (after [Tatsumura *et al.*, 2007]).

3.4. *Summary*

EOT scaling into the sub-nm regime has been accomplished by various approaches. La-based higher-k materials and silicate IL with HfO_2 showed aggressive EOT values (0.5–0.8 nm), but with large EWF shifts toward the Si E_c, limiting their application to nFET. Further exploration for pFET-compatible higher-k materials is needed. Meanwhile, IL scavenging is a promising approach to extend Hf-based high-k dielectrics to future nodes. Mobility-EOT trends in the literature suggest that short-channel performance improvement is attainable with aggressive EOT scaling via IL scavenging or La-silicate formation. However, extreme IL scaling is accompanied with severe penalty in reliability. Therefore, highly precise IL thickness control in an ultra-thin IL regime (<0.5 nm) will be the key technology to satisfy both performance and reliability requirements for future CMOS devices.

4. EWF Engineering in HKMG Stacks

4.1. *Work function/threshold voltage control in HKMG stacks*

The vacuum work function of a metal or metallic alloy used as a gate electrode in a MOSFET is a materials parameter that determines the flatband voltage, which in turn determines the V_t of the MOSFET [see Schroder 1998]. However, the V_t of a device may include other components besides the work function of the electrode such as fixed oxide charge [Schroder, 1998], charged dipoles in the dielectric [Narayanan *et al.*, 2006b] and charged point defects in the dielectric (such as oxygen vacancies) [Shiraishi *et al.*, 2004]. In addition, even in the absence of the aforementioned components, depending on the ionicity of the dielectric, the extracted electrical work function can be different than the vacuum workfunction based on the so called pinning parameter [Yeo *et al.*, 2002]. All of these effects are combined together in the term EWF, which is used to describe the extracted work function from an MOSCAP/MOSFET. Typically high performance and low power devices on bulk Si substrates or PDSOI require electrode effective work functions in the range of 4.1–4.3 eV for nMOSFETs and between 4.9 and 5.2 eV for pMOSFETs.

4.2. *V_t control in gate first integration scheme*

The GF integration scheme requires the gate electrode to be thermally stable, i.e., able to withstand junction activation temperatures >1000°C for a few seconds. However, typical elemental metals that have low work functions are highly reactive and thermally unstable (above 600°C) [Gusev *et al.*, 2006]. On the other hand, high work function metals tend to be structurally stable at these elevated temperatures but exhibit EWFs which are much lower (towards the middle of the Si band-gap) than their vacuum work function and this effect will be discussed in detail later in this section. For nMOSFETs, alloy electrodes, such as TaN [Narayanan *et al.*, 2006a], TaC [Schaeffer *et al.*, 2004] and TaSiN [Samavedam *et al.*, 2002] have been extensively evaluated due to their thermal stability and low work fucntions. However, most of these alloys tend not to scale T_{inv} aggressively [Narayanan *et al.*, 2006a] and in addition are not able to provide EWFs near the Si E_c. On the other hand, a near mid Gap EWF electrode PVD TiN provides the optimal tradeoff between electron mobility and T_{inv} and can scale $T_{inv} < 1.4$ nm with high mobilities of 200 cm^2/Vs at 1 MV (Figure 8).

To obtain near band-edge EWFs for nMOSFETs, the IBM group introduced nanoscale dielectric capping layers consisting of oxides and nitrides of gp. IIA and gp. IIIB elements, most notably, La and Mg, sandwiched between the Hf based dielectrics and a TiN electrode [Narayanan *et al.*, 2006b]. The subsequent dielectric modification of the SiO_2/HfO_2 bilayer at elevated GF processing temperatures resulted in (a) large V_t shifts towards the Si E_c, close to those exhibited by SiON/Poly-Si Band-Edge (BE) devices, and (b) equal or lower T_{inv}, compared to uncapped HfO_2/TiN stacks (Figure 9), with minimal impact to mobility [Narayanan *et al.*, 2006b]. The

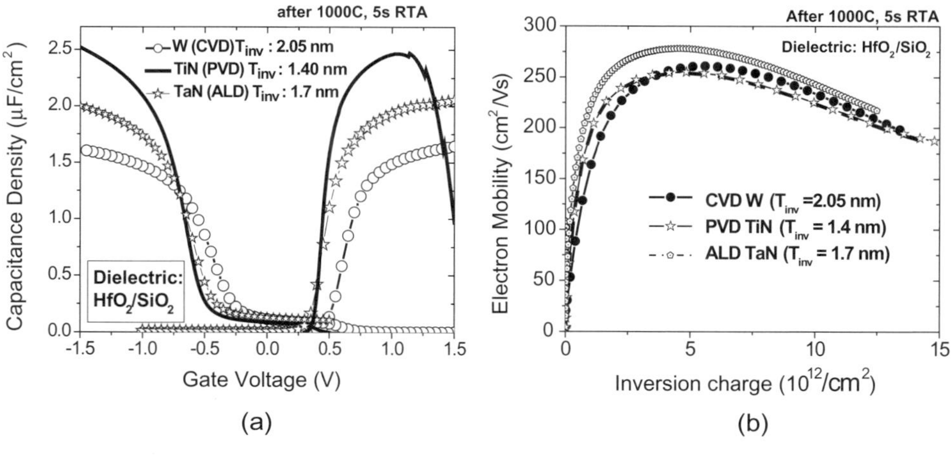

Fig. 8. (a) Inversion split C–V characteristics and (b) Electron mobility curves of self-aligned nFETs with SiO_2/HfO_2/metal gate stacks after a 1000°C, 5-s S/D RTA which clearly shows that TiN metal electrodes provide the best mobility/T_{inv} tradeoff.

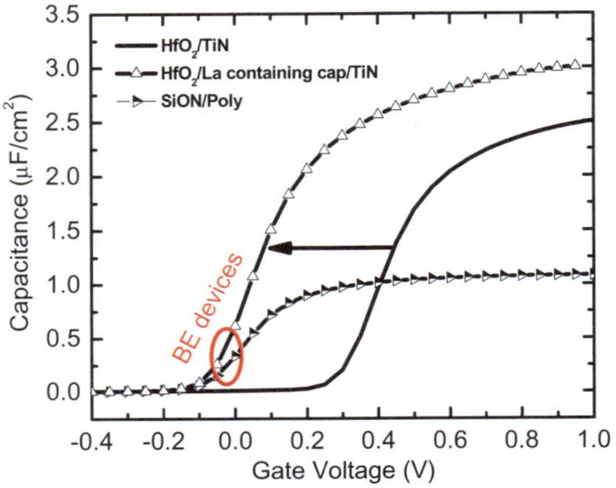

Fig. 9. nFET Split CVs of HfO_2/La containing cap layers/TiN gate stacks showing the concomitant V_t shift towards Si conduction band edge and reduction in T_{inv}@ (Vt + 0.8 V) ranging from 1.37 nm to 1.2 nm.

reason behind the voltage shifts due to the cap layers is attributed to the formation of dipoles [Narayanan *et al.*, 2006b], [Guha *et al.*, 2007]. It was argued in these papers that cations that are strongly electropositive (relative to Hf), such as group IIIB elements Sc, La, and Er or group IIA elements Mg, Ba, and Sr, will shift the V_t toward negative voltages, whereas a more electronegative cation such as Al will shift the V_t in a positive direction [Jung *et al.*, 2005]. One problem with this postulation is that the elemental electronegativity is essentially identical for Mg and Hf and cannot explain the relative V_t shifts that are observed. To explain the shifts observed, one has to invoke the differences in group electronegativity (using the Sanderson criterion) between the capping and high-k layers [Schaeffer *et al.*, 2007]. The EWF of the SiO_2/HfO_2 stack is determined by the charge neutrality level in the system. Therefore, the insertion and subsequent mixing of the capping layers with underlying $SiO(N)/HfO_2$ dielectrics result in the formation of a dipole because of the differences in group electronegativity [Jagannathan *et al*, 2008]. It has also been shown that the dipolar V_t shift is obtained only when the cation element such as La or Al is in intimate contact with the interfacial SiO_2 [Copel *et al*, 2009], [Kamimuta *et al.*, 2007], [Chudzik *et al.*, 2012], suggesting unequivocally that the dipole is activated only when the cations intermix with SiO_2. Thus, a dielectric capping layer with a higher group electronegativity than HfO_2 (such as Al_2O_3, AlN, TiO_2) would result in a capping layer–induced shift that is positive with respect to HfO_2, while a negative shift would be induced for a dielectric capping layer with a group electronegativity smaller than HfO_2 (such as La_2O_3, MgO) with SiO_2

critical for the activation of the dipole. An alternative explanation for the origin of the dipole in terms of the difference in oxygen density between the capping layer and SiO_2 has been postulated by Kita *et al.* [Kita *et al.*, 2009], though recent *ab initio* simulations have provided stronger theoretical backing for the group electronegativity theory [Lin *et al.*, 2009]. Other groups studied modulation of band alignments from adding substitutional cations to SiO_2/HfO_2 interfaces by using *ab initio* simulations and explained the experimental trend [Luo *et al.*, 2011], [Van Der Geest *et al.*, 2012].

Thermally stable high work function electrodes tend to exhibit mid-gap EWFs (and unacceptably high V_ts) after full GF processing and this has remained a fundamental problem in the realization of near band-edge pMOSFETs. The large shift towards mid gap is attributed to the formation of charged oxygen vacancies [Shiraishi *et al.*, 2004], [Cartier *et al.*, 2005] (Figure 10). It has been shown that the oxygen vacancy defect levels within the HfO_2 are located closer to the Si E_c, [Robertson *et al.*, 2007], while the high work function electrode is aligned to the Si E_v. In the presence of high temperatures, there is a large thermal driving force that results in the formation of positively charged oxygen vacancies while giving up electrons to the metal electrode and thereby creating a dipole that pulls the EWF towards midgap [Cartier *et al.*, 2005]. This reaction continues until defect level of charged vacancies is aligned Fermi level of the electrode. Substrate Si oxidation is also a critical step in the overall kinetics of oxygen vacancy formation [Akasaka *et al.*, 2006]. The SiOx IL layer when its greater than a critical thickness, can prevent the formation of these charged vacancies and thereby have a direct impact on the threshold voltage. Indeed, the formation of oxygen vacancies is believed to be the root cause for the much discussed

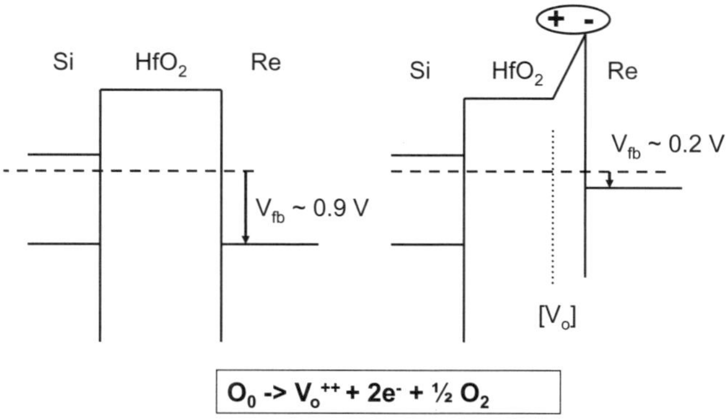

Fig. 10. Schematics explaining the impact of oxygen vacancies, [Vo], and of 'dipole' formation due to electron transfer from the vacancies to the metal (after [Cartier *et al.*, 2005]).

Table 3. T_{inv} change and V_t shift for optimized SiGe channel with reference to control Si channel (after [Krishnan *et al.*, 2011]).

Channel Material	Normalized T_{inv}	V_t shift (mV)
Si	1.00	—
Optimized SiGe	1.03	300

roll-off effect of V_t [Bersuker *et al.*, 2010] with decreasing IL thickness. On the other hand, Demkov provided an alternative theory with oxidation of high work function metals being the driving force of the formation of oxygen vacancies [Demkov, 2006].

In order to circumvent the oxygen vacancy formation issue and enable appropriate V_ts for high performance and low power CMOS, a SiGe epitaxial channel (cSiGe) was introduced selectively only on the pMOSFET channel area [Winstead *et al.*, 2007], [Harris *et al.*, 2007]. Due to the valence band offset between SiGe and Si, the EWF can be increased by more than 300 mV and thereby reduce the V_t of the pMOSFET to more appropriate values [Krishnan *et al.*, 2011] (Table 3). In addition, higher channel mobility and better NBTI reliability [Franco *et al.*, 2011] are distinct advantages for SiGe channel that are important advantages for continued technology scaling. On the other hand it is well known that the presence of Ge catalyses silicon oxidation and hence the interfacial thickness between the nFET Si channel and pFET SiGe channel is different and can result on T_{inv} deltas >2Å under the same processing conditions [Krishnan *et al.*, 2011]. Careful process optimization with appropriate cleans and anneals have been demonstrated in a planar dual channel high performance GF CMOS integration scheme [Narasimha *et al.*, 2012] with extendability to fully depeleted architectures like FinFETs [Hashemi *et al.*, 2013] and ETSOI [Cheng *et al.*, 2012].

4.3. V_t *control in gate-last integration scheme*

For GL integration, the absence of high temperature processing (i.e. $>500°C$) after metal electrode deposition implies that EWF near the Si Ec for nMOSFET and the Si Ev for pMOSFET are achieved by utilizing the intrinsic vacuum work functions of gate electrode materials. Inherently, the GL integration provides higher flexibility in the choice of metal gate electrode materials to achieve near band-edge EWF, but the choice of work function materials is often limited by their compatibility with the CMOS integration scheme, including their ability to deposit in trench like structures created from the dummy gate removal and ability to be chemically mechanically polished for planarization. Thus, only a limited number of materials have become industry viable candidate electrodes. Aluminum or Al-containing alloys are widely used as nMOSFET work function metal while TiN is often utilized as pMOSFET work function metal [Hinkle *et al.*, 2012], [Chudzik *et al.*, 2012]. Figure 11 shows examples of work function tuning in low thermal budget replacement metal gate systems that contain Al (or Al-containing alloys) and TiN for n and pMOSFETs, respectively. By proper process optimization, work function modulation is possible down from 4.1 eV, which is close to nMOSFET band edge, and up to 5.1 eV, which is near pMOSFET band edge. Despite the capability of controlling work function in a wide range with the Al containing alloy–TiN system, this system still has various challenges and concerns. Work function control toward the nMOSFET band edge is often coupled with increased gate leakage current that is likely due to lack of thermal stability of Al (or Al-containing alloys) causing Al spiking and interaction with the high-k dielectric to create excessive oxygen vacancies. Another challenge of the Al-containing alloy–TiN system is to obtain pMOSFET bandedge work functions. Since the TiN work function typically ranges between 4.8–4.95 eV [Cartier *et al.*, 2009], [Ando *et al.*, 2013], a further increase in work function toward the pFET band edge requires careful optimization of the stack with additional innovative treatments. Recently Ando *et al.*

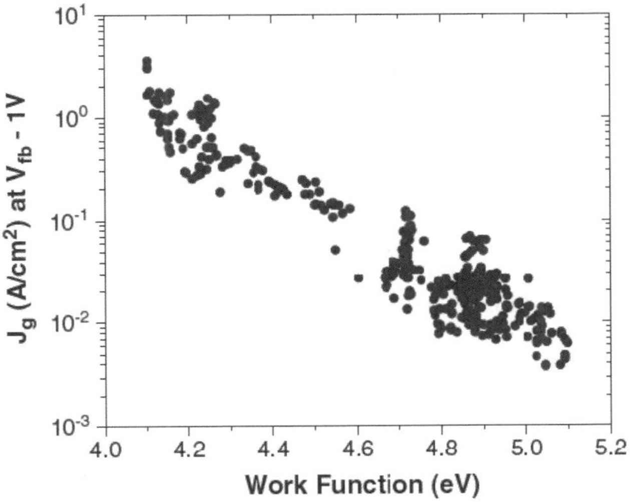

Fig. 11. Work function tuning in TiN and Al (or Al alloy) electrodes (after [Chudzik *et al.*, 2012]).

[Ando *et al.*, 2013], have also shown that the pFET EWF can also get shifted towards mid-gap in a GL integration scheme, analogous to the effect seen in GF processing, if the thickness of the SiOx IL falls below a critical thickness.

4.4. *Summary*

The Si CMOS industry using tremendous materials innovations has introduced HKMG technology into high volume manufacturing in two very different incarnations; GF and GL schemes. As the industry looks forward to ETSOI and FINFETs, the work function requirements are relaxed towards quarter gap EWF, but especially for FinFET architectures, all ALD deposition techniques for high-k and metal gates would be required which makes pathfinding for these materials even more challenging.

5. HKMG for FinFETs

5.1. *Introduction to FinFETs*

The genesis of FinFETs was in Hisamoto *et al*'s [Hisamoto *et al.*, 1989] publication on "Delta FETs", which revealed an improved three-dimensional (3D) way of increasing gate control while maintaining high mobilities. Since then, there have been giant strides that have been made in realizing 3D structures for mass production, with some introduction into the marketplace and several other announcements [Auth *et al.*, 2012], [Wu *et al.*, 2013]. A schematic view of a FinFET structure is shown in Figure 12. FinFETs represent a radical diversion from conventional planar transistor structures in that they enable channels that are entirely confined *geometrically*. While there are versions of planar technology (FDSOI or UTBB) [Cheng *et al.*, 2009] that allow for this geometrical scaling, some reports indicate better short-channel control and sub-threshold swing with FinFETs and larger active areal density of transistors.

As the industry moves towards scaled channel lengths, the drain and the source are brought closer together allowing for their depletion areas to be closer together — reducing the effective channel length and increasing the Drain-Induced-Barrier-Lowering (DIBL). In other words, the gate control becomes poorer and offstate leakage paths from source to drain that are farther from the gate start dominating. The primary methods of improving DIBL and increasing channel control are (a) Increasing the channel doping (reduces channel depletion width) (b) Increasing the halo implant doping (c)

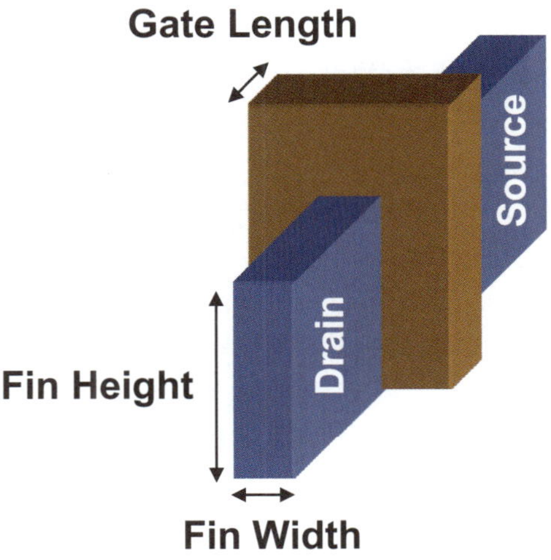

Fig. 12. Schematic view of FinFET structure.

Reducing EOT. The first two techniques both have several substantial detrimental impacts.

- Carrier mobility degradation causing on-state current loss.
- Increasing the dopant density in the channel area also has a direct impact on device threshold voltage variability.
- Reducing EOT has physical limits due to reliability concerns [Linder, 2013]
- High doping density in the substrate also places stringent requirements on Metal Gate Work Function, leading to further reliability complications.

FinFETs are a way of geometrically confining the carriers so that the onus on gate control and depletion widths is smaller:

- Gate control happens from two sides instead of one, leading to improved short channel gate control.
- Substrate doping is eliminated, leading to significantly better device control.
- Improved carrier mobility due to substrate doping elimination.
- Elimination of substrate doping allows metal gate work functions to be higher than in conventional bulk or PDSOI planar structures: Better reliability [Wang *et al.*, 2013], [Ramey *et al.*, 2014]

FinFETs, however, also have several disadvantages that need to be addressed in order to make them competitive with conventional planar technologies.

- Substantial process complexity increase leading to higher cost and new materials/tooling challenges (addressed in an upcoming section).
- Need to isolate the channel area from everything around it: This isolation is automatic in SOI FinFET, while in "bulk" FinFETs, the isolation needs to be achieved by creating reverse-biased diodes using doping differentials.
- All gate oxide and metal gate deposition techniques need to be "conformal" to the fin — leaving CVD and ALD as the processes of choice (higher cost).
- Elimination of the substrate doping has reduced the ease of creating suite of leakage devices (low to high) using V_t tuning.

5.2. *Challenges associated with HKMG on FinFET*

HKMG stack consists of high-k dielectric film and a combination of barrier, capping and work function metal films. Variations in thickness and composition of these layers become sources of variability of devices and in turn directly modulate device performance and reliability of transistors. In 3D RMG architecture in advanced technology nodes, HKMG stacks need to be formed in extremely scaled physical dimension of narrow gate length with high aspect ratio. More importantly, due to the 3D nature of FinFET where HKMG stacks form on three sides of Si fin surfaces, atomic-scale control of thickness, uniformity and conformality of metal gate films and their interaction with various crystallographic orientations around 3D FIN surfaces are extremely critical.

- **V_t setting by metal work function. Materials choice is important**

Unlike the traditional way of using channel doping to set V_t in conventional planar HKMG CMOS devices, in a 3D FinFET architecture, benefit of fully-depleted or lightly-doped channel for improving short channel control can be maximized by performing V_t adjustment by the gate work function determined by the metal electrode. Wide range of metals or metal compounds has been evaluated to meet the work function requirement and to ensure compatibility with high-k gate dielectrics. However, only a few are actively considered as candidates in advanced HKMG technologies due to the limitation of thin film process capability and stability with high-k and compatibility to complex device fabrications as reviewed in the previous section. The choice of metal gate materials is even more restricted in 3D FinFET architectures due to process challenges of achieving uniform conformal films around 3D FIN structures. Also, although V_t variability can be substantially reduced by reducing substrate doping in FinFET devices, the work function variation of metal gate remains as a major source of V_t variability [Endo *et al.*, 2010], [Lin *et al.*, 2012]. Therefore, extremely tight control of metal gate film thickness and composition is required to minimize V_t variability.

Since the thickness control of HKMG layers on 3D FinFET substrate is crucial, ALD is of great interest to provide with the required control and film conformailty. The choice of ALD work function metals, however, is often limited due to availability of ALD precursors along with process challenges to meet desired conformality and thickness control and diffusion resistance. Table 4 shows some of the resulting ALD TiN film properties deposited using either metal-organic or halide based precursors, at different temperatures and using different tooling. The final properties of the films differ dramatically depending on the method and precursor, and in turn impacting the final V_t of the devices.

Enabling multiple V_t device suites is required especially for SoC applications. In an undoped FinFET architecture, achieving multiple V_t devices is often challenged due to the lack of suitable metal gate material choices as well as process limitation to achieve various work functions and the compatibility with CMOS processing. However, it can be achieved by complex HKMG work function engineering, such as capping layer [Veloso *et al.*, 2008], metal gate engineering [Auth *et al.*, 2012], [Ando *et al.*, 2014], [Ragnarsson *et al.*, 2014], WF metal optimization/treatment [Togo *et al.*, 2014].

- **Film conformality and controls**

Table 4. ALD TiN film properties as a function of process conditions and precursors (after [Chudzik *et al.*, 2014]).

	Single Wafer MO TiN	LT Furnace TiCl4 TiN	HT Furnace TiCl4 TiN	Single Wafer TiCl4 TiN
Precursors	TDMAT&NH3	TiCl4&NH3	TiCl4&NH3	TiCl4&NH3
Proc. Temp.	340C	390C	450C	400C
Resistivity@50Å (um-cm)	~700	300–350	200	NA
Oxygen content (SIMS)	0.5%	0.08%	0.08%	0.05%

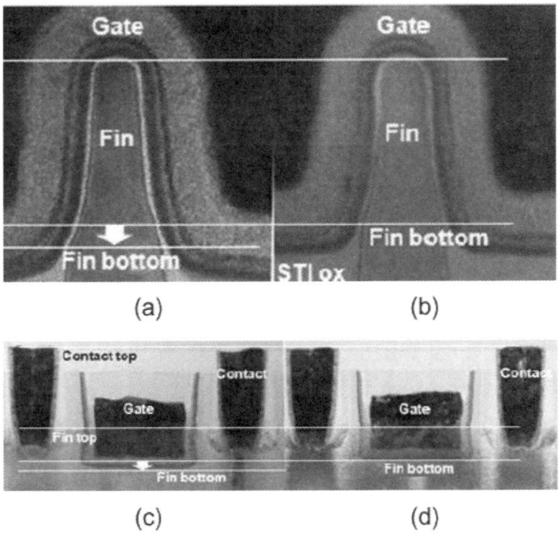

Fig. 13. Fin with STI oxide extra recess (a) and w/o extra recess (b) on XTEM images across fin. (c) and (d) are XTEM images across gate w/ and w/o STI oxide extra recess respectively (after [Togo *et al.*, 2014]).

For FinFET RMG HKMG application, fully conformal deposition of HKMG stacks along 3D FINs especially in high aspect ratio trenches becomes extremely challenging. Film uniformity around FINs (FIN side-walls as well as the top surface) needs to be tightly controlled (Figure 13). More importantly, the metal gate system often consists of more than one thin film layer. Key elements in metal gate stacks such as the work function and barrier layer films play critical roles in defining the V_t.

In the modern IC fabrication, a single work function metal film is often shared by devices with various structural densities and lengths in which same thickness of the film is required. CVD films are well known to have film thickness variation in structures with different structural density (pattern loading). One of the biggest advantages of ALD vs. CVD process is the reduction of this pattern dependent film thickness variation. However, some of the ALD processes often consist of a CVD-like component and can result in this type of microloading problem although it is at much lesser degree. In Figure 14, un-optimized ALD process (a) causes non-conformal and non-uniform film thickness through the depth of the trenches which, in turn, can result in variation in device performance, while through optimization near 100% conformality can be achieve (b). In order to further suppress microloading even with ALD films, eliminating un-intended CVD-like components and enabling complete ALD process is critical.

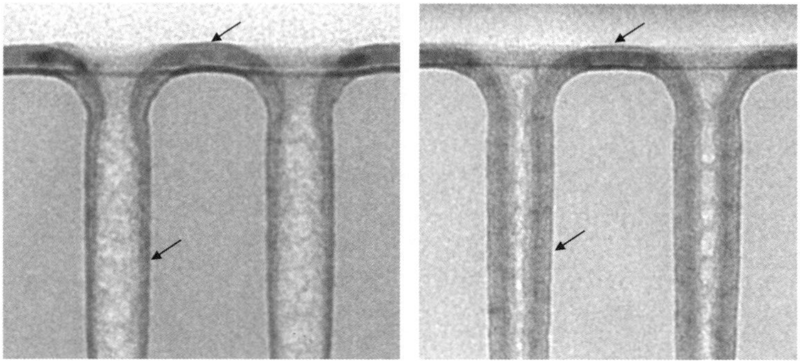

(a) (b)

Fig. 14. Examples of (a) poor and (b) good ALD film conformality in high aspect ratio trenches.

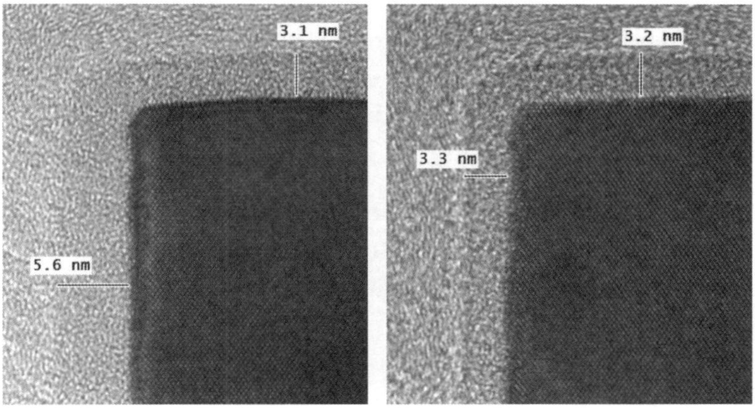

(a) (b)

Fig. 15. Film thickness on different Si crystallographic orientation of (a) thermal SiO_2 and (b) ALD SiO_2.

- **Dielectric options and interaction with FinFET**

In high-k gate oxide system, SiO_2 (or SiO_xN_y) is formed between high-k dielectric and Si substrate to prevent high-k dielectrics from interacting with Si as well as to provide high voltage I/O device options. Furthermore, unlike planar Si substrates in which the oxide forms on monolithic single crystal orientation, thermal oxidation on different silicon orientations around FinFET structures (top versus sidewall) results in differential growth rates, which makes thermal oxide potentially less controllable for FinFET structures with different crystallographic Si orientation (Figure 15). Although thermal SiO_2 creates highly reliable gate oxides or interlayer oxides, the thickness dependence on different FinFET surfaces needs to be controlled to minimize the impact on variability in device performance and reliability. ALD SiO_2 provides uniform and conformal deposition on 3D FinFET surfaces. However, the quality of as deposited SiO_2 is typically not as good as thermal SiO_2. By careful optimization of ALD SiO_2, conformal films and required dielectric reliability can be achieved [Siddiqui *et al.*, 2013].

- **Gate resistance (high aspect ratio)**
  The RMG architecture, combined with FinFETs, presents some unique challenges for metallization: a delicate optimization of parameters like work function, T_{inv} and gate resistance is required to maximize the performance out of the transistor.

A typical RMG architecture comprises of three layers: Metal Electrode, Barrier Layer and a low resistance fill. Between the various layers exist interfaces that may have additional resistance that may add to the overall gate resistance. The strategy to reduce gate resistance involves (while maintaining EWF, Reliability and EOT targets untouched) the following

- Employ lowest resistivity materials where possible.
- Reduce the thickness of high resistance layers.
- Manage integration to ensure that number of layers is kept minimal where possible.
- Reduce exposure to oxygen to reduce interface oxides to ensure minimal interface resistance.

Any typical metal starts being significantly more resistive with a reduction in thickness below a certain value due to increased scattering from grain boundaries, surface scattering [Mayadas and Schatzkes, 1970] and increased contribution of interface scattering. Optimizing the device for good gate resistance involves making sure that the low resistance fill layer is kept to a thickness greater than the Mayadas Schatzkes limit (essentially minimize all other layers till this condition is met). Finally, oxygen levels can be minimized by managing queue times and eliminating exposure to ambient where possible. A thorough optimization process will typically minimize the penalty in going to short channel lengths. FinFETs increase the complexity of this optimization by having a stringent requirement of conformality of films that is enabled by ALD, which is known to have contaminants such as Carbon, Halogens and Oxygen that typically increase resistivities. An example of gate resistance improvement at short

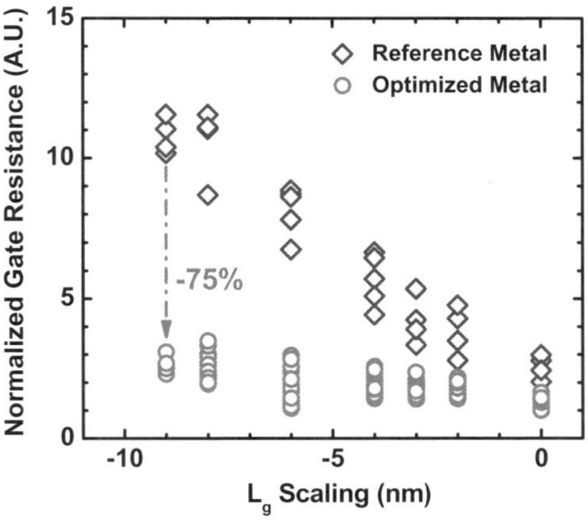

Fig. 16. Normalized gate resistance as a function of L_g scaling for optimized and reference stacks (after [Ando *et al.*, 2014]).

gate length (<25 nm) via such optimization of metal gate stacks is shown in Figure 16.

- **HKMG metrology challenges on FinFET**

Subtle variation of high-k and metal gate thickness results in variability in device parametric and limits a capability of fine tuning transistor performance. Advanced HKMG fabrication in high volume development and manufacturing environment, therefore, requires tight HKMG process control and monitoring, and it is necessary to have in-line metrology solutions for accurate HKMG film thickness/composition characterization. Industry has been utilizing in-line metrology solutions such as ellipsometry, X-ray photoelectron spectroscopy (XPS) and X-ray fluorescence (XRF) to monitor HKMG film thickness and composition. Such techniques typically can only be performed on a large planar measurement structure, which enables only indirect measurement of HKMG films since the actual devices form on a complex geometry in much smaller ground rules. Therefore, the measurement from the large planar pads often might not represent the true HKMG thickness in the device areas, resulting in an offset between the actual device area and measurement structures. This becomes even more challenging in RMG FinFET architectures due to the additional complexity of the 3D natures. In-line metrology capability of characterizing the HKMG thickness and composition on complex 3D structures become crucial for HKMG FinFET architectures. Scatterometry-based metrology has been drawing attention owing to the capability of probing information of three-dimensional shape of complicated RMG structures and thickness on

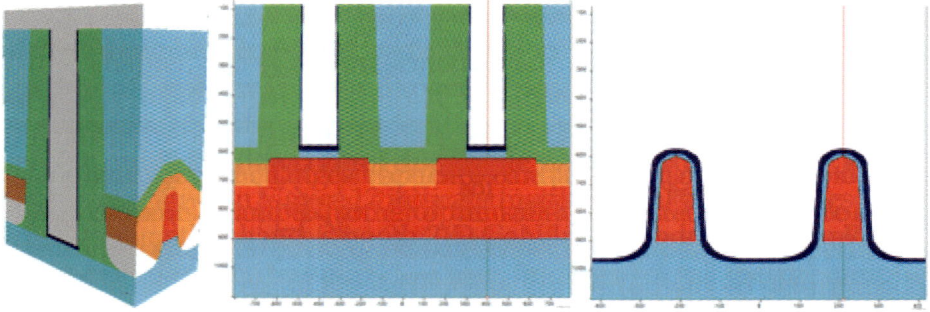

Fig. 17. Example of a scatterometry model used for the analysis of high-k on FinFET structures.

FinFET [Dziura *et al.*, 2008]. A typical example of the scatterometry model for the analysis of HKMG on FinFET structures is shown in Figure 17. Although the scatterometry-based metrology can provide a capability of measuring shape of complex 3D structures and HKMG stack thicknesses formed in the structures, accuracy and capability of characterizing uniformity of film thickness remain as challenges.

In order to ensure controlled HKMG process in high volume production of FinFET devices, suitable in-line metrology solutions with adequate accuracy and throughput have to be established.

6. HKMG for III-V Channel MOSFETs

6.1. *Introduction*

While thin films on silicon are the focus of the present volume, a brief discussion of high-k dielectrics on other semiconductors may be instructive. In particular, a comparison to III-V compound semiconductors currently under investigation as a channel material for future logic FETs will highlight the remarkably high quality of the SiO_2/Si interface, for which there is no analog in the III-V materials system.

Higher electron mobility in III-V materials such as $In_xGa_{1-x}As$ (for short, InGaAs) than in Si- and Ge- based materials provides the main motivation for efforts to develop III-V compound semiconductor FETs. Such III-V channels could operate at lower voltages while still maintaining or improving device performance. However, gate stack quality poses a tremendous challenge, as native oxide formation disrupts the III-V crystal structure, generating electronic defects. Unlike for Si(Ge) channels, preventing III-V oxidation during high-k deposition is therefore essential. We summarize three main approaches to III-V interface control. The first approach is based on direct high-k deposition onto the bare InGaAs surface, while the remaining two approaches rely on the insertion of a silicon or indium phosphide capping layer to protect and passivate the InGaAs. The impact of such caps on interface trap density and scalability will be discussed.[1]

6.2. *Dielectric/III-V interface issues and the need for oxygen control*

Despite decades of research on gate oxides for III-V semiconductors basic questions remain regarding the impact of processing and materials on gate stack quality, and regarding the very nature of the defect states. To explain the high density of trap states often observed in III-V gate stacks, various defect models have been proposed [Hasegawa, 2008]. It is evident that the compound nature of III-V materials poses a challenge during oxide/III-V interface formation, as III and/or V atoms are easily displaced from their original lattice sites at this interface, in particular when the substrate is oxidized either intentionally or unintentionally. This results in, e.g., vacancies or antisites inside the semiconductor, or in like-atom bonds and dangling bonds at the interface, many of which induce electronic states in the III-V band gap [Robertson, 2009]. Exact III:V stoichiometry should therefore be maintained when forming the interface, which is extremely difficult to achieve when forming a 'native' oxide since there is often a tendency to preferentially oxidize one of the III-V constituents. We note that this is in marked contrast to the compound semiconductor Si_xGe_{1-x}, for which arbitrary stoichiometry and random ordering are permitted, and for which interfacial SiO_2 is in fact desirable, just like on silicon.

In view of these difficulties, many of the recently proposed approaches to III-V gate stack formation are based on minimizing oxidation. One such approach, first developed in the mid-1990s, is based on initial deposition of a thin (e.g., ~1 nm) Ga_2O_3 layer onto an atomically clean, freshly grown or As-uncapped GaAs surface using a specially designed molecular beam epitaxy (MBE) process in ultra-high vacuum (see [Passlack, 2005] and references therein). Under these highly controlled conditions, it is possible to deposit the metal oxide without oxidizing the GaAs itself, thus preventing gap states from being formed [Winn *et al.*, 2007].

---

[1]This section is adapted with permission from [Frank, 2014]. Copyright 2014, The Electrochemical Society.

While MBE is attractive for its unmatched degree of control over film composition, structure, and impurities, atomic layer deposition (ALD) is more compatible with conformality requirements, e.g. for FinFETs, and is more easily scaled up to a chip manufacturing environment. It was therefore the advent of ALD-grown high-k dielectrics for Si channel devices that spurred the recent resurgence of III-V gate stack research.

In the remainder of this section, we will compare direct high-k/III-V interfaces formed by ALD to high-k/cap/III-V stacks using silicon or epitaxial InP as a cap layer. Certain parallels to the role of interfacial SiO_2 layers in high-k/Si stacks will become apparent.

6.3. *Direct ALD high-k deposition: High-k/III-V*

In direct high-k deposition onto III-V surfaces by ALD (Figure 18a), native III-V oxides can sometimes be removed rather than formed, e.g. during the popular Al_2O_3 process using cyclic exposures to $Al(CH_3)_3$ (trimethylaluminum, TMA) and H_2O [Frank *et al.*, 2005]. This phenomenon has been termed 'interfacial self-cleaning' [Hinkle *et al.*, 2008]. It requires a strong thermodynamic driving force for the metal precursor to oxidize. For example, the popular HfO_2 precursor $HfCl_4$ is much less reactive than TMA as quantified by their standard enthalpies of formation (–885 vs. –74 kJ/mol in the gas phase [Lide, 2006]), while the metal oxides have similar stability based on their Gibbs free energies of formation per oxygen atom (–527 kJ/mol for Al_2O_3 vs. –544 kJ/mol for HfO_2 [Lide, 2006]). Indeed, $HfCl_4$ has been observed to leave native GaAs oxides intact [Frank *et al.*, 2005]. The self-cleaning phenomenon suggests that it may be feasible to deposit high-k oxides via ALD without ever oxidizing an atomically clean III-V surface, thus leaving its bonding structure intact.

We note that ALD-induced self-cleaning has never been reported for high-k deposition onto oxidized Si or SiGe substrates. This is related to the higher thermodynamic stability of SiO_2 compared to III-V oxides (Table 5), making reduction unfavorable even using a highly reactive precursor such as TMA. On the other hand, atomically sharp HfO_2/Ge interfaces have been fabricated by ALD [Gusev *et al.*, 2004], likely related to the low stability of GeO_2 (Table 5).

Recently, $In_xGa_{1-x}As$ has emerged as the leading candidate material for n-channel III-V MOSFET devices, largely based on favorable tradeoffs between band gap and electron mobility. Most studies currently focus on $In_{0.53}Ga_{0.47}As$ which can be grown lattice-matched on InP substrate wafers. Since the thermodynamic tendency for In to oxidize is lower than for Ga

Table 5. Gibbs free energy of oxide formation per O atom for bulk phases of typical oxides on Si, SiGe, and III-V surfaces [Lide, 2006; Jung and Hudon, 2012; Chase, 1998].

Oxide	Gibbs Free Energy of Oxide Formation per O Atom [kJ/mol]
SiO_2	–428
Ga_2O_3	–333
In_2O_3	–277
P_2O_5 (hexagonal)	–272
GeO_2	–261
As_2O_5	–156

(Table 4), self-cleaning observed on GaAs is also expected to be achievable on InGaAs. This has been confirmed experimentally both for the TMA/H_2O process [Jevasuwan, 2012] and for HfO_2 deposition using the reactive metal precursor $Hf(NCH_3C_2H_5)_4$ (tetakis(ethylmethylamido)hafnium, TEMAH) [Chang, 2006]. With minimum queue time in air from InGaAs wet clean to the ALD chamber, a density of interface traps (D_{it}) for Al_2O_3/InGaAs of less than 10^{12} cm^{-2} eV^{-1} has been achieved [O'Connor, 2011]. However, further D_{it} reduction would be desirable for optimum device quality, motivating the search for alternate methods to preventing III-V oxidation.

6.4. *Capping layers: High-k/(Si,InP)/III-V*

An alternate, highly effective approach to preventing III-V oxidation is based on initial *in-situ* deposition of a thin semiconductor capping layer onto an atomically clean III-V surface. Silicon layers, often amorphous, as well as epitaxial indium phosphide layers have been utilized.

Silicon caps [Hasegawa, 2000], typically ~1–1.5 nm thick (Figure 18b), often improve gate stack quality on GaAs over conventional direct dielectric deposition. This was first demonstrated in 1988 using Si-based dielectrics [Tiwari and Wright, 1988; Hasegawa *et al.*, 1988; Fountain *et al.*, 1988] and verified later with high-k dielectrics [Koester *et al.*, 2006; Akazawa and Hasegawa, 2010]. A beneficial impact of atomic hydrogen on interface trap density has been observed with both types of gate stacks [Wang *et al.*, 1993; Hinkle *et al.*, 2007]. For InGaAs channels, Si capping has similar benefits [Fountain *et al.*, 1988; Wang *et al.*, 1992]. With high-k gate dielectrics, midgap D_{it} as low as 2×10^{11} cm^{-2} eV^{-1} [Akazawa and Hasegawa, 2010] and EOT below 1 nm [El Kazzi *et al.*, 2012] have been reported. This EOT is competitive with what is being practiced in state-of-the-art silicon channel devices.

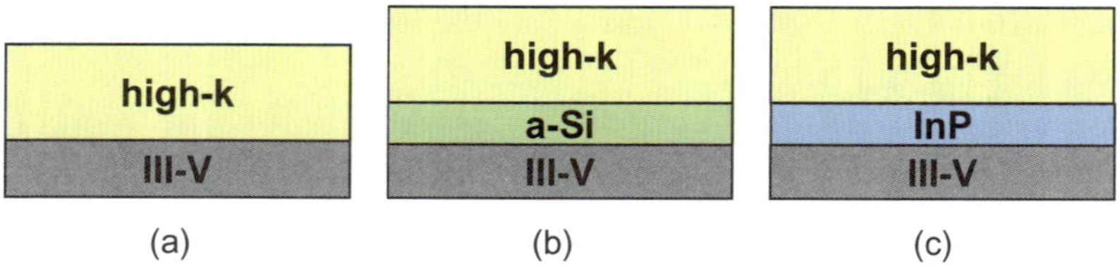

Fig. 18. Three ALD high-k/III-V interface options. For simplicity, optional additional passivating elements (e.g., S or N) or dielectric layers (e.g., SiO_2 or Si_3N_4) are omitted. Adapted with permission from [Frank, 2014]. Copyright 2014, The Electrochemical Society.

Remaining challenges include the potential for up-diffusion of III-V atoms into and through the silicon layer [Marchiori *et al.*, 2014].

As an alternative to Si caps, a thin epitaxial InP layer (Figure 18(c)) was recently found to have a beneficial impact on high-k gate stack quality on InGaAs as judged from reduced frequency dispersion in capacitance-voltage data and improved nFET performance [Zhao *et al.*, 2009; Radosavljevic *et al.*, 2009]. This scheme takes advantage of the higher band gap of InP (1.34 eV) compared to InGaAs (e.g. 0.74 eV for 53% In and 0.58 eV for 70% In) and a favorable band alignment. The band offset tends to keep carriers in the InGaAs, i.e it results in a buried channel device, ensuring that high-k/III-V interface defects are remote from the inversion layer. Their carrier mobility impact should thus be reduced. In addition, high-k/InP interfaces may have lower D_{it} than high-k/InGaAs interfaces [Radosavljevic *et al.*, 2009], perhaps due to lower interface stress [Houssa *et al.*, 2011]. However, the benefit of InP capping for gate stack quality is not undisputed, as carrier spillover into the InP cap and to traps at the high-k/InP interface may still occur [Urabe *et al.*, 2011]. Creating a buried channel device may seem unattractive in the context of EOT scaling. However, for a typical InP cap thickness of 1-2 nm, the relative permittivity of InP of ~12.6 [Lide, 2006] implies that such a cap layer is expected to add only 0.3–0.6 nm to the total EOT of the gate stack, similar to the EOT impact of a typical interfacial SiO_2 layer in high-k/Si devices.

7. Summary and Conclusion

Unlike on silicon-based channels, native oxide formation on III-V channels must be prevented for optimum gate stack quality. By employing highly reactive metal precursors for metal oxide ALD, substrate oxidation can be minimized. Even if present initially, native oxides can be removed from III-V surfaces in this manner, since their native oxides are thermodynamically less stable than on Si or SiGe. However, optimum protection of InGaAs from oxygen may require capping layers such as Si or InP. Due to the higher permittivity of Si and InP compared to SiO_2, this can be achieved with an EOT impact that is similar to that of SiO_2 in high-k/Si(Ge) devices. More work will be required to determine whether or not these high-k/cap/III-V systems exhibit similar universal mobility and reliability trends as the high-k/SiO_2/Si(Ge) system. Also, achieving fin conformality of such capping layers is a crucial requirement for their successful implementation.

References

Akasaka, Y., Nakamura, G., Shiraishi, K., Umezawa, N., Yamabe, K., Ogawa, O., Lee, M., Amiaka, T., Kasuya, T., Watabnabe, H., Chikyo, T., Ootsuka, F., Nara, Y. and Nakamura, K. (2006), *Jpn. J. Appl. Phys.*, **45** p. L1289.

Akazawa, M. and Hasegawa, H. (2010), *Appl. Surf. Sci.*, **256**, p. 5708.

Ando, T., Frank, M. M., Choi, K., Choi, C., Bruley, J., Hopstaken, M., Copel, M., Cartier, E., Kerber, A., Callegari, A., Lacey, D., Brown, S., Yang, Q. and Narayanan, V. (2009), *Tech. Dig. Int. Electron Devices Meet.*, p. 423.

Ando, T., Shimura, T., Watanabe, H., Hirano, T., Yoshida, S., Tai, K., Yamaguchi, S., Iwamoto, H., Kadomura, S., Toyoda, S., *et al.* (2009), *Appl. Phys. Express*, **2** p. 071402:1.

Ando, T., Copel, M., Bruley, J., Frank, M. M., Watanabe, H. and Narayanan, V. (2010), *Appl. Phys. Lett.*, **96** p. 132904:1.

Ando, T., Hirano, T., Tai, K., Yamaguchi, S., Yoshida, S., Iwamoto, H., Kadomura, S. and Watanabe, H. (2010), *Jpn. J. Appl. Phys.*, **49**, p. 016502:1.

Ando, T. (2012), *Materials*, **5**(3), p. 478.

Ando, T., Cartier, E. A., Bruley, J., Choi, K. and Narayanan, V. (2013), *IEEE Electron Device Lett.*, **34**(6), p. 729.

Ando, T., Kannan, B., Kwon, U., Lai, W. L., Linder, B. P., Cartier, E. A., Haight, R., Copel, M., Bruley, J., Krishnan, S. A. and Narayanan, V. (2014), *Symp. VLSI Tech. Dig. Tech. Pap.* p. 54.

Arnaud, F., Thean, A., Eller, M., Lipinski, M., Teh, Y. W., Ostermayr, M., Kang, K., Kim, N. S., Ohuchi, K., Han, J-P., Nair, D. R., Lian, J., Uchimura, S., Kohler, S., Miyaki, S., Ferreira, P., Park, J.-H., Hamaguchi, M., Miyashita, K., Augur, R., Zhang, Q., Strahrenberg, K., ElGhouli, S., Bonnouvrier, J., Matsuoka, F., Lindsay, R., Sudijono, J., Johnson, F. S., Ku, J. H., Sekine, M., Steegen, A. and Sampson, R. (2009), *Tech. Dig. Int. Electron Devices Meet.*, p. 651.

Arimura, H., Brown, S. L., Callegari, A., Kellock, A., Bruley, J., Copel, M., Watanabe, H., Narayanan, V. and Ando, T. (2011), *IEEE Electron Device Lett.*, **32** p. 288.

Auth, C., Allen, C., Blattner, A., Bergstrom, D., Brazier, M., Bost, M., Buehler, M., Chikarmane, V., Ghani, T., Glassman, T., Grover, R., Han, W., Hanken, D., Hattendorf, M., Hentges, P., Heussner, R., Hicks, J., Ingerly, D., Jain, P., Jaloviar, S., James, R., Jones, D., Jopling, J., Joshi, S., Kenyon, C., Liu, H., McFadden, R.,

McIntyre, B., Neirynck, J., Parker, C., Pipes, L., Post, I., Pradhan, S., Prince, M., Ramey, S., Reynolds, T., Roesler, J., Sandford, J., Seiple, J., Smith, P., Thomas, C., Towner, D., Troeger, T., Weber, C., Yashar, P., Zawadzki, K. and Mistry, K. (2012), *Symp. VLSI Tech. Dig. Tech. Pap.*, p. 131.

Auth, C., Cappellani, A., Chun, J.-S., Dalis, A., Davis, A., Ghani, T., Glass, G., Glassman, T., Harper, M., Hattendorf, M., Hentges, P., Jaloviar, S., Joshi, S., Klaus, J., Kuhn, K., Lavric, D., Lu, M., Mariappan, H., Mistry, K., Norris, B., Rahhal-orabi, N., Ranade, P., Sandford, J., Shifren, L., Souw, V., Tone, K., Tambwe, F., Thompson, A., Towner, D., Troeger, T., Vandervoorn, P., Wallace, C., Wiedemer, J. and Wiegand, C. (2008), *Symp. VLSI Tech. Dig. Tech. Pap.*, p. 128.

Bersuker, G., Park, C. S., Wen, H. C., Choi, K., Sharia, O., and Demkov, A. A. (2010), *IEEE Trans. Electron Devices.*, **57**, p. 2047.

Buchanan, D. A. (1999), *IBM J. Res. Develop.*, **43** p. 245.

Cartier, E., McFeely, F. R., Narayanan,V., Jamison, P., Linder, B. P., *et al.* (2005), *Symp. VLSI Tech. Dig. Tech. Pap.*, p. 230.

Cartier, E., Steen, M., Linder, B. P., Ando, T., Iijima, R., Frank, M., Newbury, J. S., Kim, Y. H., McFeely, F. R., Copel, M., Haight, R., Choi, C., Callegari, A., Paruchuri, V. K. and Narayanan, V. (2009), *Symp. VLSI Tech. Dig. Tech. Pap.*, p. 42.

Cartier, E., Kerber, A., Ando, T., Frank, M. M., Choi, K., Krishnan, S., Linder, B., Zhao, K., Monsieur, F., Stathis, J. and Narayanan, V. (2011), *Tech. Dig. Int. Electron Devices Meet.*, p. 18.4.1.

Cesana, G. (2012), Workshop Fully Depleted SOI http://www.soiconsortium.org/fully-depleted-soi/presentations/february-2012/1%20-%20Giorgio%20Cesana%20%2028%20&%2020nm%20FDSOI%20Technology%20Platforms.pdf

Cheng, K., Khakifirooz, A., Loubet, N., Luning, S., Nagumo, T., Vinet, M., Liu, Q., Reznicek, A., Adam, T., Naczas, S., Hashemi, P., Kuss, J., Li, J., He, H., Edge, L., Gimbert, J., Khare, P., Zhu, Y., Zhu, Z., Madan, A., Klymko, N., Holmes, S., Levin, T.M., Hubbard, A., Johnson, R., Terrizzi, M., Teehan, S., Upham, A., Pfeiffer, G., Wu, T., Inada, A., Allibert, F., Nguyen, B., Grenouillet, L., Le Tiec, Y., Wacquez, R., Kleemeier, W., Sampson, R., Dennard, R. H., Ning, T. H., Khare, M., Shahidi, G. and Doris, B. (2012), *Tech. Dig. Int. Electron Devices Meet.*, p. 419.

Chudzik, M., Krishnan, S., Kwon, U., Khare, M., Narayanan, V., Ando, T., Cartier, E., Bu, H., and Paruchuri, V. (2012), *High-k Gate Dielectrics for CMOS Technology* Eds., He, G., and Sun, Z., (Wiley) p. 533.

Cho, H.-J, Seo, K.-I., Jeong, W. C., Kim, Y.-H., Lim, Y. D., Jang, W. W., Hong, J.G., Suk, S. D., Li, M., Ryou, C., Rhee, H. S., Lee, J. G., Kang, H. S., Son, Y. S., Cheng, C.L., Hong, S. H., Yang, W. S., Nam, S. W., Ahn, J. H., Lee, D. H., Park, S., Sadaaki, M., Cha, D. H., Kim, D. W., Sim, S. P., Hyun, S., Koh, C. G., Lee, B. C., Lee, S. G., Kim, M. C., Bae, Y. K., Yoon, B., Kang, S. B., Hong, J. S., Choi, S., Sohn, D. K., Yoon, J. S. and Chung, C. (2011), *Tech. Dig. Int. Electron Devices Meet.*, p. 350.

Chen, X., Samavedam, S., Narayanan, V., Stein, K., Hobbs, C., Baiocco, C., Li, W., Jaeger, D., Zaleski, M., Yang, H. S., Kim, N., Lee, Y., Zhang, D., Kang, L., Chen, J., Zhuang, H., Sheikh, A., Wallner, J., Aquilino, M., Han, J., Jin, Z., Li, J., Massey, G., Kalpat, S., Jha, R., Moumen, N., Mo, R., Krishnan,

S., Wang, X., Chudzik, M., Chowdhury, M., Nair, D., Reddy, C., Teh, Y. W., Kothandaraman, C., Coolbaugh, D., Pandey, S., Tekleab, D., Thean, A., Sherony, M., Lage, C., Sudijono, J., Lindsay, R., Ku, J. H., Khare, M. and Steegen, A. (2008), *Symp. VLSI Tech. Dig. Tech. Pap.*, p. 88.

Choi, K., Jagannathan, H., Choi, C., Edge, L., Ando, T., Frank, M., Jamison, P., Wang, M., Cartier, E., Zafar, S., Bruley, J., Kerber, A., Linder, B., Callegari, A., Yang, Q., Brown, S., Stathis, J., Iacoponi, J., Paruchuri, V. and Narayanan, V. (2009), *Symp. VLSI Tech. Dig. Tech. Pap.*, p. 138.

Copel, M., Guha, S., Bojarczuk, N., Cartier, E., Narayanan, V. and Paruchuri, V. (2009), *Appl. Phys. Lett.*, **95**, p. 212903:1.

Cheng, K., Khakifirooz, A., Kulkarni, P., Kanakasabapathy, S., Schmitz, S., Reznicek, A., Adam, T., Zhu, Y., Li, J., Faltermeier, J., Furukawa, T., Edge, L. F., Haran, B., Seo, S.-C., Jamison, P., Holt, J., Li, X., Loesing, R., Zhu, Z., Johnson, R., Upham, A., Levin, T., Smalley, M., Herman, J., Di, M., Wang, J., Sadana, D., Kozlowski, P., Bu, H., Doris, B. and O'Neill, J. (2009), *Symp. VLSI Tech. Dig. Tech. Pap.* p. 212.

Chudzik, M. P., Krishnan, S., Dai, M., Siddiqui, S., Shepard, J. and Kwon, U. (2014), *ECS Trans.*, **60**(1), p. 513.

Chase, M. W., Jr. (1998), *NIST-JANAF Thermochemical Tables, Fourth Edition, Part II, Cr-Zr*, in *Journal of Physical and Chemical Reference Data Monograph No. 9*, p. 1810, The American Institute of Physics, Woodbury, NY.

Chang, C.-H.,Chiou, Y.-K., Chang, Y.-C., Lee, K.-Y., Lin, T.-D., Wu, T.-B., Hong, M. and Kwo, J. (2006), *Appl. Phys. Lett.*, **89**, p. 242911.

Demkov, A. A. (2006), *Phys. Rev. B*, **74**, p. 085310.

Dziura, T. D., Bunday, B., Smith, C., Hussain, M. M., Harris, R., Zhang, X. and Price, J. M. (2003), *Proc. of SPIE* Vol. 6922 69220V-1.

Edge, L. F., Vo, T., Paruchuri, V. K., Iijima, R., Bruley, J., Jordan-Sweet, J., Linder, B. P., Kellock, A. J., Tsunoda, T. and Shinde, S. R. (2011), *Appl. Phys. Lett.*, **98**, p. 122905:1.

Endo, K., O'uchi, S., Ishikawa, Y., Liu, Y., Matsukawa, T., Sakamoto, K., Tsukada, J., Yamauchi, H. and Masahara, M. (2010), *IEEE Electron Device Lett.*, **31**(6), p. 546.

Franco, J., Kaczer, B., Eneman, G., Roussel, P. J., Grasser, T., Mitard, J., Ragnarsson, L. , Cho, M., Witters, L., Chiarella, T., Togo, M., Wang, W.-E., Hikavyy, A., Loo, R., Horiguchi, N., and Groeseneken, G. (2011), *Tech. Dig. Int. Electron Devices Meet.*, p. 445.

Frank, M. M., Wilk, G. D., Starodub, D., Gustafsson, T., Garfunkel, E., Chabal, Y. J., Grazul, J. and Muller, D. A. (2005), *Appl. Phys. Lett.*, **86**, p. 152904.

Frank, M. M., Marchiori, C., Bruley, J., Fompeyrine, J. and Narayanan, V. (2011), *Microelectron. Eng.*, **88** p. 1312.

Frank, M. M., Zhu, Y., Bedell, S. W., Ando, T., Narayanan, V. (2014), *ECS Trans.*, **61**(1), p. 213.

Fischetti, M. V., Neumayer, D. A. and Cartier, E. A. (2001), *J. Appl. Phys.*, **90** p. 4587.

Fountain, G. G., Hattangady, S. V., Vitkavage, D. J., Rudder, R. A. and Markunas, R. J. (1988), *Electron. Lett.*, **24**, p. 1134.

Greene, B., Liang, Q., Amarnath, K., Wang, Y., Schaeffer, J., Cai, M., Liang, Y., Saroop, S., Cheng, J., Rotondaro, A. *et al.* (2009), *Symp. VLSI Tech. Dig. Tech. Pap.*, p. 140.

Gusev, E. P., Shang, H., Copel, M., Gribelyuk, M., D'Emic, C., Kozlowski, P. and Zabel, T. (2004), *Appl. Phys. Lett.*, **85** p. 2334.

Gusev, E. P., Narayanan, V. and Frank, M. M. (2006), *IBM J. Res. Dev.*, **50**, p. 387.

Guha, S., Paruchuri, V. K., Copel, M., Narayanan, V. and Wang, Y. Y. (2007), *Appl. Phys. Lett.* **90** p. 092902.

Hasegawa, H., Akazawa, M., Matsuzaki, K., Ishii, H. and Ohno, H. (1988), *Jpn. J. Appl. Phys.*, **27**, p. L2265.

Hasegawa, H. (2000), *Thin Solid Films*, **367**, p. 58.

Hasegawa, H. (2008), *Appl. Surf. Sci.*, **254**, p. 8005.

Harris, H. R., Kalra, P., Majhi, P., Hussain, M., Kelly, D., Oh, J., He, D., Smith, C. , Barnett, J., Kirsch, P. D., Gebara, G., Jur, J., Lichtenwalner, D., Lubow, Abigail, Ma, T. P., Sung, G., Thompson, S., Lee, B. H., Tseng, H.H. and Jammy, R. (2007), *Symp. VLSI Tech. Dig. Tech. Pap.*, p. 154.

Hashemi, P., Kobayashi, M., Majumdar, A., Yang, L. A., Baraskar, A., Balakrishnan, K. , Kim, W., Chan, K., Engelmann, S. U., Ott, J. A., Bedell, S.W., Murray, C. E., Liang, S., Dennard, R. H., Sleight, J. W., Leobandung, E. and Park, D.-G. (2013), *Symp. VLSI Tech. Dig. Tech. Pap.*, p. T18.

Hinkle, C., L., Galatage, R. V., Chapman, R. A., Vogel, E. M., Alshareef, H. N., Freeman, C., Christensen, M., Wimmer, E., Niimi, H., Li-Fatou, A., Shaw, J. B. and Chambers, J. J. (2012), *Appl. Phys. Lett.*, **100**, p. 153501.

Hisamoto, D., Kaga, T., Kawamoto, Y. and Takeda, E., (1989), *Tech. Dig. Int. Electron Devices Meet.*, p. 833.

Hinkle, C. L., Sonnet, A. M., Vogel, E. M., McDonnell, S., Hughes, G. J., Milojevic, M., Lee, B., Aguirre-Tostado, F. S., Choi, K. J., Kim, J. and Wallace, R. M. (2007), *Appl. Phys. Lett.*, **91**, p. 163512.

Hinkle, C. L., Sonnet, A. M., Vogel, E. M., McDonnell, S., Hughes, G. J., Milojevic, M., Lee, B., Aguirre-Tostado, F. S., Choi, K. J., Kim, H. C., Kim, J. and Wallace, R. M. (2008), *Appl. Phys. Lett.*, **92**, p. 071901.

Hobbs, C., Fonseca, L., Dhandapani, V., Samavedam, S., Taylor, B., Grant, J., Dip, L., Triyoso, D., Hegde, R., Gilmer, D., Garcia, R., Roan, D., Lovejoy, L., Rai, R., Hebert, L., Tseng, H., White, B. and Tobin, P. (2003), *Symp. VLSI Tech. Dig. Tech. Pap.*, p. 9.

Houssa, M., Scarrozza, M., Pourtois, G., Afanas'ev, V. V. and Stesmans, A. (2011), *Appl. Phys. Lett.*, **98**, p. 141901.

Hubbard, K. J. and Schlom, D. G. (1996), *J. Mater. Res.*, **11** p. 2757.

Hwang, C. S., Park, S. O., Cho, H.-J., Kang, C. S., Kang, H.-K., Lee, S. I. and Lee, M. Y. (1995), *Appl. Phys. Lett.*, **67**, p. 2819.

Jagannathan, H., Narayanan, V. and Brown, S. L. (2008), *ECS Trans.* **16**, p. 19.

James, D. (2014), *Tech. Dig. 25th Annual SEMI Advanced Semiconductor Manufacturing Conference*, p. 431.

Jevasuwan, W., Urabe, Y., Maeda, T., Miyata, N., Yasuda, T., Yamada, H., Hata, M., Taoka, N., Takenaka, M. and Takagi, S. (2012), *Materials*, **5**, p. 404.

Jung, H. S., Lee, J. H., Han, S. K., Kim, Y. S., Lim, H. J., *et al.* (2005), *Symp. VLSI Tech. Dig. Tech. Pap.*, p. 232.

Jung, H.-S., Lee, J.-H, Han, S. K., Kim, Y.-S., Lim, H. J., Kim, M. J., Doh, S.J., Yu, M. Y., Lee, N.-I., Lee, H.-L., Jeon, T.-S., Cho, H.-J., Kang, S. B., Kim, S. Y., Park, I. S.,

Kim, D., Baik, H. S. and Chung, Y. S. (2005), *Symp. VLSI Tech. Dig. Tech. Pap.*, p. 232.

Jung, I.-H. and Hudon, P. (2012), *J. Am. Ceram. Soc.*, **95**, p. 3665.

Kamimuta,Y., Iwamoto, K., Nunoshige, Y., Hirano, A., Mizubayashi, W., *et al.* (2007), *Tech. Dig. Int. Electron Devices Meet.*, p. 341.

Kazzi, M.El, Czornomaz, L., Rossel, C., Gerl, C., Caimi, D., Siegwart, H., Fompeyrine, J. and Marchiori, C. (2012), *Appl. Phys. Lett.*, **100**, p. 063505.

Kim, S. K., Choi, G.-J., Lee, S. Y., Seo, M., Lee, S. W., Han, J. H., Ahn, H.-S., Han, S. and Hwang, C. S. (2008), *Adv. Mater.*, **20**, p. 1429.

Kim, H., McIntyre, P. C., Chui, C. O., Saraswat, K. C. and Stemmer, S. (2004), *J. Appl. Phys.*, **96** p. 3467.

Kirsch, P. D., Quevedo-Lopez, M. A., Krishnan, S. A., Lee, B. H., Pant, G., Kim, M.J., Wallace, R. M. and Gnade, B. E. (2006), *Appl. Phys. Lett.*, **89**, p. 242909:1.

Kita, K., Kyuno, K. and Toriumi, A. (2005), *Appl. Phys. Lett.*, **86**, p.102906:1.

Kita, K. and Toriumi, A., *Appl. Phys. Lett.* **94**, p. 132902.

Koester, S. J., Kiewra, E. W., Sun, Y., Neumayer, D. A., Ott, J. A., Copel, M., Sadana, D. K., Webb, D. J., Fompeyrine, J., Locquet, J.-P., Marchiori, C., Sousa, M. and Germann, R. (2006), *Appl. Phys. Lett.*, **89**, p. 042104.

Krishnan, S., Kwon, U., Moumen, N., Stoker, M. W., Harley, E. C.T., Bedell, S., Nair, D., Greene, B., Henson, W., Chowdhury, M., Prakash, D. P., Wu, E., Ioannou, D., Cartier, E. , Na, M.-H., Inumiya, S., McStay, K., Edge, L., Iijima, R., Cai, J., Frank, M., Hargrove, M., Guo, D., Kerber, A., Jagannathan, H., Ando, T., Shepard, J., Siddiqui, S., Dai, M., Bu, H. , Schaeffer, J., Jaeger, D., Barla, K., Wallner, T., Uchimura, S., Lee, Y., Karve, G., Zafar, S., Schepis, D., Wang, Y., Donaton, R., Saroop, S., Montanini, P., Liang, Y. , Stathis, J., Carter, R., Pal, R., Paruchuri, V., Yamasaki, H., Lee, J.-H., Ostermayr, M. , Han, J.-P., Hu, Y., Gribelyuk, M., Park, D.-G., Chen, X., Samavedam, S., Narasimha, S., Agnello, P., Khare, M., Divakaruni, R., Narayanan, V. and Chudzik, M. (2011), *Tech. Dig. Int. Electron Devices Meet.*, p. 634.

Li, J. and Ma, T. P. (1987), *J. Appl. Phys.*, **62**, p. 4212.

Lide, D. R. (2006), *CRC Handbook of Chemistry and Physics*, 87th ed., CRC, Boca Raton, FL.

Lin, L. and Robertson, J., *Appl. Phys. Lett.* **95**, p. 012906.

Lin, C.-H., Kambhampati, R., Miller, R. J., Hook, T. B., Bryant, A., Haensch, W., Oldiges, P., Lauer, I., Yamashita, T., Basker, V., Standaert, T., Rim, K., Leobandung, E., Bu, H. and Khare, M. (2012), *Symp. VLSI Tech. Dig. Tech. Pap.*, p. 15.

Linder, B. P. (2013), *VLSI-TSA Tech. Dig.* p. JS22.

Linder, B. P., Cartier, E., Krishnan, S., Tian, C. E. and Narayanan, V. (2014), *Micro- and Nanoelectronics: Emerging Device Challenges and Solutions (Devices, Circuits, and Systems)*, Ed. Brozek, T. (CRC Press) p. 138.

Litta, E. D., Hellström, P-E. and Östling, M. (2013), *Technical Digest of ESSDERC*, p. 155.

Luo, X., Bersuker, G. and Demkov, A. A. (2011), *Phys. Rev. B*, **84**, p. 195309.

Marchiori, C., Frank, M. M., Bruley, J., Narayanan, V. and Fompeyrine, J. (2011), *Appl. Phys. Lett.*, **98** p.052908-1.

Marchiori, C., Kazzi, M.El, Czornomaz, L., Pierucci, D., Silly, M., Sirotti, F., Abel, S., Uccelli, E., Sousa, M. and Fompeyrine, J. (2014), *J. Phys. D: Appl. Phys.*, **47**, p. 055101.

Mayadas, A. F. and Shatzkes, M. (1970), *Phys. Rev. B*, **1**, p. 1382.

Mistry, K., Allen, C., Auth, C., Beattie, B., Bergstrom, D., Bost, M., Buehler, M., Cappelani, A., Chau, R., Choi, C. H., Ding, G., Fischer, K., Ghani, T., Grover, R., Han, W., Hanken, D., Hattendorf, M., He, J., Hicks, J., Huessner, D., Ingerley, D., Jain, P., James, R., Jong, L., Joshi, S., Kenyon, C., Kuhn, K., Lee, K., Liy, H., McIntyre, B., Moon, P., Neiynck, J., Parker, C., Parsons, D., Pipes, L., Prince, M., Ranade, P., Reynolds, T., Sandford, J., Schifren, L., Sebastian, J., Seiple, J., Simon, D., Sivakumar, S., Smith, P., Thomas, C., Troeger, T., Vanderyoon, P., Williams, S. and Zawadzi, K. (2007), *Tech. Dig. Int. Electron Devices Meet.*, p. 247.

Mise, N., Tonomura, O., Sekiguchi, T., Horii, S., Itatani, H., Ogawa, A., Saito, T., Sakai, M., Takebayashi, Y., Yamazaki, H. and Torii, K. (2009), *Tech. Dig.*

Monaghan, S., Cherkaoui, K., O'Connor, É., Djara, V., Hurley, P. K., Oberbeck, L., Tois, E., Wilde, L. and Teichert, S. (2009), IEEE Electron Device Lett., **30** *p. 219.*

Int. Electron Devices Meet., p. 267.

Narasimha, S., Chang, P., Ortolland, C., Fried, D., Engbrecht, E., Nummy, K., Parries, P., Ando, T., Aquilino, M., Arnold, N., Bolam, R., Cai, J., Chudzik, M., Cipriany, B., Costrini, G., Dai, M., Dechene, J., DeWan, C., Engel, B., Gribelyuk, M., Guo, D., Han, G., Habib, N., Holt, J., Ioannou, D., Jagannathan, B., Jaeger, D., Johnson, J., Kong, W., Koshy, J., Krishnan, R., Kumar, A., Kumar, M., Lee, J., Li, X., Lin, C-H., Linder, B., Lucarini, S., Lustig, N., McLaughlin, P., Onishi, K., Ontalus, V., Robison, R., Sheraw, C., Stoker, M., Thomas, A., Wang, G., Wise, R., Zhuang, L., Freeman, G., Gill, J., Maciejewski, E., Malik, R., Norum, J. and Agnello, P. (2012), *Tech. Dig. Int. Electron Devices Meet.*, p. 52.

Narayanan, V., Paruchuri, V. K., Bojarczuk, N. A., Linder, B. P., Doris, B., Kim, Y. H., Zafar, S., Stathis, J., Brown, S., Arnold, J., *et al.* (2006), *Symp. VLSI Tech. Dig. Tech. Pap.*, p. 224.

Narayanan, V., Maitra, K., Linder, B. P., Paruchuri, V. K., Gusev, E. P., Jamison, P., Frank, M. M., Steen, M. L., La Tulipe, D., Arnold, J. *et al.* (2006), *IEEE Electron Device Lett.*, **27** p. 591.

O'Connor, É., Monaghan, S., Cherkaoui, K., Povey, I. M. and Hurley, P. K. (2011), *Appl. Phys. Lett.*, **99**, p. 212901.

Ootsuka, F., Katakami, A., Shirai, K., Watanabe, T., Nakata, H., Kitajima, M., Aoyama, T., Eimori, T., Nara, Y., Ohji, Y. and Tanjyo, M. (2008), *IEEE Trans. Electron Devices*, **55** p. 1042.

Ota, H., Hirano, A., Watanabe, Y., Yasuda, N., Iwamoto, K., Akiyama, K., Okada, K., Migita, S., Nabatame, T. and Toriumi, A. (2007), *Tech. Dig. Int. Electron Devices Meet.*, p. 65.

Packan, P., Akbar, S., Armstrong, M., Bergstrom, D., Brazier, M., Deshpande, H., Dev, K., Ding, G., Ghani, G., Golonzka, O., Han, W., He, J., Heussner, R., James, R., Jopling, J., Kenyon, C., Lee, S-H., Liu, M., Lodha, S., Mattis, B., Murthy, A., Neiberg, L., Neirynck, J., Pae, S., Parker, C., Pipes, L., Sebastian, J.,

Seiple, J., Sell, B., Sharma, A., Sivakumar, S., Song, B., Amour, A. St., Tone, K., Troeger, T., Weber, C., Zhang, K., Luo, Y. and Natarajan, S. (2009), *Tech. Dig. Int. Electron Devices Meet.*, p. 659.

Panasonic press release, Sep 15, (2010), http://panasonic.co.jp/corp/news/official.data/data.dir/en100915-2/en100915-2.html

Passlack, M. (2005), *J. Vac. Sci. Technol. B*, **23**, p. 1773.

Pawlak, M. A., Kaczer, B., Wang, W.-C., Kim, M.-S., Popovici, M., Swerts, J., Tomida, K., Opsomer, K., Schaekers, M., Vrancken, C., Govoreanu, B., Belmonte, A., Demeurisse, C., Debusschere, I., Altimime, L., Afanas'ev, V. V. and Kittl, J. A. (2011), *Symp. VLSI Tech. Dig. Tech. Pap.*, p. 168.

Radosavljevic, M., Chu-Kung, B., Corcoran, S., Dewey, G., Hudait, M. K., Fastenau, J. M., Kavalieros, J., Liu, W. K., Lubyshev, D., Metz, M., Millard, K., Mukherjee, N., Rachmady, W., Shah, U. and Chau, R. (2009), *Tech. Dig. Int. Electron Devices Meet.*, p. 319.

Radosavljevic, M., Dewey, G., Fastenau, J. M., Kavalieros, J., Kotlyar, R., Chu-Kung, B., Liu, W. K., Lubyshev, D., Metz, M., Millard, K., Mukherjee, N., Pan, L., Pillarisetty, R., Rachmady, W., Shah, U. and Chau, R. (2010), *Tech. Dig. Int. Electron Devices Meet.*, p. 126.

Robertson, J. (2008), *J. Appl. Phys.*, **104** p. 124111.

Robertson, J. (2009), *Appl. Phys. Lett.*, **94**, p. 152104.

Saito, S., Hisamoto, D., Kimura, S. and Hiratani, M. (2003), *Tech. Dig. Int. Electron Devices Meet.*, p. 797.

Samavedam, S. B., La, L. B., Smith, J., Dakshina-Murthy, S., Luckowski, E., *et al.* (2002), *Tech. Dig. Int. Electron Devices Meet.*, p. 433.

Schroder, D. K. (1998), *Semiconductor Material and Device Characterization*, 2nd Edition (John Wiley and Sons).

Schaeffer, J. K., Capasso, C., Fonseca, L. R. C., Samavedam, S., Gilmer, D. C., *et al.* (2004), *Tech. Dig. Int. Electron Devices Meet.*, p. 287.

Schaeffer, J. K., Gilmer, D. C., Capasso, C., Kalpat, S., Taylor, B., *et al.* (2007), *Microelectron Eng.*, **84**, p. 2196.

Shang, H., Jain, S., Josse, E., Alptekin, E., Nam, M. H., Kim, S. W., Cho, K. H., Kim, I., Liu, Y., Yang, X., Wu, X., Ciavatti, J., Kim, N. S., Vega, R., Kang, L., Meer, H. V., Samavedam, S., Celik, M., Soss, S., Utomo, H., Ramachandran, R., Lai, W., Sardesai, V., Tran, C., Kim, J. Y., Park, Y. H., Tan, W. L., Shimizu, T., Joy, R., Strane, J., Tabakman, K., Lalanne, F., Montanini, P., Babich, K., Kim, J. B., Economikos, L., Cote, W., Reddy, C., Belyansky, M., Arndt, R., Kwon, U., Wong, K., Koli, D., Levedakis, D., Lee, J.W., Muncy, J., Krishnan, S., Schepis, D., Chen, X., Kim, B. D., Tian, C., Linder, B. P., Cartier, E., Narayanan, V., Northrop, G., Menut, O., Meiring, J., Thomas, A., Aminpur, M., Park, S. H., Lee, K. Y., Kim, B. Y., Rhee, S. H., Hamieh, B., Srivastava, R., Koshy, R., Goldberg, C., Pallachalil, M., Chae, M., Ogino, A., Watanabe, T., Oh, M., Mallela, H., Codi, D., Malinge, P., Weybright, M., Mann, R., Mittal, A., Eller, M., Lian, S., Li, Y., Divakaruni, R., Bukofsky, S., Kim, J. D., Sudijono, J., Neumueller, W., Matsuoka, F. and Sampson, R. (2012), *Symp. VLSI Tech. Dig. Tech. Pap.*, p. 129.

Shiraishi, K., Yamada, K., Torii, K., Akasaka, Y., Nakajima, K., *et al.* (2004), *Jpn. J. Appl. Phys.*, **43**, p. L1413.

Siddiqui, S., Chowdhury, M. M., Brodsky, M. J., Rahim, N., Dai, M., Krishnan, S., Fugardi, S., Wu, E., Chou, A., Narasimha, S., Li, J., Mcstay, K., Linder, B., Maciejewski, E., Rettmann, R., Mittl, S., Kwon, U., Narayanan, V., Henson, W., Schepis, D. and Chudzik, M. (2013), *ECS Trans.*, **53**(3), p. 137.
Suzuki, M., Tomita, M., Yamaguchi, T. and Fukushima, N. (2005), *Tech. Dig. Int. Electron Devices Meet.*, p. 433.
Takagi, S., Toriumi, A., Iwase, M. and Tango, H. (1994), *IEEE Trans. Electron Devices*, **41**, p. 2257.
Tatsumura, K., Goto, M., Kawanaka, S. and Kinoshita, A. (2007), *Tech. Dig. Int. Electron Devices Meet.*, p. 1.
Tomida, K., Kita, K. and Toriumi, A. (2006), *Appl. Phys. Lett.*, **89**, p.142902:1.
Tiwari, S. and Wright, S. L. (1988), *IEEE Electron Device Lett.*, **9**, p. 488.
Togo, M., Joshi, M., Meer, H. V., Liu, Y., Yong, C., Liu, B., He, X., Wu, X., Mun, S. Y., Zhang, X., Konduparthi, D., Lian, J., Bohra, G., Tong, W. H., Xiao, C. Y., Triyoso, D., Banghart, E., Pandey, S. M., Wei, A., Pal, R., Carter, R., Nam, M. H., Eller, M. and Samavedam, S. (2014), *Symp. VLSI Tech. Dig. Tech. Pap.*, p. 140.
Urabe, Y., Yasuda, T., Ishii, H., Itatani, T., Miyata, N., Yamada, H., Fukuhara, N., Hata, M., Takenaka, M. and Takagi, S. (2011), *Microelectron. Eng.*, **88**, p. 1076.
Van Der Geest, A. G., Blaise, P. and Richard, N. (2012), *Phys. Rev. B*, **86**, p. 085320.
Veloso, A., Witters, L., Demand, M., Ferain, I., Son, N. J., Kaczer, B., Roussel, Ph. J., Simoen, E., Kauerauf, T., Adelmann, C., Brus, S., Richard, O., Bender, H., Conard, T., Vos, R., Rooyackers, R., Elshocht, S. Van, Collaert, N., Meyer, K. De, Biesemans, S. and Jurczak, M. (2008), *Symp. VLSI Tech. Dig. Tech. Pap.*, p. 14.
Wang, Z., Mui, D. S.L., Demirel, A. L., Biswas, D., Reed, J. and Morkoç, H. (1992), *Appl. Phys. Lett.*, **61**, p. 1826.
Wang, Z., Lin, M. E., Biswas, D., Mazhari, B., Teraguchi, N., Fan, Z., Gui, X. and Morkoç, H. (1993), *ppl. Phys. Lett.*, **62**, p. 2977.
Wang, M., Muralidhar, R., Stathis, J. H., Linder, B. P., Jagannathan, H. and Faltermeier, J. (2013), *IEEE Electron Device Lett.*, **34**(7), p. 837.
Yeo, Y. C., King, T. J. and Hu, C. H. (2002), *J. Appl. Phys.*, **92**, p. 7266.
Wang, J., Tateshita, Y., Yamakawa, S., Nagano, K., Hirano, T., Kikuchi, Y., Miyanami, Y., Yamaguchi, S., Tai, K., Yamamoto, R., Kanda, S., Kimura, T., Kugimiya, K., Tsukamoto, M., Wakabayashi, H., Tagawa, Y., Iwamoto, H., Ohno, T., Saito, M., Kadomura, S. and Nagashima, N. (2007), *Symp. VLSI Tech. Dig. Tech. Pap.*, p. 46.
Winstead, B., Taylor, W. J., Verret, E., Loiko, K., Tekleab, D., Capasso, C., Foisy, M., and Samavedam, S.B. (2007), *IEEE Electron Device Lett.*, **28**(8), p. 719.
Winn, D. L., Hale, M. J., Grassman, T. J., Sexton, J. Z., Kummel, A. C., Passlack, M. and Droopad, R. (2007), *J. Chem. Phys.*, **127**, p. 134705.
Wu, S.-Y., Lin, C. Y., Chiang, M. C., Liaw, J. J., Cheng, J. Y., Yang, S. H., Liang, M., Miyashita, T., Tsai, C. H., Hsu, B. C., Chen, H. Y., Yamamoto, T., Chang, S. Y., Chang, V. S., Chang, C. H., Chen, J. H., Chen, H. F., Ting, K. C., Wu, Y. K., Pan, K. H., Tsui, R.F., Yao, C. H., Chang, P. R., Lien, H. M., Lee, T. L., Lee, H. M., Chang, W., Chang, T., Chen, R., Yeh, M., Chen, C. C., Chiu, Y. H., Chen, Y. H., Huang, H. C., Lu, Y. C., Chang, C.W., Tsai, M. H., Liu, C. C., Chen, K. S., Kuo, C. C.,

Lin, H. T., Jang, S. M. and Ku, Y. (2013), *Tech. Dig. Int. Electron Devices Meet.*, p. 224.

Yamaguchi, T., Iijima, R., Ino, T., Nishiyama, A., Satake, H. and Fukushima, N. (2002), *Tech. Dig. Int. Electron Devices Meet.*, p. 621.

Yamaguchi, S., Tai, K., Hirano, T., Ando, T., Hiyama, S., Wang, J., Hagimoto, Y., Nagahama, Y., Kato, T., Nagano, K., Yamanaka, M., Terauchi, S., Kanda, S., Yamamoto, R., Tateshita, Y., Tagawa, Y., Iwamoto, H., Saito, M., Nagashima, N. and Kadomura, S. (2006), *Symp. VLSI Tech. Dig. Tech. Pap.*, p. 192.

Yang, S. H., Sheu, J. Y., Ieong, M. K., Chiang, M. H., Yamamoto, T., Liaw, J. J., Chang, S. S., Lin, Y. M., Hsu, T. L., Hwang, J. R., Ting, J. K., Wu, C. H., Ting, K. C., Yang, F. C., Liu, C. M., Wu, I. L.,Chen, Y. M., Chent, S. J., Chen, K. S., Cheng, J. Y., Tsai, M. H., Chang, W., Chen, R., Chen, C. C., Lee, T. L., Lin, C. K., Yang, S. C., Sheu, Y. M., Tzeng, J.T., Lu, L. C., Jang, S. M., Diaz, C. H. and Mii, Y. J. (2011), *Custom Integrated Circuits Conference*, p. 1.

Zhao, H., Chen, Y.-T., Yum, J. H., Wang, Y., Goel, N. and Lee, J. C. (2009), *Appl. Phys. Lett.*, **94**, p. 193502.

Chapter 8

Materials for DRAM Memory Cell Applications

Uwe Schroeder
NaMLab gGmbH/TU Dresden, Germany

Kyuho Cho
Samsung Electronics, Semiconductor Research Center, Hwasung, Korea

Stefan Slesazeck
NaMLab gGmbH/TU Dresden, Germany

Semiconductor memory is one of the key technologies driving the success of Si-based information technology within the last five decades. The most prominent representative memory type, the dynamic random access memory (DRAM) was patented in 1967[1] and was introduced into the market by Intel Corporation in 1972. Until the year 2001 and the realization of the 110 nm technology node, DRAM was the driving force on the lithography shrink roadmap, before NAND FLASH took over that role. Hence, the development of the DRAM technology was long time the forerunner for the exponentially growing large-scale integration and promoted similar advances in logic chips. One of the reasons of the success of the DRAM is its simple cell structure, which consists of only one transistor (1T) and one capacitor (1C), where the information is stored in form of a charge.

1. DRAM Operation

1.1. *Basic operation*

Figure 1 depicts the comparatively simple structure of a 1T-1C DRAM cell. In this cell, the information is stored as electrical charge on the storage capacitor *C*. *C* can be electrically connected to the bitline (BL) via the access transistor T. This means T acts as a switch and its gate is controlled by the wordline (WL), which is arranged perpendicular orientated to the BL. Hence, each individual DRAM cell can be addressed by one WL and one BL. The information is written to the capacitor *C* by switching on the access transistor T when applying a high voltage V_{WLH} to WL. *C* is then charged or discharged, applying a bitline voltage V_{BLH} or $\mathrm{V_{BLL}}$ to the BL, respectively. When *C* becomes disconnected from the BL when forcing a low voltage V_{WLL} to the gate of transistor T the remaining charge on the capacitor *C* represents then the binary information. A charged capacitor denotes a logical '1', whereas a discharged capacitor represents a logical '0'.

The memory state is read using an external circuit — the so called sense amplifier SA, which is connected to each of the bitlines. This SA commonly consists of a cross-coupled inverter circuitry (flip flop) to translate the charge back into digital information. Typically, two bitlines are connected to the two terminals of the SA in order to maximize the signal to noise margin during read operation.

For reading out the information from the DRAM cell, first an equalizing signal of $V_{\mathrm{BLH}}/2$ is applied to these two differential inputs. The operation voltage as well as the ground terminal of the sense amplifier

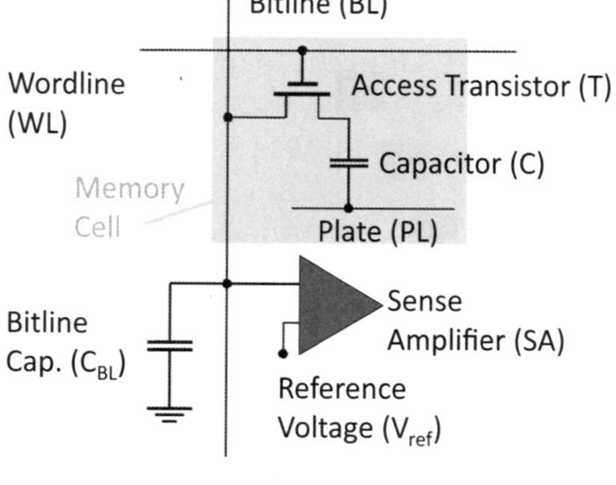

Fig. 1. DRAM memory cell within a memory array.

are disconnected. The flip flop is now in an instable state. Subsequently, the selected WL is activated by applying V_{WLH} so that cell capacitor C and corresponding BL are connected via the access devices. The stored charge on the corresponding capacitor C is redistributed between the cell capacitance, C, and the bitline capacitance C_{BL}. This leads to a voltage change on the active BL, whereas the voltage on the passive BL stays at the pre-charge level of $V_{\mathrm{BLH}}/2$. When the sense-amplifier becomes activated by connecting the operation voltage and ground terminal, the flip flop swings into the stable state according to the sign of the voltage change at the active BL. If the capacitor was empty, the voltage on the active BL drops to a lower values than $V_{\mathrm{BLH}}/2$ ending in the flip flop reaching the stable state '0'. If the capacitor was charged, the bitline becomes more positive and the flip flop falls into the '1' state. The state of the flip flop can be read by additional digital logic. The subsequent use of the read information is described in section 1.2. The read operation of the DRAM cell is called 'destructive' readout, since the information which was stored in the DRAM cell is destroyed. Hence, after each readout, the information has to be restored for a subsequent read operation. The refresh operation is performed by keeping the WL activated during the whole readout process. With the flip flop swinging to a stable state, the information in the cell can be rewritten. If the capacitor was uncharged, the flip flop moves to the '0' state by pulling the bitline with the activated cell to 0 V. If the capacitor was charged, i.e. in '1' state, the bitline is pulled to V_{BLH}. This means the capacitor is recharged according to its previous state. When turning off the access transistor, the information is restored in the cell as it was before the read operation. In order to reduce the electrical stress over the capacitors dielectric and to minimize the leakage currents, the plateline (PL) is always kept at $V_{\mathrm{BLH}}/2$. Hence, the effective voltage over the capacitor is switched between $\pm V_{\mathrm{BLH}}/2$ instead of switching between 0 V and full level V_{BLH}.

Unfortunately, leakage paths exist through the capacitors dielectric as well through the non-ideal transistor. The most prominent leakage mechanisms in the transistor are the junction leakage at the so called node side of the transistor (the terminal where the capacitor is connected to) and the channel leakage. Those leakage currents might charge or discharge C over time. In order to prevent the complete loss of the information, a certain charge on C and a corresponding minimum voltage change of the BL has to be guaranteed, which is still detectable by the sense amplifier during the read operation. Therefore, the charge of each memory capacitor has to be refreshed periodically. For the 16 Gb generation, the refresh time is at 64 or 32 ms depending on the DRAM type.[2] A special built-in counter takes care of the refresh for those rows that are not selected by the working program during the given refresh period. In case of cutting the operation voltage of the DRAM, the information is lost. Therefore, the DRAM is a volatile memory component.

In general, by selecting a WL, all the select-transistors in this row are active so that all cells in this row (1024 or 2048, depending on the chip design) can be read simultaneously. So, each BL has to be connected to one sense amplifier at the array edge. This is the reason why sense amplifier layouts have to be optimized to fit into the very dense pitch which is given by the BLs. The number of cells which can be connected to one BL depends mainly on the available cell capacitance C, BL capacitance per cell and the minimum signal on the BL which can be detected by the sense amplifier.

1.2. *Access transistor*

The planar devices with symmetric junctions used since the beginning of DRAM were not shrinkable below 120 nm without modification. In order to switch off the array devices sufficiently to guarantee a channel leakage well below 1 fA, a negative V_{WLL} was mandatory (see section 1.1). The resulting high electrical fields in the gate-to-junction overlap region lead to an increased junction leakage, the so called gate-induced drain leakage (GIDL). In order to minimize that leakage current, the pn-junctions had to be designed very soft and the boron channel doping concentration of the devices had to be kept below $\approx 2 \cdot 10^{17}/\mathrm{cm}^3$. Low channel doping in turn decreases the threshold voltage of the access devices, resulting in even more negative gate voltages. Hence in order to minimize the leakage currents, a well-balanced dopant profile was mandatory. Further constraints were given by the performance of the access devices. These have to drive a current in the range of $\approx 20\,\mu$A when switched on, allowing to read and rewrite the DRAM cells in an access time well below 10 ns. In order to improve the device performance, the more uncritical pn-junction with respect to leakage currents at the BL-side was doped more heavily and a single-sided so called halo implant was used to increase the BL-sided channel doping. The result was an unsymmetrical device design.

For groundrules below the 70 nm generation, the channel length of the access devices reached a distance at which the potential drop between source and drain lead to an unacceptably high electrical field at the capacitor-side

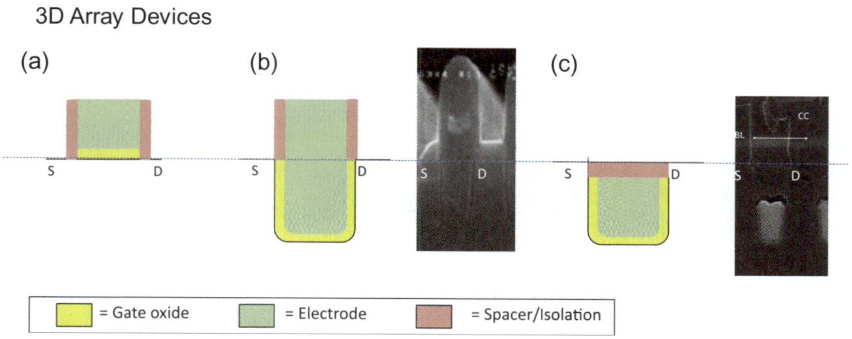

Fig. 2. Access transistor overview for different DRAM technology nodes. (a) Schematic view of a planar transistor in comparison to 3D array devices: (b) recessed channel device and (c) buried word line (BWL) device. For the BWL device, even the electrode is buried in the Si substrate.[71] The dotted line is the wafer surface.[3,4]

pn-junction. Therefore, the so-called 3D devices had to be introduced. In contrast to planar devices, the 3D devices are either formed with a vertical channel [3] or a recess channel [4] as shown in Figure 2(b). Up to that time, the WL typically was integrated in the gate stack module. This means, it ran as poly-gate with tungsten cap over the array. Now additional processing steps had to be introduced for the formation of the 3D devices. Furthermore, with the introduction of the so called "buried wordline" concept, the wordline was shifted below the silicon surface and the gate stack layer[71] was used to integrate the BL (Figure 2(c)). In order to guarantee a fast array access and sufficiently low RC-delay on the WL, a metal-gate had to be introduced. The main advantages of the curved channel below the silicon surface were the spatial separation of both source drain junctions, the possibility to locally increase the channel doping (no increasing electric field at capacitor-sided junction) and the threshold voltage increase due to the concave channel shape. As a result, further margin was yielded for the device design in the attempt to find the best compromise between junction leakage and sub-threshold channel leakage.

1.3. *Capacitor specifications*

The cell charge must be sufficiently high to compensate for leakage in the memory cell. Accordingly, the storage capacitance C needed to be kept constant and could only be slowly decreased during DRAM scaling when other factors were improved e.g. the sensitivity of the sense amplifier (Figure 3). For the 8 Gb chip, C is approx. 20 fF per capacitor. C can be

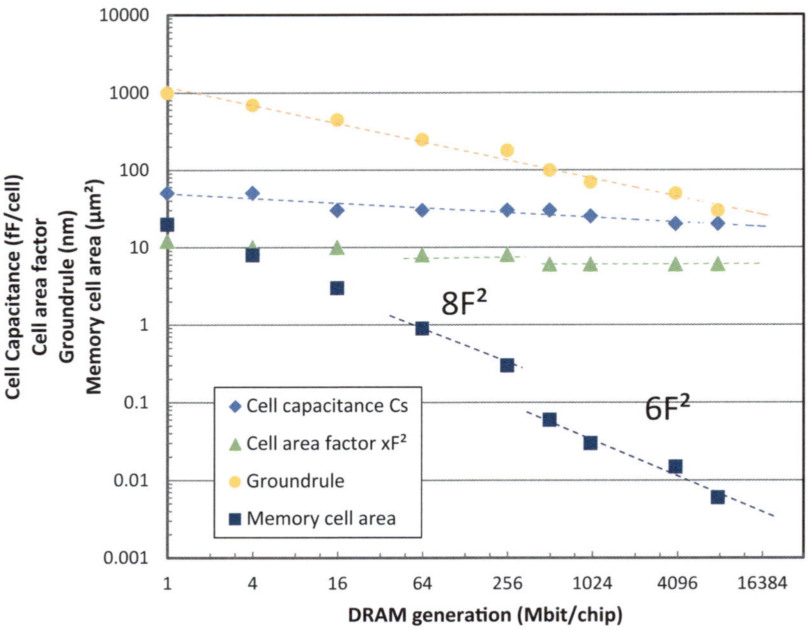

Fig. 3. DRAM roadmap trends during the last 30 years for cell storage capacitance, memory cell area, cell area factor and groundrule.[5,6,8]

calculated by the simple equation (Eq. 1)

$$C = \varepsilon_0 \cdot \varepsilon_{r\,SiO_2} \cdot \frac{A}{CET} \; with \; CET = \frac{\varepsilon_{r\,SiO_2}}{\varepsilon_r} \cdot d \; and \; \varepsilon_{r\,SiO_2} = 3.9 \qquad (1)$$

Here, ε_r is the relative permittivity, ε_0 is the vacuum permittivity, A is the capacitor area (here the average surface area of top and bottom electrode), and d is the physical thickness of the dielectric. Eq. 1 also defines the capacitance equivalent thickness (*CET*) with respect to the relative permittivity of SiO_2, ($\varepsilon_{r,SiO2} = 3.9$) — the dielectric, which was used for the first DRAM capacitors. Accordingly, the cell capacitance is determined by two main parameters: the cell capacitor area and the capacitance equivalent thickness. As can be seen from eq. 1, for high density DRAM with smaller and smaller cell area (Figure 3), there are only a few options to keep C constant when the cell area decreases:

- reduce *CET* by enhancing the ε_r-value of the dielectric or by shrinking of the physical thickness d.
- keep A constant by increasing the aspect ratio of the structure with decreasing cell area.

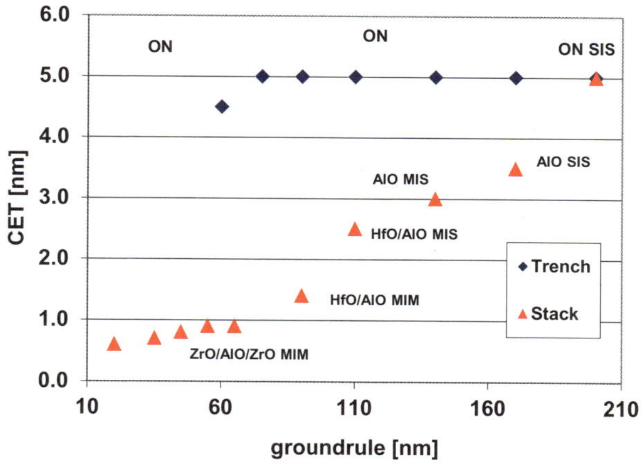

Fig. 4. CET values of capacitor dielectric for trench and stack capacitors for different technology nodes.[8–17]

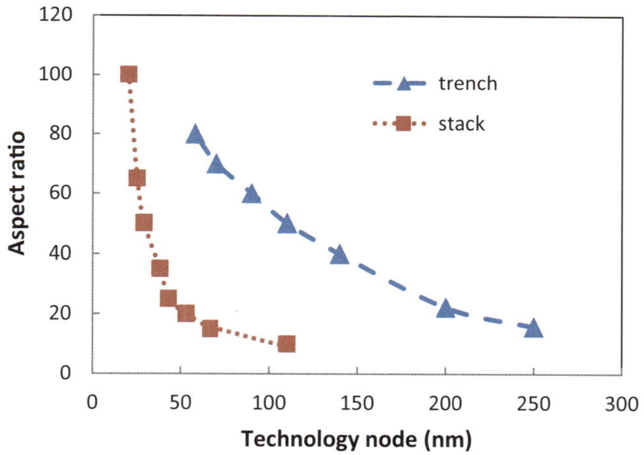

Fig. 5. Aspect ratio of capacitor structure vs. technology node.[7,17]

During the evolution of capacitors in the past DRAM generations, both options have been exercised (see Figures 4 and 5).

SiO_2 with a dielectric constant of 3.9 was used as the capacitor dielectric material for the first DRAM generations. With higher integration density, the dielectric was replaced by a mixture of Si oxide and nitride layers with higher effective dielectric constant ($\varepsilon_r \approx 7$), the so-called ON dielectric.[8] Below the 200 nm groundrule, stack capacitor technology in this conventional design approached its limits: the thickness of the (Si) oxide/nitride (ON) compound layer allowed no further thinning beyond $\approx$5 nm because of unacceptably high tunneling leakage currents (see section 2.3). A search for new materials started and industry followed in general the same trend. As a first replacement dielectric material Al_2O_3 was introduced by most of the companies (see Figure 4).[9–11] There are three factors favoring the material, namely its higher dielectric constant ($\varepsilon_r \approx 11$) compared to ON, low leakage current density i.e. high band gap, and also the availability of atomic layer deposition (ALD) processes, along with the ALD equipment suitable for a manufacturing environment. A first fully functional MIS (metal-insulator-semiconductor structure) 1 Gb DRAM chip using Al_2O_3 deposited by ALD on hemispherical Si grains (HSG, for surface area enhancement) with TiN top electrodes[10] was demonstrated. The typical capacitance increase compared to SIS ON (CET $\approx$ 5 nm) was in the order of 200% altogether for HSG introduction, 30% for Al_2O_3 dielectric (CET $\approx$ 3.5 nm), and 10% for depletion reduction by a change from doped poly-Si to TiN top electrode.

As a next step, DRAM manufacturers developed layered or "laminate" dielectrics, in order to gain maximum advantage from each of the two different dielectric materials used in these laminates. For example, amorphous Al_2O_3/HfO_2 structures have been demonstrated by Samsung.[12] The apparent advantage is that Al_2O_3 has a high band gap of $\approx$8 eV, while amorphous HfO_2 has a higher dielectric constant ($\varepsilon_r \approx 22$) at a lower band gap of $\approx$6 eV.[13] MIS Al_2O_3/HfO_2 capacitors showed CET values of $\approx$2.5 nm, which compensated the area loss due to the fact that hemispherical grains for surface area enhancement were no longer feasible in conjunction with this dielectric as ALD HfAlO showed no conformal growth on the grains in a high aspect ratio structure. Further capacitance gain was achieved by introducing a TiN bottom electrode, which reduced the CET value down to $\approx$1.2 nm for a TiN/HfAlO/TiN metal-insulator-metal (MIM) capacitor.

For most of the stack capacitor companies the HfAlO was the last amorphous dielectric. In 2006 the introduction of crystalline tetragonal ZrO_2 (CET $\approx$ 0.8 nm, $\varepsilon_r \approx 40$) with an amorphous Al_2O_3 layer in the center of the ZrO_2 film (ZAZ) to optimize breakdown behavior and reliability was announced (Figure 6).[14,15] In parallel to stack capacitor based DRAM, also memory cells with capacitor underneath the select transistor (trench capacitor) was developed.[16] The main advantage was that the surface area of the capacitor could be easier increased by etching the structure deeper into the Si-substrate. Accordingly, the ON based dielectric material could be used down to 70 nm technology node.[7] However, below the 58 nm node, the trench technology could not be extended because it was difficult to keep

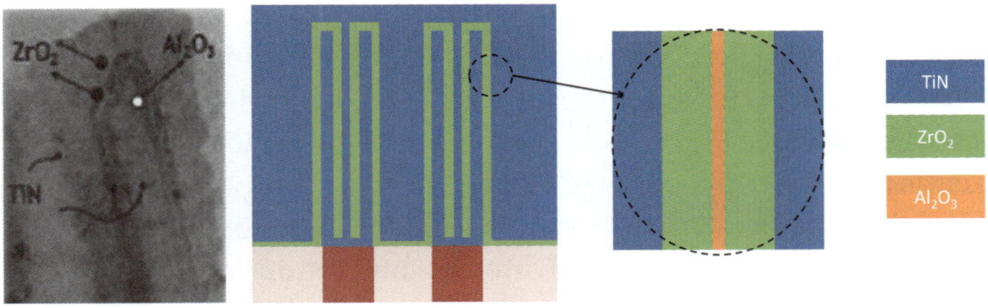

Fig. 6. DRAM capacitor with TIT ZAZ structure: (left) SEM image of typical top layer stack (©2006 IEEE. Reprinted, with permission, from Ref. 14) (middle) schematic overview of capacitor structure (right) detailed overview of MIM structure.

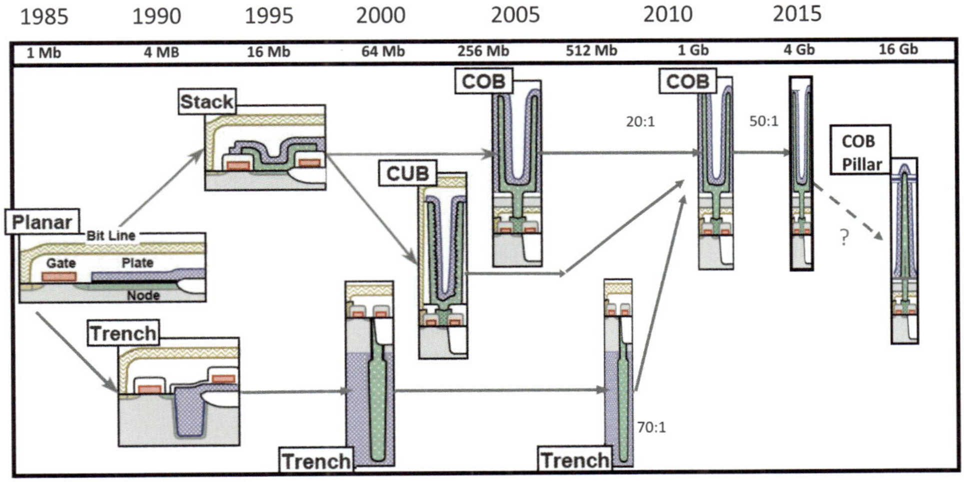

Fig. 7. Schematic overview of different cell concepts.[18]

the capacitor area A_s constant. High-ε_r dielectric materials were difficult to implement due to the high thermal budget of subsequent processing steps (e.g. transistor dopant activation anneals). Accordingly, only stack capacitors with ZAZ dielectric were used up to current 18 nm technology nodes (Figures 4 and 6).[17]

In the beginning of DRAM fabrication, only planar capacitors were processed (see Figure 7). Starting with the 4 Mb generation, the required capacitor area (A_S) was too large for the available cell area (A_{cell}) and the cell capacitance could not be achieved by a planar geometry. Consequently, the third dimension had to be used in the form of trenches and later also of posts in a stacked geometry (see Figure 7). An increasing height of the cell capacitor was maintained the capacitor area, although the area of the DRAM cell was continuously decreased in high density DRAM. As a consequence, the aspect ratio of DRAM capacitor was increased up to 80 for the trench technology[7] and up to 50 for current stack capacitor devices below 30 nm technology node (see Figure 4).[17] For the last 58 nm trench technology, the ε_r value was about five times lower compared to the stack case and accordingly, the aspect ratio of the structure was also about a factor of five higher than in the stack structure (see Figure 5). Different integration concepts, having the capacitor under (CUB: capacitor under bitline) and over the bitline (COB: capacitor over bitline) of the memory array, were verified. Below 18 nm technology the current capacitor integration concept (cylinder structure, see section 2.3) might be too large for the remaining cell area and pillar concepts might be the only feasible way to form the capacitor. Overall, high aspect ratio capacitor structures require a deep contact hole etch as well as conformal dielectric and electrode depositions (see section 2.6). ALD processes enable the required conformal deposition. In addition, the prevention of leaning and bending is critical when the storage node height is increased. Therefore, below 50 nm stack capacitor technology node, a support structure was successfully adapted to high aspect ratio storage node to stabilize the movement of the capacitor top region[19,20] (see Figure 16.5 in comparison to Figure 16.6).

2. Materials for Capacitor Stacks

2.1. *High-permittivity dielectrics*

As discussed in section 1, a dielectric material with dielectric constant ε_r and a high band gap E_g is required to fabricate a capacitor with sufficient capacitance and minimal charge loss due to leakage current. Looking at published data of oxides for most elements of the periodic table,[13] only a small number of binary oxides have a dielectric constant above 20 (see Figure 8). Since a plot of the dielectric constant versus the band gap follows a hyperbolic behavior (see Figure 12) an even smaller number of materials exhibit a band gap close to or beyond 4 eV: Al doped TiO_2, ZrO_2, and HfO_2. In addition, perovskite type dielectrics like $SrTiO_3$ (STO) show even better dielectric constants and low CET values at target leakage current (see Figure 9).

For other binary dielectrics like Nb_2O_5 and Ta_2O_5 the crystallization temperature is higher than 550°C to reach the reported dielectric constants larger than 40. This temperature is too high to be integrated in a typical DRAM process flow. Here, values below 500°C would be feasible.

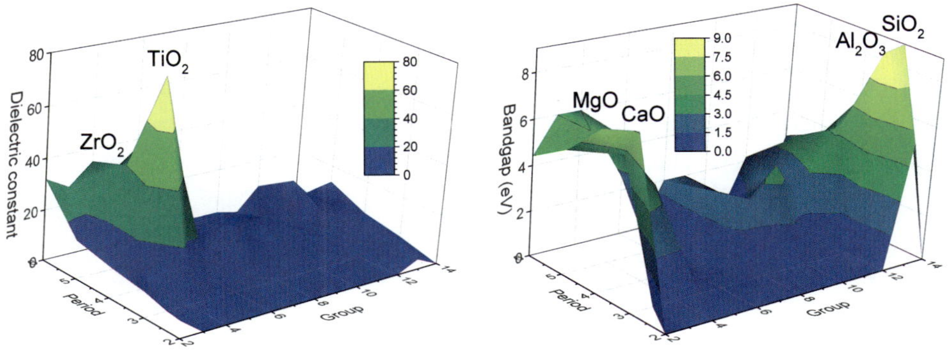

Fig. 8. Overview of dielectric constant and band gap for oxide materials of the periodic table [values mainly from Robertson *et al.*[13]]

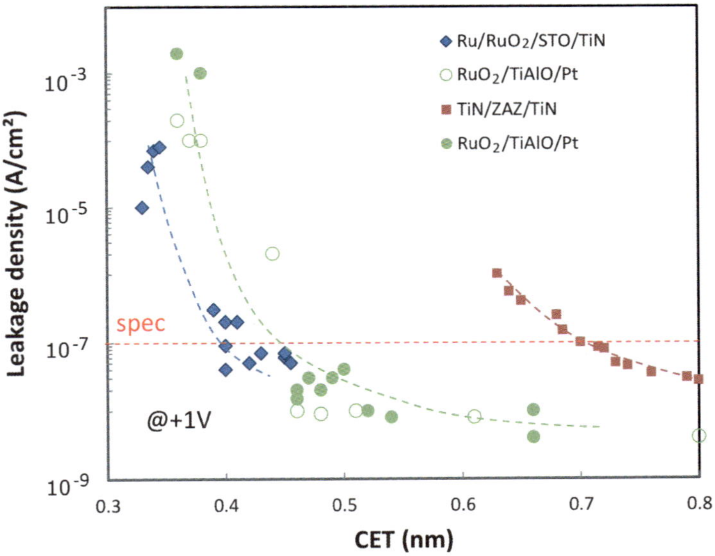

Fig. 9. Leakage current density versus CET value for DRAM capacitor dielectrics as reported in literature for the dielectric materials: doped TiO_2,[22,24] ZrO_2[23] and STO.[21] The leakage current specification limit for DRAM applications is at $1 \cdot 10^{-7}$ A/cm^2.

Having a closer look at the reported literature values for CET plotted as a function of leakage current density (see Figure 9), STO[21] shows the best CET results at target leakage specification ($1 \cdot 10^{-7}$ A/cm^2). These results indicate a slight improvement for $SrRuO_3$ compared to pure Ru electrodes. For $SrRuO_3$ the lattice constants are very similar to STO which reduced dead layer effects.[5,21] The good results of a STO dielectric are followed by Al doped TiO_2[22] and ZrO_2 with Al_2O_3[23] or SrO_2[53] interlayer.

Table 1. Material properties of high work function electrodes.

	TiN	Ru	Ir	Pt
Effective Work Function ϕ [eV]	4.2–4.7[25,26]	4.8[27]	5.6[28]	5.7[33]
Resistivity ρ [$\mu\Omega$.cm][29]	21.7	7.1	4.7	10.5

For a decision on a new dielectric for future DRAM generations important parameters are:

- STO and Al doped TiO_2 would require cost intensive noble metal electrodes with higher work function values (see section 2.2).
- In addition, the physical thickness of the overall MIM capacitor is critical. Due to the lower E_g values of STO and Al doped TiO_2 the overall physical thickness of the dielectric has to be higher than for ZrO_2. This could be critical for future scaled capacitors structures and development is ongoing to reduce the stack height.

2.2. *High work function electrodes*

Different electrode materials could be used for a DRAM capacitor. For best capacitor performance a material with low resistivity, high work function, and high conduction band offset to the adjacent dielectric would be beneficial. In addition, the electrode should be semiconductor compatible, cheap, and a deposition into high aspect ratio structures needs to be possible. Typical material properties of different reported electrodes are summarized in Table 1. For many years, TiN was the material of choice due to the high work function, high thermal stability and good reliability performance. In addition, the process of TiN is relatively inexpensive due to the low precursor cost ($TiCl_4$ and NH_3). For future DRAM capacitors also noble metal electrodes are of interest due to their high work function, high conduction band offset values, and low resistivity. But here, a conformal deposition in high aspect ratio structures and capacitor reliability still need to be proven. Furthermore, the high cost of current noble metal precursors needs to be reduced.

With shrinking capacitor dimensions, the dielectric thickness becomes thinner and electrode effects on the leakage currents become more and more important since the leakage current of the capacitor becomes mainly determined by the dielectric/electrode interface. In addition, the resistance of the electrode is important considering the high operating speed of

DRAMs. These two important properties of the electrode material (Table 1) are briefly discussed below.

2.2.1. *Work function*

According to the Schottky-Mott rule, the conduction band offset *CBO* of a metal insulator contact is determined by the electron affinity χ of the dielectric material and the work function ϕ of the electrode material (Eq. 2) (see Figure 11).

$$CBO = \phi - \chi \tag{2}$$

The Schottky-Mott rule does not consider the interface reaction between electrode and dielectric. Therefore, the observed conduction band offset value of the metal/dielectric contact is usually different from calculated material parameters. Considering an interface formation, the Schottky-Mott rule can be modified as:

$$CBO = \phi - \chi + e \cdot D_{int} \tag{3}$$

(with $e \cdot D_{\mathrm{int}}$: term induced by the interface dipole)

There have been several approaches to determine the interface dipole value systematically. A pinning factor S and the addition of a density of interface state N were suggested to calculate the conduction band offset.[31,32]

$$\begin{gathered} CBO = S(\phi - \phi_s) + (\phi_s - \chi) \\ S = [1 + (e^2 N\delta/\varepsilon_{int}\varepsilon_0)]^{-1} \end{gathered} \tag{4}$$

(with e: elemental charge, N: density of interface states, δ: interface thickness, $\varepsilon_{\mathrm{int}}$: relative permittivity of the interface, ε_0: permittivity of free space, ϕ_{s}: interface energy level of the dielectric).

It was also found that higher- ε_r dielectric materials showed a larger deviation from Schottky-Mott rule due to a small S value (see Figure 10). However, no universal formula has been suggested to successfully calculate the conduction band offset yet.

Regardless of such a deviation, high work function electrode materials increase the *CBO* of the metal insulator contact which results in a lower capacitor leakage current. Replacing the low-ε_r SiON ($\varepsilon_{\mathrm{r}} \approx 7$) to the higher-$\varepsilon_r$ ZrO_2 dielectric ($\varepsilon_{\mathrm{r}} \approx 40$) at 80 nm technology node decreased the barrier height of the capacitor inevitably since the electron affinity of ZrO_2 ($\chi \approx 2.2$ eV) is higher than the one of SiON ($\chi \approx 1.5$). Therefore, TiN was chosen as a higher work function material ($\phi \approx 4.7$ eV) to replace the doped

Si ($\phi \approx 4.0$ eV) electrode to maintain the barrier height. The usage of TiN as an electrode has been successfully extended for 10 years to the current 18 nm technology node. Further scaling of the DRAM capacitor requires higher ($\varepsilon_r > 40$) dielectric constant materials. Unfortunately, the electron affinity of higher dielectric constant materials is higher than ZrO_2 (e.g TiO_2: 3.3 eV[34]). Therefore, higher work function electrode materials are necessary to maintain the barrier height. There has been tremendous research effort to integrate an noble metal based electrode such as Ru ($\phi \approx 4.8$ eV), RuO_2 ($\phi \approx 5.0$ eV[35]), Ir ($\phi \approx 5.6$ eV), Pt ($\phi \approx 5.7$ eV)[72] into DRAM capacitors. Ru and RuO_2 were most intensively studied because Ru is relatively easy to etch and the cost of Ru precursors is relatively low among the noble metal electrode materials.

Even though the work function of the noble metal is higher than the one of TiN, which is an advantage for further scaling of the dielectric, several technical issues remain to be solved. First, a conformal deposition process for the thin noble metal is one of the major challenges for integrating the material into the capacitor. Unlike oxide and nitride materials a pure metal deposition process does not have a self-limiting surface reaction. Conformal thin inert noble metal films in high aspect ratio capacitor require appropriate ALD precursors and processes. Considering the high cost of the noble metals, process cost should be minimized by development of a low cost precursors and a fast deposition process.

In addition, the *CBO* does not improve as expected because of two reasons: As described in Eq. 3 pinning due to interface dipoles is lowering the *CBO*.[36] Furthermore, noble metal atoms can easily diffuse into the dielectric material, which causes traps at the interface and degrades the barrier height of the capacitor[37] (see Figure 11). Accordingly, new integration schemes are necessary to prevent these effects. The experimental determination of the work function and the *CBO* is described in section 2.4.

2.2.2. *Resistance*

To store information in a DRAM capacitor the charge needs to be transferred within the electrode to the dielectric layer interface. Possible write and read times are determined by the RC-delay of the cell (see section 1). If the resistance of the electrode is too high, the RC-delay reduces the charge transport within the electrode material, which lowers the effective cell capacitance. During capacitor scaling the thickness of the capacitor electrodes was continuously decreased, which causes a higher RC-delay since the resistivity of the thin film electrode is higher than the one of

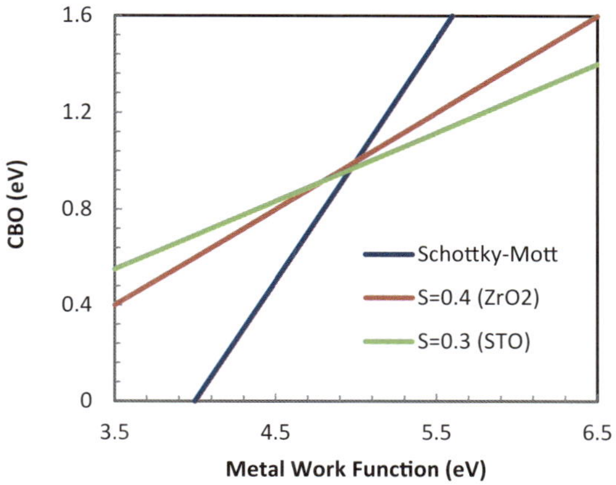

Fig. 10. Metal work function vs. barrier height: prediction from CBO models that assume a fixed separation between interface states and metal.[30]

the bulk material.[38] In addition, the top electrode is thinner than the bottom electrode, which further impacts the RC-delay. The main reason for this difference in electrode thickness is the integration flow described in section 2.6 (see Figure 16). Here, a certain thickness is required for the bottom electrode to guarantee the stability of the whole structure and the top electrode only fills remaining gaps in the structure.

$$R = \rho \cdot \frac{l}{A} \tag{5}$$

The resistance R of a thin TiN wire depends on the resistivity ρ, the length l and area A of a wire (Eq. 5). The resistivity of a film can be reduced by increasing the film density and improving the conformality of the film. The resistivity of inert noble metal electrodes is several times lower than that of TiN (Table 1.) Therefore, applying inert metal electrodes to DRAM capacitors can effectively reduce the electrode resistance.[72]

2.3. *Leakage mechanisms and traps*

The charge on the storage capacitor is decreased by leakage currents. Therefore, DRAM capacitors need to be recharged periodically, which is called "refresh". Typical refresh times are 64 or 32 ms depending on the DRAM type. The leakage current of the DRAM capacitor is determined by several parameters such as dielectric thickness, barrier height, trap energy level and trap density (see Figure 11).

Electrodes have a major impact on the leakage current since they supply both types of charge carriers: electrons and holes. When the dielectric is

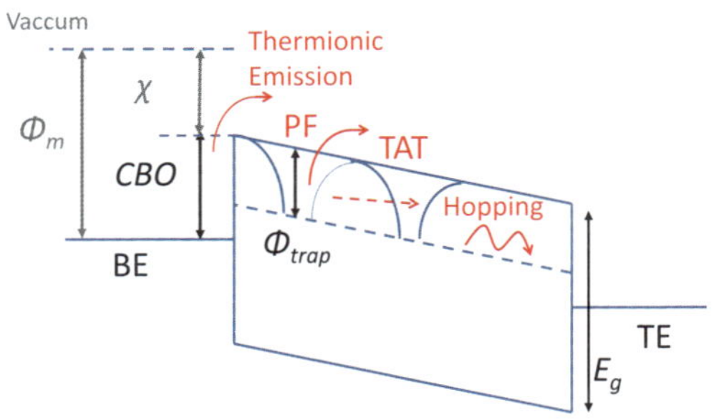

Fig. 11. Schematic overview of different leakage mechanisms for DRAM capacitors.

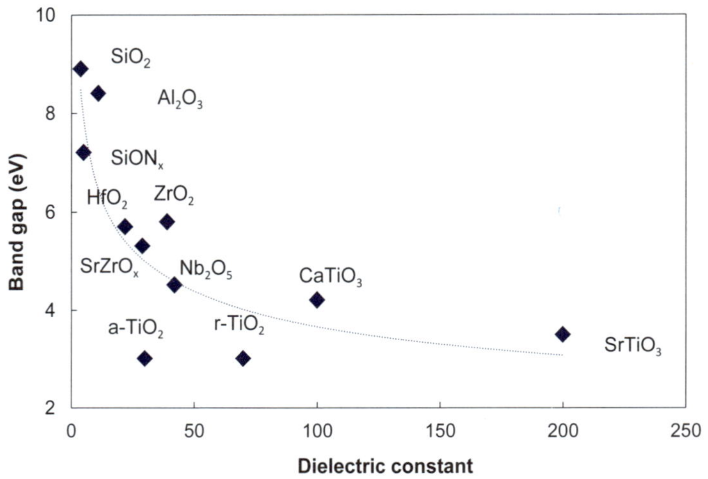

Fig. 12. Empirical trend between the band gap E_g and the dielectric constant ε_r for various dielectrics.[13] The dashed line is a fitting curve.

in contact to the electrodes, a band structure is formed where charged carriers like electrons need to overcome an energy barrier (Schottky barrier or conduction band offset *CBO*). This barrier is a dielectric metal interface material property and dependent to the metal workfunction ϕ and the affinity χ of the dielectric as described in Eq. 2. The band gap E_g was empirically found to show an inversely proportional dependence on the dielectric constant (see Figure 12).[39] Accordingly, materials with high dielectric constant have a low band gap and likely also a lower CBO and need to be deposited with a much higher thickness to compensate for the higher leakage current.

Overall, the leakage current can be divided into three major components: thermionic emission, tunneling, and hopping mechanisms.[40] Thermionic emission effects could be injection from the Fermi-level of the electrode to the conduction band of the dielectric (Schottky emission: Eq. 6):

$$j = \alpha \cdot T^2 \exp\left(-q\left(\phi_{trap} - \frac{\sqrt{\frac{qF}{\pi\varepsilon}}}{k_B T}\right)\right) \tag{6}$$

(where j is the current density, α is a constant, q is the charge of electron, φ_{trap} is the trap depth without external field, F is the electric field in the dielectric, k_B is the Boltzmann constant, and T is the temperature.

Or by Poole-Frenkel emission of trapped carriers from the trap site to the conduction band (Eq. 7):

$$j = \beta \cdot (qN_{PF}\mu)F \exp\left(\frac{E_A}{T}\right) \text{ with } E_A = -q/k_B \cdot \left(\phi_{trap} - \sqrt{\frac{qF}{\pi\varepsilon_0\varepsilon_\infty}}\right) \tag{7}$$

(where $j, q, \varphi_{\text{trap}}, k_B, T, F$ see above, N_{PF} is the trap density, μ is the electron mobility, ε_0 is the vacuum permittivity, and ε_∞ is the dynamic permittivity) Both effects are field enhanced and increase with temperature.

On the other hand, bulk transport effects need to be discussed: Tunneling can occur from the Fermi-level of the electrode to adjacent trap sites or between different trap locations. This effect follows for a square barrier (Eq. 8):

$$j = \sigma V; \sigma = \frac{q^2\sqrt{2m^* q\phi_e}}{4\pi\hbar^2\omega} \exp\left(-\frac{2\omega}{\hbar}\sqrt{2m^* q \cdot CBO}\right) V \ll CBO \tag{8}$$

or:

$$j = \sigma F^2 \exp\left(-\frac{4\sqrt{2m^*}}{3q\hbar F}(q \cdot CBO)^{\frac{3}{2}}\right) \quad with\ \sigma = \frac{q^3}{16\pi^2\hbar} \cdot q \cdot CBO \tag{9}$$

($j, q, \varphi_{\text{trap}}$, F see above, m^* the effective tunneling mass, ω the relevant width of the barrier, $\hbar$ the Planck constant) for higher applied fields for a triangular shaped barrier according to the Fowler-Nordheim equation (Eq. 9). As a last mechanism, hopping to nearest neighbors is mentioned and further mechanisms are described in literature.[40] Depending on the transport mechanism, different procedures are described to obtain e.g. the trap depth. As one example, the extraction of parameters for Poole-Frenkel emission is described in Fig. 13.[41] Here, the logarithm of the leakage current density versus one over temperature is plotted for a ZAZ capacitor with

TiN electrode. The data should follow a straight line and a trap depth of 0.8–1.4 eV can be calculated.[42]

2.4. *Capacitor band structure*

The knowledge of the DRAM capacitor band diagram can be used to model leakage currents [43] and to further optimize the film stack for future DRAM generations. The band structure has been estimated by several experimental methods. Terraced oxide measurements are performed to determine the work function of an electrode material on a dielectric.[44] A SiO_2 layer is grown on a Si substrate and a stepped structure is formed by wet chemical etching. On this surface, a thin dielectric (eg. ZrO_2) and metal layer pads (e.g. TiN) are deposited. After measurement of the flatband voltage shift for different dielectric thickness, values the work function of the metal can be extracted.[44] In addition, internal photoemission (IPE) spectroscopy enables

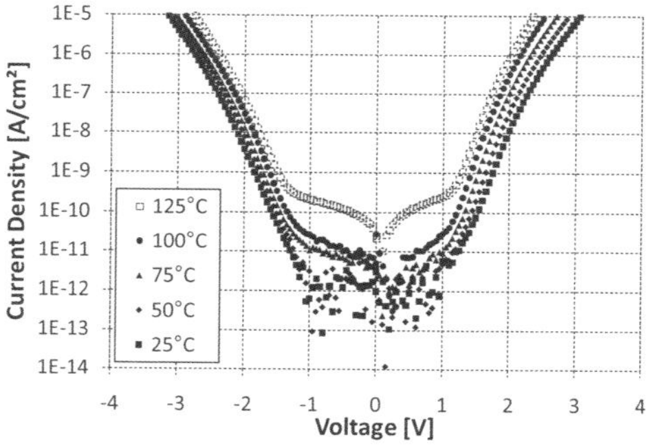

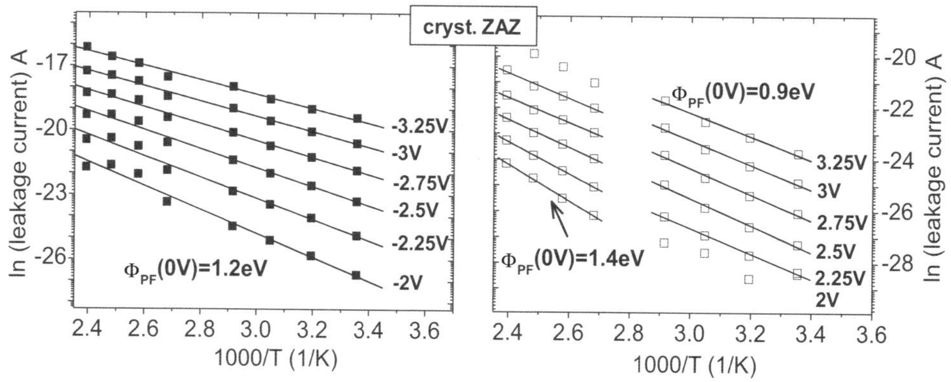

Fig. 13. (Top) Leakage current density of a 10 nm ZAZ capacitor dielectric with TiN top and bottom electrode for different bias voltage and different substrate temperatures (©2014 IEEE. Reprinted, with permission, from Ref. 41). Values for low bias voltages are limited by resolution of measurement setup. (Bottom) Extraction of trap levels assuming a Poole-Frenkel leakage mechanism.[42]

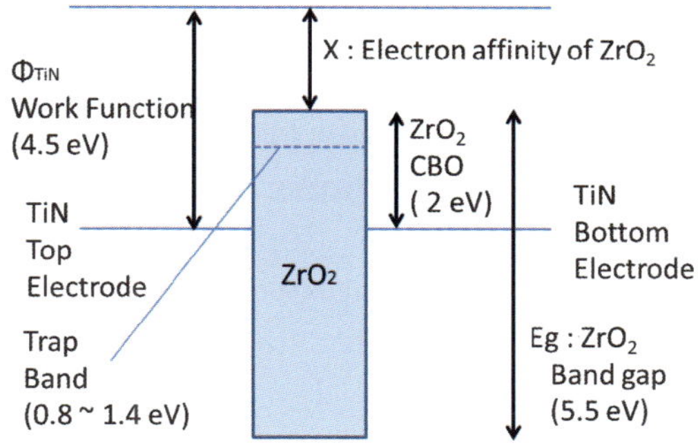

Fig. 14. Band diagram for a MIM capacitor with TiN electrodes and a ZAZ dielectric.[67]

straightforward CBO barrier determination as the minimal energy required to excite an electron from the TiN Fermi-level to the conduction band of the dielectric.[45] Furthermore, the band gap of the dielectric is characterized by photoconductivity (PC) measurements.

The band diagram of a MIM capacitor with a $ZrO_2/SrO/ZrO_2$ very similar to a currently used ZAZ dielectric was summarized as Figure 14[67] The *CBO* between the Fermi-level of the TiN electrode and the conduction band of ZrO_2 is about 2 eV and the band gap of ZrO_2 about 5.5 eV. Both numbers are lower compared to values reported in literature for bulk materials.[42] This can be related to diffusion of electrode atoms into the dielectric, interface oxidation of the electrode material and to effects caused by property changes in thin films. In addition, trap levels at 0.8–1.4 eV below the ZrO_2 conduction band and a TiN work function of $\approx$4.5 eV were determined.[53]

The band structure of a MIM capacitor with Ru electrode and $SrTiO_3$ dielectric was evaluated by Kaczer *et al.*[46] Again, the IPE method was used to evaluate the conduction band offset and band gap of the dielectric materials. A conduction band offset of $\approx$1.5 eV and a band gap of 3.2 eV was determined. Considering a *CBO* of $\approx$ 2 eV for ZAZ based capacitor with TiN electrodes, the *CBO* value for a $Ru/SrTiO_3$ interface needs to be improved to obtain a maximum advantage from their high dielectric constant. Sr rich $SrTiO_3$ was suggested to suppress the leakage current of the $Ru/SrTiO_3$ system.[47]

Moreover, the band structure of Al doped TiO_2 with a RuO_2 electrode was studied. It was suggested that the RuO_2 electrode was inevitable to deposit the rutile high dielectric constant phase of TiO_2. Furthermore, Al doping improved the conduction band offset of the RuO_2/TiO_2 interface, which reduced the leakage currents effectively.[48] However, due to a low E_g value of TiO_2 (3.1 eV), the conduction band offset of RuO_2/Al doped TiO_2 was estimated as ≈1.5 eV, which is also smaller than the one of a ZAZ capacitor with TiN electrodes. To compensate for these low *CBO* values, IrO_2 electrodes were deposited on Al doped TiO_2. Here, the higher work function values resulted in a *CBO* improvement and a decrease in leakage current for the Al doped TiO_2 capacitor stack.[49]

2.5. *Capacitance equivalent thickness scaling*

As discussed in section 1.3, a CET downscaling of DRAM capacitor can be achieved by two different approaches. On one hand, a decrease of the physical thickness of the current ZrO_2 dielectric is evaluated and on the other hand, new higher- ε_r dielectric materials could be applied to the DRAM capacitor.

Figure 15 shows a plot of the leakage current as a function of the CET value indicating an exponential increase of the leakage current for lower CET values caused by the thickness scaling of the ZrO_2 dielectric. According to the ITRS roadmap[51] a leakage current density of $1 \cdot 10^7$ A/cm^2 is required at a bias voltage of 0.5 V(V_{BLH}). This corresponds to about 1 fA/memory cell for a 3D capacitor when the surface area is not exactly known. At target leakage specification, typical CET values of 0.7 nm are reached for

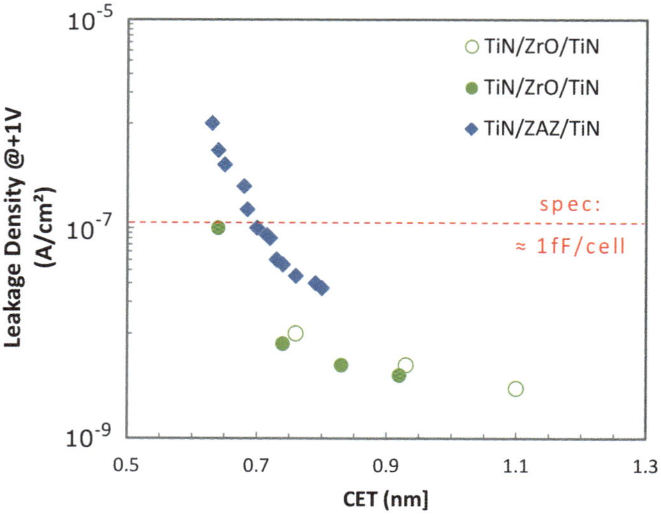

Fig. 15. Leakage current density versus capacitance equivalent thickness for different ZrO_2 based dielectric with TiN electrodes in a MIM capacitor configuration (square dots[23]; open dots;[50] closed dots[18]).

a crystalline pure ZrO_2 dielectric.[50] This material does not meet the 10 year lifetime criteria for DRAM capacitors due to the crystalline structure of the ZrO_2 dielectric. Here, ZrO_2 grains span the whole distance between bottom and top electrode and the leakage current is increased along the grain boundaries.[52] Typically, an amorphous Al_2O_3 interlayer is introduced in the center of the ZrO_2 dielectric to separate the ZrO_2 layer into two crystalline films with smaller grains size and an overall improved leakage and reliability behavior.[15] Due to the lower ε_r- value of the introduced Al_2O_3 the CET value is slightly increased which resulted in a similar CET/leakage performance compared to pure ZrO_2. In the last years the ZAZ process was improved and currently CET values below 0.7 nm at target leakage current can be reached. Furthermore, studies are ongoing to search for an interlayer material with higher dielectric constant than Al_2O_3 and similar electrical properties.[53]

As explained in paragraph 2.3 different leakage mechanisms could be responsible for conduction through the dielectric material. Temperature dependent leakage current measurements could give an indication of the main conduction paths. For the ZAZ capacitor stack a bias dependent leakage current mechanism is reported.[42] Assuming a Poole-Frenkel conduction mechanism, trap levels in the order of 0.8–1.4 eV are extracted between 1.5 to 3 MV/cm and related to oxygen vacancy driven trap sites[54] (see paragraph 1.2). The leakage current at higher field was typically attributed to Fowler-Nordheim tunneling.[56]

More aggressive CET scaling has been achieved by higher-*k* ($\varepsilon_r > 40$) dielectric materials such as Al doped TiO_2 and $SrTiO_3$ with noble metal electrode (see Figure 9). A CET value below 0.5 nm was obtained for Al doped TiO_2 with RuO_2 top and bottom electrode.[54] Even better results were achieved for MIM capacitors with $SrTiO_3$ dielectric and Ru or $SrRuO_3$ electrodes. A CET value below 0.4 nm was successfully shown (see Figure 9) and recent leakage current simulations using conduction band offset and oxide tunneling mass as a key parameter suggested that $SrTiO_3$ might be the only high-*k* dielectric candidate which is able to attain CET values below 0.3 nm.[21]

2.6. *3D capacitor fabrication*

Unlike planar structure MIM capacitor, formation of 3D structure DRAM capacitors requires additional process step and technology.

The integration flow for a DRAM capacitor was kept stable since the 50 nm technology but the aspect ratio of the structure drastically increased.

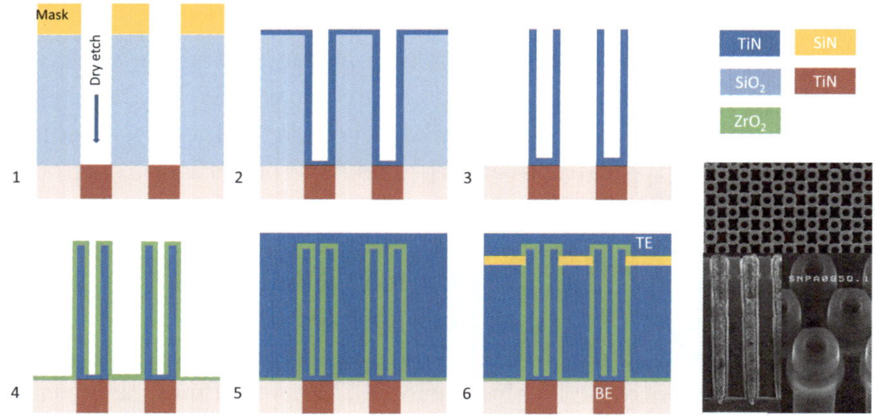

Fig. 16. Integration flow for DRAM stack capacitor fabrication; SEM: (top) top view of capacitor with support structure (bottom left) cross section similar to Figure 16.6 (bottom right) top view of capacitor under certain angle (©2013 IEEE. Part of figure reprinted, with permission, from Ref. 4).

In a first step (Figure 16.1), a mask is deposited on a micrometer thick mold SiO_2 layer before very precise trenches are dry etched into the SiO_2. These trenches need to be placed exactly on top of the contact structures and require an almost 90° taper angle. Both steps are already very challenging for the according process equipment.

After dry etching of trenches a chemical vapor deposition (CVD) of titanium nitride into a trench structure is performed (Figure 16.2). Typically, titanium chloride ($TiCl_4$) and ammonia (NH_3) are used as precursors. The mold oxide material is removed by a wet chemical process (Figure 16.3) followed by the ZrO_2-based dielectric deposition onto the very brittle TiN bottom electrode cup-like structure (Figure 16.4). Due to the layer by layer growth of the ALD processing very conformal films can be deposited in these high aspect ratio structures. For ZAZ film deposition typical precursors are $Zr[N(CH_3)C_2H_5]_4$ or $CpZr[NMe_2]_3$ for Zr, $Al(CH_3)_3$ (TMA) for Al and ozone as the oxygen source at deposition temperatures of about 300°C.[55] Finally, the CVD TiN top electrode is deposited onto the ZrO_2 dielectric (Figure 16.5). After fabrication, each capacitor in the array has a separate bottom contact to the select transistor and a common top electrode. To stabilize the cup shaped bottom electrode and to prevent bending and leaning a supporting SiN-based structure was implemented in the processing sequence starting with 50 nm technology node (Figure 16.6). For future DRAM capacitor structures the cell area is decreasing further and even higher aspect ratio structures are necessary to compensate the area loss. As shown in Figure 5, the aspect ratio might increase to 100:1. This will drastically enhance the surface area of the capacitor and accordingly, also of a 300 mm wafer and impact process specifications. Since the electrode layer might need to be thinned to relax the scaling requirements for future devices the resistivity within the electrode layer will become increasingly more critical.

As described in literature,[56] during the ZrO_2 ALD deposition process the underlying TiN bottom electrode is slightly oxidized by the usage of the ozone precursor and a TiO_xN_y is formed at the interface. During top electrode deposition only a nitridation of the dielectric is occurring and oxygen is diffusing into the TiN during subsequent anneal processing steps. Hence, the top electrode is less oxidized and a slightly asymmetric capacitor stack is formed. Since the leakage current is partially impacted by the interfaces these structural differences are leading to a non-symmetric leakage current behavior.

The high aspect ratio DRAM capacitor is fabricated on top of the transistor and further back end metal lines are processed on the capacitor. Compatibility to current DRAM integration scheme is essential for the DRAM capacitor. From all possible integration parameters, a conformal physical thickness of dielectric material and electrode are one of the most important process parameters for capacitor integration. The ratio between thinnest and thickest physical thickness in 3D structure is called as step coverage. The step coverage of dielectric material is important for both the cell capacitance and the leakage current. Due to the high aspect ratio of a DRAM capacitor structure, the thickness of the dielectric layer at the top of the structure is thicker than that on the bottom. Precursor and reactant concentration differences between top and bottom during the deposition are considered to be the origin of this difference.[57] The region with the thinnest dielectric — most likely at the bottom of the structure determines the leakage current of the storage capacitor. Hence, a minimum bottom dielectric thickness should be maintained to prevent high leakage currents and early breakdown of the device. On the contrary, a thicker dielectric at the top reduces the overall cell capacitance. Therefore, the target would be to improve the step coverage for best capacitance and leakage current performance. This becomes more and more challenging for smaller capacitor structures with higher aspect ratio. Novel atomic layer deposition processes were successfully developed for Al_2O_3 and ZrO_2 deposition to maintain conformal step coverage down to 18 nm DRAM generations on a complete 300 mm wafer.[58] Physisorption and chemical decomposition of precursors and incomplete chemisorption or chemical reaction are considered as the main reasons for conformality degradation.[59,60] Further improvements are necessary below the 18 nm node to be able to continue to achieve these step coverage requirements.

A poor conformality of the TiN electrode at the bottom of the structure degrades the mechanical strength. In addition, a reduced thickness of TiN in the top region causes resistance degradation (see section 2.2.2). Again, novel deposition process needs to be developed to guarantee further scaling of the capacitor.

These issues also impact the introduction of possible new dielectric or high work function electrode materials. Although perovskite type high-k materials such as $SrTiO_3$ have been successfully applied to bulk capacitors starting in middle of 20th century,[61] they are still not implemented in DRAM capacitor despite of the tremendous development efforts.[73] Poor step coverage of the current ALD processes for $SrTiO_3$ and the Ru electrode are some of the typical issues for device integration.[62]

3. Reliability of the Capacitor

The stored charge on the capacitor electrodes decreases over time due to leakage via the capacitor dielectric and the select transistor structure within the memory cell (see Figure 1). According, the charge needs to be restored in 64 or 32 ms (refresh time depends on the DRAM type). For an accurate sensing of the memory state the sense amplifier requires a maximum charge drop of less than 10% within the refresh time. For this reason, potential detractors will be discussed in the following chapter.

3.1. *Relaxation currents*

Dielectric relaxation is ability of a capacitor to completely discharge to zero. As depicted in Figure 18, the leakage current density in a capacitor is dominated during the initial time period by relaxation currents. For most dielectrics, they follow the empirical Curie-von Schweidler relation with a time dependence of the relaxation current density $j_{\mathrm{relax}}(\mathrm{t})$[63] (see Eq. 10):

$$j_{relax}(t) = j_0 \left(\frac{t}{t_0}\right)^{-\alpha} \quad \text{with } 0.5 < \alpha < 1 \qquad (10)$$

The time dependence was typically explained by a sequence of "Debye" relaxations ($j_i(t) \propto \exp(-t/\tau_{\mathrm{i}})$) with very different relaxation times τ_{i}. Since there is no microscopic physical model, the charge loss is

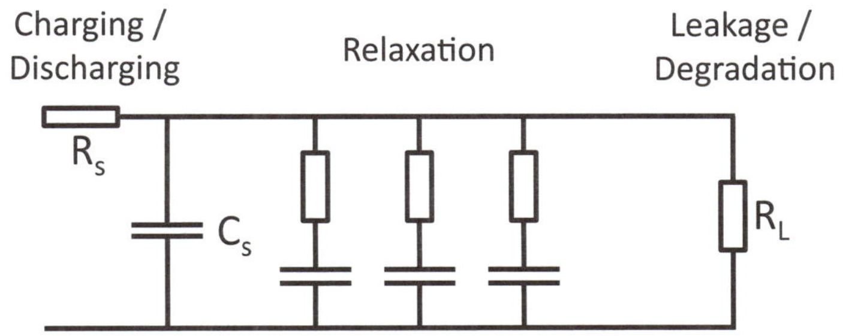

Fig. 17. Schematic diagram of equivalent current for dielectric relaxation.

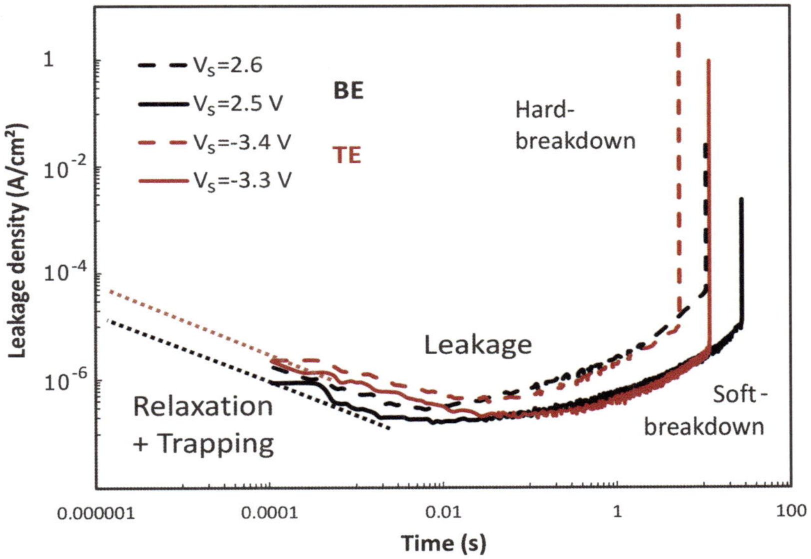

Fig. 18. Time evolution of current response after charging a MIM capacitor with ZAZ dielectric.

described in a model of parallel RC elements with different time constants (see Figure 17).[64]

After this initial relaxation phase (typically above the ms range) a steady state leakage current is visible before resistance degradation of the device (see section 3.2), i.e. the high resistivity of the insulator decreases gradually (soft breakdown) or nearly instantaneously by several orders of magnitude (hard breakdown) (see Figure 18). Both are unacceptable for DRAM operation because of possible lifetime restrictions.

The relaxation current can be determined by charging the capacitor for a defined amount of time, discharging the device for a short time followed by a measurement of the relaxation current. These relaxation currents degrade

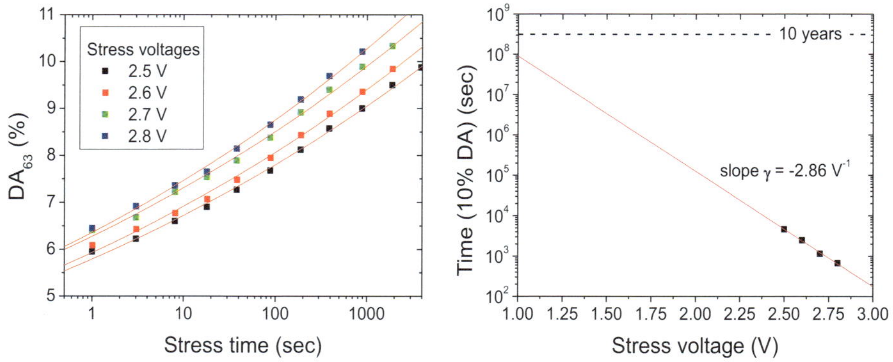

Fig. 19. (Left) Dielectric relaxation for different stress times and stress voltages. (Right) Extrapolation of dielectric relaxation to conditions at which a DRAM capacitor is used.

over time and the behavior was characterized for planar ZAZ capacitors at accelerated stress conditions (see Figure 19). For an extrapolation, the time is plotted when a capacitor showed a loss of 10% due to relaxation at a certain voltage. This time can be extrapolated to 10 years and typical conditions where the capacitor is used ($V_{BLH} \sim 0.5\,V$). Hence, relaxation is uncritical for a capacitor when the device is used at room temperature and with charging pulses in the μs range. This phenomenon might become more critical at elevated temperatures or for shorter switching times.

3.2. *Dielectric resistance degradation and breakdown*

The above mentioned constant leakage current regime is followed by an increasing leakage current and subsequently by a slow (soft) or abrupt (hard) breakdown (see Figure 20). A clear distinction of this different breakdown behavior is visible for amorphous versus crystalline ZrO_2 caused by different amounts of defects in the layer.[41] These determine the lifetime and reliability of the capacitor. The lifetime of a single capacitor can be defined as the time for which the leakage current has increased ten times compared to the steady state minimum. Overall, the benchmark for the lifetime of capacitors is typically 10 years at an operation voltage of $\approx 0.5\,V$ ($= V_{BLH}$). Data for the lifetime evaluation has to be collected in accelerated tests for a statistical relevant set of production type stack capacitors. Here, higher stress temperatures and voltages are used to avoid extremely long testing times. A statistical distribution of a larger number of breakdown events is plotted in a Weibull graph (see Figure 21 (left)). The most likely breakdown time is extracted at the time when 63% of the devices failed. This time can be plotted for different stress voltages and extrapolated to

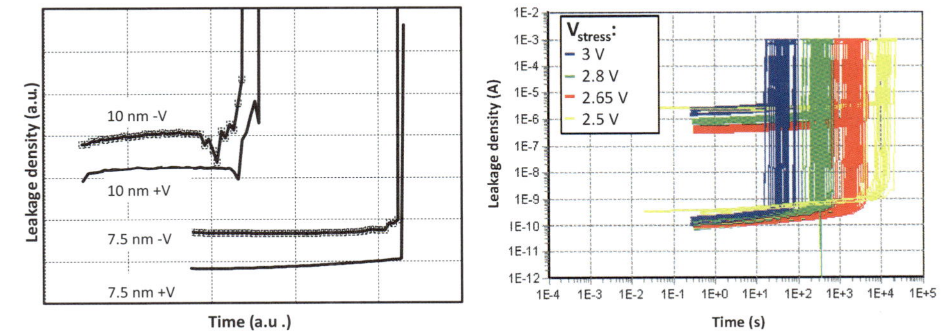

Fig. 20. (Left) Leakage current increase over time caused by dielectric breakdown of a 10 nm thick crystalline ZAZ layer (soft + hard breakdown) compared to a 7.5 nm amorphous layer (only hard breakdown).[19] (Right) breakdown behavior for different bias voltages applied to a production type ZAZ based DRAM capacitor structure.[65]

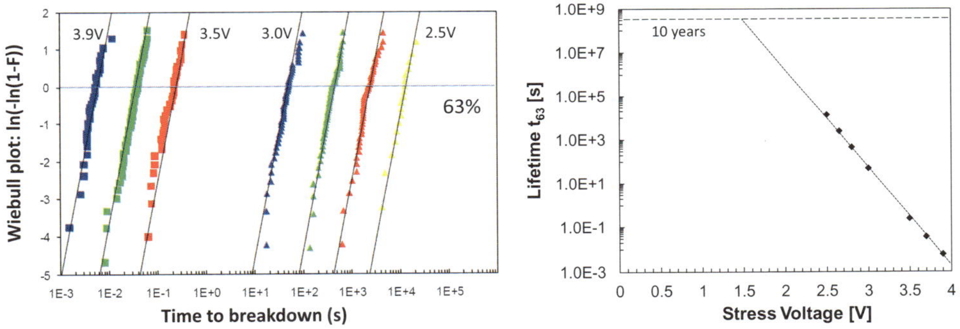

Fig. 21. (Left) Weibull plot of a breakdown lifetime distribution in Figure 20: the failure rate is plotted vs. the lifetime of the device. (Right) The lifetime of a device (from Figure 21 left) that broke down after 63% of similar devices showed a breakdown event (see line) is plotted versus the applied stress voltage. Data points are extrapolated to use voltage of 0.5 V.

operation conditions of $V_{\mathrm{BLH}} \approx 0.5\,\mathrm{V}$ (see Figure 21 (right)). Since devices are stressed at elevated temperatures, this should be taken into account for the determination of the overall lifetime. The temperature dependence of the lifetime can be described by Arrhenius behavior, i.e. the underlying mechanism is thermally activated. The extracted thermal activation energies for ZrO_2 thin films vary between 0.7 eV and 1.6 eV.[41] Overall, ZAZ dielectric layers in TiN electrode based capacitor configurations meet the reliability criteria for current DRAM specifications.[41]

AC stress measurements are reported for MIM capacitors with a thin SiO_2 instead of Al_2O_3 interlayer in the middle of the ZrO_2 dielectric.[66] A rectangular 100 Hz signal was applied, varying between the constant

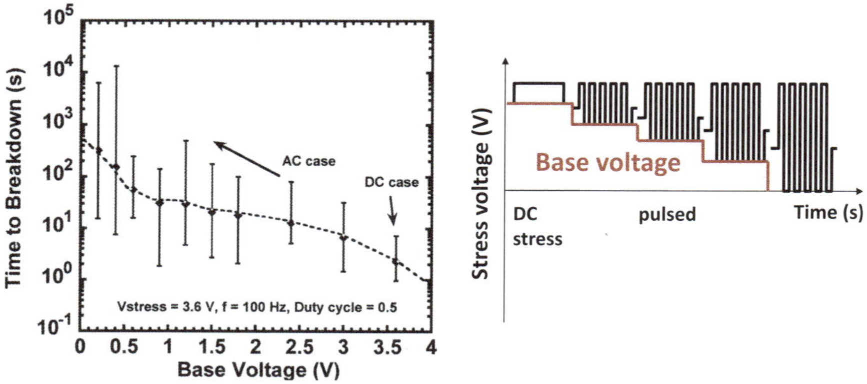

Fig. 22. (Left) Time to breakdown versus base voltage as applied in AC stress condition. The lifetime of the capacitors are improving with lower base voltage. (Right) Schematic overview of pulsing scheme (© 2009 IEEE. Reprinted, with permission, from Ref. 66).

stress voltage of 3.6 V and the base voltage (Figure 22 (right)). The base voltage was lowered in several steps. At each step, a certain number of devices were allowed to breakdown and the average lifetime is extracted (Figure 22 (left)). Overall, for lower base bias conditions an improvement by two to three orders of magnitude in time-to-breakdown is observed. This is very likely due to a reduction of the charge build-up in the dielectric. Furthermore, the frequency dependence of the time-to-breakdown was studied[67] and an improvement for higher test frequencies was found (see Figure 23). The breakdown time increased by two orders of magnitude since charge carriers in the dielectric are more and more unable to follow the higher frequencies and less damage can occur in the dielectric. Since refresh times are in the millisecond range and the clock speed of processors is in the GHz range, a considerable reliability and lifetime margin can be gained under DRAM AC operating bias conditions with this frequency range.

4. Scaling Challenges for DRAM Capacitors

The groundrule for DRAM has been reduced continuously down to 18 nm generations, which resulted in different benefits such as high speed, low power consumption and lower production cost. However, beyond 18 nm generation, there are many technological challenges for further shrinking of the technology. Innovative approaches are necessary to overcome the technological challenges. It is commonly discussed that the scaling limit of DRAM is determined by the capacitor scaling limit.[23] Therefore, the expectation of a scaling limit for the DRAM capacitor implies a scaling limit

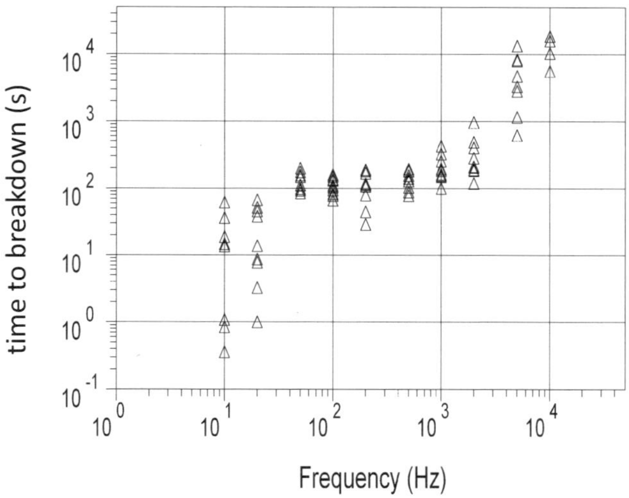

Fig. 23. (Left) Frequency dependence of time-to-breakdown (tbd). The measurements were performed at stress voltage of 3.8 V on ZrO_2 based capacitors.[67]

of future DRAM technologies. A minimum capacitance needs to be obtained to guarantee the functionality of every new DRAM generation. The aspect ratio of the capacitor has been increased and the equivalent oxide thickness has been decreased continuously.

The key question in the discussions on a DRAM capacitor scaling limit is the required minimum cell capacitance to operate DRAM. In fact, the question cannot be answered without discussing the device architecture (e.g. length of the wordlines) and the sensing circuits, which determine the ability to sense the charge of the cell capacitance. Improvements in sensing resulted in a reduction of the required capacitance value from 30 fF/cell in 2000 to 25 fF/cell in 2009 down to 20 fF/cell for the current 18 nm DRAM generations (see Figure 3). Therefore, it is hard to predict the scaling limit of DRAM capacitor without knowledge of the minimum cell capacitance requirement. In the near future, 10 fF/cell might be sufficient to operate DRAM device. If a minimum capacitance value of 10 fF/cell is feasible, further DRAM scaling might be easier. Therefore, a new device architecture or new processes which could improve the capacitor sensing ability can extend the shrinking limits in the roadmap for future DRAM generations.

The second key topic for the DRAM capacitor scaling limit is the maximum cell capacitance that can be manufactured in a future DRAM. The cell capacitance is determined by the CET value and the capacitor area (see chapter 1.3). Up to now, new material and patterning technology made it possible to reduce the CET value and increase the aspect ratio of 3D capacitor

(Figures 4 and 5). Innovative future technologies could reduce the CET value dramatically. For example, it was reported that a $SrTiO_3/Ba_xSr_{1-x}TiO_3$ super lattice can increase the dielectric constant by several orders of magnitude ($\varepsilon_r > 30,000$).[68] Even though the fabrication of a super lattice in a 3D structure is still a technical challenge, perovskite structured $SrRuO_3$ electrodes as a template for super lattice growth of $SrTiO_3/Ba_xSr_{1-x}TiO_3$ could be a possible solution.[69] Considering the worldwide efforts and the high speed of technology evolution, it is also hard to predict the limit of CET scaling and the highest possible aspect ratio of future DRAM structures.

The third and more obvious key question concerning the DRAM capacitor scaling limit is "What is the minimum thickness of DRAM capacitor?" Each 3D cell capacitor in a cup structure is composed of two bottom electrodes, four layers of dielectrics and four layers of top electrodes (10 layers, see Figure 16). The maximum available space for this film stack decreases for smaller groundrule. Each electrode and dielectric layer requires a certain thickness for best performance in the capacitor structure (see Figure 15 for the trend of leakage current density for different dielectric thickness). If the overall thickness requirement for the DRAM capacitor is thicker than the available space determined by the groundrule, the physical thickness would be the limiting factor for future scaling. One possibility to overcome this issue would be the introduction of a pillar capacitor with five layers (one bottom electrode, two dielectric layers, two top electrode layers) instead of a cup structure with ten layers (see Figure 7). This would lead to a drastic increase of the aspect ratio of the capacitor to compensate for the reduced capacitor footprint. But even for the pillar structure a limitation in the available space is critical. The limit will only be shifted to smaller groundrules. Therefore, a thickness scaling of the dielectric layer and of the electrode is the key challenge to overcome DRAM capacitor scaling limit.

In these discussions, only optimal parameters are mentioned so far. In addition, variations in the lithography of the capacitor pattern layout and non-perfect taper angles of the capacitor structure caused during dry etching of the initial trench structure to form the capacitor (see Figure 16) hinder an optimal structure formation. Furthermore, the conformality of the dielectric and the electrode layer is slightly reduced due to the high aspect ratio of the capacitor structure. Hence, the allowed film thickness for each dielectric or electrode layer might be lower than maximum space allowed by the groundrule.

The final key question about the DRAM capacitor scaling limit is "What is the advantage of DRAM capacitor scaling?" It should be noticed that the main purpose of the device scaling is to reduce production cost. Device scaling has increased the number of chips on the same wafer, which has reduced the cost per chip. If the cost of the scaling is higher than the cost reduction, the scaling has no economic advantage anymore. In this case, the driving force for the scaling would diminish. Device capacitor scaling is accompanied by a high aspect ratio etch, a conformal deposition process, thickness scaling, and a new high-k/high work function material. All of them increase the production cost of the DRAM.[70] Therefore, new, cost effective technologies are necessary to extend the DRAM capacitor scaling limit. Otherwise, the first scaling limit for the DRAM capacitor will not be a physical limit, but a cost limit.

Acknowledgements

The authors gratefully acknowledge Steve Knebel for fruitful discussions and electrical characterization and Tony Schenk for carefully reviewing the chapter and checking the symbols and formulas.

References

1. Dennard, R. H. (1967), United States Patent 3387286, July 14.
2. NEC 256 Mb SDRAM (μPD45256441) Datasheet (1998).
3. Weis, R. *et al.* (2001), *IEEE IEDM conference proceedings.*
4. Kim, K. *et al.* (2005), *IEEE IEDM conference proceedings*, 333.
5. Waser, R., *Nanoelectronics and Information Technology*, 3rd edition (Wiley VCH), chapter 27.
6. Advances in Solid State Circuit Technologies, Chap. 15.
7. Amon, J., Kieslich, A., Heineck, L. *et al. IEEE IEDM conference proceedings* (2004).
8. Kotecki, D. (1997), *Integrated Ferroelectr*, **16**, 1.
9. Kim, Y. K. *et al.* (2000), *IEEE IEDM conference proceedings*, 369.
10. Park, I.-S. *et al.* (2000), Symposium on VLSI Technology Digest of Technical Papers 42.
11. Kim, Y. K. *et al.* (1998), Symposium on VLSI Technology Digest of Technical Papers 52.
12. Lee, J. H. *et al.* (2002), Symposium on VLSI Technology Digest of Technical Papers, 114.
13. Robertson, J. (2013), *Journal of Vacuum Science & Technology A*, **31**, 050821.
14. Kil, D. S. *et al.* (2006), Symposium on VLSI Technology Digest of Technical Papers.
15. Park, Y. K. *et al.* (2007), Symposium on VLSI Technology Digest of Technical Papers.

16. Gutsche, M. *et al.* (2001), *IEEE IEDM conference proceedings.*
17. Hong, S. J. *et al.* (2010), *IEEE IEDM conference proceedings.*
18. Yoon, K. R., *et al.* (2005), SSDM conference, Kobe, 188.
19. EE Times Asia, 09.Apr.2009, Unlock Micron's 50 nm DRAM Technology.
20. Chang, K. B. *et al.* (2013), Simulation of Semiconductor Processes and Devices (SISPAD).
21. Pawlak, M. A. *et al.* (2011), *Appl. Phys. Lett.* **98**, 182902 (2011) or Pawlak, M. A. *et al.* Symposium on VLSI Technology Digest of Technical Papers 168.
22. Han, J. H. *et al.* (2011), *Applied Physics Letters* **99**, 022901.
23. Park, S. K. Future Memory Technology Workshop, Sematech, Seoul 2011.
24. Hudec, B. *et al.* (2013), ECS Transaction **50**, 13, 79.
25. Fillot, F. *et al.* (2005), Microelectronic Engineering, Volume 82, Issues 3–4, December.
26. Lima, L. P. B. *et al.* (2012), Microelectronic Engineering, 92.
27. Park, K. J. *et al.* (2005), *Appl. Phys. Lett.* **86**, 051903.
28. Speight, James. Lange's Handbook of Chemistry, 16th ed.; McGraw-Hill Professional: Boston, MA, 2004; p 1:132.
29. CRC Handbook, Handbook of Chemistry & Physics 95th edition (2014).
30. Tung, R. T. (2014), *Appl. Phys. Rev.* **1**, 011304.
31. Schmid, P. E. (1985), *Helv. Phys. Acta* **58**, 371.
32. Inglesfield, J. E. (1982), *Rep. Prog. Phys.* **45**, 223–284.
33. Cowley, W. and Sze, S. M. (1965), *J. Appl. Phys.* **36**, 3212.
34. Ru, H. *et al.* (1997), *J. Chem. Phys.* **107**(20), 22 November.
35. Park, C. S., *et al.* (2010), *Electrochem. Solid-State Lett.* **13**, 4.
36. Robertson, J. (2000), *J. Vac. Sci. Technol. B* **18**, 3.
37. Tapajna, M., *et al.* (2004), The Fifth International Conference on Advanced Semiconductor Devices and Microsystems.
38. Yokota, K. *et al.* (2004), *J. Phys. D: Appl. Phys.* **37**, 1095.
39. Mise, N. *et al.* (2009), *IEEE IEDM conference proceedings.*
40. Braun, D. *et al.* (2003), Polymer Devices 2622.
41. Knebel, S. *et al.* (2014), *IEEE Transactions on Device and Materials Reliability*, **14**, 1.
42. Zhou, D. *et al.* (2010), *Journal of Applied Physics* **108**, 124104.
43. Jegert, G. *et al.* (2011), *Applied Physics Letters* **96**, 062113; or *J. Appl. Phys.* **109**, 014504.
44. Wen, H. C. *et al.* (2006), *IEEE Electr. Dev. Lett.*, **27**, 7.
45. Afanas'ev, V. V. *et al.* (2007), *Journal of Applied Physics*, **102**, 081301.
46. Kaczer, B. *et al.* (2013), *Journal of Vacuum Science & Technology B*, **31**, 01A105.
47. Popovici, M. *et al.* (2014), *Appl. Phys. Lett.*, **104**, 082908.
48. Kim, S. K. *et al.* (2008), *Adv. Mater.*, **20**, 1429–1435.
49. Kim, S. K. (2011), *Physica Status Solidi (RRL)*, **5**, 8, 262–264.
50. Berthelot, A. *et al.* (2006), *ESSDERC conference proceedings* 343.
51. ITRS Roadmap, http://www.itrs.net/
52. Martin, D. *et al.* (2011), *J. Vac. Sci. Technol. B*, **29**, 1.
53. Knebel, S. (2015), *J. Appl. Phys.* **117**, 224102 and Pesics, M. (2016), *J. Appl. Phys.* **119**, 06401.

54. Besuker, G. *et al.* (2008), *IEEE IEDM conference proceedings*, 791.
55. Londergan, A. *et al. ECS Transaction, Atomic layer deposition application*, 4.
56. Weinreich, W. *et al.* (2009), *J. Vac. Sci. Techn. A* **31**, 01A123 (2013); or INFOS Conference, Cambridge.
57. Gordon, R. G. *et al.* (2003), *Chemical Vapor Deposition*, **9**, 2.
58. Arabindra Das, EE Times, 25. Mar. 2014
59. Kim, J. Y. *et al.* (2007), *J. Appl. Phys.*, **101**, 073502.
60. Elliott, S. D. (2012), *Semicond. Sci. Technol.* **27**, 074008.
61. Tanaka, T. (1954), Bulletin of the Institute for Chemical.
62. Hwang, C. S. *et al.* (2005), *J. Electr. Soc.* **152**, 6.
63. Jonscher, A. K. (1996), Universal Relaxation Law, Chelsea Dielectrics Press Ltd, London.
64. Reisinger, (2001), *IEEE IEDM conference proceedings.*
65. Knebel, S. (2009), Diploma thesis, University of Applied Science, Zwickau and Knebel, S. *et al.* (2014), *Trans. Dev. Mat. Ref.* **14**(1) 154.
66. Agaiby, R. *et al.* (2009), *IEEE Electr. Dev. Lett.*, **30**, 4.
67. Knebel, S. *et al.* submitted to IEEE TDMR (unpublished).
68. Tsurumi, T. *et al.* (2004), *Science and Technology of Advanced Materials*, **5**, 425–429.
69. Han, J. H. (2012), *Chem. Mater.* **24**, 4686–4692.
70. Yoon, J. H. (2013), Flash memory summit.
71. Schloesser, T. *et al.* (2008), *IEEE IEDM Conference Procedings.*
72. Pesic, M. *et al.* (2016), *Sol. State Electr.* **115** (133).
73. Popovici, M. *et al.* (2015), *Microelectr. Eng.* **147**(108).

Chapter 9

Ferroelectric Oxides on Silicon

Alexander A. Demkov and Agham-Bayan Posadas
Department of Physics,
The University of Texas at Austin, USA

In this chapter, we discuss efforts at integrating thin films of ferroelectric materials on silicon and other semiconductors for the purpose of memory and logic device applications. We discuss the phenomenology and microscopic origin of ferroelectricity with its signature property of a non-volatile switchable macroscopic dipole moment. We then describe the most common device architectures for implementing non-volatile memory on a semiconductor platform using ferroelectric materials and their benefits and shortcomings. Finally, we briefly go over the materials properties and device implementation of ferroelectric materials that have had some degree of success in integration with semiconductors. This review is not meant to be exhaustive but to just give the reader an idea of what has already been done and what remains to be done in the field of ferroelectrics on semiconductors.

1. Ferroelectricity

1.1. *Phenomenology*

A material is said to be ferroelectric if it possesses a non-zero intrinsic electric polarization P even in the absence of an external electric field E (called the remnant polarization P_r), and that the direction of this polarization can be reversed by applying a sufficiently large external electric field. This minimum electric field that must be applied in order to switch the polarization is known as the coercive field E_c and must be smaller than the breakdown field of the material. In this way, ferroelectric materials usually exhibit hysteresis when the polarization is plotted as a function of applied electric field (Figure 1), in analogy with the magnetization hysteresis exhibited by ferromagnets. In fact, it is this analogous behavior with ferromagnetism where the term **ferro**electricity comes from and not from being associated with iron. Another important characteristic of most ferroelectric materials is the existence of a phase transition temperature or Curie temperature T_C, above which the material is paraelectric (although in a few cases, the material melts before reaching T_C). The phase transition can either be an order-disorder transition or a displacive structural transition.[1] Other more exotic mechanisms of ferroelectricity have been found[2–7] but the two main ones are dominant in ferroelectric materials used in device applications. In order-disorder transitions, the lattice sites change from random occupation to a specific super structure. In displacive transitions, slight distortions in the atomic positions occur that can usually be assigned to one or a few phonon modes that soften. The phase transition usually involves a crystal space group change from a centrosymmetric one to a non-centrosymmetric (and polar) one as the material is cooled through the Curie temperature. The polar nature of the ferroelectric crystals necessarily implies that all ferroelectrics are also both pyroelectric and piezoelectric.

Historically, ferroelectricity has been known as Seignette-electricity or Rochelle-electricity, following the discovery by the American physicist Joseph Valasek, of the anomalous dielectric properties of Rochelle salt.[8] Rochelle salt, or potassium sodium tartrate ($KNaC_4H_4O_6{\cdot}4H_2O$), was first prepared in 1655 by Pierre Seignette, a pharmacist in the French city of La Rochelle.[9] It was one of the first materials discovered to exhibit piezoelectricity. Debye was the first to use the analogy with magnetism

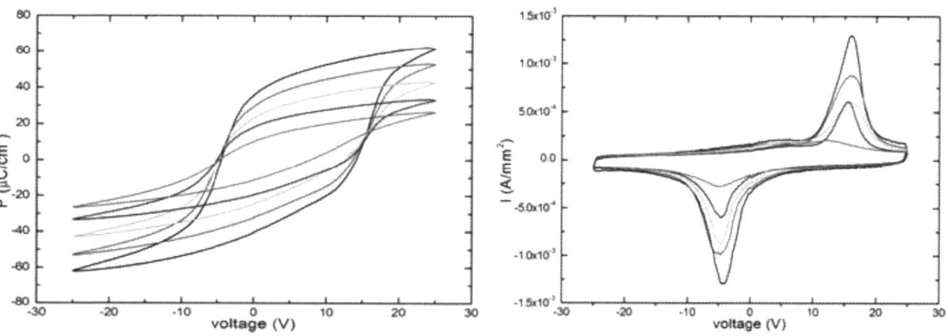

Fig. 1. Polarization vs. electric field for various ferroelectric thin film superlattices (left). The corresponding current response during the voltage sweeps (right). The current response peaks during the polarization switching. With kind permission from Springer Science+Business Media: Physics of Ferroelectrics: A Modern Perspective, Modern Physics of Ferroelectrics: Essential Background, vol. 105, 2007, p. 3, Rabe *et al.*), Figure 1.

to describe piezoelectricity, and he gave the temperature at which the dielectric constant diverges Curie's name. But it was Schrödinger who coined the term *"ferroelektrisch"* or *ferroelectric*, as the state where permanent dielectric polarization exists without applying an external electric field. The analogy between ferromagnets and ferroelectrics was the subject of Valasek's doctoral work at the University of Minnesota.[8]

At the phenomenological level, ferroelectricity is most naturally described in the context of the phase transition theory proposed by Landau.[10] The central concept of this theory is that the so-called order parameter η has a value of zero in the high-symmetry phase above the transition temperature and a finite value below (see Figure 2). The physical significance of the order parameter is the measure of order. The emerging order could be structural as in structural phase transitions, magnetic as in a ferromagnet, or electric as in a ferroelectric. Mathematically, the order parameter could be a number, a vector, or even a tensor. For example, in the case of ferroelectricity, the polarization vector $\vec{P}$ or one of its components is a natural choice. Landau postulated that the Gibbs free energy depends on the order parameter η just as it does on temperature and pressure and that at the equilibrium value of the order parameter, the free energy has a minimum:

$$\frac{\partial\Phi(P,T,\eta)}{\partial\eta} = 0$$

$$\frac{\partial^2\Phi(P,T,\eta)}{\partial\eta^2} > 0$$

In the high symmetry phase, the minimum is reached for $\eta = 0$ and in the low symmetry phase, for some finite value η. The fashion in which the order parameter reaches its finite value depends on the type of the phase

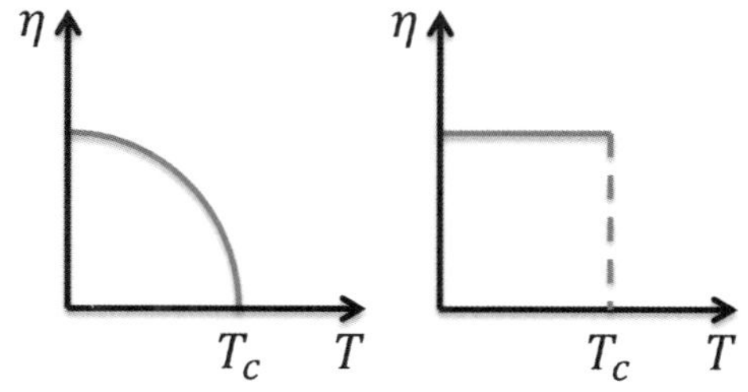

Fig. 2. Left: in a second order phase transition the order parameter emerges gradually below the transition temperature T_C. Right: In a first order transition, the order parameter changes abruptly from zero in a high-symmetry phase to a specific value in the low symmetry phase.

transition. For first order transitions the change is abrupt, and for second order transitions it is gradual (see Figure 2). If we limit our discussion to second order transitions, then in the low symmetry phase near the transition temperature the order parameter is small and the Gibbs free energy can be expanded in a Taylor series in the order parameter. For $\eta = 0$ to minimize free energy above the transition temperature (high symmetry phase), and for the system to remain isotropic, we have to assume that η and η^3 do not appear in the expansion:

$$\Phi(T, P, \eta) = \Phi_0(T, P) + A(T, P)\eta^2 + B(T, P)\eta^4 + \cdots$$

Ignoring the dependence of the free energy on pressure or strain, we assume that near the phase transition the second order term coefficient is proportional to $(T - T_c)$ and therefore changes sign at the transition temperature (see Figure 3), and that the fourth order term is a positive constant. Then we have:

$$\Phi(T, \eta) = \Phi_0(T) + \frac{1}{2}\alpha(T - T_c)\eta^2 + \frac{1}{4}\beta\eta^4 + \cdots$$

Above the transition temperature, the order parameter η is zero; and below T_c using the equilibrium condition (the minimum of free energy), it is

$$\eta_0^2 = -\frac{\alpha(T - T_c)}{\beta}$$

Therefore, the equilibrium free energy above and below the transition temperature T_c is

$$\Phi(T) = \begin{cases} \Phi_0(T) & \text{for } T > T_c \\ \Phi_0 - \dfrac{[\alpha(T - T_c)]^2}{4\beta} & \text{for } T < T_c \end{cases}$$

In the case of a ferroelectric it is important to include the free energy dependence on the electric field. Assuming that the order parameter describes a shift of a group of atoms and ρ is the effective density of the

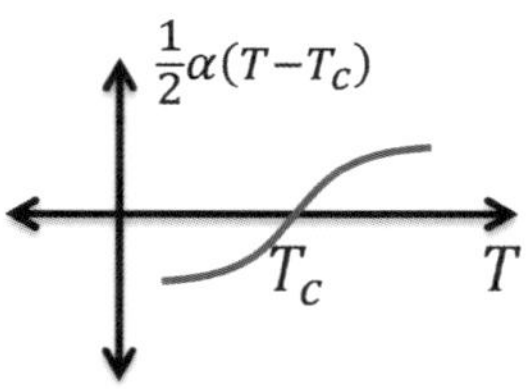

Fig. 3. Behavior of the coefficient of the quadratic term near the transition temperature T_c.

ionic charge, $E_j dP_j$ amounts to $-\rho\eta E$, where we have dropped the Cartesian index for simplicity. Thus the free energy expansion assumes the following form (Figure 4):

$$\Phi(T, \eta, E) = \Phi_0 + \frac{1}{2}\alpha(T - T_c)\eta^2 + \frac{1}{4}\beta\eta^4 - \rho\eta E$$

The system polarization and susceptibility can now be found as

$$P = -\frac{\partial \Phi}{\partial E} = \rho\eta_0$$

and

$$\chi_{jj}^{p,T} = \frac{\partial P}{\partial E} = \rho\frac{\partial \eta_0}{\partial E}$$

Thus in the absence of an electric field, below the transition temperature, the system has non-zero spontaneous polarization:

$$P_s = \pm\rho\left(\frac{\alpha\,(T - T_c)}{\beta}\right)^{1/2}$$

The susceptibility $\chi = \frac{\partial P}{\partial E}$ can be determined by first minimizing the system free energy in the presence of an electric field:

$$\frac{\partial \Phi}{\partial \eta} = \alpha(T - T_c)\eta_0 + \beta\eta_0^3 - \rho E = 0,$$

and then taking the derivative of that expression with respect to electric field:

$$\frac{\partial \eta_0}{\partial E} = \frac{\rho}{\alpha(T - T_c) + 3\beta\eta_0^2}$$

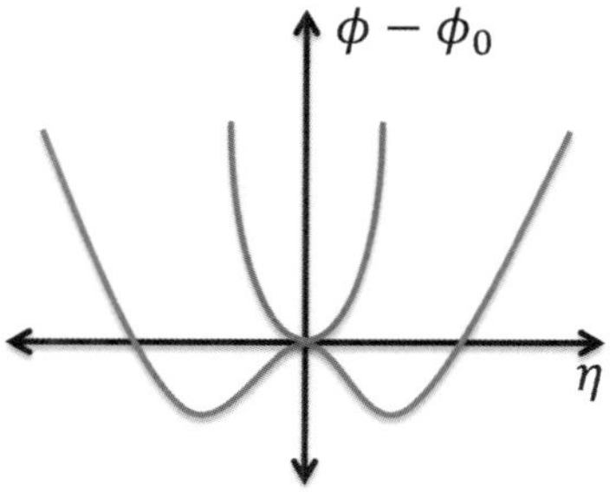

Fig. 4. Dependence of the free energy on the order parameter above and below the transition temperature T_C.

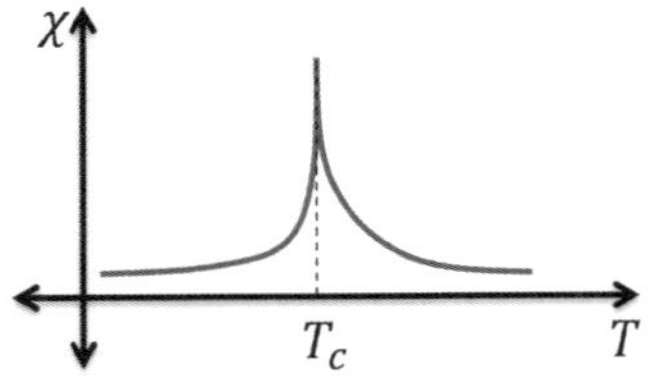

Fig. 5. Susceptibility as a function of temperature for a ferroelectric.

Then the dielectric constant is

$$\varepsilon_{jj}^{P,T} \approx 4\pi\chi_{jj}^{P,T} = \begin{cases} \dfrac{4\pi\rho^2}{\alpha(T - T_c)} & \text{for } T > T_c \\ \dfrac{-2\pi\rho^2}{\alpha(T - T_c)} & T < T_c \end{cases}$$

which is essentially the Curie-Weiss law with the Curie-Weiss constant $C_{C-W} = 4\pi\rho^2/\alpha$ (see Figure 5).

1.2. *Effective microscopic Hamiltonian*

Though the theory presented above provides significant insight, it is macroscopic and not materials-specific as it is essentially based on symmetry and constraints it imposes on the free energy. To describe specific materials, a microscopic theory needs to be developed. It would be desirable to start with an *ab initio* atomistic picture and work all the way up to predicting the temperature of the ferroelectric transition. Unfortunately, this is a rather complicated problem. A workable path towards such a theory turned out to be the introduction of an effective or model Hamiltonian based on the relevant microscopic degrees of freedom, and then deriving the free energy expression from it. The parameters of such a model Hamiltonian can be computed from first principles.[11,12]

To illustrate this two-step approach, let's consider a simplified two-dimensional version of $BaTiO_3$ (see Figure 6).[a] For simplicity, assume that Ba and O atoms do not move at all and that Ti atoms can only move up or down (along the c-axis of the tetragonal cell). The role of Ba and O is to provide an effective local potential, which has to have the form of a double well for the non-centrosymmetric phase to be the ground state. Then locally, in the *i*-th unit cell we can write the energy of the Ti atom as a function of

[a]This is just an illustration. The ferroelectric transition in $BaTiO_3$ is of the order-disorder type rather than displacive, and Ti atoms move mostly along the [111] direction.

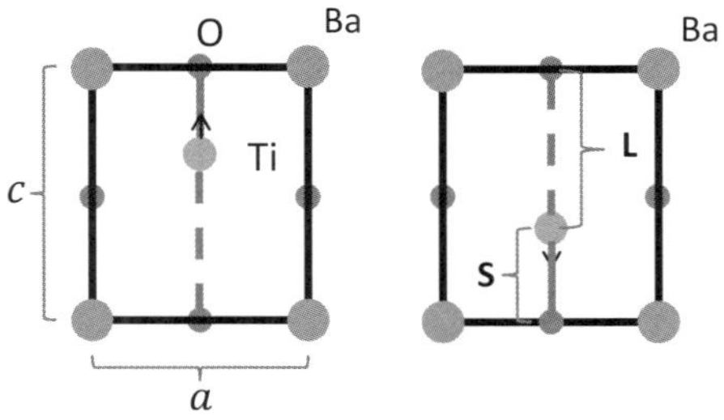

Fig. 6. A two-dimensional schematic of the crystal symmetry lowering due to displacement of Ti atoms indicated with arrows. Displacements up and down are equivalent and result in two distinct Ti-O bond lengths, one short (S) and one long (L).

its displacement from the center of symmetry u_i as

$$\frac{1}{2}au_i^2 + \frac{1}{4}bu_i^4$$

The neighboring Ti atoms can know about each other, thus a term proportional to $(u_i - u_j)^2$ needs to be added. Summing over the entire crystal, the effective Hamiltonian is

$$H = \sum_i \left(\frac{1}{2}au_i^2 + \frac{1}{4}bu_i^4\right)^2 + \frac{1}{2}c\sum_{i,j}(u_i - u_j)^2$$

Physically, this is a set of Einstein oscillators that have quartic anharmonicity, and in addition are coupled to each other. This is just a simple model that we expect to have a phase transition from a centro-symmetric or ***paraelectric*** to non-centro-symmetric or ***ferroelectric*** state. The connection between this Hamiltonian and the Landau free energy can be established by using a mean-field approximation.[13] The derivation is beyond the scope of this chapter, but the main ingredients are as follows. It is convenient to switch to the reciprocal space description with $u_j = \sum_j u_k e^{ik\cdot(ja)}$ where the sum runs over all lattice sites (a is the lattice vector) and periodic boundary conditions are used. Then a long wavelength limit is assumed. The order parameter η is then taken to be the thermodynamic average $\langle u_l \rangle$ in the limit of $k = 0$. Salje and co-workers have shown using this approach that the Landau expression for the free energy is obtained.[13] The power and utility of this method is that it relates the coefficients in the Landau expression to material-specific constants used to generate the model Hamiltonian. At this point, it is useful to say a few words about why the effective potential should look the way it does in Figure 7.

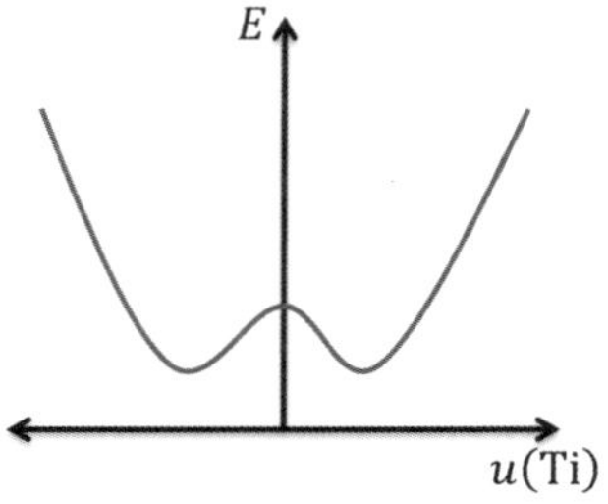

Fig. 7. The effective local potential energy due to the atomic displacement.

In $BaTiO_3$, the loss of the center of inversion and emergence of a local dipole can be traced to a peculiar covalency of the titanium-oxygen bond. First principles calculations indeed show that the lowest energy structure is obtained when the Ti and O sub-lattices are shifted. Consider a Ti^{4+} ($3d^0$) ion residing at the center of an octahedron formed by oxygen. In the simplest possible picture, the question of forming a dipole is reduced to shortening/strengthening one of the Ti-O bonds at the expense of stretching/weakening the other (see Figure 6). Assuming the total energy can be split into elastic and electronic contributions, we can readily write the cost of shifting a Ti atom off-center as $\frac{1}{2}ku^2$ where u is the displacement, and k the effective elastic constant. The change of the electronic energy can be estimated as follows. Taking the oxygen p level to be at zero energy, the energy gap between the titanium d and oxygen p levels to be Δ_{pd}, and the hopping integral between them to be t_{pd}, the energy of the bonding state is $\frac{-t_{pd}^2}{2\Delta}$ with respect to the oxygen p level (in this model the bonding state is fully occupied and the antibonding state is empty corresponding to $3d^0$ nature of $BaTiO_3$). Now consider the hopping integral change for the stretched bond $\tilde{t}_{pd} = t_{pd} - \lambda u$ and for the shortened bond $\tilde{t}_{pd} = t_{pd} + \lambda u$ (to first order in the displacement). Then, the overall energy change for the two bonds is a net gain of $\frac{-\lambda^2 u^2}{\Delta}$, and as long as it is larger than $\frac{1}{2}ku^2$ the distortion is favored. This is schematically shown in Figure 7.

1.3. *Experimental measurement of hysteresis*

The original method for measuring ferroelectric hysteresis is the so-called Sawyer-Tower circuit.[14] This involves having two capacitors in series: a ferroelectric capacitor and a regular capacitor with known capacitance. Because the charge on the plates of capacitors in series has to be the same, the potential across the known capacitor can be directly related to the charge across the ferroelectric and hence the polarization. This works reasonably well for macroscopic ferroelectrics or very thick (μm level) films that have virtually no leakage. For thin films, however, that are not totally insulating due to tunneling, Schottky injection, grain boundaries, etc., the Sawyer-Tower method makes it hard to distinguish between a leaky non-linear dielectric and a true ferroelectric. A better method is to simultaneously measure the current (displacement + resistive) through the ferroelectric as voltage is applied. A rapid rise and fall in the current should accompany polarization switching. This allows one to separate switching from non-switching current and integration of the switching current should yield the polarization change ($2P_r$). To more clearly remove artifacts, a method known as PUND — positive up, negative down — is utilized (Figure 8). This involves a series of voltage pulses (usually a setting pulse followed by a sequence of two positive and two negative pulses or vice versa) and the measurement of the transient current in the material resulting from each pulse. This method is now implemented in commercial ferroelectric testers such as those by aixACCT or Radiant Technologies.[15,16]

It should be pointed out that the presence of an electrical hysteresis does not necessarily mean the material is ferroelectric. First, there are

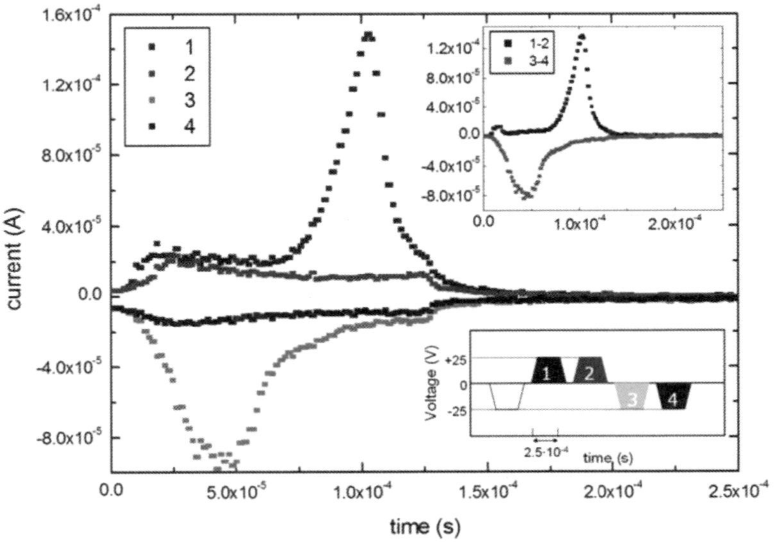

Fig. 8. Result of a PUND measurement on a ferroelectric sample. The lower inset shows the sequence of voltage pulses applied. The upper inset shows the difference between pulses 1 and 2 and pulses 3 and 4 showing only the switching current. The integral of the switching current is the polarization. With kind permission from Springer Science+Business Media: Physics of Ferroelectric: A Modern Perspective, Modern Physics of Ferroelectrics: Essential Background, vol. 105, 2007, p. 5, Rabe *et al.*), Figure 2.

certain materials and heterostructures that would exhibit such behavior, for example, electrets where mobile ions/charged defects produce the hysteretic behavior, and p-n junctions. Electrical hysteresis can also arise due to measurement artifacts, e.g. due to trapping/detrapping of charges injected from the electrodes, ionic electro-migration, or even simple leakage through a regular dielectric where rounded loops can appear in polarization vs. electric field plots. One way to distinguish charge trapping from true ferroelectric switching in metal-ferroelectric-semiconductor capacitors is the direction of the hysteresis (clockwise for p-type and counter-clockwise for n-type), which is often (but not always) reversed in the case of charge trapping. The exact shape of the hysteresis curve is determined by a combination of the domain structure and the intrinsic ferroelectricity. For device applications, a square-like, symmetric hysteresis is best as that means the switching is rapid and the two polarization states are very distinct. This is what one would get for a single domain, single crystalline, non-leaky ferroelectric material.

2. Ferroelectric Memory Devices

In this section, we briefly describe the main types of electronic devices that can be fabricated when integrating ferroelectric materials onto a semiconductor platform. Because of the non-volatile polarization of a ferroelectric, non-volatile logic and memory devices are the most obvious device applications. Figure 9 shows the three main types of ferroelectric memory devices that have been developed and fabricated on silicon. We discuss each type of device below. In addition, we discuss the concept of negative capacitance that ferroelectrics can exhibit and how this may be used to significantly improve the switching speed of MOSFETs. A more detailed treatment of ferroelectric memories can be found in the texts by Ishiwara and Okuyama[17] and by Scott.[18]

2.1. *Ferroelectric random access memories (FeRAM)*

The idea of incorporating ferroelectric thin films into semiconductors for the purpose of non-volatile memory applications dates back from the late 1960s to early 1970s when both integrated circuits and technological advances in thin film deposition were still in their infancy.[19–21] However, materials processing techniques for ferroelectric materials did not really become viable until the mid- to late 1980s[22,23] when complex oxide thin film deposition techniques became sufficiently mature, spurred on by

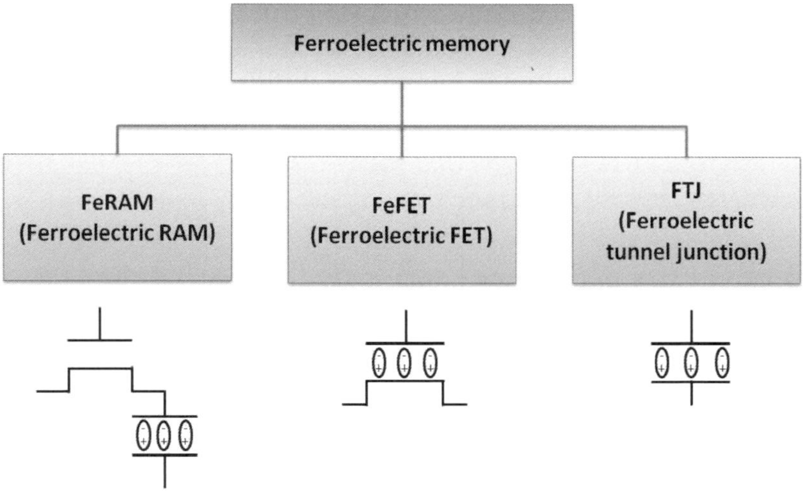

Fig. 9. Ferroelectric memory devices can be classified into three types: ferroelectric RAM (FeRAM), ferroelectric FET (FeFET), and ferroelectric tunnel junction (FTJ). FeRAM devices consist of a ferroelectric capacitor grown on Si that is electrically coupled to and controlled by a normal FET. FeFET devices have the same architecture as a standard FET but with the gate dielectric replaced by a ferroelectric material. FTJs utilize the ferroelectric as a non-volatile resistive switch in a two-terminal configuration.

the discovery of high-Tc superconducting oxides. The earliest claim of a working ferroelectric memory integrated with silicon was reported in 1988 by Eaton *et al.*[24] The device consisted of two ferroelectric capacitors based on lead zirconate titanate (PZT) coupled to a 6-transistor static random access memory (SRAM) cell. In this so-called ferroelectric random access memory (FeRAM or FRAM) design, the bit state (whether "1" or "0") is stored in the state (up or down) of the remanent polarization in the ferroelectric capacitor. Readout of the bit state is accomplished by measuring the displacement current when the bit is switched one way, which also makes the readout destructive necessitating the need to re-write the bit after each read operation. Even so, FeRAM was technologically attractive because it had fast read/write cycle times (40 ns), very low power consumption during read/write, and high endurance ($>10^{15}$ cycles), while at the same time having non-volatile (>10 years) data retention.[25] Performance-wise, FeRAM combines the best of DRAM and Flash memory.

The main disadvantage of FeRAM comes from the need to have a sufficiently large capacitor in order for the total switched charge to be measurable and also from the need to have an extra control transistor. This results in a large feature size as well as a large cell size, which means a low memory density, limiting their current applications. FeRAM also suffers from a slightly reduced endurance compared to DRAM due to the need for destructive readout (i.e. device has to be switched every time it is read) although they still have write cycles more than 10^9 times that of Flash. From the original 6T-2C design by Ramtron in 1987, improved circuit design led to a smaller 2T-2C 4 kb FeRAM cell that performed sufficiently well to be mass produced in 1992.[24,26] Since then, various research groups in both industry and academia worked on the problem and there has been rapid growth in FeRAM capacity, which is now being mass produced at the 64 to 128 Mb level using a 1T-1C architecture. According to the International Technology Roading for Semiconductors (ITRS), FeRAM feature sizes are currently at 130 nm with a cell area of $22F^2$. Using 3D architecture, this is expected to improve to 45 nm feature size with cell area of $13F^2$ by 2026. Improvements in read/write endurance came as a result of better materials quality as well as the development of alternative materials. PZT was the material of choice in the early days of FeRAM, but because it was initially used in the form of a ceramic thin film with random crystallographic orientation endurance was limited to at most 10^{12} cycles. This led to the study of other, alternative ferroelectric oxides for use in FeRAM including $SrBi_2Ta_2O_9$ (SBT)[27,28] and $Bi_4Ti_3O_{12}$ (BIT).[29,30] Highly (111)-textured PZT films can now also be readily made, further improving both the endurance and data retention of the devices.[31] While FeRAM is a mature and well-developed technology now and is even superior in most aspects (power consumption, endurance, retention time) to Flash memory, it is still way behind in terms of data density. Improvement in device density requires further reductions in the ferroelectric capacitor area and also reductions in the thickness of the ferroelectric layer to accommodate the lower core voltages being used by transistors, requiring better quality ferroelectric materials to be grown.[32] These device size reductions, however, should not affect retention time and cycling endurance, while still maintaining sufficient signal margin between the up and down polarization states. For further information regarding FeRAM, the reader is referred to the many excellent specialized texts and reviews that have already been published.[17,18,33–35]

2.2. *Ferroelectric field effect transistors (FeFET)*

FeRAMs utilize a ferroelectric capacitor (metal-ferroelectric-metal sandwich) fabricated in close proximity to and in electrical contact with a standard MOS transistor. The capacitor stores the data in a non-volatile fashion while the transistor is used for reading the memory state and switching the capacitor. In contrast, when one talks about the ferroelectric

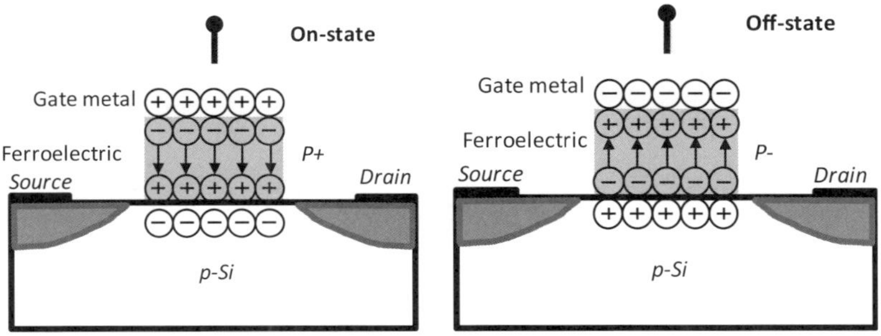

Fig. 10. Schematic of a (1T) non-volatile ferroelectric field effect transistor and its ideal mode of operation, illustrated on p-type Si. When the ferroelectric polarization is pointing down (+P), holes are driven away and an inversion layer is created resulting in a large source-drain current (on state). When the polarization is pointing up (−P), holes are accumulated and there is no source-drain current (off state). Because the two polarization states are ideally non-volatile, the transistor state is also non-volatile.

field effect transistor or FeFET, this involves using a ferroelectric material as the gate dielectric in a standard MOS transistor configuration (Figure 10). Hence, the ideal structure is that of a metal gate electrode on top of a ferroelectric material, which is grown directly on the transistor channel material (usually Si), resulting in what is often called a metal-ferroelectric-semiconductor (MFS) transistor. In contrast to the 1T-1C design of FeRAM, a FeFET is a 1T device, with the transistor performing the data storage, switching, and reading tasks of the entire memory device.[36] The operating principles of the FeFET are shown schematically in Figure 10. When the ferroelectric polarization is directed upward, for example, the Si channel is in weak accumulation resulting in a nearly zero source-drain current since electrons cannot be injected from the source into the channel (off state). On the other hand, when the ferroelectric polarization is directed downward, the Si channel is switched to strong inversion and electrons can now freely flow between source and drain (on state). Because the polarization is nonvolatile, i.e. there is a large remanent polarization even in the absence of an external electric field, the state of the transistor is retained. The main benefits of the FeFET over FeRAM architecture are the significantly faster switching times, non-destructive readout, scalability and the capability of performing both non-volatile logic and memory functions in the same device. The main problem with the FeFET is not in its design but in its technological implementation. To date, data retention characteristics of FeFETs reported in the literature have been very poor and no commercial device has been produced that can be considered non-volatile.[33,37,38] There are many reasons for this lack of data retention with the two major ones being leakage currents, especially for scaled devices that use thin ferroelectric layers, and depolarization due to the lack of sufficient screening on the semiconductor side of the gate stack.[39] Another stumbling block is that it has not been possible to deposit a single crystalline, single domain ferroelectric material directly on silicon without forming some kind of interfacial layer, which is usually detrimental to device performance. FeFETs are currently listed in the ITRS as part of "Emerging Research Devices". If the technological issues with FeFET fabrication can be overcome, FeFETs are expected to be able to achieve 22 nm feature size with cell areas of $4F^2$, much smaller than the projections for scaled down FeRAM. It is also expected to have read/write time of 2.5 ns and endurance of $>10^{16}$ cycles (both at least 10 times better than FeRAM technology) with the same property of non-volatility (>10 year retention time).[40] However, it is the retention time that is the current primary problem plaguing FeFET device technology,[41–43] with the best reported value being only about 3.5 months.[44] Table 1 summarizes the key device parameters for a FeFET including their projected and demonstrated values. Below we discuss the historical development of the FeFET and its future prospects based on new materials and new deposition processes.

The concept of using a ferroelectric material as a nonvolatile transistor was first patented in 1957, soon after the transistor itself was developed.[45] The first successful fabrication of a FeFET on silicon was reported by Wu in 1974.[46] The source and drain were formed using high temperature diffusion with an SiO_2 diffusion mask with a source-drain distance of 76 um. The SiO_2 was then etched away and the ferroelectric material $Bi_4Ti_3O_{12}$ was deposited using RF sputtering to a thickness of 3.6 um. A gold gate electrode was then

Table 1. Demonstrated and projected key device parameters for a ferroelectric FET according to the International Technology Roadmap for Semiconductors, 2012 edition.[25]

	Value	
Device parameter	Demonstrated	Best projected [38, 40]
Feature size	28 nm [86]	22 nm
Cell area	$4F^2$ [87]	$4F^2$
Write/Erase time	10 ns [88]	2.5 ns
Retention time	3.5 mo [44]	10 yr
Endurance	10^{12} cycles [89]	10^{16} cycles
Write operating voltage	4 V [38]	<0.9 V

deposited on top and patterned using photolithography. Contact windows were then opened in the bismuth titanate layer to access the source and drain regions by etching in concentrated HCl. This first FeFET on silicon showed hysteresis in the capacitance vs. voltage measurement and the transistor on and off states were claimed to be stable for at least 2 hours. Because of the ferroelectric film thickness needed to obtain sufficient measurable out-of-plane polarization, the operating voltages were also quite large, on the order of 20 to 100 V. Since then, a tremendous amount of research has gone into trying to optimize the FeFET using various combinations of materials and device layouts.[33,36,41,47–49] However, to date, there are still no commercially viable FeFET devices and the field continues to attract a lot of interest.[41,50–73]

The difficulty in achieving a non-volatile FeFET vs. that of FeRAM is due to the requirement of intimate integration between a ferroelectric oxide and silicon. In the case of FeRAM, the ferroelectric capacitor is on the order of 0.1 um away from the MOSFET and the two essentially act as distinct but interacting devices. There is also no depolarization in FeRAM as the ferroelectric is usually sandwiched between two metal electrodes where complete screening is usually possible. In a FeFET, the ferroelectric oxide needs to be in direct or almost direct contact with the MOSFET channel such that the polarization of the ferroelectric is able to influence the carriers in the channel. The crucial device parameters such as threshold voltage, on/off ratio, and memory window are strongly affected by defects at the interface, such as interfacial reaction layers, impurities due to interdiffusion, dislocations, and dangling bonds. Hence, one must treat the ferroelectric-semiconductor interface itself as the device. The technological inability to make a clean ferroelectric-semiconductor interface is the main reason why FeFETs have not performed as expected with respect to retention time. Because of this inability to directly deposit a ferroelectric oxide on silicon without disrupting the interface, most studies done on FeFETs have utilized the incorporation of an insulating buffer layer between the ferroelectric and silicon — the so-called MFIS transistor or metal-ferroelectric-insulator-semiconductor transistor. In the 1990s, buffer layers of SiO_2, Si_3N_4 and CeO_2 were studied.[33,34] However, they were not a tremendous help in the problem of retention time. With the development of high-k gate dielectrics (HfO_2 and $HfAlO_x$), more recent work has focused on using them as the insulating buffer. The best observed retention times on the order of months were in a system where the gate stack consisted of Pt/SBT/high-k dielectric/Si.[41,44,74]

The origin of the depolarization field in FeFETs is the incomplete compensation of the screening charge on the semiconductor side of the device. This can result in the gradual switching of ferroelectric domains to the opposite polarity leading to 180° domains, eventually resulting in a net zero polarization.[73,75,76] Charge leakage is the other main cause of loss of retention either due to defects in the ferroelectric (grain and domain boundaries) or Schottky injection through the ferroelectric into the semiconductor.[39,77] Working FeFETs are primarily characterized by measuring the hysteresis in capacitance vs. voltage (C-V) characteristics and, more importantly, in the observation of threshold voltage switching hysteresis (Figure 11). Hysteretic C-V is usually insufficient to show non-volatile switching even after eliminating other sources of hysteresis such as uncompensated traps and ionic motion. Carefully measured clockwise/anticlockwise C-V hysteresis on p-/n-type semiconductors, eliminating other sources of hysteresis, can only confirm that the gate insulator is ferroelectric. A more definite way of showing ferroelectric modulation of the semiconductor is hysteresis in the source-drain current vs. gate voltage curve. The shift in the threshold voltage or in the C-V curve is known as the memory window, which must remain sufficiently large over long time periods (10 years is the typical metric used in the semiconductor industry to be considered non-volatile). The development of epitaxial $SrTiO_3$ (STO) growth on Si, Ge, and GaAs show much promise for enabling the nearly

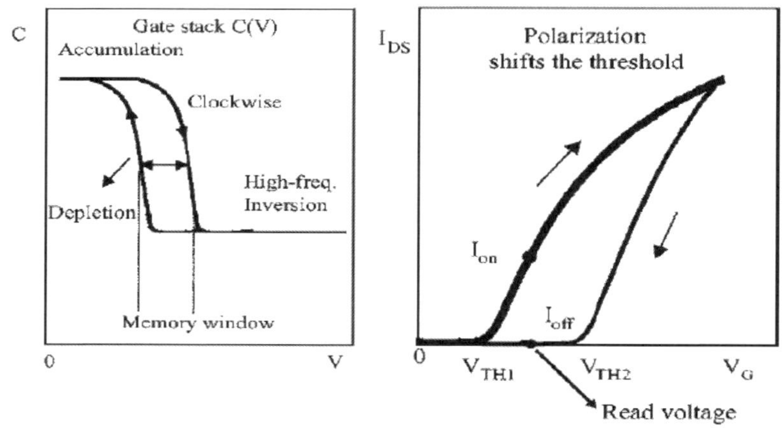

Fig. 11. Typical capacitance vs. gate voltage and source-drain current vs. gate voltage characteristics of a ferroelectric FET. The hysteretic behavior of the polarization also results in hysteresis in C-V and I_{DS}-V_g curves. The reading voltage is a value in between the threshold voltages of the on and off states. Reprinted with permission from Ref. 34. Copyright 2006, AIP Publishing LLC.

direct integration of ferroelectric oxides with semiconductors.[78–84] Because single crystalline, single domain ferroelectrics can now be potentially grown on these semiconductor substrates, depolarization due to domain back-switching will become less important (square hysteresis loops)[77] and FeFETs may yet have hope as a technology. Another new development is the discovery of ferroelectricity in doped HfO_2[85] that may also facilitate growing heterostructures where the ferroelectric is in intimate contact with the semiconductor. We discuss some of these recent developments in Sec. 3.

2.3. *Negative capacitance FETs*

Power dissipation in MOSFETs is one of the leading technological bottlenecks in further improvements in scaling and performance of integrated circuits. One obvious way to reduce power consumption is to reduce the supply voltages used for transistor operation. However, simply reducing the source-drain voltage would result in a large decrease in the on/off ratio. One way to overcome this is to be able to increase the subthreshold slope. The maximum subthreshold slope is dictated by a combination of the operating principles of a FET and the laws of thermodynamics of electron excitation ($= e/kT \ln 10$).[90] At room temperature it has a value of ~17 decades of current change per volt (more commonly stated as 60 mV/decade). There is a large body of work in trying to circumvent this limitation using new device architectures.[91]

A modification to the FeFET concept described in the previous section has been proposed as a possible way to overcome this thermodynamic limit.[92] Defining capacitance as the derivative of the charge transferred to the surfaces of the ferroelectric with respect to the voltage applied ($C = dQ/\mathrm{d}V$), in an unscreened poled ferroelectric, the amount of net surface charge can decrease in response to a voltage that would normally increase the surface charge. In other words, it is possible for the differential capacitance in a ferroelectric to be negative under certain conditions. It should be noted, however, that the condition in which this is true is thermodynamically unstable and some means of momentarily stabilizing this state is necessary. This can, theoretically, be achieved by combining a regular positive capacitance in series with the negative capacitance from the ferroelectric as shown in Figure 12. If the two capacitances are almost exactly compensated, the total gate-to-channel capacitance becomes very large resulting in what is effectively a large change in the channel charge

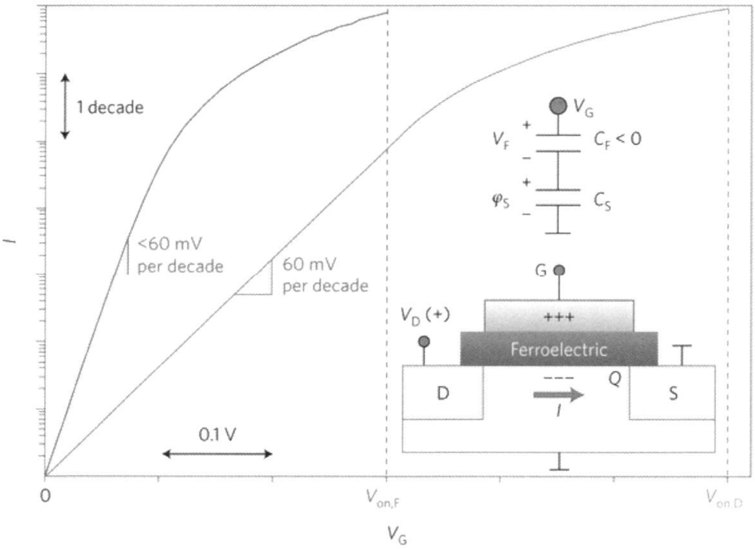

Fig. 12. Schematic of a negative capacitance FET. Because the transient capacitance of a ferroelectric can be made to be negative under certain conditions, by inserting a ferroelectric in the gate stack of a FET, the series combination of the ferroelectric capacitance and the semiconductor capacitance could potentially allow for exceeding the subthreshold slope limit, which is thermodynamically limited to be 60 mV/decade at room temperature, allowing for significantly faster switching times. Reprinted by permission from Macmillan Publishers Ltd: Nature Materials 14, 137, copyright 2015.

density for a small change in the gate voltage.[92] The negative capacitance therefore works as a kind of voltage amplifier.[93]

Negative capacitance has been measured as a transient state by Khan *et al.*[94] In their work, the put a ferroelectric capacitor consisting of epitaxial PZT on $SrRuO_3$ conductive oxide grown on STO substrate, with a top contact made of Au patterned into a 30 um square. To be able to measure the negative capacitance, a resistor was placed in series with the ferroelectric capacitor to provide lag time for the ferroelectric switching. As a result of this, the resulting PE hysteresis loops acquire some curvature while it is switching and certain regions have a negative slope (Figure 13).

One major problem with implementing the negative capacitance effect in a field effect transistor is that the series capacitance due to the semiconductor is not constant as it changes from depletion (off state) to inversion (on state), with the capacitance changing by as much as a factor of ten. This is a problem for the need to match the ferroelectric capacitance in order to achieve the voltage amplification effect. Either the effect is lost or hysteretic behavior occurs.[95,96] This can be partly alleviated by inserting an ultrathin

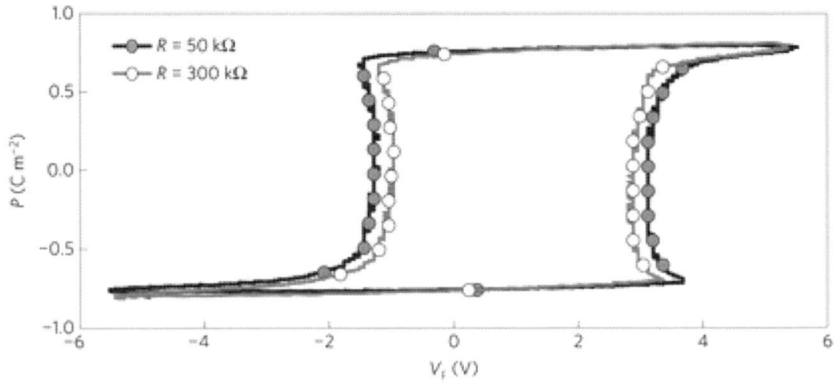

Fig. 13. Experimental P-E hysteresis loops in epitaxial PZT on conductive oxide bottom electrode with a series resistance of 50 or 300 kΩ. In both cases, the ferroelectric switching is delayed resulting in portions of the P-E loop having a negative slope limit. The differential capacitance in those regions of the curve is negative. Reprinted by permission from Macmillan Publishers Ltd: Nature Materials 14, 184, copyright 2015.

dielectric layer between the ferroelectric and the semiconductor. However, there is a second potential issue in that the switched charge of a ferroelectric is typically very large, about 2-20 uC/cm^2.[97] This amount of sheet charge is sufficient to break down many dielectrics so some means of partly screening the polarization charge, possibly using degenerately doped semiconducting oxides would be needed. A floating gate was proposed by Rusu *et al.*[98] to solve this capacitance mismatch, and also partly screens the large polarization switching charge. In their work, they were indeed able to show a subthreshold slope steeper than the thermodynamic limit although with hysteretic behavior. However, this is not suitable for highly scaled logic FETs as the gate can float to any potential depending on leakages and the history of the device, which would cause shifts in the switching potential.[97]

A novel way of reducing the capacitance mismatch without using a floating gate is the so-called quantum metal FeFET concept (QMFeFET) proposed by Frank *et al.*[97] Instead of a floating gate, a very thin (quasi-2D) conductive layer with a relatively low carrier density is inserted between the ferroelectric and the semiconductor, with the conductive layer in electrical contact with the source region (Figure 14). Basically, this "quantum metal layer" screens out most of the polarization charge of the ferroelectric. The conductive layer can still sustain a small field that the semiconductor can sense, allowing for more reasonable charge swings in the channel. Using an effective mass model, Frank *et al.* show that an idealized QMFeFET is capable theoretically of producing a subthreshold voltage slope of

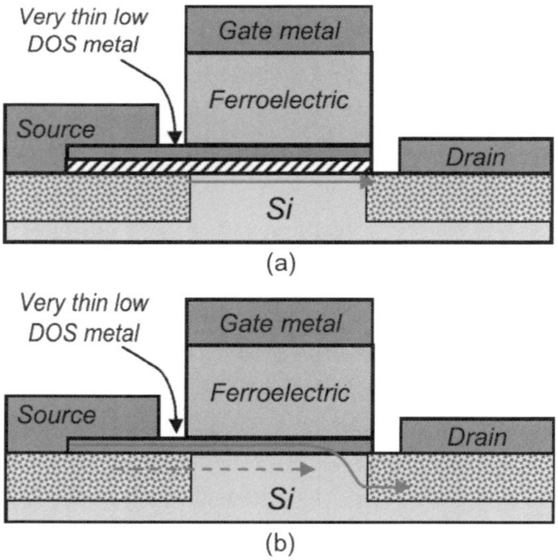

Fig. 14. Schematic diagram of the quantum metal ferroelectric FET (QMFeFET) proposed by Frank *et al.* [97]. The low carrier density 2D metallic layer eliminates the electrical mismatch between the ferroelectric and the semiconductor. Two variants (with and without a thin high-k dielectric layer) are shown. Figure taken from Ref. 97 with permission.

2 mV/decade.[97] It is suggested that a thin (2 nm) layer of n-doped $SrTiO_3$ (1×10^{21} cm^{-3} carrier density) may be a suitable candidate for the quantum metal layer.

2.4. *Ferroelectric resistive switches*

Using the ferroelectric material as a tunneling barrier between two conductive regions results in a device concept known as the ferroelectric tunnel junction.[33,99–109] The key mechanism of the device is that the direction of the ferroelectric polarization affects either the Schottky barrier height or the available density of states on either side of the barrier. These two will affect the tunneling probability and hence the tunneling current. Since tunneling is very sensitive to these parameters, large on/off ratios can in principle be achieved.[106] A FTJ requires a ferroelectric layer to be thin enough for significant electron tunneling, yet still remain ferroelectric with well-defined polarization states. In many cases, the ideal ferroelectric thickness is very close to the fundamental limits for the existence of ferroelectricity. FTJs have been demonstrated in all-perovskite oxide heterostructures, where very high quality ultrathin ferroelectric films down to 2 nm can be achieved.[110,111] Working ferroelectric tunnel junctions were reported in 2012 by Chanthbouala *et al.*[111] using 0.5 um diameter Au/Co top electrodes, 2-nm-thick $BaTiO_3$ BTO tunnel barrier, and (La, Sr) M_nO_3 (LSMO) conductive oxide bottom electrode, all grown on a $NdGaO_3$ substrate. Straining the BTO compressively helps maintain the ferroelectricity of the BTO layer down to such small thicknesses. Chanthboula *et al.* reported on/off ratios of over 100 with the mechanism being modulation of the tunnel barrier height. On/off ratios of more than 10^4 were demonstrated by Wen *et al.* in 2013 for 2.8 nm BTO tunnel barriers on Nb-doped STO substrate with 30 um diameter Pt top electrodes, showing modulation of the effective tunneling barrier width from the formation of a space charge layer in the Nb-STO (Figure 15).[110] Integration on Si, however, is not yet well-established. Li *et al.* demonstrated indirect integration on Si, utilizing the ability to grow epitaxial $SrTiO_3$ on Si to grow a stack consisting of an LSMO bottom electrode, 3.2 nm BTO ferroelectric tunnel barrier, and a conductive atomic force microscope (AFM) tip as the top electrode (Figure 16).[112] A more conventional Si-integrated FTJ device structure was recently reported by Guo *et al.* using a 5 um Pt top electrode. They show ferroresistive switching with an on/off ratio of 10^2.[113] A slightly different approach is to use Schottky injection instead of direct tunneling.[114–118] This allows for thicker ferroelectric layers to be used for resistive switching. Ngai *et al.* showed that an epitaxial BTO (40 nm)/ $Ba_1\text{-}_2Sr_2TiO_3$ (BST) (20 nm)/Ge structure with Pt top electrode has hysteretic

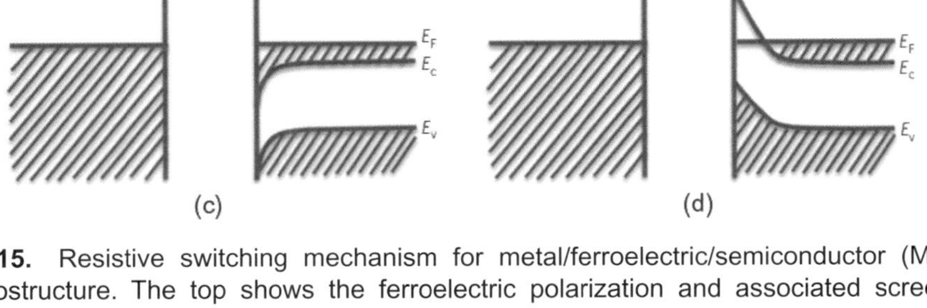

Fig. 15. Resistive switching mechanism for metal/ferroelectric/semiconductor (M/F/S) heterostructure. The top shows the ferroelectric polarization and associated screening charges. The bottom shows the corresponding energy band diagram. The left side shows the low resistance state while the right side shows the high resistance state as a result of the formation of a depletion layer in the semiconductor. Reprinted by permission from Macmillan Publishers Ltd: Nature Materials 12, 617, copyright 2013.

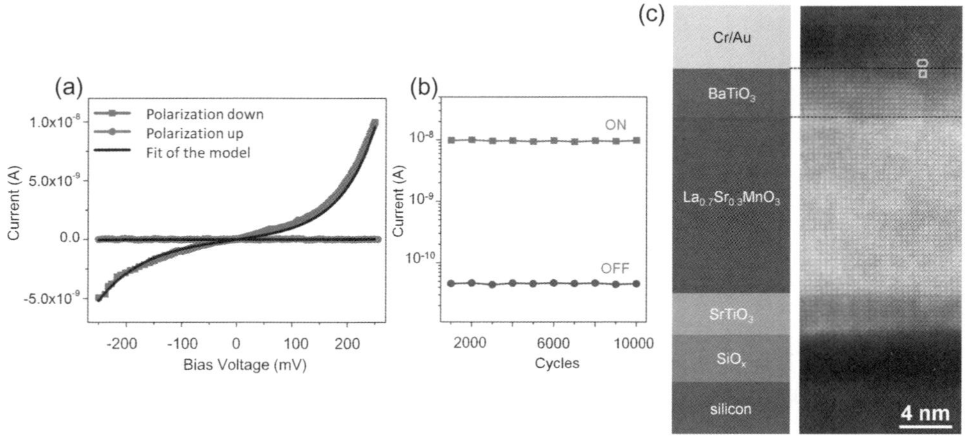

Fig. 16. (a) Current vs. voltage characteristics of silicon-integrated FTJ for the two polarization states of the BTO tunnel barrier. (b) Tunneling current (at 250 mV bias) for the on and off state as a function of cycling, (c) STEM image of the FTJ device grown on Si. Figures taken fron Ref. 112, copyright 2014 by John Wiley and Sons, reprinted with permission.

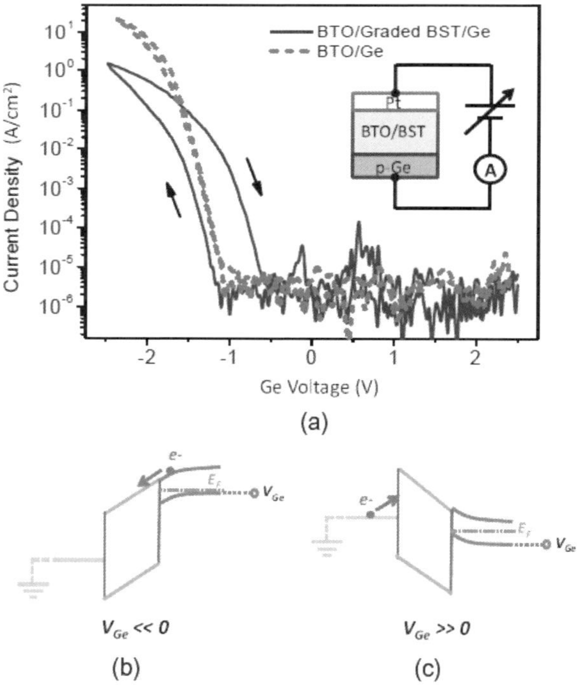

Fig. 17. (a) Current vs. voltage characteristics for Pt/BTO/BST/Ge diode structure. Threshold voltage modulaton of about 0.5 V is demonstrated. (b) Energy band diagram showing mechanism of hysteretic behavior. Reprinted with permission from Ref. 119, copyright 2014, AIP Publishing LLC.

I-V characteristics depending on the polarization direction of the BTO layer (Figure 17). Their device utilizes the large Schottky barrier between BTO and Pt to prevent electron flow in the off state, and the nearly zero conduction band offset between Ge and BTO to easily allow for electron flow in the on state.[119]

3. Ferroelectric Materials on Silicon and Other Semiconductors

In this section we discuss some of the historical and recent developments in the integration of ferroelectric materials on Si and other semiconductors, either as a textured polycrystalline film or an epitaxial thin film. We focus only on those materials that are technologically viable for implementation in nonvolatile memory applications. We also briefly discuss the ferroelectric polymer PVDF and its potential integration on Si, which may be useful for low cost applications.

3.1. $BaTiO_3$

$BaTiO_3$ (BTO) has the simplest crystal structure among the ferroelectric oxides. At room temperature, the crystal structure is a slight polar distortion from the cubic perovskite structure (Figure 18). It consists of a Ti atom surrounded octahedrally by oxygens with Ba ions in the voids in

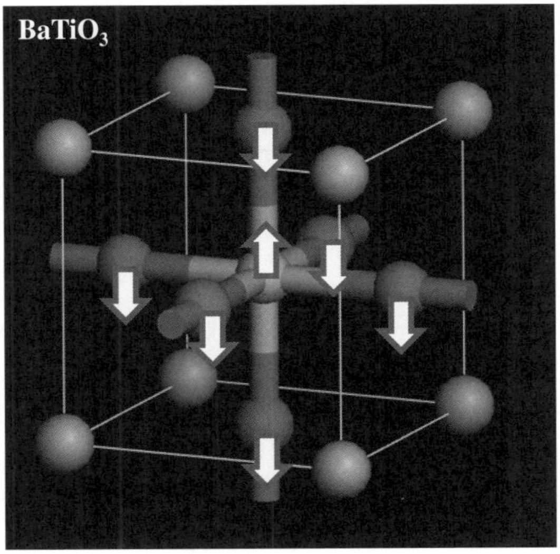

Fig. 18. Schematic of a unit cell of $BaTiO_3$. Ba atoms are represented by the balls at the corners, a Ti atom by a ball at the center and oxygen atoms by balls in the center of each face. Arrows represent the displacement of O and Ti sub-lattices in the ferroelectric state.

between octahedra. Because the Ba ions are large, chemical bonding and geometric constraints cause the Ti to shift slightly either up or down from its centrosymmetric position to a configuration with lower energy. This bistability of the Ti position produces the switchable permanent dipole moment characteristic of ferroelectricity. $BaTiO_3$ loses its ferroelectricity when it is heated above its Curie temperature (120°C) where it becomes truly cubic. At room temperature BTO has space group P4 mm with lattice parameters a = 3.992 Å and c = 4.036 Å with a remanent polarization of 27 uC/cm^2. BTO has still more structural phase transitions at lower temperatures leading to a ferroelectric orthorhombic phase below 5°C and a ferroelectric rhombohedral phase below −90°C. In the room temperature ferroelectric phase, the net polarization is along the 001 direction. For the lower temperature ferroelectric phases, the polarization points along the 110 and 111 directions relative to the original cubic paraelectric phase. In addition to the ferroelectric memory applications of BTO, it also exhibits a strong electro-optic effect as demonstrated by its very large effective Pockels coefficient (r_{eff} = 148 pm/V), making the material potentially useful for integrated Si nanophotonics.[120,121] Further details of the use of BTO as an electro-optic material are discussed in the chapter by Abel *et al.*

The recently developed ability to grow layers of transition metal oxides with atomic precision by means of physical vapor deposition has opened up a possibility of monolithic integration of these oxides on semiconductors. Monolithic integration of BTO on Si is enabled by molecular beam epitaxy (MBE) through the use of buffer layers, including BaO,[82] $Ba_{0.7}Sr_{0.3}TiO_3$,[84] and most commonly, $SrTiO_3$ (STO).[80,120,122] Here we review the recent progress in integrating ferroelectric films with Si, Ge, and GaAs and their potential applications in electronics and nanophotonics. A more complete review was recently published by Mazet *et al.*[123]

3.1.1. *BTO on Si*

The epitaxial integration of BTO on Si (001) presents several challenges, particularly for applications requiring that the ferroelectric polarization be pointing out of plane. The lattice mismatch between BTO and Si is 4% (and even larger at typical growth temperatures) with the BTO compressed in-plane. While the direction of the mismatch is favorable for producing *c*-axis oriented BTO, the relatively large mismatch usually results in a very high defect density that can degrade the electrical properties of the film. Further, there has not yet been any successful direct growth of BTO on Si as no suitable wetting interlayer that will enable direct epitaxy has been discovered. Another challenge is the large thermal expansion mismatch between BTO and Si, with BTO having a thermal expansion coefficient about three times larger than Si. While BTO films have been grown on Si using various buffers,[124–127] these typically produce *a*-axis oriented films as a result of the thermal expansion mismatch, which causes the BTO to experience tensile stress while cooling down through its Curie temperature. To resolve this problem, a combination of a suitable buffer that reduces both lattice and thermal expansion mismatch, and slow cool down to reduce stress is necessary to obtain *c*-axis oriented films on Si. Such an approach was first demonstrated by Vaithyanathan *et al.*[84]

One implementation of that approach with a thinner buffer layer is to use a fully relaxed layer of STO grown on Si as a pseudosubstrate for the subsequent BTO deposition.[80] The lattice mismatch between BTO and relaxed STO still induces compressive strain that stabilizes out of plane orientation of the BTO film with a significantly reduced lattice mismatch of 2.2%, as well as a reduced thermal expansion mismatch such that BTO now experiences favorable compressive stress during cool down (the BTO thermal expansion coefficient is slightly smaller than the STO thermal expansion coefficient). By using this approach of tailoring the heterostructure layers and their interfaces, lattice and thermal expansion mismatch issues can be overcome and epitaxial *c*-axis oriented BTO films on Si with thicknesses up to 20 nm can be grown.

After the growth of 6 nm of relaxed STO on Si by means of the Sr Zintl template layer (Figure 19(a)), the substrate is heated at 20°C/min to 700°C while simultaneously ramping the O_2 pressure to 5×10^{-6} Torr. When the target O_2 pressure and substrate temperature are reached, BTO growth by means of alternating monolayer dosing of Ba and Ti is initiated, with Ba

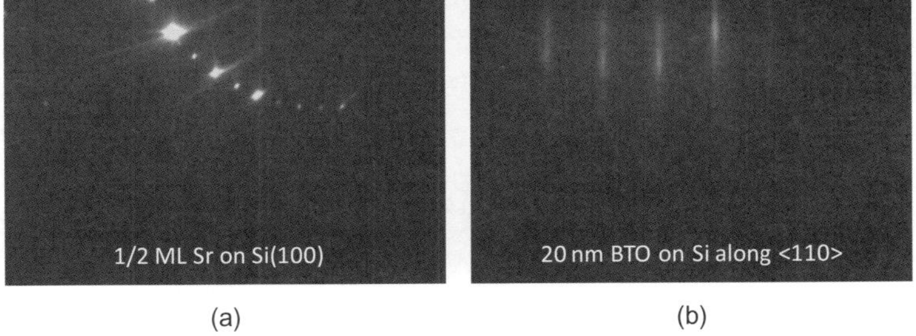

(a) (b)

Fig. 19. RHEED pattern for (a) Sr Zintl template on Si along Si ⟨110⟩ showing a sharp 2 × reconstruction pattern with an atomically flat surface; (b) 20 nm BTO grown on STO/Si along the ⟨110⟩ azimuth showing a streaky 1× pattern.

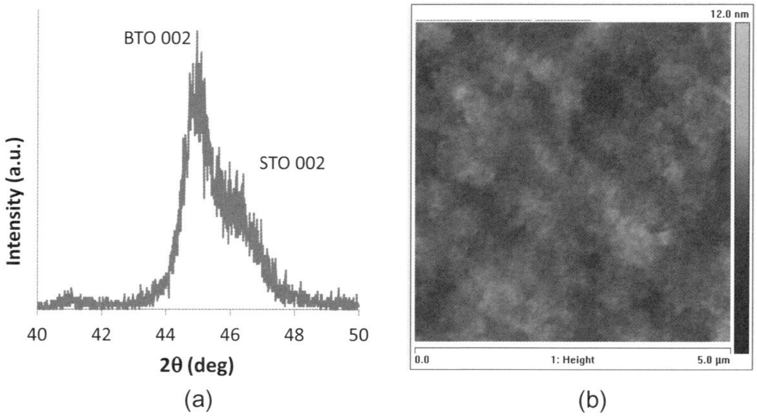

Fig. 20. (a) X-ray diffraction pattern of 10 nm BTO on 6 nm STO on Si showing c-axis orientation of BTO. (b) Atomic force microscopy (AFM) of stoichiometric BTO film grown on STO/Si over a $5 \times 5\,\mu m^2$ area with a root mean square roughness of $\sim$0.5 nm.

first and Ti last. The entire growth process is monitored *in-situ* by reflection high-energy electron diffraction (RHEED) to ensure high crystallinity and stoichiometry. Figure 19(b) shows a RHEED pattern after growth of a 20 nm BTO film on STO/Si. After the desired thickness of BTO is grown, the sample is cooled at 5°C/min in O_2 to room temperature. After growth, X-ray diffraction and atomic force microscopy measurements are performed on the BTO films. An X-ray diffraction $2\theta - \theta$ scan around the BTO (002) peak is shown in Figure 20(a), and an atomic force microscope image of a BTO film after growth is shown in Figure 20(b).

BTO has also been grown epitaxially on Si using atomic layer deposition (ALD). Ngo *et al.* have grown *c*-axis oriented epitaxial BTO at 225°C using ALD with barium bis(triisopropylcyclopentadienyl) on 1.6-nm STO-buffered Si(001).[128] The BTO films have a good degree of crystallinity and no amorphous layer was observed at the STO/Si interface. A 5-min vaccum anneal at 600°C significantly improved BTO crystallinity. Annealing using a relatively high temperature ramping rate (30°C/min) could destroy the tetragonal structure, while annealing using a lower ramping rate (5°C/min) is able to maintain the *c*-axis oriented tetragonal structure. The dielectric constant of the BTO/STO stack on Si was estimated to be $\sim$660 from capacitance-voltage measurements. We expect the ALD method of growing *c*-axis oriented epitaxial BTO on STO-buffered Si(001) at low temperature to be a potential way to fabricate the negative capacitance gate oxide structure on Si with no amorphous layer between STO and Si in the near future.

3.1.2. *BTO on Ge*

For BTO growth on Ge(001),[79] the Ge surface is first prepared using a combination of wet-etching, oxygen plasma cleaning and annealing, resulting in a (2 × 1) surface reconstruction, low surface roughness, and clear surface signal in the angle resolved photoemission spectrum.[129] Prior to BTO growth, 1/2 monolayer of barium is deposited to prevent germanium from oxidizing during the initial growth of the first 5 unit cells of STO at 200°C. After increasing the temperature to 600°C to crystallize the STO, BTO is deposited on the STO/Ge template at a temperature of 650°C.

Films were characterized by X-ray diffraction (XRD) on using a Philips X'Pert double-crystal diffractometer. Figure 21 shows symmetric $\theta - 2\theta$ scans of a 160-Å-thick BTO film grown on Ge, both with and without a 20-Å-thick STO buffer layer. The BTO peaks correspond to a single orientation with an out-of-plane spacing of 4.06 Å for the film with the STO buffer, and 3.98 Å for the film without the buffer. This result clearly demonstrates the *c*-axis orientation for the STO-buffered sample compared with the *a*-axis orientation for the BTO layer grown directly on Ge. Rocking curve scans around the BTO 002 Bragg peak typically show a full-width at half-maximum of 0.5–0.8°.

The ferroelectric properties of the BTO layer were tested using band excitation-piezoresponse spectroscopy measurements (BEPS).[130] We

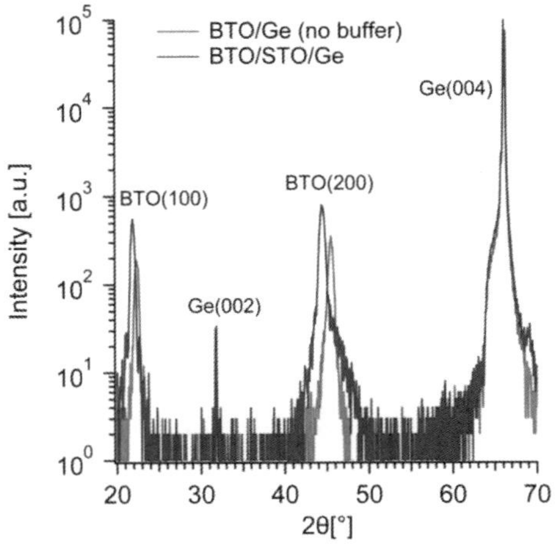

Fig. 21. X-ray diffraction $2\theta - \theta$ scan of 16 nm $BaTiO_3$ epitaxially grown on Ge(001) substrate, with and without a 2 nm $SrTiO_3$ buffer layer. Without the buffer, the d-spacing corresponds to the *a* axis of BTO directed out of plane. With the STO buffer, the *c* axis of BTO is directed out of plane. Figure taken from Ref. 79.

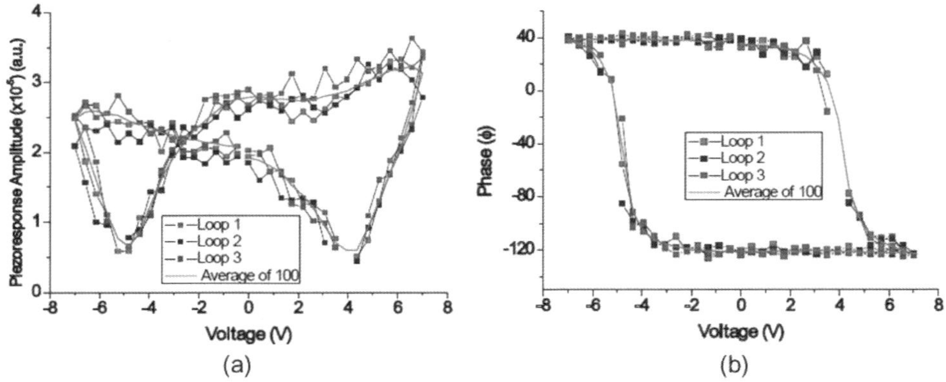

Fig. 22. Hysteresis loops of BTO/Ge captured using BEPS. Selected amplitude and phase loops are shown in (a) and (b), respectively, with the average loop shown as a solid line in olive, showing ferroelectric behavior. Figure taken from Ref. 79.

performed a BEPS measurement on the BTO film across a 10×10 grid to yield a data set of 100 measurements in total. Three selected responses at individual points from the 100 measurements, after fitting to a simple harmonic oscillator model, are shown in Figure 22(a, b). The average of 100 measurements is shown as a solid olive line. The amplitude curves clearly follow the expected butterfly-like hysteresis typical of ferroelectrics, while the phase loop appears to show a change somewhat $<180°$.

3.1.3. *BTO on GaAs*

Recently, Contreras-Guerrero *et al.* have demonstrated high crystallinity ferroelectric $BaTiO_3$ films grown on n-type GaAs by MBE utilizing a two unit cell $SrTiO_3$ buffer to ensure out-of-plane polarization.[83] Once the GaAs wafer was introduced into the oxide chamber, it was heated to a temperature of 300°C. 1/2 ML Ti was then deposited onto the c(4×4) As-stabilized GaAs surface. During the Ti deposition, the reconstruction switched from c(4×4) to (2×1). Two unit cells of STO was then deposited on the Ti/GaAs surface prior to the deposition of $BaTiO_3$. Growth of the BTO layer was initiated by opening the Ba, Ti, and molecular oxygen sources at a substrate temperature of 500°C and an oxygen pressure of 1×10^{-7} mbar. They observed a streaky RHEED pattern with low background intensity, indicative of a high quality commensurate crystalline layer throughout the BTO growth.

Ferroelectricity was investigated using piezoresponse force microscopy (PFM). To characterize the domain structure, vertical-PFM was used at room temperature in ambient. The usual PFM characterization of writing a square within a square of opposite polarization was utilized. The center square and outer ring were patterned with −3 V, while the inner ring was patterned with +3 V. The results of piezoresponse force spectroscopy confirm that the polarization is switchable. The piezoresponse amplitude is reported to be about 5 pm/V, with a coercive voltage of 1-2 V. The pattern was quite stable, and was easily re-imaged after one hour. Similar to the case of BTO on Ge, the as-grown ferroelectric appeared to be pre-poled.

3.2. *Pb(Zr,Ti)O*$_3$ *(PZT)*

PZT or lead zirconate titanate (discovered in 1949) was the first commercially viable ferroelectric material considered for making ferroelectric memory devices. The properties of PZT are tunable by virtue of the fact that it is a solid solution of two perovskite materials: $PbTiO_3$ and $PbZrO_3$. $PbTiO_3$ is a ferroelectric with a tetragonal crystal structure while $PbZrO_3$ is antiferroelectric with an orthorhombic crystal structure. The phase diagram of PZT is shown in Figure 23.[131] Around a composition of 47% Ti is the so-called morphotropic phase boundary, which is the boundary between a tetragonal structure with space group P4 mm and a rhombohedral structure with space group of either R3m and R3c, depending on temperature. More recent structural studies of PZT showed that the boundary is not a line but a narrow composition region where the structure of the material has monoclinic symmetry.[132,133] Most functional PZT films utilize compositions in the range of 48% $PbTiO_3$ to 100% $PbTiO_3$. In this composition range, PZT has the same crystal structure as $PbTiO_3$ with the polarization along the crystallographic [001] direction. The more $PbZrO_3$ is mixed into the solid solution, the stronger the piezoelectric effects become at the expense of the size of the ferroelectric polarization. Piezoelectricity reaches its maximum effect at the morphotropic phase boundary, which is where the crystal structure adopts the monoclinic structure. For applications emphasizing piezoelectricity, compositions at or close to the morphotropic phase boundary are used. For applications emphasizing ferroelectricity, Ti-rich compositions (80-100% $PbTiO_3$) are preferred. Tetragonal PZT has P4 mm symmetry in its ferroelectric phase with a = 3.899·Å and c = 4.153 Å (data for $PbTiO_3$). Figure 24 shows the variation in the room temperature lattice constants of PZT as a function of composition. It transforms to its cubic, paraelectric phase at temperatures ranging from 350 (50% Zr) to 490°C (pure $PbTiO_3$) depending on the Zr content. The remanent polarization of $PbTiO_3$ is relatively large with a value of 75 uC/cm^2.[134] The remanent polarization is directed either along [001] or along [111], depending on which side of the morphotropic phase boundary the material is in (Figure 25). Near

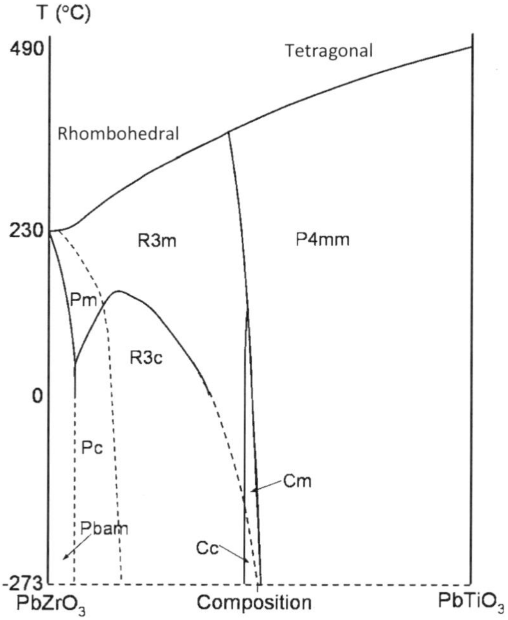

Fig. 23. Phase diagram of ferroelectric perovskite $Pb(Zr,Ti)O_3$. The Ti-rich side is tetragonally distorted with space group P4mm. The Zr-rich side is rhombohedrally distorted with spacegroup R3c (low temperature) or R3m (high temperature). For $PbZrO_3$, the structure is orthorhombic (Pbam) with antiferroelectric ordering. Small regions of monoclinic phases appear at the major phase boundaries, including the morphotropic phase boundary near the center of the composition axis. Reprinted with permission from Woodward *et al.*, Phys. Rev. B 72, 104110 (2005). Copyright 2005 by the American Physical Society.

the morphotropic phase boundary, PZT is highly sensitive to strain and is strongly piezoelectric with piezoelectric coefficient d_{33} reaching values as high as 200 pm/V. The partial subsitution of other cations such as La for Pb has been found to further increase the piezoelectric response.[31]

Because Pb and Si react readily to form lead silicides, some kind of buffer layer is needed in order to integrate PZT onto silicon. This reaction also results in extreme lead deficiency in the ferroelectric film often leading to the formation of secondary phases. For the past several decades, the fabrication of metal-ferroelectric-metal capacitor structures on a silicon substrate has been the main thrust of research into PZT thin film integration with Si. Such a structure is suitable for FeRAM-type (1T1C) memory devices. One of the more commonly used substrates is platinized Si, which consists of a stack of relatively thick films of Pt, TiO_2, and SiO_2 on single crystal Si. The Pt usually grows as a highly 111-oriented polycrystalline film with strong in-plane texture as well. The TiO_2 layer serves as both a diffusion barrier and adhesion layer. Depending on the growth conditions, PZT grown on platinized Si is either strongly (111) textured or (001) textured. A large number of groups too numerous to list in this short review is able to routinely deposit textured polycrystalline PZT on platinized Si substrates.[135–137] Because of the problem of fatigue in Pt/PZT/Pt capacitors as a result of oxygen electromigration, there has also been a lot of work in using conductive oxide bottom electrodes which has experimentally been shown to drastically reduce or even eliminate fatigue.[138,139] There are two main categories of conductive oxide bottom electrodes: rutile-type oxides (IrO_2 and RuO_2)[140] and perovskite-type oxides ((La,Sr)CoO_3, (La,Sr)MnO_3, $LaNiO_3$, $SrRuO_3$).[141]

For FeFET applications, one requires a much closer contact between PZT and Si. Fortunately, a way to epitaxially grow $SrTiO_3$ directly on Si was reported by McKee *et al.* in 1998.[142] Because single crystal STO is the most widely used substrate for the high quality epitaxial growth of PZT, the ability to grow high quality epitaxial layers of 100-oriented STO on Si opened up a route for the epitaxial integration of PZT on Si.[29,42,43,45] Growth of STO on Si involves the deposition of 0.5 monolayer of Sr on the clean Si surface. This allows for mild oxidation resistance allowing for STO to nucleate on Si. Such STO/Si templates are now routinely used by several groups to grow epitaxial PZT on Si. The STO layer prevents Si oxidation, Pb diffusion, and helps stabilize the (001) PZT out of plane orientation.

Because the crystal structure of PZT is a simple distorted perovskite (Figure 18), thin films of PZT can be grown to very high quality. With the advent of better physical vapor deposition methods such as PLD and sputtering, single crystalline, epitaxial PZT thin films with c-axis orientation could be readily grown on a variety of commercially available single crystal perovskite oxide substrates. This ability led to a large amount of research into the intrinsic properties of PZT and $PbTiO_3$ and their scaling down to very small sizes.[143–145] Efforts to integrate it epitaxially on Si[146–155] and other semiconductors[156–158] followed soon after. The ability to integrate epitaxial PZT onto Si, in particular with the development of the direct epitaxy of $SrTiO_3$ on Si, allows one to fabricate high performance ferroelectric capacitors on a silicon substrate. Such methods may even ultimately allow for direct epitaxy of PZT on Si for making FeFET-type devices.[39]

3.3. *$SrBi_2Ta_2O_9$ (SBT)*

In the 1980s, when thin film deposition technology was not yet at the level to routinely achieve epitaxial oxides on silicon, most ferroelectric films utilized for fabricating ferroelectric devices on silicon were polycrystalline.

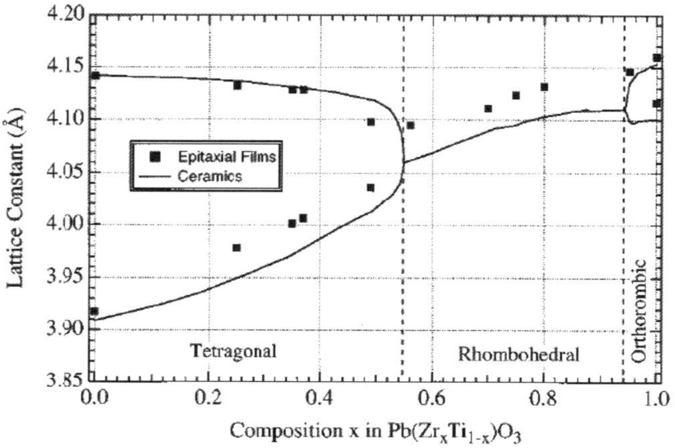

Fig. 24. Variation of the room temperature lattice parameters of PZT with composition for epitaxial MOCVD-grown films. The maximum tetragonal distortion occurs in $PbTiO_3$ with a c/a ratio of 1.065. Reprinted with permission from C.M. Forster *et al.*, J. Appl. Phys. 81, 2349 (1997). Copyright 1997, AIP Publishing LLC.

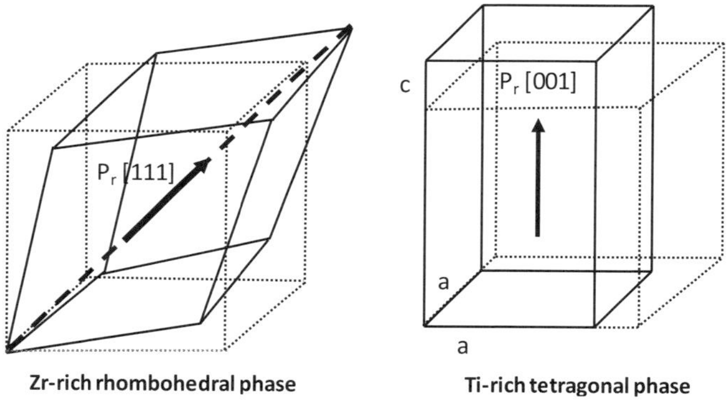

Fig. 25. PZT has a distorted perovskite structure with a polar direction. In Zr-rich PZT, the polar distortion occurs along the [111] direction of the undistorted cube resulting in a rhombohedral phase. In Ti-rich PZT, the polar distortion is along the [001] direction of the undistorted cube resulting in a tetragonal phase.

While polycrystalline PZT had a fairly sizable switchable polarization even in polycrystalline form, it suffers from a phenomenon known as polarization fatigue, where repeated switching when used with certain electrodes resulted in the gradual loss of switchable polarization.[159] Partly due to this, a second ferroelectric material, strontium bismuth tantalate (with chemical formula $SrBi_2Ta_2O_9$ or SBT), became popular in the early 1990s for use in making FeRAM devices on silicon.[27,160,161] The crystal structure of SBT is one of the so-called Aurivillius phases[162] and is a derivative of the

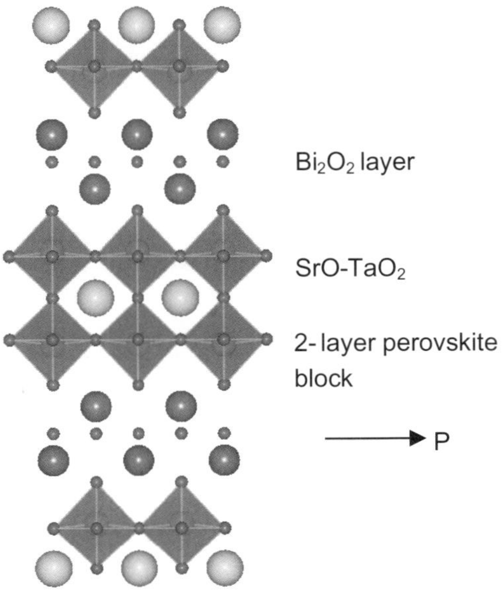

Fig. 26. Crystal structure of Aurivillius phase ferroelectric $SrBi_2Ta_2O_9$ (SBT). The structure consists of perovskite blocks separated by Bi_2O_2 layers. The ferroelectric polarization is along the a-axis, parallel to the layers.

simple perovskite structure of PZT where two TaO_2 layers (equivalent to the $(Zr,Ti)O_2$ layer in PZT) are separated by a SrO layer (equivalent to the PbO layer in PZT). However, this perovskite functional block is separated from each other by a rocksalt-like bilayer of Bi_2O_2. This results in a very long c-axis (the stacking direction) equivalent to two sets of TaO_2-SrO-TaO_2 and Bi_2O_2 displaced diagonally in the ab-plane. The low temperature ferroelectric crystal structure (Figure 26) has space group $A2_1am$ with lattice parameters $a = 5.531$ Å, $b = 5.534$ Å, and $c = 24.984$ Å. The remanent polarization in SBT is due to the asymmetry in the cation displacements of the Bi_2O_2 and SrO layers in the ab-plane. Thus the polarization in a single crystal of SBT points along the a-axis with no net polarization along the c-axis. Ferroelectricity in SBT was first reported by Smolenskii *et al.* in the early 1960s.[163] Figure 27 shows the difference in ferroelectric hysteresis loops measured on an SBT single crystal along the [001] and [110] directions. One can clearly see the absence of a remanent polarization along the c-axis but a well-saturated hysteresis loop can be observed in the obliquely oriented film.[164]

Randomly oriented polycrystalline SBT grown on SiO_2/Si exhibits a sufficient net out-of-plane polarization to be usable in FeRAM devices. Such SBT films do not display fatigue behavior and shows lower coercivity

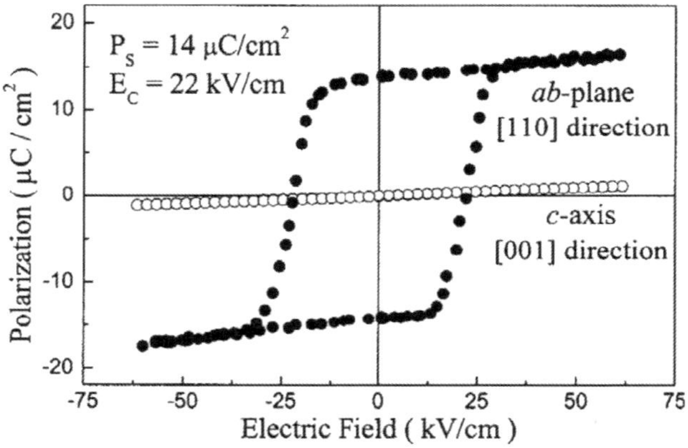

Fig. 27. Polarization vs. electric field for SBT single crystal measured along the ab-plane or along the c-axis. Ferroelectric hysteresis is observed only along the ab-plane. Reprinted with permission from H. Amorin *et al.*, Appl. Phys. Lett. 85, 5667 (2004). Copyright 2004, AIP Publishing LLC.

allowing for lower voltage operation but with a much lower switchable polarization compared to polycrystalline PZT. However, one issue with SBT is that it actually prefers to grow with the c-axis out of plane, which has no net polarization. Further, film growth of SBT requires a much higher crystallization temperature compared to PZT (~100–200°C higher than the 500°C crystallization temperature of PZT), which may be an issue with integration on silicon. Thus, a tremendous amount of work was done to try and grow a-axis oriented SBT with lower growth temperatures.[165–168] High quality polycrystalline SBT films with a-axis preferred orientation showed no fatigue after 10^{12} switching cycles while also showing good retention time, low coercivity and low leakage.[27,169] The Curie temperature of SBT is typically reported to be 335°C and its dielectric constant is ~180 at room temperature.[164,170–172] The fatigue-free behavior of SBT is ascribed to the ability of the charged $(Bi_2O_2)^{2+}$ layers to adjust their position to compensate for any space charge developing in the ferroelectric layer[173] as well as to provide oxygen ions for oxygen vacancy control making it suitable for devices where the electrodes are Pt.

Most modern FeRAM devices are fabricated using either SBT or PZT.[174] SBT is the material of choice of Matsushita (Panasonic)[174] for their smart cards. Matsushita has produced 500 million units of SBT thin-film integrated ferroelectrics and is using a 0.13-um stacked 4-level-metal design.[174] More recently, significant progress in improving the retention time of FeFET devices has been reported using a combination of SBT films with high-k dielectric materials in a MFIS structure.[89,175–180] The first major breakthrough in the retention time for a fully functioning FeFET on Si was reported by Ishiwara at the Tokyo Institute of Technology in 2007[175] using 400 nm SBT with 8 nm HfO_2 and Pt electrodes. They reported an on/off ratio of $>10^3$ after 30 days using relatively large (5 um) gate lengths. For scaled down devices, the best SBT-based FeFET performance was reported by Sakai *et al.*[176] using a Hf-Al-O high-k dielectric. They reported on/off ratio $>10^5$ after 37 days for long channel devices with 600 nm SBT and 7 nm Hf-Al-O layers. Sakai *et al.* also demonstrated good progress in scaled down FeFET devices.[177] Using transistors with gate lengths of 0.56 um, they were able to show on/off ratios of 10^5 that were unchanged after more than a day. They report an endurance of at least 10^8 cycles and a memory window of 0.9 V for an operating voltage of ± 5 V. While SBT-based long channel FeFETs appear to have sufficient retention and endurance, much more work is needed to scale down the devices for high-density, low-voltage, non-volatile operation.

3.4. $Bi_4Ti_3O_{12}$

Another problem related to ferroelectrics on silicon is the issue of imprint. This is the phenomenon where over time the hysteresis loop of a ferroelectric film shifts toward one voltage polarity and can ultimately result in a preferred polarization direction. The degree of imprint also seems to be dependent on temperature but the detailed mechanism resulting in imprint is still not yet fully settled.[181–184] Both PZT and SBT suffer from imprint effects and so in the 1990s, research into alternative ferroelectric materials continued. A potential material that appears to exhibit very little imprint was reported in 1999 by Park *et al.*[185,186] They doped the well-known ferroelectric material $Bi_4Ti_3O_{12}$ with lanthanum or neodymium on the bismuth site.[185] The parent compound is also an Aurivillius phase but with 3 TiO_2 layers separated from each other by a BiO monolayer and with the entire 5-layer functional block again separated from other blocks by a Bi_2O_2 bilayer (Figure 28).[162] Bismuth titanate films were difficult to grow with sufficient quality and suffered from the same issues as SBT. However, partial substitution of Bi by La or Nd showed a preference to grow with a-axis orientation on certain substrates resulting in large switchable out of plane polarizations and with much lower processing temperatures than SBT, but with an increased coercive field. While the material appeared as a whole better than either polycrystalline PZT or SBT, most groups researching ferroelectric memories opted to stay with PZT and SBT since the improvements were not sufficiently drastic.

Within the 3-layer perovskite-like blocks of bismuth titanate, the Ti ions are at the center of an octahedron whose corners are the oxygen atoms. These oxygen atoms serve as the link for sharing between adjacent octahedra through the formation O-Ti-O linear chains. Bi ions serve as electron donors and spacers for the TiO_6 octahedral framework. Similar to SBT, bismuth titanate also has orthorhombic symmetry with the ferroelectric polarization lying along the a-axis. Its lattice parameters at room temperature are a = 5.410, b = 5.448, and c = 32.84 Å.[162] Because of the layered structure, bismuth titanate possesses highly anisotropic properties.[185,187] The partial substitution of Bi for La or Nd, ideally substituting into the Bi site, yields the materials named bismuth lanthanum titanate ($Bi_{4-x}La_xTi_3O_{12}$, abbreviated as BLT) and bismuth neodymium titanate ($Bi_{4-x}Nd_xTi_3O_{12}$, abbreviated as BNT). These doped bismuth titanate materials allows one to improve fatigue behavior, enhance the remanent polarization, and decrease the processing temperature of the material.[186,188] A description of the basic ferroelectric device-related properties of bismuth titanate can be found in Funakubo *et al.*[190] Table 2 compares the properties of textured polycrystalline films of PZT, SBT, and BLT.[185]

3.5. *Doped* HfO_2

Both HfO_2 and ZrO_2 were studied extensively in the last decade in terms of their structural and electrical properties as well as methods of thin film deposition for their application as a replacement for SiO_2/Si_3N_4 as the

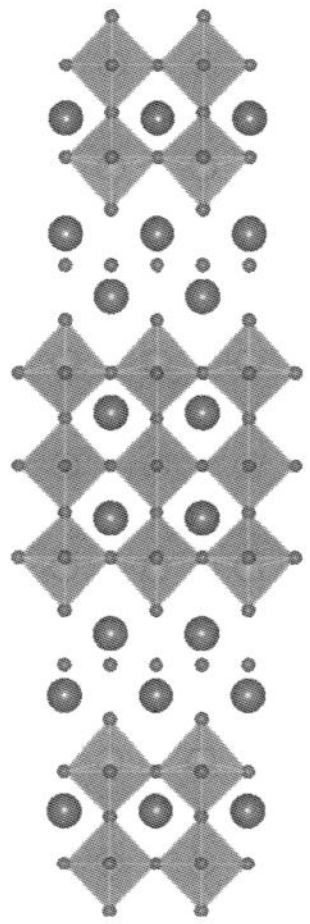

Fig. 28. Crystal structure of Aurivillius phase ferroelectric $Bi_4Ti_3O_{12}$. The structure consists of TiO_2-BiO perovskite blocks separated by Bi_2O_2 layers. The ferroelectric polarization is along the *a*-axis, parallel to the layers.

Table 2. Comparison of the properties of different types of thick film polycrystalline ferroelectric capacitors. PZT = $Pb(Zr,Ti)O_3$, SBT = $SrBi_2Ta_2O_9$, and BLT = $(Bi,La)_4Ti_3O_{12}$. Adapted from Ref. 185.

	PZT	SBT	BLT
Switchable polarization (uC/cm^2)	20–70	4–16	16–20
Fatigue resistance (on Pt electrodes)	Poor	Good	Good
Leakage current at 5 V (A/cm^2)	10^{-7}	10^{-9}	10^{-7}
Processing temperature (°C)	500–600	750–850	650–700

gate dielectric in sub-45 nm CMOS technology. Recently, it was reported by Boscke *et al.*[85] that in HfO_2 doped with a few mol% of SiO_2, an additional intermediate orthorhombic phase exists within the commonly known sequence of structural phase transitions of HfO_2 of monoclinic to tetragonal to cubic crystal structure. This orthorhombic phase was confirmed to be non-centrosymmetric and to exhibit ferroelectricity as demonstrated by hysteretic P-E, C-V, and PFM measurements.[85,191] The ferroelectric behavior exists in a narrow composition range of 3–4 mol%, with antiferroelectric behavior occurring at somewhat higher concentrations and a return to a linear dielectric at concentrations over 8.5 mol% (Figure 29).[192] The ferroelectric phase is obtained only when the doped HfO_2 is grown amorphous and then capped (usually with TiN) before crystallization.[85] It was soon discovered that other cations are also able to stabilize this ferroelectric orthorhombic phase in a narrow composition range. The cations that are confirmed experimentally to form the ferroelectric phase in HfO_2 include Y, Al, Gd, Sr, and La.[191–195] The existence of a non-centro symmetric orthorhombic phase and ferroelectricity in doped HfO_2 was shown to be possible using ab initio calculations.[191,196,197]

While the remanent polarization of doped HfO_2 is typically smaller than that of single crystal PZT ($\sim$15 uC/cm^2 vs 60 uC/cm^2), from the point of view of device fabrication and performance, HfO_2 may be more advantageous. One such advantage of HfO_2 is its unusually high coercive field. Typical values for Si-doped HfO_2 are in the range of 1–2 MV/cm [198] compared to a value of $\sim$100 kV/cm for crystalline PZT and even lower for SBT or BTO. The high coercive field allows for aggressive gate length scaling and at the same time drastically reduces polarization loss due to internal depolarization fields when coupled to a semiconductor. This has been experimentally demonstrated through the observation of ferroelectric hysteresis in a 5 nm HfO_2 ferroelectric capacitor grown on Si.[198] The high

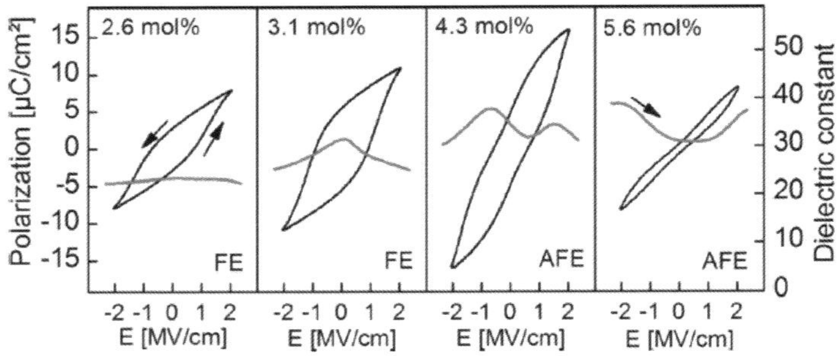

Fig. 29. Evolution of the polarization (black) and dielectric constant (red) as a function of applied electric field of a capacitor structure with Si-doped HfO_2 as the dielectric. Ferroelectric hysteresis begins to emerge at 2.6% Si and transforms into antiferroelectricity at 4.3% Si. Reprinted with permission from Ref. 85. Copyright 2011, AIP Publishing LLC.

coercivity is not always beneficial though. The dielectric breakdown field of HfO_2 is only about a factor of two larger than its coercive field and there is a much greater chance for device failure due to dielectric breakdown. In practice, this has limited the endurance of ferroelectric capacitors based on HfO_2 to about 10^9 switching cycles.[199,200] The high coercivity also puts limits on the minimum voltage needed to switch the ferroelectric and may be a negative factor for low voltage operation. The main advantage of HfO_2 from a device fabrication perspective is its CMOS compatibility and mature fabrication process. Atomic layer deposition of HfO_2 is well-established and is at a stage where the process is routinely used in manufacturing. It is stable in contact with Si under a wide range of conditions and has a sufficiently large band offset to both valence and conduction bands of Si unlike titanate perovskite ferroelectrics.

This manufacturability advantage may ultimately make doped HfO_2 to be the material of choice for the fabrication of ferroelectric field effect transistors for non-volatile logic, which is held back from development because of CMOS-compatibility and scaling issues. Both PZT and BTO require STO buffers in order to be integrated onto Si, making the fabrication more complicated since the growth of crystalline STO on Si may require the use of molecular beam epitaxy. Also, in order to have a sufficiently large memory window in the device as well as stable polarization when grown on Si, the ferroelectric oxide must be sufficiently thick (on the order of at least several tens of nm) in the case of titanate ferroelectrics. HfO_2 can achieve the same memory window using an order of magnitude thinner ferroelectric layer. However, as mentioned above, the closeness of the coercive field to the breakdown field in HfO_2 results in a much more limited endurance for HfO_2-based FeFETs. It remains to be seen which

Table 3. Comparison of the properties of doped HfO_2 compared to the more commonly used ferroelectric ceramics of $SrBi_2Ta_2O_9$ (SBT) and $Pb(Zr,Ti)O_3$ (PZT). Table adapted from Ref. 198.

	PZT	SBT	Doped HfO_2
Typical thickness	70 nm	25 nm	5-30 nm
Processing temperature	500–600°C	750°C	450–1000°C
Remanent polarization	20–70 uC/cm^2	4–16 uC/cm^2	1–40 uC/cm^2
Coercive field	~50 kV/cm	10–100 kV/cm	1–2 MV/cm
Breakdown field	0.5–2 MV/cm	~2 MV/cm	4–8 MV/cm
Dielectric constant	~1300	150–250	~30
ALD capability	limited	limited	mature
CMOS compatibility	Pb and O diffusion	Bi and O diffusion	stable

drawback is more detrimental. A comparison of the ferroelectric and device processing characteristics of PZT, SBT and doped HfO_2 is shown in Table 3.

A CMOS-compatible FeFET using ferroelectric HfO_2 was fabricated by Muller *et al.* in 2012.[201] The ferroelectric gate consisted of 10 nm of Si:HfO_2 grown on Si with a TiN top electrode. There is a very thin (~1 nm) SiO_2 interfacial layer between the Si and the HfO_2. They observed non-volatile switching down to a 20 ns pulse width with a memory window of 1.2 V. They show retention of the polarization up to 10^6 seconds and extrapolate a residual 10-year memory window of 0.6 V. The same group later studied the scalability of such a device and found that HfO_2-based FeFETs can be readily scaled down to 28 nm gate length. However, the devices are still limited by their low endurance (~10^4 cycles).[198] Because of its CMOS-compatibility and scaling advantage, HfO_2 may be more ideally suited for achieving negative capacitance effects in a FET.[92,202] An enhancement in the subthreshold slope for a transistor using ferroelectric HfO_2 as a gate dielectric has been reported with a swing as steep as 5 mV/decade.[203] The ability to use ultrathin ferroelectric layers while providing robust ferroelectricity is also beneficial for applications such as FTJs fabricated directly on Si.

3.6. $Pb_5Ge_3O_{11}$

Lead germanate with empirical formula $Pb_5Ge_3O_{11}$ is a ferroelectric material that has a crystal structure based on the mineral apatite. The structure consists of alternating layers of isolated single GeO_4 tetrahedra and isolated double Ge_2O_7 tetrahedra with the spaces in between filled by Pb^{2+} ions (Figure 30).[204] Ferroelectricity in this material was first reported by Iwasaki *et al.* in 1971[205] and confirmed by dielectric, electro-optic, and piezoelectric measurements.[206–208] The material has trigonal P3 symmetry in its ferroelectric phase and P-$\bar{6}$ symmetry in its paraelectric phase, with the spontaneous polarization pointing along the c-axis and having a Curie temperature of 178°C. Because polarization can only develop along a single axis, unlike in the perovskites, the material can only have 180° domains. The polarization results from a combination of Ge-O tetrahedral rotations as well as displacements of the Pb^{2+} ions with their extended 6s lone pairs. The lattice parameters for the ferroelectric phase are a = 10.190 Å and c = 10.624 Å. In terms of electrical properties related to device performance, $Pb_5Ge_3O_{11}$ has small values for the dielectric constant and the remanent polarization. The relatively low dielectric constant (~30) makes it suitable for FeFETs while still maintaining a modest polarization that is capable of charge modulation while also keeping depolarization fields low. Decent quality films of $Pb_5Ge_3O_{11}$ were reported in the 1990s using pulsed laser deposition,[209,210] sol-gel,[211] and metal-organic chemical vapor deposition.[212]

Ferroelectric transistors on silicon were reported in 2001 by T. Li *et al.*[213] Because of the difficulty in directly interfacing Pb with Si, a MFMIS structure was utilized. The relatively low dielectric constant of $Pb_5Ge_3O_{11}$ allows for significantly lower programming voltages compared to high k materials like PZT and SBT for the same thickness. The c-axis oriented films grown by Li *et al.* using MOCVD had a remanent polarization of 3 uC/cm^2 and a coercive field of 100 kV/cm. The required processing temperatures were also low (520°C) making it very compatible with CMOS gate last processing.[214] $C - V$ measurements show a memory window of 2–3 V at a programming voltage of 3 V. Transistors having gate length as small as 0.3 um were demonstrated with on/off ratio close to 10^9. The 1T memory devices show very good memory characteristics and decent retention properties. The memory window is observed to rapidly decrease from an initial value of 2 V to about 1.2 V in an hour, likely due to polarization loss as a result of defects. However, the memory window retains its value after this initial drop over the course of more than 5 days because of the low depolarization field. Using semiconducting InO_x as a buffer instead of a dielectric layer, Li *et al.* managed to improve retention by preventing the threshold voltage shift of the off state. They project that the memory window they measured for 10 days would still have a sufficiently large value of 0.7 V even after 10 years (Figure 31).[215]

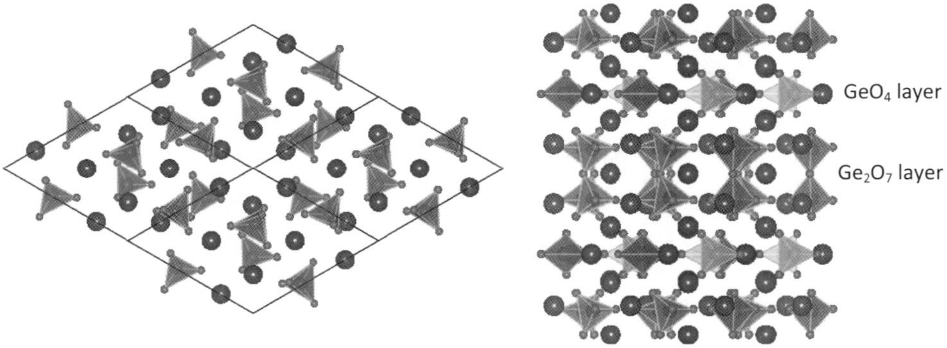

Fig. 30. Structure of ferroelectric lead germanate based on the apatite crystal structure. The left panel shows a view along the c-axis showing the trigonal symmetry of the unit cell. The structure consists of stacks of GeO_4 single tetrahedra and Ge_2O_7 double tetrahedra stacked alternately. The stacking is shown on the right. The Pb atoms (black spheres) occupy the spaces in between the Ge-O tetrahedra and serve as electron donors. The polarization is along the *c*-axis.

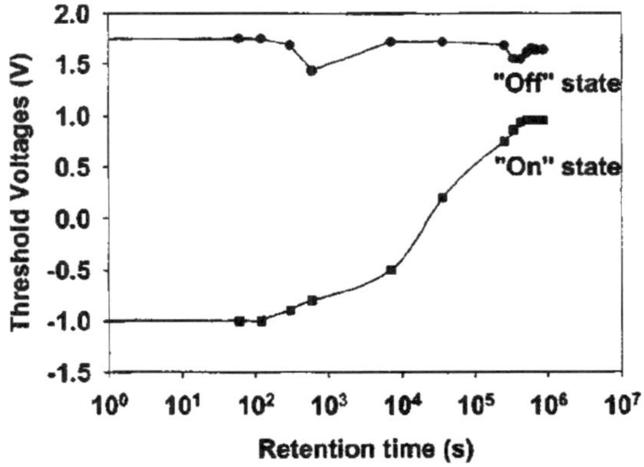

Fig. 31. Retention properties of 0.6 um 1T FeFET based on lead germanate. There is a gradual change in the on state threshold voltage of a few days days but a memory window of 0.7 V is expected to remain over a 10 year projection. Reprinted with permission from Ref. 239. Copyright 2005, AIP Publishing LLC.

3.7. *Ferroelectric polymers*

Ferroelectricity can also be exhibited by non-oxide material. One of the more interesting ones is a class of ferroelectric polymers, based on polyvinylidene fluoride (PVDF).[216] The basic monomer has composition [-CH_2CF_2-]. The two hydrogen and two fluorine atoms are alternately bonded to the carbon atoms of the polymer backbone as shown in Figure 32. When

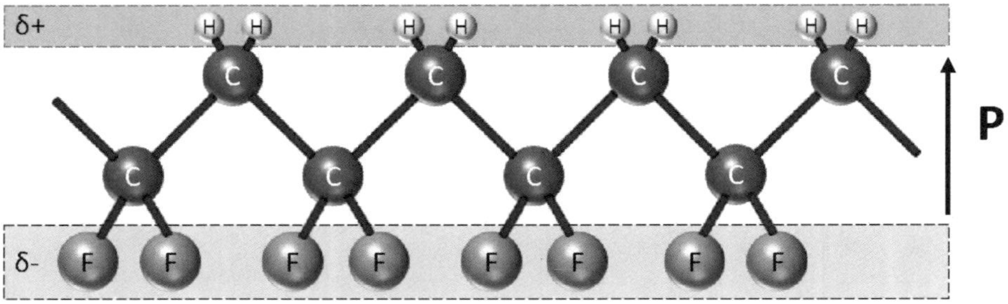

Fig. 32. Molecular structure of the β phase of PVDF. All the H atoms are on one side of the carbon backbone while all the F atoms are on the opposite side. The larger electronegativity of F compared to results in the formation of an electric polarization.

Table 4. Comparison of the ferroelectric properties of PVDF with the two most commonly used ferroelectric ceramic materials (PZT = $Pb(Zr,Ti)O_3$ and SBT = $SrBi_2Ta_2O_9$) in the form of FeRAM devices. Data taken from Ref. 217.

	PZT FRAM (Ramtron)	SBT FRAM (Celis)	PVDF FET
Remanent Polarization (uC/cm^2)	35	15–25	8
Leakage current (nA/cm^2)	100	100	—
Coercive field (kV/cm)	500–700	300–500	50–5000
Breakdown field (MV/cm)	1	3.8	10
Dielectric constant	300–1300	150–250	10
Switching speed (ns)	—	370	2000
Typical thickness (nm)	500	800	1–1000
Endurance	$>10^{10}$	$>10^{10}$	—
Retention time	10 yr	10 yr	>1 yr

the molecule is configured such that all the H atoms are on one side and all the F atoms are on the other, a strong dipole moment develops perpendicular to the direction of the backbone of the polymer as a result of the high electronegativity of fluorine. Such a configuration where all the H and F ends are all in a trans arrangement and the molecule is stretched out is known as the β phase. The β phase can be stabilized by the addition of co-polymers of trifluoroethylene (TrFE). The combination of 70% PVDF with 30% TrFE allows for direct crystallization of the ferroelectric β phase from the melt or from solution. Ferroelectricity in PVDF arises from the electric dipoles that can be inverted by the application of an external electric field primarily through rotations of the C-C bonds in the backbone.

An advantage of PVDF is that it is chemically inert and does not require high temperature processing. As with many polymers, the fabrication methods (e.g. solvent-based methods) for PVDF are relatively cheap compared with the thin film deposition systems required for high quality oxide thin film ferroelectrics. 70:30 PVDF-TrFE polymers exhibit a Curie temperature of about 110°C with a melting point of about 150°C. The remanent polarization is about 8 uC/cm^2. One disadvantage is that the switching from one polarization state to another is quite slow, on the order of microseconds at best and requires large electric fields. A typical coercive field for PVDF is in the range of 1-5 MV/cm, one to two orders of magnitude more than the titanate perovskite ferroelectrics. The large coercive field naturally leads to trying to reduce the film thickness to be able to use lower voltages. Most PVDF films aimed for transistor applications are grown by the Langmuir-Blodgett technique or by vacuum evaporation.[217] A Si-based MFIS capacitor using PVDF as the ferroelectric was reported in 2003 by Ducharme *et al.*[218] In this work, a 170 nm Langmuir-Blodgett film was grown on 100 nm SiO_2 on Si. The device showed a memory window of 2.8 V when the gate voltage was cycled between +25 V and −25 V. The capacitance changed by a factor of 8 between the "on" and "off" states. Using thinner SiO_2 (10 nm) and a thinner PVDF layer (36 nm), they reduced the operating voltage to 3 V for a memory window of 1 V.[219] A similar result was reported by S. Fujisaki[220] using a 77:23 composition of PVDF:TrFE and a 100 nm thick film on Si with 3 nm high-k Ta_2O_5 dielectric instead of SiO_2. They reported a memory window of 2.9 V for a voltage sweep of +/−4 V. One major problem with these devices is the very poor retention time of about 15 min.[219] Scaling of thickness of PVDF devices is currently limited by the polymer film surface roughness (~20 nm). Comparison of the basic ferroelectric properties of PVPF-based devices with oxide ferroelectric devices is shown in Table 4.

An actual FeFET structure incorporating PVDF was reported in 2010 by Ducharme using a 35 nm thick PVDF layer with a 10 nm SiO_2 dielectric layer on lightly doped p-type Si. Source and drain regions were defined by As ion implantation. Au was used as the top gate electrode. The device exhibited excellent transistor characteristics for gate voltages from 0 to 3.5 V. However, the transistor showed extremely short retention times of about 10 s. State retention in FeFETs can be limited by instability in the ferroelectric polarization or by charge leakage and trapping in the insulating layers, or both. In the case of PVDF-based transistors, it is charge leakage as a result of the short circuit condition when the device is in the off state. In open circuit conditions, the polarization state is preserved for at least several hours.[221] PVDF on Si has been used for fabricating tunnel FETs[222] and negative capacitance devices[223] with a demonstrated 53 mV/decade swing with an on/off ratio of 10^3.

4. Conclusions

We provided a broad overview of the applications of ferroelectric thin films that are integrated onto semiconductors. Further research will undoubtedly increase both the number of ferroelectric materials integrated with semiconductors and the variety of functional devices that can be fabricated using these materials, along with processes that allow for the effective creation of these devices. Because of the permanent but switchable electric polarization, applications involving low-power and non-volatile memory are the most obvious candidates. We described three main types of ferroelectric memory device structures and their principles of operation as well as their benefits and shortcomings. We also described several classes of ferroelectric materials that have been used to implement these device structures. It is clear that ferroelectric memories have great potential to become the next generation memory technology capable of replacing both hard drives and flash memory. With the continued development of thin film deposition processing of crystalline oxides, many of the existing materials science issues preventing further development of these ferroelectric devices is expected to be overcome in the near future. Ferroelectric materials, especially $BaTiO_3$, are additionally useful for optoelectronic device applications. This is discussed in more detail in the chapter by Abel *et al.*

Acknowledgments

In the process of writing the chapter, we have benefited greatly from many discussions with Chungwei Lin, Andrew O'Hara, Kurt Fredrickson and Kristy Kormondy.

References

1. Lines, M. E. and Glass, A. M., *Principles and Applications of Ferroelectrics and Related Materials*, (Clarendon, Oxford, 1977).
2. Subramanian, M. A., He, T., Chen, J. Z., Rogado, N. S., Calvarese, T. G. and Slight, A. W. (2006), *Adv. Mater.*, **18**, 1737.
3. Ikeda, N., Ohsumi, H., Ohwada, K., Ishii, K., Inami, T., Kakurai, K., Murakami, Y., Yoshii, K., Mori, S., Horibe, Y. and Kito, H. (2005), *Nature*, **436**, 1136.

4. van Aken, B. V., Palstra, T. T. M., Filippetti, A. and Spaldin, N. A. (2004), *Nature Materials*, **3**, 164.
5. Levanyuk, A. P. and Sannikov, D. G. (1974), *Sov. Phys. Usp.*, **17**, 199.
6. Kimura, T., Goto, T., Shintani, H., Ishizaka, K., Arima, T. and Tokura, Y. (2003), *Nature*, **426**, 55.
7. Kenzelmann, M., Harris, A. B., Jonas, S., Broholm, C., Schefer, J., Kim, S. B., Zhang, C. L., Cheong, S.-W., Vajk, O. P. and Lynn, J. W. (2005), *Phys. Rev. Lett.*, **95**, 087206.
8. Valasek, J. (1922), *Phys. Rev.*, **19**, 478.
9. Jona, F. and Shirane, G. *Ferroelectric Crystals* (Dover Publications, 1993).
10. Strukov, B. A. and Lavanyuk, A. P. (1998), *Ferroelectric Phenomena in Crystals*, Springer, Berlin.
11. Zhong, W., Vanderbilt, D. and Rabe, K. M. (1995), *Phys. Rev. B*, **52**, 6301.
12. King-Smith, R. D. and Vanderbilt, D. (1994), *Phys. Rev. B*, **49**, 5828.
13. Salje, E. K. H., Wruck, B. and Thomas, H. (1991), *Z. Phys. B — Cond. Matter*, **82**, 399.
14. Sawyer, C. B. and Tower, C. H. (1930), *Phys. Rev.*, **35**, 269.
15. Aixacct Technologies website: http://www.aixacct.com
16. Radiant Technologies website: http://www.ferrodevices.com
17. *Ferroelectric Random Access Memories*, of Topics in *Applied Physics*, Vol. 93, edited by Ishiwara, H. and Okuyama, M., (Springer-Verlag, Berlin, 2004).
18. Scott, J. F., *Ferroelectric Memories*, Springer Series in *Advanced Microelectronics* (Springer-Verlag, Berlin, 2000).
19. Chapman, D. W. (1969), *J. Appl. Phys.*, **40**, 2381.
20. Francombe, M. H. (1972), *Thin Solid Films*, **13**, 243.
21. Sharma, B. S., Vogel, S. F. and Prentky, P. I. (1973), *Ferroelectrics*, **5**, 69.
22. Blum, J. B. and Gurkowich, S. R. (1985), *J. Mater. Sci.*, **20**, 4479.
23. Budd, K. D., Dey, S. K. and Payne, D. A. (1988), *Br. Ceram. Proc.*, **36**, 107.
24. Eaton, S. S., Butler, D. B., Parris, M., Wilson, D. and McNeillie, H. (1988), *Dig. Tech. Pap.-IEEE Int. Solid-State Circuits Conf.*, **130**, 329.
25. The International Technology Roadmap for Semiconductors (ITRS), *Emerging Research Devices*, 2011 http://www.itrs.net/
26. Evans, J. and Womack, R. (1988), *IEEE J. Solid-State Circuits*, **23**, 1171.
27. de Araujo, C. A. P., Cuchiaro, J. D., MacMillan, L. D., Scott, M. C. and Scott, J. F. (1995), *Nature*, **374**, 627.
28. Sumi, T. *et al.* (1994), Dig. Tech. Pap.-*IEEE Int. Solid-State Circuits Conf.*, p. 268.
29. Choi, E. *et al.* (2004), *Integr. Ferroelectr.*, **66**, 107.
30. Sinharoy, S., Buhay, H., Lampe, D. R. and Francombe, M. H. (1992), *J. Vac. Sci. Technol. A*, **10**, 1554.
31. Izyumskaya, N. *et al.* (2007), Processing, Structure, Properties, and Applications of PZT Thin Films. *Critical Reviews in Solid State and Materials Sciences*, **32**, 111–202.
32. Auciello, O., Araujo, C. A. P. de and Celinska, J. in *Emerging Non-Volatile Memories* (eds. Hong, S., Auciello, O. and Wouters, D.) 3–35 (Springer US, 2014).

33. Takeuchi, K. and Chen, A. in *Emerging Nanoelectronic Devices* (eds. Chen, A., Hutchby, J., Zhirnov, V. and Bourianoff, G.) 110–122 (John Wiley & Sons Ltd, 2014).
34. Setter, N. *et al.* (2006), *J. Appl. Phys.*, **100**, 051606.
35. Ramesh, S. Aggarwal and Auciello, O. (2001), *Mater. Sci. Eng. R*, **32**, 191.
36. Ishiwara, H. in *Ferroelectric Random Access Memories* (eds. Ishiwara, P. H., Okuyama, P. M. and Arimoto, D. Y.) 233–254 (Springer Berlin Heidelberg, 2004).
37. Thomas, R., Scott, J. F., Bose, D. N. and Katiyar, R. S. (2010), *J. Phys.: Condens. Matter*, **22**, 423201.
38. Ishiwara, H. (2009), *Current Applied Physics*, **9**, S2–S6.
39. Pan, X. and Ma, T. P. (2011), *Appl. Phys. Lett.*, **99**, 013505.
40. Fitsilis, M., Mustafa, Y. and Waser, R. (2005), *Integrated Ferroelectrics*, **70**, 29–44.
41. Arimoto, Y. and Ishiwara, H. (2004), *MRS Bull.* **29**, 823.
42. Ishiwara, H. (2000), *Mater. Res. Soc. Symp. Proc.*, **596**, 427.
43. Kohlstedt, H. and Ishiwara, H., in *Nanoelectronics and Information Technology:Advanced Electronic Materrial and Novel Devices*, edited by R. Waser (Wiley-VCH, Weinbeim, 2000), Chap. 14, p. 387.
44. Kaneko, Y., Tanaka, H., Ueda, M., Kato, Y. and Fujii, E. (2011), *IEEE Transactions on Electron Devices*, **58**, 1311–1318.
45. Ross, I. M. (1957), U.S. Patent No. 2, 791, 760.
46. Wu, S. Y. (1974), *IEEE Trans. Electron Devices*, **ED-21**, 499.
47. Ishiwara, H. (2001). *J. Semicond. Technol. and Sci.*, **1**, 1–14.
48. Hoffman, J., Pan, X., Reiner, J. W., Walker, F. J., Han, J. P., Ahn, C. H. and Ma, T. P. (2010), *Advan. Materi.*, **22**, 2919.
49. Cheng, C.-H. and Chin, A. (2014), *IEEE Electron Device Letters*, **35**, 138–140.
50. Miller, S. L. and Mcwhorter, P. J. (1992), *J. Appl. Phys.*, **72**, 5999.
51. Nicollian, E. H. and Brews, J. R. (1982), *MOS Physics and Technology* (New York: Wiley).
52. Thomas, R., Pradhan, D. K., Melgarejo, R. E., Saavedra-Arias, J. J., Karan, N. K., Palai, R., Murari, N. M. and Katiyar, R. S. (2008), *ECS Trans.*, **13**, 363.
53. Gerber, A., Kohlstedt, H., Fitsilis, M., Waser, R., Reece, T. J., Ducharme, S. and Rije, E. (2006), *J. Appl. Phys.*, **100**, 024110.
54. Kohlstedt, H. and Ishiwara, H. (2003), *Nanoelectronics and Information Technology* ed Waser Rainer (Weinheim: Wiley–VCH).
55. Ko, E., Pak, J., Nam, K. and Park, G. (2005), *J. Korean Phys. Soc.*, **46**, 269.
56. Sakamaki, K., Hirai, T., Uesugi, T., Kishi, H. and Tarui, Y. (1999), *Japan. J. Appl. Phys.*, **38**, L451.
57. Choi, H. S., Park, K. S., Hur, J.-S. and Choi, I.-H. (2001), *J. Korean Phys. Soc.*, **39**, S228.
58. Hirai, T., Teramoto, K., Nagashima, K., Koike, H. and Tarui, Y. (1995), *Japan. J. Appl. Phys.*, **34**, 4163.
59. Shih, W.-C., Kang, K.-Y. and Lee, J. Y.-M. (2007), *Appl. Phys. Lett.*, **91**, 232908.
60. Liu, C.-H., Wu, J.-M. and Wu, L.-J. (2006), *Appl. Phys. Lett.*, **88**, 122901.
61. Park, B.-E. and Ishiwara, H. (2004), *Integr. Ferroelectr.*, **62**, 141.

62. Senzaki, J., Kurihara, K., Nomura, N., Mitsunaga, O., Iwasaki, Y. and Ueno, T. (1998), *Japan. J. Appl. Phys.*, **37**, 5150.
63. Juan, P.-C., Hu, Y.-P., Chiu, F.-C. and Lee, J. Y.-M. (2005), *J. Appl. Phys.*, **98**, 044103.
64. Chen, H.-Y., Wu, J.-M., Huang, H.-E. and Bor, H.-Y. (2007), *Appl. Phys. Lett.*, **90**, 112907.
65. Takahashi, M. and Sakai, S. (2005), *Japan. J. Appl. Phys.*, **44**, L800.
66. Thomas, R., Melgarejo, R. E., Murari, N. M., Pavunny, S. P. and Katiyar, R. S. (2009), *Solid State Commun.*, **149**, 2013.
67. Thomas, R., Ehrhart, P., Luysberg, M., Boese, M., Waser, R., Roeckerath, M., Schubert, J., Elshocht, S. V. and Caymax, M. (2007), *Electrochem. Soc.*, **154**, G147.
68. Thomas, R., Bhakta, R., Milanov, A., Patil, U., Devi, A. and Ehrhart, P. (2007), *Chem. Vapour Depos.*, **13**, 98.
69. Thomas, R., Ehrhart, P., Luysberg, M., Boese, M., Waser, R., Wagner, M., Schubert, J., Elshocht, S. V. and Caymax, M. (2006), *Appl. Phys. Lett.*, **89**, 232902.
70. Park, B. E, Takahashi, K. and Ishiwara, H. (2004), *Appl. Phys. Lett.*, **85**, 4448.
71. Takahashi, K., Aizawa, K., Park, B.-E. and Ishiwara, H. (2005), *Japn. J. Appl. Phys.*, **44**, 6218.
72. Takahashi, M., Horiuchi, T., Wang, S., Li, Q.-H. and Sakai, S. (2008), *J. Vac. Sci. Technol.* B **26**, 1585.
73. Dawber, M., Rabe, K. M., and Scott, J. F. (2005), *Rev. Mod. Phys.*, **77**, 1083.
74. Takahashi, K., Manabe, K., Morioka, A., Ikarashi, T., Yoshihara, T., Watanabe, H. and Tatsumi, T. (2004), Abstracts of the International Conference Solid State Devices and Materials, Tokyo, Paper No. 1–2.
75. Batra, I. P., Würfel, P. and Silverman, B. D. (1973), *Phys. Rev. Lett.*, **30**, 384.
76. Metha, R. R., Siverman, B. D. and Jacobs, J. T. (1973), *J. Appl. Phys.*, **44**, 3379.
77. Ma, T. P. and Han, J.-P. (2002), *IEEE Electron Device Letters*, **23**, 386–388.
78. Ponath P. *et al.* (2015), *Nat. Commun.*, **6**, 6067.
79. Dubourdieu C. *et al.* (2013), *Nat. Nanotechnol.*, **8**, 748.
80. Niu, F. and Wessels, B. W. (2007), *J. Vac. Sci. Technol. B*, **25**, 1053.
81. Mazet, L. *et al.* (2014), *J. Appl. Phys.*, **116**, 214102.
82. McKee, R. A., Walker, F. J. and Chisholm, M. F. (2001), *Science*, **293**, 468.
83. Contreras-Guerrero, R., Veazey, J. P., Levy, J. and Droopad, R. (2013), *Appl. Phys. Lett.*, **102**, 012907.
84. Vaithynathan, V. *et al.* (2006), c-axis oriented BaTiO3 films on Si, *J. Appl. Phys.*, **100**, 024108.
85. HfO2 FE (2011), *Appl. Phys. Lett.*, **99**(10), 102903.
86. Muller, J. *et al.* (2012), *Symp. On VLSI Tech.*, pp. 25–26.
87. Zhang, X., Takahashi, M., Takeuchi, K. and Sakai, S. First 64 kb ferroelectric-NAND flash memory array with 7.5 V program, 10^8 endurance and long data retention, *Inter. Conf. Solid State Dev. and Mater.* (SSDM), pp. 975–976 (2011).
88. Kaneko, Y., Nishitani, Y., Ueda, M., Tokumitsu, E. and Fujii, E. (2011), *Appl. Phys. Lett.*, 99, 182902.
89. Sakai, S. and Ilangonvan, R. (2004), *IEEE Electron Device Letters*, **25**, 369–371.
90. Taur, Y. and Ning, T. (1998), *Fundamentals of Modern VLSI Devices*. Cambridge, U.K.: Cambridge Univ. Press.

91. Theis, T. N. and Solomon, P. M., In quest of the 'next switch': Prospects for greatly reduced power dissipation in a successor to the silicon field-effect transistor, *Proc. IEEE*, **98**(12), pp. 2005–2014, Dec. (2010)
92. Salahuddin, S. and Datta, S. (2008), *Nano Lett.*, **8**(2), pp. 405–410.
93. Zhirnov, V. V. and Cavin, R. K. (2008), *Nat. Nano.*, **3**, 77–78.
94. Khan, A. I. *et al.* (2015), *Nat. Mater.*, **14**, 182–186.
95. Jimenez, D., Miranda, E. and Godoy, A. (2010), *IEEE Trans. Electron Devices*, **57**(10), pp. 2405–2409, Oct.
96. Khan, A. I., Yeung, C. W., Hu, C. and Salahuddin, S. (2011), "Ferroelectric negative capacitance MOSFET: Capacitance tuning & antiferroelectric operation," in *Proc. IEEE IEDM*, Dec. pp. 255–258.
97. Frank, D. J. *et al.* (2014), *IEEE Transactions on Electron Devices*, **61**, 2145–2153.
98. Rusu, A., Salvatore, G., Jimenez, D. and Ionescu, A. M., Metal-ferroelectric-metal-oxide-semiconductor field effect transistor with sub-60 mV/decade subthreshold swing and internal voltage amplification, in *Proc. IEEE IEDM*, Dec. (2010), pp. 395–398.
99. Esaki, L., Laibowitz, R. B. and Stiles, P. J. (1971), *IBM Tech. Discl. Bull.*, **13**, 2161.
100. Kohlstedt, H., Pertsev, N. A. and Waser, R. (2002), *Mater. Res. Soc. Symp. Proc.*, **688**, 161.
101. Qu, H., Yao, W., Garcia, T., Zhang, J., Sorokin, A. V., Ducharme, S., Dowben, P. A. and Fridkin, V. M. (2003), *Appl. Phys. Lett.*, **82**, 4322.
102. Contreras, J. R., Kohlstedt, H., Poppe, U., Waser, R., Buchal, C. and Pertsev, N. A. (2003), *Appl. Phys. Lett.*, **83**, 4595.
103. Wolf, R., Blom, P. W. M. and Krijn, M. P. C. U.S. Patent No. 5,541,422 (1996).
104. Indlekofer, M. and Kohlstedt, H. (2005), *Europhys. Lett.*, **72**, 282.
105. Zhuravlev, M. Y., Sabirianov, R. F., Jaswal, S. S. and Tsymbal, E. Y. (2005), *Phys. Rev. Lett.*, **94**, 246802.
106. Kohlstedt, H., Pertsev, N. A., Contreras, J. R. and Waser, R. (2005), *Phys. Rev. B*, **72**, 125341.
107. Garcia, V. *et al.* (2009), *Nature*, **460**, 81–84.
108. Maksymovych, P. *et al.* (2009), *Science*, **324**, 1421–1425.
109. Gruverman, A. *et al.* (2009), *Nano Lett.*, **9**, 3539–3543.
110. Wen, Z., Li, C., Wu, D., Li, A. and Ming, N. (2013), *Nature Materials*, **12**, 617.
111. Chanthbouala, A. *et al.* (2012), *Nat Nano*, **7**, 101–104.
112. Z. Li *et al.* (2014), *Adv. Mater.*, **26**, 7185.
113. Guo, R. *et al.* (2015), *Scientific Reports*, **5**.
114. Blom, P. W. M., Wolf, R. M., Cillessen, J. F. M. and Krijn, M. P. C. M. (1994), *Phys. Rev. Lett.*, **73**, 2107.
115. Gotoh, K., Tamaura, H., Takauchi, H. and Yoshida, A. (1996), *Jpn. J. Appl. Phys.*, Part 1 **35**, 39.
116. Okano, M. and Watanabe, Y. (2000), *Appl. Phys. Lett.*, **76**, 233.
117. Meyer, R. and Kohlstedt, H. (2006), *IEEE Transactions on Ultrasonics, Ferroelectrics, and Frequency Control*, **53**, 2340–2348.
118. Zhao, W. and Jena, D. (2004), *Appl. Phys. Lett.*, **96**, 2095.
119. Ngai, J. H., Kumah, D. P., Ahn, C. H. and Walker, F. J. (2014), *Appl. Phys. Lett.*, **104**, 062905.

120. Abel, S., Stöferle, T., Marchiori, C., Rossel, C., Rossell, M. D., Erni, R., Caimi, D., Sousa, M., Chelnokov, A., Offrein, B. J. and Fompeyrine, J. (2013), *Nat. Commun.*, **4**, 1671.
121. Xiong, C., Pernice, W. H. P., Ngai, J. H., Reiner, J. W. Kumah, D., Walker, F. J. Ahn, C. H. Tang, H. X. (2014), *Nano Lett.*, **14**, 1419.
122. Niu, F. and Wessels, B. W. (2007), *J. Vac. Sci. Technol. B Microelectron. Nanom. Struct.*, **25**, 1053.
123. Mazet, L., Yang, S. M., Kalinin, S. V., Schamm-Chardon, S. and Dubourdieu, C. (2015), *Sci. Technol. Adv. Mater.*, **16**, 036005.
124. McKee, R. A., Walker, F. J., Conner, J. R., Specht, E. D. and Zelmon, D. E. (1991), *Appl. Phys. Lett.*, **59**, 782.
125. Lee, M.-B., Kawasaki, M., Yoshimoto, M. and Koinuma, H. (1995), *Appl. Phys. Lett.*, **66**, 1331.
126. Yu, Z., Ramdani, J., Curless, J. A., Overgaard, C. D., Finder, J. M., Droopad, R., Eisenbeiser, K. W., Hallmark, J. A., Ooms, W. J. and Kaushik, V. S. (2000), *J. Vac. Sci. Technol.*, **18**, 2139.
127. Meier, A. R., Niu, F. and Wessels, B. W. (2006), *J. Cryst. Growth*, **294**, 401.
128. Ngo, T. Q., Posadas, A. B., McDaniel, M. D., Hu, C., Bruley, J., Yu, E. T., Demkov, A. A. and Ekerdt, J. G. (2014), *Appl. Phys. Lett.*, **104**, 082910.
129. Ponath, P., Posadas, A. B., Hatch, R. C. and Demkov, A. A. (2013), *J. Vac. Sci. Technol. B*, **31**, 031201.
130. Jesse, S., Baddorf, A. P. and Kalinin, S. V. (2006), *Appl. Phys. Lett.*, **88**, 062908.
131. Woodward, D. I., Knudsen, J. and Reaney, I. M. (2005), *Phys. Rev. B*, **72**, 104110.
132. Noheda, B., Cox, D. E., Shirane, G., Gonzalo, J. A., Cross, L. E. and Park, S.-E. (1999), *Appl. Phys. Lett.*, **74**, 2059.
133. Noheda, B., Cox, D. E., Shirane, G., Guo, R., Jones, B. and Cross, L. E. (2001), *Phys. Rev. B*, **63**, 014103.
134. Landolt-Bornstein: *Numerical Data and Functional Relationships in Science and Technology, Group III*, vol. 16, Part A (Springer, Berlin, 1981).
135. Lin, A., Hong, X., Wood, V., Verevkin, A. A., Ahn, C. H., McKee, R. A., Walker, F. J. and Specht, E. D. (2001), *Appl. Phys. Lett.*, **78**, 2034.
136. Gruverman, A. (1999), *Appl. Phys. Lett.*, **75**, 1452.
137. Nagarajan, V., Aggarwal, S., Gruverman, A., Ramesh, R. and Waser, R. (2005), *Appl. Phys. Lett.*, **86**, 262910.
138. Stolichnov, I., Tagantsev, A., Setter, N., Cross, J. and Tsukada, M. (1999), *Appl. Phys. Lett.*, **74**, 3552.
139. Al-Shareef, H., Auciello, O. and Kingon, A. (1994), *J. Appl. Phys.*, **77**, 2146.
140. Nakamura, T., Nakao, Y., Kamisawa, A. and Takasu, H. (1994), *Appl. Phys. Lett.*, **65**, 1522.
141. Cross, J. S., Fujiki, M., Tsukada, M., Matsuura, K., Otani, S., Tomotani, M., Kataoka, Y., Kotaka, Y. and Goto, Y. (1999), *Integr. Ferroelectr.*, **25**, 605.
142. McKee, R. A., Walker, F. J. and Chisholm, M. F. (1998), *Phys. Rev. Lett.*, **81**, 3014.
143. Lichtensteiger, C., Triscone, J.-M., Junquera, J. and Ghosez, P. (2005), *Phys. Rev. Lett.*, **94**, 047603.
144. Lichtensteiger, C. and Triscone, J.-M. (2004), *Integr. Ferroelectri.*, **61**, 143–148.

145. Fong, D. D., Stephenson, G. B., Streiffer, S. K., Eastman, J. A., Auciello, O., Fuoss, P. H. and Thompson, C. (2004), *Science*, **304**, 1650.
146. Chirila, C. *et al.* (2015), *J. Mater. Sci.*, **50**, 3883–3894.
147. Yin, S. *et al.* (2012), *Thin Solid Films*, **520**, 4572–4575.
148. Shi, Y. *et al.* (2014), *J. Appl. Phys.* 115, 214108.
149. Borowiak, A. S. *et al.* (2012), *Thin Solid Films*, **520**, 4604–4607.
150. Dekkers, M. *et al.* (2009), *Appl. Phys. Lett.*, **95**, 012902.
151. Kim, D. M. *et al.* (2006), *Appl. Phys. Lett.*, **88**, 142904.
152. Fujito, K., Wakiya, N., Mizutani, N. and Shinozaki, K. (2005), *Jpn. J. Appl. Phys.*, **44**, 6900.
153. Gilmore, W. M. *et al.* (2003), *J. Mater. Rese.*, **18**, 111–114.
154. Liu, B. T. *et al.* (2002), *Appl. Phys. Lett.*, **80**, 4801–4803.
155. Wang, Y. *et al.* (2002), *Appl. Phys. Lett.*, **80**, 97–99.
156. Louahadj, L. *et al.* (2013), *Appl. Phys. Lett.*, **103**, 212901.
157. Xiao, B. *et al.* (2007), *Appl. Phys. Lett.*, **91**, 182908.
158. Paisley, E. A. *et al.* (2013), *J. Appl. Phys.*, **113**, 074107.
159. Tagantsev, A. K., Stolichnov, I., Colla, E. and Setter, N. (2001), *J. Appl. Phys.*, **90**, 1387.
160. Scott, J. F., Ross, F. M., Paz, de Araujo, C. A., Scott, M. C. and Huffman, M. (1996), *MRS Bulletin*, **21**, 33.
161. Auciello, O., Scott, J. F. and Ramesh, R. (1998), *Phys. Today*, **51**, 22.
162. Aurivillius, B. (1949), *Arkiv Kemi*, **1**, 499.
163. Smolenski, G. A., Isupov, V. A., Arganovskaya, A. A. and Popov, S. N. (1960), *Fiz. Tverd. Tela*, **2**, 2906.
164. Amorín, H., Shvartsman, V. V., Kholkin, A. L. and Costa, M. E. V. (2004), *Appl. Phys. Lett.*, **85**, 5667.
165. Lee, H. N., Zakharov, D. N., Reiche, P., Uecker, R. and Hesse, D. in *Ferroelectric Thin Films X* (eds. Gilbert, S. R., TrolierMcKinstry, S., Miyasaka, Y. and Wouters, D. J.) **688**, 291–296 (Materials Research Society, 2002).
166. Watanabe, T. *et al.* (2002), *Jpn. J. Appl. Phys.*, **41**, L1478–L1481.
167. Miu, D., Martinez, J. C., Wiehl, L., Raitieri, R. and Adrian, H. (2005), *Mater. Lett.*, **59**, 1243–1247.
168. Iijima, R. (2001), *Appl. Phys. Lett.*, **79**, 2240–2242.
169. Amanuma, K., Hase, T. and Miyasaka, Y. (1995), *Appl. Phys. Lett.*, **66**, 221.
170. Subbarao, E. C. (1962), *J. Phys. Chem. Solid.*, **23**, 665.
171. Ko, J.-H., Hushur, A., Kojima, S., Sih, B. C. and Ye, Z.-G. (2002), *Appl. Phys. Lett.*, **81**, 4043.
172. Kamba, S., Pokorny, J., Porokhonsky, V., Petzelt, J., Moret, M. P., Garg, A., Barber, Z. H. and Zallen, R. (2002), *Appl. Phys. Lett.*, **81**, 1056.
173. Park, B. H., Kang, B. S., Bu, S. D., Noh, T. W., Lee, J. and Jo, W. (1999), *Nature (London)*, **401**, 682.
174. Scott, J. F. (2007), *Science*, **315**, 954–959.
175. Ishiwara, H. and Ishiwara, H. (2007), Recent progress in ferroelectric random access memory technology. In *Symposium I — Materials and Processes for Nonvolatile Memories II*, **997**.

176. Sakai, S. and Takahashi, M. (2010), *Materials*, **3**, 4950–4964.
177. Van Hai, L., Takahashi, M. and Sakai, S. (2010), *Semicond. Sci. Technol.*, **25**, 115013.
178. Zhang, W., Takahashi, M. and Sakai, S. (2013), *Semicond. Sci. Technol.*, **28**, 085003.
179. Han, J.-P. (2004), Ferroelectric 1-T memory device — will it be viable for nonvolatile memory applications? In *Symposium D — Materials and Processes for Nonvolatile Memories*, **830**.
180. Tokumitsu, E., Fujii, G. and Ishiwara, H. (1999), *Appl. Phys. Lett.*, **75**, 575–577.
181. Al-Shareef, H. N. Dimos, D., Warren, W. L. and Tuttle, B. A. (1996), *J. Appl. Phys.*, **80**, 4573.
182. Arlt, G. and Neumann, H. (1988), *Ferroelectrics*, **87**, 109.
183. Grossmann, M., Lohse, O., Bolten, D., Boettger, U., Waser, R., Hartner, W., Kastner, M., and G. Schindler (2000), *Appl. Phys. Lett.*, **76**, 363.
184. Gruverman, A., Rodriguez, B. J., Kingon, A. I., Nemanich, R. J., Cross, J. S. and Tsukada, M. (2003), *Appl. Phys. Lett.*, **82**, 3071.
185. Park, B. H., Kang, B. S., Bu, S. D., Noh, T. W., Lee, J. and Jo, W. (1999), *Nature (London)*, **401**, 682.
186. Park, B. H., Hyun, S. J., Bu, S. D., Noh, T. W., Lee, J., H.-D. Kim, Kim, T. H. and Jo, W. (1999), *Appl. Phys. Lett.*, **74**, 1907.
187. Subbarao, E. C. (1961), *Phys. Rev.*, **122**, 804.
188. Joshi, P. C. and Krupanidhi, S. B. (1992), *J. Appl. Phys.*, **72**, 5517.
189. Shiosaki, T. and Shimizy, M. (1998), *J. Kor. Phys. Soc.*, **32**, S1316.
190. Funakubo, H., Watanabe, T., Morioka, H., Nagai, A., Oikawa, T., Suzuki, M., Uchida, H., Kouda, S. and Saito, K. (2005), *Mat. Sci. Eng. B*, **118**, 23.
191. Muller, J. *et al.* (2013), Ferroelectric hafnium oxide: A CMOS-compatible and highly scalable approach to future ferroelectric memories. In *Electron Devices Meeting (IEDM), 2013 IEEE International* 10.8.1–10.8.4.
192. Schroeder, U. *et al.* (2014), *Jpn. J. Appl. Phys.*, **53**, 08LE02.
193. Müller, J. *et al.* (2011), *J. Appl. Phys.*, **110**, 114113.
194. Mueller, S. *et al.* (2012), *Adv. Funct. Mater.*, **22**, 2412–2417.
195. Mueller, S. *et al.* (2012), *ECS J. Solid State Sci. Technol.*, **1**, N123–N126.
196. Müller, J. *et al.* (2012), *Nano Lett.*, **12**, 4318–4323.
197. Clima, S. *et al.* (2014) *Appl. Phys. Lett.*, **104**, 092906.
198. Müller, J., Polakowski, P., Müller, S. and Mikolajick, T. (2014), *ECS Trans.*, **64**, 159–168.
199. Müller, S., Müller, J., Schröder, U. and Mikolajick, T. (2013), *IEEE T. Device Mater. Res.*, **13**, 93–97.
200. Müller, J., Yurchuk, E., Schlosser, T., Paul, J., Hoffmann, R., Müller, S., Martin, D., Slesazeck, S., Polakowski, P., Sundqvist, J., Czernohorsky, M., Seidel, K., Kucher, P., Boschke, R., Trentzsch, M., Gebauer, K., Schroder, U. and Mikolajick, T. (2012), in *Symposium on VLSI Technology (VLSIT)*, p. 25–26.
201. Muller, J. *et al.* (2012), *IEEE Electron Device Letters*, **33**, 185–187.
202. Krowne, C. M., Kirchoefer, S. W., Chang, W., Pond, J. M. and Alldredge, L. M. B. (2011), *Nano Letters*, **11**(3), 988–992.
203. Cheng, C. H. and Chin, A. (2014), *Electron Device Letters, IEEE*, **35**(2), 274–276.

204. Baikie, T. *et al.* (2010), *Acta Cryst B*, **66**, 1–16.
205. Iwasaki, H., Sugii, K., Yamada, T. and Niizeki, N. (1971), *Appl. Phys. Lett.*, **18**, 444.
206. Yamada, T., Iwasaki, H. and Niizeki, N. (1972), *J. Appl. Phys.*, **43**, 771.
207. Iwasaki, H., Niyazawa, S., Koisumi, H., Sugii, K. and Niizeki, N. (1972), *J. Appl. Phys.*, **43**, 4907.
208. Ochida, N., Saku, T., Iwasaki, H. and Onuki, K. (1972), *J. Appl. Phys.*, **43**, 4933.
209. Krupanidhi, S. B., Roy, D., Maffei, N. and Peng, C. J. (1991), Proceedings of 3rd International Symp. on Integrated Ferroelectrics, 100. C. J. Peng etc. 1992.
210. Peng, C. J., Roy, D. and Krupanidhi, S. B. (1992), *Appl. Phys. Lett.*, **60**, 827.
211. Lee, J. J. and Dey, S. K., *Appl. Phys. Lett.*, **60**. 2487.
212. Li, T., Zhang, F. and Hsu, S. T. (1999), *Appl. Phys. Lett.*, **74**, 296.
213. Li, T., Hsu, S. T., Ulrich, B., Ying, H., Stecker, L., Evans, D., Ono, Y., Maa, J.-S. and Lee, J. J. (2001), *Appl. Phys. Lett.*, **79**, 1661.
214. Li, T., Hsu, S. T., Ulrich, B. D., Stecker, L., Evans, D. R. and Lee, J. J. (2002), *IEEE Electron Device Lett.* **23**, 339.
215. Li, T., Hsu, S. T., Ulrich, B. D. and Evans, D. R. (2005), *Appl. Phys. Lett.*, **86**, 123513.
216. Bune, A. V., Fridkin, V. M., Ducharme, S., Blinov, L. M., Palto, S. P., Sorokin, A. V., Yudin, S. G. and Zlatkin, A. (1998), *Nature*, 391, 874.
217. Ducharme, S. (2005), *IEEE Trans. on Dev. and Mat. Rel.*, **5**(4).
218. Reece, T. J., Ducharme, S., Sorokin, A. V. and Poulsen, M. (2003), *Appl. Phys. Lett.*, **82**, 142–144.
219. Gerber, A. *et al.* (2006), *J. Appl. Phys.*, **100**, 024110.
220. Fujisaki, S., Ishiwara, H. and Fujisaki, Y. (2007), *Appl. Phys. Lett.*, **90**, 162902.
221. Reece, T. J., Gerber, A., Kohlstedt, H. and Ducharme, S. (2010), *J. Appl. Phys.*, **108**, 024109.
222. Ionescu, A. M. *et al.* (2010), *IEEE Trans. Electron Devices*, **57**, 3518–3524.
223. Jo, J. *et al.* (2015), *Nano Lett.*, **15**, 4553–4556.

Chapter 10

Electro-Optically Active Oxides on Silicon for Photonics

Stefan Abel and Jean Fompeyrine
IBM Research — Zurich
Säumerstrasse 4
CH-8803 Rüschlikon
Switzerland
sab@zurich.ibm.com

1. Introduction

The class of oxide materials shows numerous properties that are appealing for many real-world applications, ranging from high-temperature superconductivity, ferromagnetism, tunable permittivities [Mikheev *et al.*, 2012, Reiner *et al.*, 2010], pyro-, ferro- and piezoelectricity [Baek *et al.*, 2011] and interfacial 2-dimensional electron gases [Kim *et al.*, 2013] to metal-insulator transitions [Yang *et al.*, 2011]. In addition, some functional oxides have interesting optical characteristics: their refractive indices can be altered by applying an electric field or by injecting a current. The present chapter discusses those electro-optical properties which are the fundament of both well-established applications such as high-speed electro-optic modulation in bulk crystals [Wooten *et al.*, 2000] and novel devices, for example efficient optical switches with very small footprints [Ryckman *et al.*, 2013]. Due to the importance for the emerging field of integrated silicon photonics and potential applications in quantum photonics [Banaszek *et al.*, 2001] and ultra-compact plasmonics [Diest *et al.*, 2009], the focus of the following discussion is set on the linear electro-optic field-effect in oxide thin films

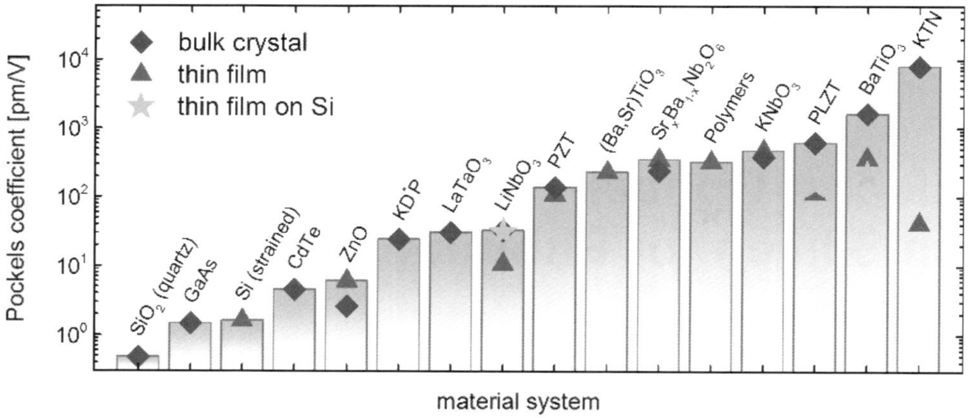

Fig. 1. Comparison of literature values for the Pockels effect in bulk crystals and thin films for various material systems [Chmielak *et al.*, 2011, Cook Jr, N.D.A, Cook Jr, N.D.B, Cuniot-Ponsard, 2011, Dalton *et al.*, 2010, Haertling and Land, 1971, Hoerman *et al.*, 2003, Huang, 1995, Kityk *et al.*, 2002, Rabiei *et al.*, 2013, Reitze *et al.*, 1993, Spirin *et al.*, 1998, Wang *et al.*, 2011, Wessels, 2007, Xiong *et al.*, 2014, Yariv and Yeh, 1984]. Data from films on silicon substrates are highlighted with a yellow star. Only largest Pockels coefficients for bulk and thin films, respectively, available in the wavelength range from 633 nm to 1550 nm are shown. The grey bars indicate the maximal value for each material system.

on silicon. Structural phase transitions in oxides that are accompanied by strong changes of the optical properties are also covered.

The linear electro-optic effect, also known as the Pockels effect, was already discovered in the 19th century, and many materials have electro-optically been investigated since then [Johnston and Weingart, 1965, Soref, 1992, Weis and Gaylord, 1985]. Oxide crystals show among the largest Pockels coefficients of all known materials (Fig. 1), which motivated the development of passive and active optical waveguiding components from oxide materials, such as linear phase shifters and acousto-optical modulators. The most prominent material platform is lithium niobate ($LiNO_3$) crystals: Passive optical waveguides are fabricated by slightly increasing the refractive index in $LiNO_3$ by interdiffusion of titanium, and active switches are realized by applying an electric field to these structures [Wooten *et al.*, 2000]. Even up to date this mature technology is state-of-the-art for high-speed electro-optical modulators [Janner *et al.*, 2009]. Due to the small index contrast in the waveguides, the bending radii are in the millimeter-range and cannot be reduced sufficiently enough to obtain ultra-compact, photonic integrated circuits (PICs).

To overcome this limitation, thin oxide films with strong Pockels coefficients have been deposited on host substrates, such as magnesium oxide and glass, where higher refractive index contrasts and thus smaller bending radii are obtained using fully etched waveguide cross sections [Wessels, 2007]. However, the effective electro-optical effect typically decreases in thin films compared to bulk crystals (Fig. 1). An example for such a decrease is found in barium titanate crystals, which have a strong linear electro-optical effect with Pockels coefficients of $r > 1500$ pm/V [Johnston and Weingart, 1965], among the largest Pockels coefficients known. Thin films of $BaTiO_3$ however show a large deviation from the bulk Pockels coefficients with typically much lower values as little as 22 pm/V for thin polycrystalline layers [Niu *et al.*, 2000, Petraru *et al.*, 2003, Tang *et al.*, 2005]. The linear electro-optical effect is even reported to completely vanish if strontium is mixed into the film [Kim *et al.*, 2003, Takeda *et al.*, 2009].

There are numerous possible origins of this reduction: Due to the generation of defects at interfaces and different thermal expansion coefficients between the film and the substrate, the crystalline symmetry and lattice constants are often altered in thin layers compared to their bulk counterparts. Because of the strong link between the Pockels effect and the microscopic crystal structure, this modification has a direct impact on the electro-optic properties. Similarly, the crystalline and ferroelectric domain structures influence the electro-optical response, as well as the presence of pores and grain boundaries. In order to approach the bulk Pockels coefficients in thin films, high-quality, single-crystalline layers would therefore be desired.

The integration of optical nonlinear oxide films with high crystalline quality on silicon is a challenging task, in particular compared to depositing them on oxide substrates: the formation of amorphous silicon oxide at the oxide/silicon interface may prevent any epitaxial and thus single-crystalline film growth and has therefore to be suppressed during the process. It is not surprising that only few examples of oxide films showing a large Pockels effect and grown on silicon substrates have been reported in literature (Fig. 1) despite the outlook for the low-cost fabrication of highly integrated photonic devices utilizing nonlinear optical effects. In addition to the difficulty of obtaining high-quality films with large Pockels coefficients, the electric field distribution is also affected in the presence of semiconducting silicon that shows a different conductivity and permittivity than oxide substrates. The higher conductivity of silicon can result in a strongly reduced overlap between the optical and the electric field, thus limiting the effective electro-optic response of the structure.

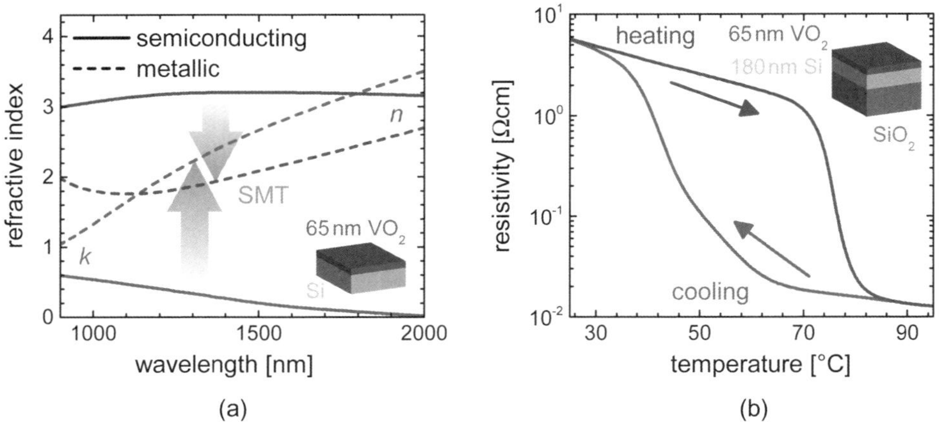

Fig. 2. (a) Real and imaginary part of the refractive index as a function of wavelength of VO_2 in the insulating and metallic phase, respectively. The green arrows indicate the change during the semiconductor-to-metal phase transition (SMT). (b) Resistivity as a function of temperature. The insets represent the sample structure. Data replotted from [Briggs *et al.*, 2010] with permission of OSA, The Optical Society.

Besides optical nonlinear materials, another class of oxides has recently come into focus for photonic applications, namely oxides that exhibit a structural transition between two crystalline phases with strongly different optical properties. A popular representative of this class of oxides is vanadium dioxide, which shows a semiconductor-to-metal phase transition (SMT) at 68°C with a large change of the electrical and optical properties (Fig. 2) [Briggs *et al.*, 2010, Verleur *et al.*, 1968]. Although it has intensively been studied for decades [Jerominek *et al.*, 1993, Verleur *et al.*, 1968] and many devices such as electronic switches and thermal sensors have been demonstrated [Yang *et al.*, 2011], VO_2 has only recently been applied to photonic structures on silicon [Briggs *et al.*, 2010, Ryckman *et al.*, 2013]. The transition between the semiconducting and the metallic state is a consequence of a change in the crystalline structure from a low-symmetry monoclinic to a higher-symmetry rutile phase [Verleur *et al.*, 1968]. Strictly speaking such transition is not a pure *electro-optic* effect since the phase transition is often triggered by temperature variations or optical stimuli. However, electric driven SMTs have already been demonstrated in all-electric devices [Ruzmetov *et al.*, 2009] and research for electro-optic devices is ongoing [SITOGA, 2014]. The present chapter therefore highlights the current developments of VO_2-based switches specifically for photonic applications while avoiding to discuss the mechanism of the SMT, which has been previously reviewed by Yang *et al.* [2011].

In the following sections, first the physical background of the electro-optic response is discussed (Sec. 2), before comparing different integration methods of nonlinear oxide layers on silicon such as epitaxial growth and wafer bonding (Sec. 3). Methods of how to characterize the Pockels effect of such thin films are presented in Sec. 4, with a detailed example of quantitatively determining the Pockels effect in barium titanate layers on silicon substrates. Finally, current and future applications are discussed (Sec. 5), and a conclusion and outlook for the field of nonlinear oxides on silicon is provided (Sec. 6).

2. Physical Background of Electro-Optic Effects

2.1. *Electro-optical effects in oxides*

Applying an electric field E to an oxide generally modifies its refractive index. In a simplified version, neglecting any tensor properties, the real part of the refractive index n is described by [Saleh and Teich, 2007]

$$n(E) = n_0 - \frac{1}{2} r\, n_0^3 E - \frac{1}{2} \xi\, n_0^3 E^2, \tag{1}$$

where r is the Pockels coefficient, ξ the electro-optic Kerr effect, and n_0 the refractive index when no electric field is applied. The second-order Kerr effect is present in all materials, while the first-order Pockels effect vanishes for centro-symmetric crystals [Nye, 1960]. Although both effects can be utilized in order to switch light [Kuhlow, 2006], the Pockels effect is often preferred compared to the Kerr effect for high-speed modulation because of the linear dependence of n on the applied electric field and the typical much larger induced phase shift, both of which is desired in many applications. The following sections therefore focuses on the first-order, linear electro-optical effect. A more detailed description of the quadratic, second-order electro-optical Kerr effect can be found for example in Saleh and Teich, 2007.

The microscopic origin of the Pockels effect is the small, field-induced distortion of the crystalline structure, which results in a perturbation of the optical properties as discussed in the next section. In addition to this distortion, charged defects can be created and chemical reactions can be triggered within the crystal as a consequence of the applied electric field or large intensities of the optical beam. These processes may alter the optical properties of the oxide in two ways: first, defects and modified chemical states alter the complex part of the refractive index and thus change the absorption characteristics, as described by the electrochromic and the photochromic effect [Irie, 2000, Mortimer, 2011]. Second, charged defects create internal electric fields, which can change the real part of the refractive index via the Pockels effect: incident photons can excite free charge carriers at impurities in the crystal, which may diffuse to other defects before recombining, leaving fixed charges at the initial impurity [Saleh and Teich, 2007]. The charges consequently create an electric field within the film. Various applications such as holographical storage and optical phase conjugation have been developed that utilize this so-called photorefractive effect as non-volatile and local variation of the refractive index. Since these applications are typically based on bulk crystals rather than on thin films on silicon, the photorefractive effect is not further discussed in this chapter. An extensive review of the physical background of photorefraction, suitable materials, and applications can be found in references Günter and Huignard, 2006, Günter and Huignard, 2007a, Günter and Huignard, 2007b, Nolte, 1995.

2.2. *Microscopic origin of the Pockels effect*

Among the different electro-optical effects, the electro-refractive effects in oxide thin films is in particular important for applications on silicon substrates (Sec. 5). On the microscopic scale, the non-linear optical response of crystalline oxides is described by two different mechanisms [Khatib *et al.*, 1993a, Kinase *et al.*, 1986]: First, the applied electric field perturbates the electron distribution in the crystal, corresponding to a change of the spontaneous local electric fields at the individual ions in the crystal. The modification of the local electric field due to the applied field impacts the electronic polarizability and thus the optical properties of the material, which can directly be calculated using a quantum-mechanical description of the individual wave-functions [Kinase *et al.*, 1986].

As a second mechanism, applying an electric field deforms the crystalline lattice due to the piezoelectric effect. The displacement of the ions varies the polarization that is induced by the electric field of the optical wave. It additionally changes the polarizability of the crystal and thus the permittivity and the refractive index [Khatib *et al.*, 1993b, Khatib *et al.*, 1993a, Kinase *et al.*, 1986]. The interaction of the mechano-optical, elasto-optical, and electro-optical effects is macroscopically described by coupling the corresponding tensor elements [Bernasconi *et al.*, 1995, Zgonik *et al.*, 1994].

Both of the two mechanisms, the change of the electronic polarizability due to the change of the local fields as well as the change of the optical properties due to the lattice deformation occur in stress-free crystals. However, if the film is mechanically clamped to a substrate, or if the applied frequency of the external electric field is above the piezoelectric resonance, a deformation of the crystalline lattice is suppressed and the effective electro-optic response is reduced, typically by a factor of ~ 2 compared to free crystals [Abarkan *et al.*, 2008, Günter and Zgonik, 1991]. Mathematically, the stress-free Pockels effect r^T can be represented as $r^T = r^s + r_a$, where r^S is the clamped Pockels effect and r_a is the electro-optic effect originating from acoustic lattice vibrations. The clamped Pockels tensor can again be split, into contributions caused by the perturbation of the electron distribution in the crystal, r_e, and contributions caused by optical phonons, r_o [Günter, 1987]. The total stress-free Pockels effect can thus be described as

$$r^T = r_a + r_o + r_e. \tag{2}$$

Using this equation, the frequency dependence of the Pockels effect and the two regimes of clamped and unclamped materials can be illustrated (Fig. 3). The stress-free Pockels effect is present up to $\sim 10^{12}$ Hz [Günter, 1987, Valdmanis, 1990], which allows the utilization of the electro-optic effect for applications operating at very high speed, such as ultra-fast electro-optic modulators.

The pure electronic contribution to the Pockels effect at even shorter time scales is equivalent to the second order non-linear susceptibility $\chi^{(2)}$, which

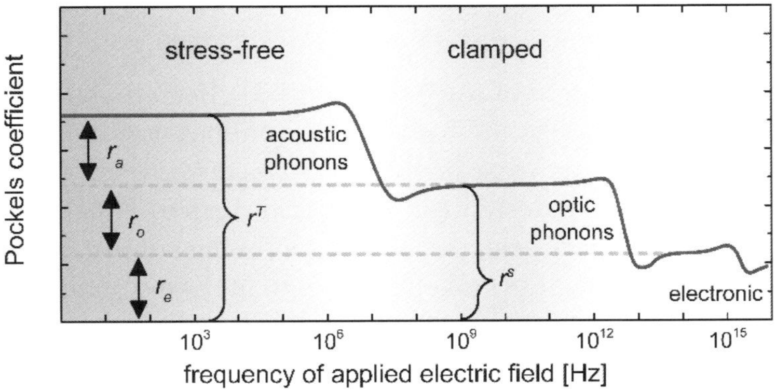

Fig. 3. Schematics of different contributions and the frequency dependence of the Pockels effect. The frequency range for the stress-free/unclamped and the clamped Pockels coefficients are highlighted in blue and yellow, respectively. Adapted from [Gnter, 1987] with permission of Taylor & Francis Ltd, http://www.tandfonline.com.

is commonly used at optical frequencies, via $r_e = -2\chi^{(2)}/n^4$ [Saleh and Teich, 2007]. $\chi^{(2)}$ is relevant for any nonlinear optical effects such as second harmonic generation or four-wave mixing, where the applied electric field E in Eq. (1) corresponds to the electric field of the optical beam. A more treatment of $\chi^{(2)}$-processes and related applications can be found in Boyd, [2003], Saleh and Teich, [2007].

In the recent decade, first principle density functional theory (DFT) studies have been carried out in order to gain better insight into the microscopic origin of the electro-optic behavior of optical nonlinear crystals, such as lithium niobate and barium titanate [Hermet *et al.*, 2009, Meng *et al.*, 2007, Veithen and Ghosez, 2002, Veithen *et al.*, 2004, Veithen *et al.*, 2005, Zhang *et al.*, 2010]. These studies separate the pure electronic contributions and a phonon-mediated electro-optic response and discuss the influence of phononic modes, the crystalline structure, temperature, and the bonding character between the individual ions on the optical properties of the crystal. New insight into the electro-optic behavior from atomistic models have been obtained, such as the difference between the strong nonlinear behavior of $BaTiO_3$ compared to structurally similar $PbTiO_3$, which is explained by the significantly larger atomic displacement of the ions along the Ti-O chains in $BaTiO_3$ [Veithen *et al.*, 2004]. Since the results of the electro-optic tensors calculated via DFT are close to experimental data [Sando *et al.*, 2014, Veithen *et al.*, 2004], first principle studies provide an excellent way to specifically design novel compounds with strong electro-optic properties.

2.3. *Phenomenological description of the Pockels tensor*

The simplified description of the electro-optical effect as shown in Eq. (1) is not sufficient to describe the response of nonlinear crystals as required for correctly interpreting experiments to characterize their Pockels effect (Sec. 4) and to properly design electro-optical devices [Abel *et al.*, 2013a, Janner *et al.*, 2009, Petraru *et al.*, 2002]. The following subsection provides a tensorial treatment of the Pockels effect, shown in the example of $BaTiO_3$. A phenomenological description of the Pockels effect is used to calculate the behavior of parameters such as the orientation of the optical axis and the refractive index, which are typically experimentally accessible.

In optical nonlinear materials the refractive index is dependent on the applied electric field E_{ext} as indicated in Eq. (1). The elements n_{ij} of the refractive index tensor as a function of the electric field are defined as [Saleh and Teich, 2007]

$$n_{ij}\left(E_{ext}\right) = n_0 - \frac{1}{2} r_{ijk}\, n_0^3\, E_{ext,l} - \frac{1}{2} \xi_{ijlm}\, n_0^3\, E_{ext,l}\, E_{ext,m} \tag{3}$$

In Eq. (3), Einstein notation is used, r refers to the Pockels tensor, ξ to the Kerr tensor, and $E_{ext,l/m}$ to different components of the applied field. The second order Kerr effect will not be considered here since it is typically much smaller than the Pockels effect in oxides with strong nonlinear properties [Günter, 1987, Zgonik *et al.*, 1994] at electric field strengths accessible in most experiments.

As can be deduced from the tensorial Eq. (3) the electro-optical response on the light depends on the wavevector k and the polarization relative to the electric field E_{ext} and the crystalline orientation, which defines the coordinate system of the equation. The impact of the wavevector and the polarization on Eq. (3) is reflected by the refractive index tensor n_{ij} [Saleh and Teich, 2007]. A visualization of Eq. (3) is shown in Fig. 4 in which the calculated optical indicatrix $\hat{\eta} = (\hat{n}^{-1})^2$ and its modification upon applying an electric field is plotted [Saleh and Teich, 2007], using the bulk material parameters of barium titanate as an example [Bernasconi *et al.*, 1995].

The polarization of light traveling through a birefringent material such as $BaTiO_3$ does not change if the polarization of the light is oriented along one of the two major axes. The ordinary and extraordinary refractive indices n_o and n_{eo}, respectively, as well as the orientation of these axes can be constructed from $\hat{\eta}$ as explained in the caption of Fig. 4. Typically, changes in n_o and n_{eo} and the rotation of the axes are experimentally detectable (Sec. 4). In order to obtain these values from Eq. (3), the optical indicatrix $\hat{\eta} = \hat{\eta}(\vec{E}_{ext})$ can be written as an eigenvalue equation using the displacement field $\vec{D}$ of the electro-magnetic light wave, the normalized wavevector $\vec{u}_k = \vec{k}/k_0$ with $k_0 = 2\pi/\lambda$ where λ is the wavelength, and the refractive index n [Saleh and Teich, 2007]:

$$\vec{u}_k \times \left(\vec{u}_k \times \hat{\eta}\left(\vec{E}_{ext}\right)\vec{D}\right) = \frac{1}{n^2}\vec{D} \tag{4}$$

The dependence of the tensor $\hat{\eta} = \hat{\eta}(\vec{E}_{ext}) = (\hat{n}^{-1}(\vec{E}_{ext}))^2$ on the electric field E_{ext} is given through Eq. (3). The eigenvalues n and eigenvectors $\vec{D}$ of $\vec{u}_k \times \hat{\eta}(\vec{E}_{ext})$ correspond to the refractive indices and the direction of the major optical axes of the crystal, respectively. The next subsection shows examples of solutions of Eq. (4) for specific sample and measurement geometries.

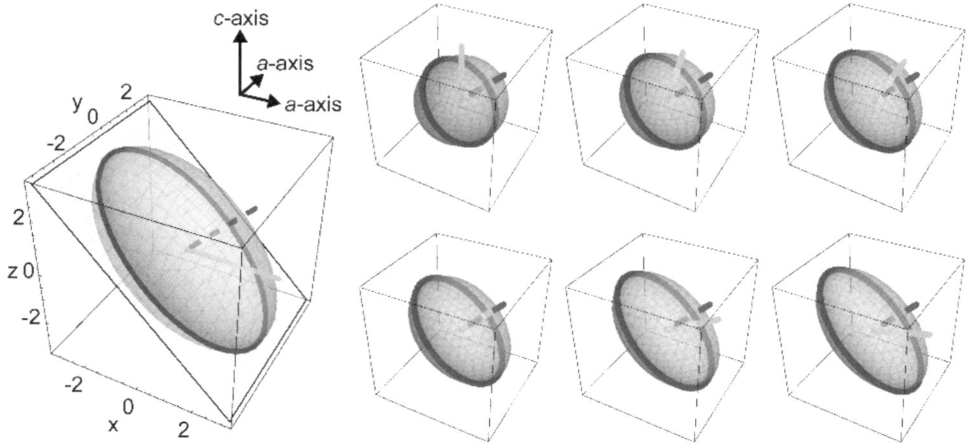

(a) (b)

Fig. 4. Three-dimensional representation of the optical indicatrix $\hat{\eta} = (\hat{n}^{-1})^2$ and the impact of an applied electric field on it, using the bulk $BaTiO_3$ Pockels tensor [Abel *et al.*, 2012]. The electric field strength is taken to be very large ($E_{ext} = 10^8$) V/m) for a clear visualization. (a) Illustration of the construction of the refractive indices for a specific orientation of the incoming light k (dashed, blue) and the electric field E_{ext} (yellow line): the short and long axis of the intersection ellipse (red) between the indicatrix and a plane perpendicular to k correspond to the ordinary and extraordinary index, respectively. The orientation of the $BaTiO_3$ crystalline axes is indicated by the coordinate system on the top. (b) Optical indicatrix for various orientations of the electric field.

2.4. *Impact of crystalline symmetry on the electro-optical response*

The electro-optic response of a crystal strongly depends on the relative orientation between the incident light, the applied electric field, and the crystalline orientation (Sec. 2.3). Consequently, the crystalline and ferroelectric domain structure directly impacts the effective electro-optical response of a crystal that is accessible during thin-film characterization and in integrated electro-optic devices. The following section describes the response of first a single domain in an epitaxial film, before including multiple, well-oriented 90°-domains. Finally, polycrystalline and amorphous films are treated.

Epitaxial films

The influence of the domain structure on the electro-optical response is illustrated in the example of a thin epitaxial film of tetragonal $BaTiO_3$ on silicon with the long, crystalline c-axis being parallel to the film surface. Due to the mismatch of the crystalline symmetry between the cubic silicon

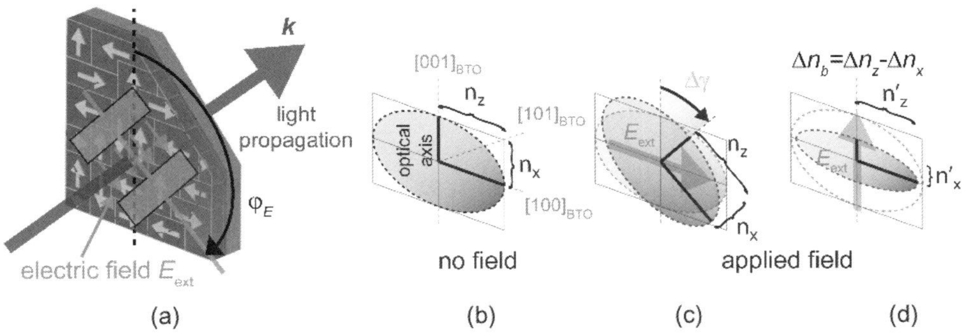

Fig. 5. (a) Example of the domain structure in ferroelectric $BaTiO_3$ on silicon. Two electrodes are used to create an in-plane electric field, which is oriented at an angle φ_E relative to the ferroelectric polarization (indicated by white arrows) of a single domain in the a-axis film (tetragonal c-axis of $BaTiO_3$ parallel to the surface). The light is transmitted perpendicular to the sample between the gap. (b) Refractive index ellipse in $BaTiO_3$, showing the refractive index n_x and n_z along different crystalline directions. (c) Applying an electric field E_{ext} rotates the optical axis by $\Delta\gamma$ and (d) changes the birefringence Δn_b depending on the orientation of E_{ext} (not to scale).

substrate and the tetragonal perovskite, 90°-domains are formed (Fig. 5(a)). This domain structure is commonly observed for ferroelectric $BaTiO_3$ films on silicon with thicknesses of >50 nm [Abel *et al.*, 2013b, Abel *et al.*, 2013a]. The changes of the birefringence in the $BaTiO_3$ layer depend on the direction of the applied electric field E_{ext}, which can experimentally be varied by fabricating electrodes with different orientations relative to the main crystalline axes ($[001]_{BTO}$ and $[100]_{BTO}$). However, even for a fixed electrode geometry, the relative orientation between E_{ext} and the ferroelectric polarization varies between different domains (Fig. 5(a)), and a different electro-optic response among the different domains is expected.

Response of a single-crystalline domain

When considering only one single domain in the film, the changes of the refractive indices $\Delta n_{o/eo} = n^0_{o/eo} - n_{o/eo}(E_{ext})$ and the birefringence $\Delta n_b = n_o(E_{ext}) - n_{eo}(E_{ext})$, as well as the rotation of the optical axis can be calculated as a function of the angle between the applied field φ_E and the ferroelectric polarization (Fig. 6). The results show the strong anisotropic behavior of the electro-optic response, and underlines the importance of considering the full electro-optic tensor when working with optical nonlinear thin films. In the case of $BaTiO_3$, the optical axis strongly rotates while the birefringence hardly changes when the electric field is perpendicular to the crystalline c-axis ($\varphi_E = \pm 90°$), as illustrated in Fig. 5(c). The opposite behavior is

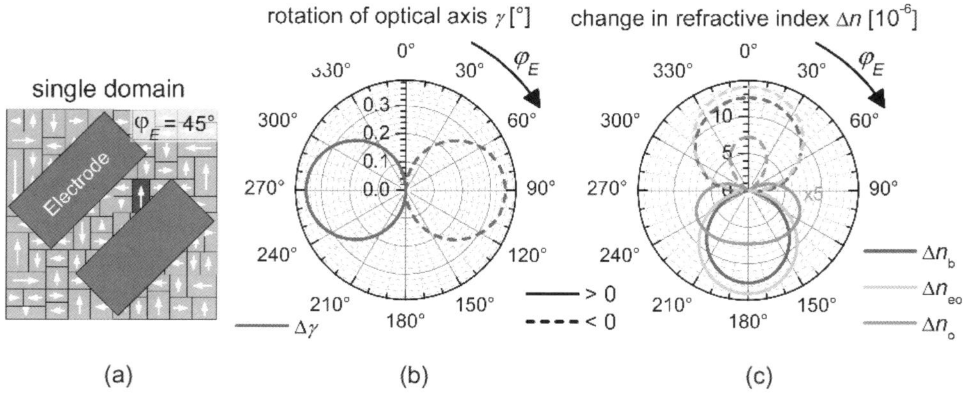

Fig. 6. Simulated response of a single ferroelectric domain in $BaTiO_3$ thin films. (a) Schematics of the domain structure shown for a specific electrode configuration with $\varphi_E = 45°$. (b) Rotation of the optical axis $\Delta\gamma$ and (c) variation of the birefringence Δn_b and the ordinary and extraordinary refractive index Δn_o (scaled by factor 5) and Δn_{eo}, respectively, as a function of the direction of the applied field. Solid and dashed lines correspond to positive and negative values, respectively. The definition of φ_E, $\Delta\gamma$ and Δn_b is illustrated in Fig. 5.

observable when the electric field is parallel to the *c*-axis ($\varphi_E = 0°/180°$, schematics in Fig. 5(d)).

Response of 180° *and* 90° *domains*

To calculate the macroscopic electro-optical properties of a multi-domain film as shown in Fig. 5(a), the different response of all domains and interference effects between the different polarization states have to be considered to describe the transmitted light. Applying an electric field at an arbitrary direction to a set of two 180° oriented domains causes a rotation of the optical axis within each domain but with opposite sign (Fig. 7(a)). Also, the change in birefringence is almost opposite between the two domains. For both domains, the polarization of the light will change after transmitting through the sample. However, the phase shift between the ordinary and extraordinary light component will be opposite, which results into the opposite sense of rotation of the transmitted, elliptically polarized light. The interference of these two polarization states, being left- and right-circular polarized, respectively, will result in a zero net-change in the polarization of the transmitted light.

By applying an electric field above the ferroelectric coercive field, domains can flip their orientation and 180° domains will be eliminated. In contrast, 90° domains are linked to a different crystallographic orientation and still exist after such a poling process. The response of two 90° domains

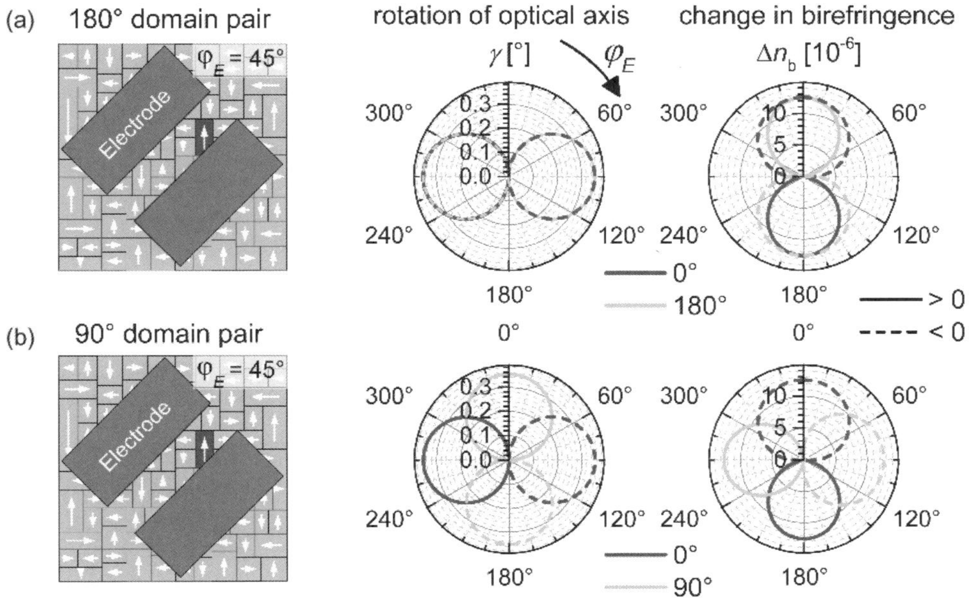

Fig. 7. Rotation of the optical axis and change in the birefringence for two domains with different ferroelectric polarization as a function of direction of the electric field in $BaTiO_3$ thin films, (a) 180° domains and (b) 90° domains. Solid and dashed lines correspond to positive and negative values, respectively. The schematics on the left shows the ferroelectric domain distribution in the example of an electrode configuration with $\varphi_E = 45^\circ$. Fig. 5 illustrates the definitions of φ_E, $\Delta\gamma$ and Δn_b.

generally shows a different magnitude when the direction of the electric field is varied (Fig. 7(b)). Thus, even if the two polarization states interfered destructively, they would not fully cancel each other out. In the case of 90° oriented domains the changes of the different polarization states even interfere constructively [Abel *et al.*, 2012], which can eventually be utilized in electro-optic devices. An example of how to experimentally observe the crystalline and ferroelectric domain structure in an electro-optic setup is given in Sec. 4.3.

Polycrystalline films

The tensorial character of individual domains in polycrystalline films with randomly oriented grains is macroscopically not visible because the overall response of such layer is the sum of many differently oriented crystalline regions. An average, isotropic response is visible. The properties of a polycrystalline film can be calculated by integrating and equally weighting the response of a single domain in all spatial angles, similarly to the procedure for two domains as described above.

On silicon substrates, oxide films often tend to be textured with a preferential rather than a fully random orientation of the domain [Wei *et al.*, 2005, Yoon and Kim, 1996]. In this case, the individual domains have to be weighted differently when calculating the net response of the film, leaving an over-all anisotropic electro-optic response [Ruglovsky *et al.*, 2006]. Therefore, a tensorial mathematical treatment is required to estimate the Pockels effect in textured films [Xiong *et al.*, 2012]. However, the anisotropy of the effective Pockels effect in the layers is typically reduced compared to single-crystalline oxides.

3. Integration of Electro-Optical Active Oxides on Silicon

As the technology of choice for integrated photonics, silicon photonics defines the boundary conditions for electro-optical active oxides to be used in potential applications. Different methods have been reported in order to obtain optical nonlinear oxide thin films of various crystalline perfections on silicon substrates. Using novel materials in active or passive optical devices is challenging, since such devices are extremely sensitive to structural and chemical defects that might impact light propagation in an undesired way. In addition and as reviewed in the previous section, electro-optical activity is a tensorial property, and the crystallographic axis of the active material should ideally be well defined with respect to the light polarization and propagation direction. The following section will therefore focus on methods to deposit (Sec. 3.1) and layer-transfer (Sec. 3.2) single crystalline layers in order to get the highest crystalline perfection as beneficial for the electro-optic response. Most of the work reported in this section deals with the deposition of perovskites as the most popular electro-optical active material, but also includes recent reports related to $LiNbO_3$ thin films.

3.1. *Epitaxial films*

Epitaxial film deposition

Epitaxy refers to the conditions where the crystalline orientation of a deposited film is determined by the crystalline structure of the underlying substrate. Heteroepitaxial processes are most relevant for this section, as the substrate and the deposited layer are of different chemical composition. In this particular case, one of the crucial steps is the building of the heteroepitaxial interface, especially the arrangement of the first deposited atoms on the host crystalline surface. Once this step is overcome, homoepitaxial processes are taking place, with the adlayer being chemically identical to the surface. Physically, the process and product of the deposition is described by the laws of thermodynamics, statistics, and quantum mechanics, as excellently reviewed by Herman *et al.* [Herman *et al.*, 2004]. At the atomic level, it involves many different processes such as adsorbtion, chemisorbtion, diffusion, reaction and reevaporation of different atomic and molecular species. This complexity provides many opportunities for scientists to design and fabricate novel crystalline heterostructures that would naturally not exist.

Additionally, the microstructure of the grown heteroepitaxial layers strongly depends on the differences in the crystalline structure of the two materials at the interface. The atoms can for example either arrange in small islands, or wet the surface and arrange regularly into a two-dimensional crystalline lattice on the substrate. In all cases, the mismatch of the lattice constants between the substrate and the film is accommodated by creating defects that will relax the adlayer. This mismatch varies with the growth temperature, according to the difference in the linear coefficient of thermal expansion. Finally, the crystal symmetry of the substrate and of the film can also be different, as for example when a perovskite is grown on a rock-salt structure. In this case, the structural mismatch can also lead to specific extended defects.

This very short description of heteroepitaxy shows that many parameters can be artificially varied during the deposition process. Understanding, controlling, and fine-tuning these numerous parameters represent the challenges to grow electro-optically active oxides on silicon.

Functional oxides on silicon

The epitaxial growth of perovskites oxides on silicon substrates is an excellent example of an artificial structure that could only be realized by thoroughly choosing the process conditions and using advanced deposition tools. As silicon immediately oxidizes in the presence of oxygen, the consequent formation of an amorphous SiO_2 layer on the surface hinders any further epitaxial growth. A first approach to achieve epitaxial growth, developed in the late 80s, relies on the reaction at high temperature between the deposited oxide and silica. It is widely used to obtain (001) oriented yttria-stabilized zirconium dioxide (YSZ) [Fukumoto *et al.*, 1988]. In this case, the interfacial silica is progressively consumed and the lattice registry between silicon and the oxide can take place. A second approach aims to start from the crystalline silicon surface and to build the oxide lattice while avoiding the formation of any silicon dioxide. Being more elegant, this method is also technologically more relevant as it enables the *direct* growth of perovskites on silicon. This research field was pioneered by McKee *et al.* [McKee *et al.*, 1991] who for the first time epitaxially deposited a very thin perovskite material, barium titanate ($BaTiO_3$), on silicon substrates. They extended their work and defined a generic process to grow strontium titanate ($SrTiO_3$) [McKee *et al.*, 1998], a common substrate when used as a bulk crystal to grow other perovskite oxides. The epitaxial growth of $SrTiO_3$ on silicon (Fig. 8(a)) relies on a surface treatment of the silicon wafers to create one monolayer thick, ordered alkaline earth metal-silicide structures such as SrSi or BaSi. This template is commensurate to silicon and less reactive towards oxygen than bare silicon. By carefully choosing the oxygen partial pressure and temperature, the heteroepitaxial growth of $SrTiO_3$ is possible, using in particular solid-phase epitaxy. A few groups worldwide have developed different version of a similar process, and $SrTiO_3$ grown on silicon is today available on large wafers up to 200 mm, with ultra-thin layers as thin as 4 nm, and with a structural quality approaching the one of compound semiconductors grown on silicon [Demkov and Posadas, 2014, Gu *et al.*, 2009, Norga *et al.*, 2006, Park *et al.*, 2009].

So far, this process has been achieved using molecular beam epitaxy (MBE). This deposition method, briefly described below, provides in the same tool a large range of oxygen partial pressures and temperatures. In particular, it enables the deposition at low temperature ($<200°C$) and low oxygen partial pressure ($<10^{-7}$Torr). This combination is critical to control the process described by McKee *et al.* [1998]. A similar process for integrating epitaxial oxides on semiconductors has been reported using germanium substrates and atomic layer deposition [McDaniel *et al.*, 2014], taking advantage of the lower stability of GeO_2 to minimize the interfacial oxidation.

Besides providing a sensitive control over the interface, MBE deposition tools also enable an accurate regulation over the stoichiometry of the grown layer. Since perovskites are ternary systems, reaching the appropriate ratio between the two cationic sublattices is important. To achieve this control, MBE systems exploit the large mean free path length of atoms in the reactors, which is in the order of several meters in the molecular beam regime. Therefore, solid materials such as metals can locally be evaporated and, in a properly-designed chamber, directed as an atomic or molecular beam towards a specific position where the sample is mounted (Fig. 8(b)). The atomic compositions at the sample surface can be controlled by blocking the beams of different sources with mechanical shutters. Sub-monolayer

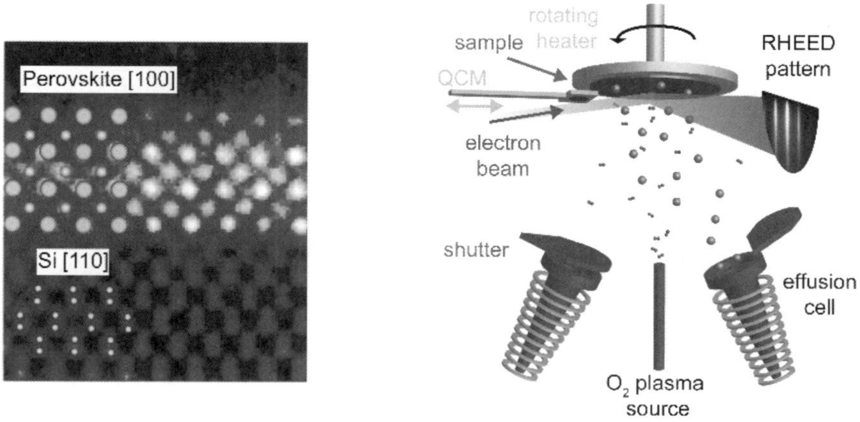

(a) (b)

Fig. 8. (a) Z-contrast transmission electron microscopy image of epitaxial $SrTiO_3$ on silicon. Reprinted from [McKee *et al.*, 1998] with permission from APS. Copyright (1988) by the American Physical Society. (b) Schematics of a MBE system: The sample is mounted on a rotating heater, onto which different atomic and molecular beams from gas and metal sources are directed. A quartz crystal microbalance (QCM) serves for calibration of the fluxes prior to the deposition. A reflection high energy electron diffraction image can be obtained using an electron beam at grazing incidence to the sample.

(ML) coverages can be well controlled by using low atomic fluxes, and the rates are typically determined using a quartz crystal micro-balance (QCM) prior to the deposition process. More advanced schemes to control the stoichiometry are also considered, and can be sorted in two set of methods: first, methods with the goal to improve the long term stability of the evaporation sources. It can be achieved by for example using quadrupole mass spectrometry to monitor the actual evaporation rate. More recently, hybrid physical-chemical routes have been successfully demonstrated [Jalan *et al.*, 2009], as the flux of chemical precursors can be better controlled than the direct evaporation of metals. The second sort of methods to control the stoichiometry follows the goal to improve the monitoring of the actual material flux, in order to compensate deviations from the targeted composition by a precise shuttering sequence [Haeni *et al.*, 2000]. Other methods have also been investigated, ranging from optical absorption to *in-situ* ion spectroscopy or X-ray fluorescence [Braun and Ploog, 2003].

Oxygen stoichiometry is also an extremely important parameter in the growth of functional oxides. This is in particular true for titanium based materials that are considered for electro-optic applications, as titanium can have a mixed valence to accommodate possible oxygen vacancies. It has been shown that only a minute amount of oxygen must be introduced during the initial stage of growth to enable the crystal registry between silicon and the oxide [Norga *et al.*, 2005]. On the contrary, highly reactive oxygen species must be provided to avoid an excessive oxygen deficiency while growing the oxide in a vacuum chamber. Oxygen can be easily introduced into the chamber as a molecular species. A more reactive form of oxygen can be generated by plasma-cracking of the molecular oxygen or by ozone generation, leading to a better oxidation of the deposited oxide. Provided a high quality, thin $SrTiO_3$ nucleation layer has been grown on the silicon substrate, other deposition techniques than MBE can be used to grow high quality perovskite oxides with high oxygen partial pressures: physical vapor deposition (PVD), metal-organic chemical vapor deposition (MOCVD), and pulsed laser deposition (PLD) have routinely been used to grow perovskites on $SrTiO_3$ substrates [Abel *et al.*, 2013b, Guiller *et al.*, 2008]. The work reported by Guiller *et al.* [2008] clearly demonstrates that a thin $SrTiO_3$ seed layer promotes the subsequent PLD growth of the ferroelectric $KTa_xNb_{(1-x)}O_3$. Attempts to grow the same material in direct contact with silicon will on the contrary result in a non-ferroelectric pyrochlore phase.

Ultimately, we can say that a path exists for the heterogeneous integration of functional perovskites with silicon. Today, the majority of the research in this area aims to exploit some of the unique properties of the crystalline oxide layer itself. Ferroelectricity has been reported in ultrathin $BaTiO_3$ layer grown on silicon [Dubourdieu *et al.*, 2013] with the aim to exploit ferroelectricity as a source for negative capacitance. Piezoelectricity can be exploited in novel switching device concepts based on the coupling between a piezoelectric and piezoresistive layer [Doll and Pruitt, 2013]. Ferro- and ferrimagnetic oxides have been obtained using a similar approach [Bachelet *et al.*, 2011, Posadas *et al.*, 2011], paving the way towards application for spin injection devices. It has also been demonstrated that 2D-electron gases observed at the interface between $LaAlO_3$ and $SrTiO_3$ could be obtained in structures grown on silicon [Park *et al.*, 2010], opening a possible path to exploit unexpected physical phenomena in new type of devices. Most of the research mentioned above aims however to find a replacement for the 50 years old silicon based field effect transistor as a switching digital device. The unique properties of some of functional perovskites can however be exploited for other purposes, for example leading to giant piezoelectricity as demonstrated in PMN-PT heterostructures [Baek *et al.*, 2011], where an enhanced piezoelectric response is obtained for single crystalline oxide grown on micro-machined silicon. As photonic technologies become increasingly important, there is a growing interest in exploiting the peculiar optical properties of crystalline oxides, both in passive and active devices. The strong linear electro-optic response of prototype compounds such as $BaTiO_3$ can for example be used in active optical devices, which will be reviewed more specifically in the following paragraphs.

Epitaxial $BaTiO_3$ layers on silicon

Owing to its excellent electro-optical bulk properties, $r_{BTO} > 1500\,\text{pm/V}$ (Sec. 1), $BaTiO_3$ is a promising candidate for active optical components in silicon photonics [Abel *et al.*, 2013a, Petraru *et al.*, 2002, Wessels, 2007]. It undergoes a transition from a ferroelectric tetragonal phase to a paraelectric cubic phase when heated above its Curie temperature of $T_c = 123°C$ [Sakayori *et al.*, 1995]. In the cubic phase the Ti atoms are octahedrally coordinated by six oxygen atoms. In the tetragonal phase atoms are displaced along the z-direction and the unit cell is elongated along the *c*-axis (Fig. 9(a)). This displacement results in a variety of functional properties such as piezoelectricity [Damjanovic *et al.*, 2002], ferroelectricity [Choi *et al.*, 2004], and electro-optic activity [Abel *et al.*, 2013a, Wessels, 2007].

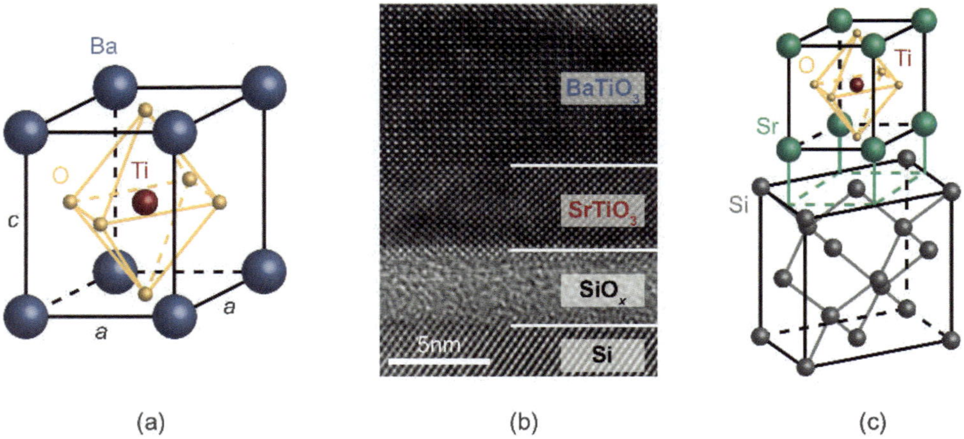

Fig. 9. (a) Tetragonal unit cell of $BaTiO_3$ with one long *c*-axis and two shorter a-axes. (b) High resolution transmission electron microscopy (HRTEM) image of $BaTiO_3$ epitaxially grown on silicon with a $SrTiO_3$ buffer layer. The amorphous interfacial SiO_x is formed during the high-temperature $BaTiO_3$ deposition and does not harm the epitaxial growth [Abel, 2014]. (c) Schematics of the epitaxial growth of a perovskite on silicon: The $SrTiO_3$ unit cell is rotated by 45° in respect to the silicon unit cell. The interface between the two layers is simplified in the illustration compared to first-principle calculations [Ashman, 2004, Hellberg, 2012].

As mentioned in section 2.2, the optical nonlinear properties originate from a displacement of the Ti atoms along the *c*-axis with respect to their centrosymmetric position in the unit cell. The functional properties of $BaTiO_3$ are therefore absent in the cubic phase [Nye, 1960]. Although the tetragonal phase is stable at room temperature in bulk $BaTiO_3$ [Kay and Vousden, 1949], thin films consisting of polycrystalline $BaTiO_3$ tend to stabilize in the cubic symmetry when the grain size is smaller than a few tens of nanometers [Jang *et al.*, 1997, Salama and Siciunas, 1972]. It is therefore important to ensure that both the grain size and the crystalline orientation is well defined with respect to the underlying silicon substrate. Epitaxial growth during the monolithical integration of $BaTiO_3$ thin films onto Si substrates is then required. As a perovskite structurally closely related to $SrTiO_3$, $BaTiO_3$ can easily be grown epitaxially on silicon, using a thin $SrTiO_3$ seed layer as described above (Fig. 9(b)) [Abel *et al.*, 2013a, Niu *et al.*, 2011]. During the epitaxial growth of $BaTiO_3$, the large lattice-mismatch of -26% is reduced to $-4-5\%$ by rotating the $BaTiO_3$ crystalline lattice by 45° [McKee *et al.*, 1991]. This rotation is imposed by the underlying $SrTiO_3$ and leads to the epitaxial relationship $[100]_{BTO}||[110]_{Si}$ (Fig. 9(c)).

Once the tetragonal structure is stabilized, it has recently been shown that the orientation of the long *c*-axis of the tetragonal lattice strongly depends on the deposition conditions. It is determined by the different thermal expansion coefficients of silicon and $BaTiO_3$ [Abel *et al.*, 2013b], and by the epitaxial strain introduced in the $BaTiO_3$ layer at the growth temperature. At high growth temperatures, fully relaxed $BaTiO_3$ layers crystallize in a cubic structure. Due to the smaller thermal expansion coefficient of Si, $BaTiO_3$ layers are subjected to biaxial tension when cooled through the Curie temperature. To relieve such biaxial tensile strain the films tend to orient with their longer *c*-axis in the plane of the substrate and the short *a*-axis out-of-plane, resulting in an in-plane ferroelectric polarization of the $BaTiO_3$ films and a 90° domain structure (Fig. 10). The growth of tetragonal $BaTiO_3$ on Si with the long *c*-axis perpendicular to the film surface has nevertheless been reported using $10-30$ nm thick $Ba_{0.7}Sr_{0.3}TiO_3$ buffer layers [Vaithyanathan *et al.*, 2006], and recently also with 4 nm thin $SrTiO_3$ buffer layers (Fig. 10) [Abel *et al.*, 2013b, Dubourdieu *et al.*, 2013, Niu *et al.*, 2011]. In this case it is believed that the large epitaxial strain present at the growth temperature is not compensated during cooling, and that a residual in-plane compressive strain forces the long *c*-axis to orient out-of-plane.

Practically, the evolution of the crystalline orientation of $BaTiO_3$ thin films as a function of their thickness has been found to provide a way to

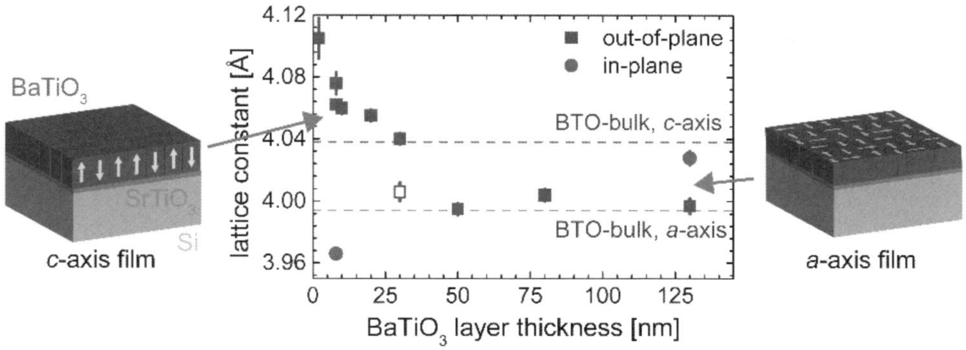

Fig. 10. Out-of-plane (red) and in-plane (blue) lattice parameter of MBE-grown $BaTiO_3$ films on $SrTiO_3$/Si templates determined via X-ray diffracton (XRD) analysis. The open red square shows a second lattice parameter visible at a thickness of 30 nm, indicating a transition regime with both *a*- and *c*-axis domains. The dashed lines correspond to the bulk lattice constants. The illustrations show a *c*-axis and an *a*-axis film, respectively, where white arrows indicate the direction of the ferroelectric polarization which is parallel to the crystalline *c*-axis [Abel, 2014, Abel *et al.*, 2013b].

adjust the electro-optical response depending on the crystalline orientation of the film (Fig. 10). The thickness is primarily determined by the optical device design: typically, layer thicknesses of ~50 nm are required for electro-optical applications in order to get a sufficiently large overlap between light and the active material [Abel *et al.*, 2013c, Petraru *et al.*, 2002, Wessels, 2007, Xiong *et al.*, 2014]. The control of the crystalline orientation gives then an additional design parameter to fully exploit the electro-optical activity of $BaTiO_3$.

3.2. *Layer transfer via bonding*

The recent development described in the previous section is limited to the direct growth of perovskites onto silicon. Many oxides whose optical properties are relevant for silicon photonics active devices are however not crystallizing in this structure. One can cite as an example $LiNbO_3$ and yttrium iron garnet (YIG) whose electro-optical and magneto-optical properties, respectively, have been long exploited in bulk crystal. $LiNbO_3$ is for example one of the most widely used materials for bulk modulators frequently used in telecommunication application [Wooten *et al.*, 2000]. It is transparent in a wide range from ultraviolet to infrared, and its non-centro-symmetricity is associated with a non-zero Pockels coefficient of about 30 pm/V. Few successful reports can be found to show that $LiNbO_3$ can be grown *c*-axis oriented on silicon [You *et al.*, 2013]. Unlike perovskite materials though, large $LiNbO_3$ wafers can be purchased from commercial suppliers, with diameter up to 6″. The difficulty to obtain high quality grown material on silicon, combined with the availability of bulk $LiNbO_3$ wafers, make bonding an attractive approach for the monolithic integration with silicon.

Generally, bonding refers to many different approaches and processes where heterogeneous integration is performed by creating adhesive forces between a substrate and the heteroepitaxial layer. It is not the purpose of this section to provide an exhaustive review [Gösele and Tong, 1998]. The research in this area can be classified in different approaches, depending whether direct or indirect bonding is performed, and whether it takes place at the wafer level or at the die or device level. In all cases though, an economically viable process can only be developed if thin layers can be transferred from a donor substrate in such way that the substrate can possibly be re-used for multiple bonding steps. For $LiNbO_3$, helium implantation has been reported as an effective method to split thin layer from a crystal [Huang *et al.*, 2014]. Upon an appropriate thermal treatment, $LiNbO_3$ slabs exfoliate from the source wafer. Common to all bonding methods, special care should be given to the difference of thermal expansion coefficients between silicon and the donor wafer. When the transferred layer is thick ($>1\mu m$) and the bonding temperature high, large differences in thermal expansion coefficients could lead to cracks in the heterostructures.

Recent work reported the use of $LiNbO_3$ platelets co-integrated with Si photonics structures. In this case, the adhesion forces between the $LiNbO_3$ platelets and the silicon wafer are provided by Benzocyclobutene (BCB) [Chen and Reano, 2012]. Platelets from tens to hundreds of micrometers can be transferred onto silicon photonics structures. BCB bonding does not need the stringent requirements that direct bonding has in terms of surface roughness. It puts however severe constraints on the post-processing of such structures and is challenging to envision as an industrial process to monolithically integrate $LiNbO_3$ with Si photonics. Direct bonding has therefore also been considered both for the transfer of pieces or full wafers. Several millimeter long, and few μm thick $LiNbO_3$ microcrystals have been dispersed on the target SOI wafer containing the Si photonic devices [Lee *et al.*, 2011]. The localization is random as in the case of BCB bonding, and an additional 1000°C anneal is required prior to bonding to recover from ion implantation damages. This relatively large temperature forces the annealing to be performed on $LiNbO_3$ substrate so that the difference in CTE does not strain the transferred material. Finally, a similar process has been reported in [Rabiei *et al.*, 2013] but at the full wafer level. Figure 11

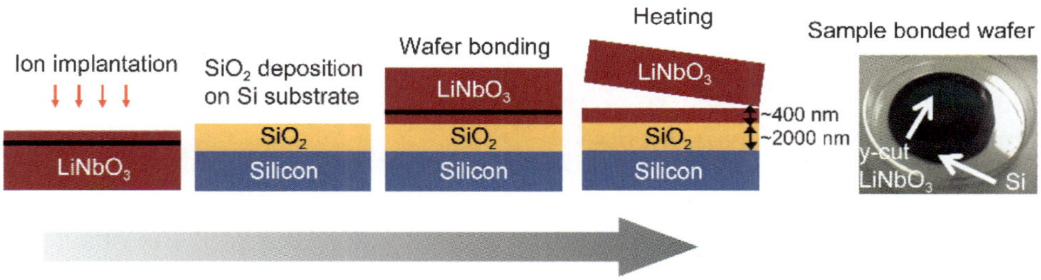

Fig. 11. (a)–(d) Direct bonding of thin $LiNbO_3$ layers on silicon using bulk $LiNbO_3$ substrates and ion implantation. (e) Photograph of a successful bonded 3" wafer LNO wafer on a 4" silicon wafer. Adapted from [Rabiei *et al.*, 2013] with permission of OSA, The Optical Society. Provided by Partow Technologies LLC.

schematically shows the direct wafer bonding process mediated by silica, which was enabled by the sub-nm roughness obtained on un-processed $LiNbO_3$ wafers. In this case, 3″ $LiNbO_3$ wafers were used for a layer transfer to 4″ SOI wafers. However, the limit for this process is only given by the availability of $LiNbO_3$ wafers and could in principle be scaled up to 6″. Up to now, all wafer bonding methods for $LiNbO_3$ result in rather thick transferred layers, intrinsically limited by the use of ion implantation for layer splitting. Thicknesses range from 0.5 μm to a few μm, and set boundaries for the post-processing and the design of hybrid Si photonic circuits containing $LiNbO_3$: First, the bonding annealing process cannot exceed 250°C due to the large difference in CTE, and second, no advanced waveguides with large optical confinement in the active oxide structures such as slot-waveguides [Abel *et al.*, 2013c, Almeida *et al.*, 2004, Xiong *et al.*, 2014] are possible.

4. Electro-Optical Characterization

Compared to bulk materials, a larger variation of nonlinear coefficients is generally reported in thin films, because the morphology and crystalline structure of the thin layers depend on the substrate and the deposition methods. Also the oxide/semiconductor interface might strongly impact the film properties for example by introducing interfacial charges that in turn create depolarization fields. Therefore, the optical nonlinear properties of thin films have to be determined specifically for each deposition process, since the estimation from previous studies or from bulk properties is generally not possible.

Light propagating through an electro-optic material experiences a phase-shift when applying an electric field, as a consequence of the modification of the refractive index (Eq. (1)). This phase shift can cause a change in the polarization or the intensity of the probing light, which is experimentally detectable. There are however two main experimental challenges in the measurements:

- The change of the refractive index n is relatively small (typically, $\Delta n \sim 10^{-3} - 10^{-5}$) which often results in only small changes of the observed signal. These small changes can be at the edge of the experimental resolution, in particular for thin films where the interaction length between light and the electro-optical material is very short.
- The tensorial nature of the electro-optical effect (Sec. 2.3) makes the determination of the Pockels and Kerr tensor complicated because the measurement values depend on the relative orientation between the crystal, the applied electric field, the direction of the propagation and the polarization of the light. Furthermore, the existence of ferroelectric domains (Sec. 2.4) has to be taken into account.

Different solutions have been reported that tackle these issues. The most common methods to electro-optical characterize thin-films are briefly compared in the following section. Comments will be made if characterization techniques are applicable for investigating optical properties of materials with semiconductor-to-metal phase transitions (SMT) such as VO_2. Due to the large optical contrast between the different crystalline phases (e.g. $\Delta n > 1$ for semiconducting and metallic VO_2 [Briggs *et al.*, 2010]), commercially available standard techniques such as ellipsometry are typically used to investigate these films. The focus of current research for oxides with SMTs is on the mechanisms and dynamics of the phase transitions using techniques such as time resolved optical spectroscopy [Cavalleri *et al.*, 2004] and THz spectroscopy [Pashkin *et al.*, 2011, Wall *et al.*, 2012], which will not be covered in the present chapter.

4.1. *Thin film electro-optical characterization techniques*

Ellipsometry

Spectroscopic ellipsometry is a wide-spread technique to characterize the optical properties of thin films. It is based on the detection of changes in the polarization of light in thin films for a broad wavelength range and different angles of incidence. The detected signal is then compared with a model of the sample in order to obtain values such as the refractive index, layer thicknesses, or interface roughnesses [Tompkins and Irene,

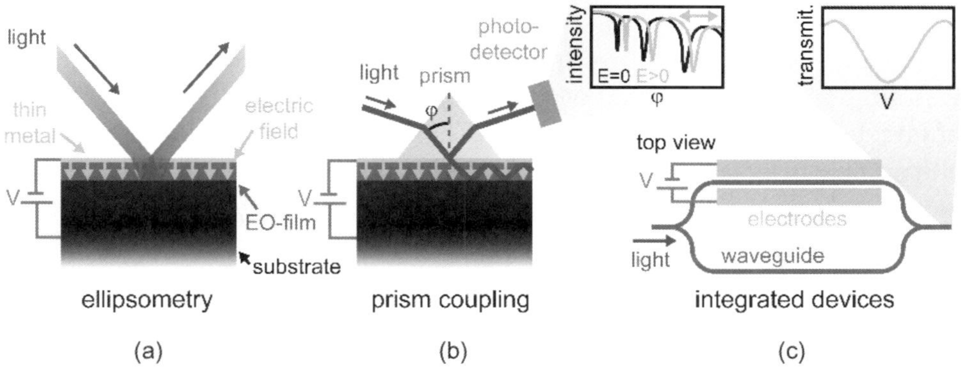

Fig. 12. Common methods to investigate the electro-optical properties of thin films. (a) Reflection methods sensitive to changes in the polarization, such as spectroscopic ellipsometry or the Teng-Man method. (b) Prism coupling to excite planar optical modes in the active layer. The reflected intensity drops for specific angles of incidence, which can be shifted by applying an electric field (see inset). (c) Active Mach-Zehnder interferometer as example for an integrated device. The transmission is modified for electro-optical structures when applying a voltage to the electrodes (see inset).

2005]. The commercially available tools can be extended to characterize electro-optical properties by applying an electric field to the sample and comparing the refractive indices as determined for different field strengths [Kang *et al.*, 2008]. Typically, thin, semitransparent metal top-electrodes and the grounded substrate are used to apply a field (Fig. 12(a)). The method suffers however from a rather low sensitivity, as a reliable detection of $\Delta n \sim 10^{-3}$ is at the limit of the technique, in particular for very thin films of only a few tens of nm. Additionally, the electro-optical effect is not implemented in the standard modeling tools which make the tensorial analysis complicated. Nevertheless, the effective Pockels coefficients of $\sim 1\,\mu$m-thick oxides have successfully been measured with spectroscopic ellipsometry [Kang *et al.*, 2008]. Due to the low sensitivity, especially for films with thicknesses below 100 nm and potentially low Pockels coefficients, more sensitive methods are often preferred for characterizing thin electro-optic oxides.

Thin oxide films with semiconductor-to-insulator phase transitions have successfully been optically characterized using spectroscopic ellipsometry (Fig. 2(a)), which can resolve the large optical contrast without special modifications of the tools [Briggs *et al.*, 2010, Kana *et al.*, 2011]. However, the phase transitions are thermally triggered and the technique serves as a method to investigate individual crystalline phases rather than obtaining their electro-optically characteristics.

Teng-Man reflectometry

A similar method like ellipsometry with a significantly better resolution was suggested by Teng and Man [1990] and Schildkraut [1990], which is often referred to as Teng-Man method. The major difference to spectroscopic ellipsometry is the modulation of the electric field in combination with a lock-in detection scheme, which increases the sensitivity by several orders of magnitude. Additionally, a laser instead of a broadband light source is used. Beside this standard configuration, many modifications such as a combination of perpendicular incidence with interdigitated electrodes for in-plane electric fields have been implemented [Ma *et al.*, 2009]. Although the Teng-Man technique is widely spread for thin film characterization [Enami *et al.*, 2007, Park, 2008], it is prone to misinterpretation of the measurement signals [Park, 2008]. Interference effects from multiple reflections, the influence of the semi-transparent electrode, and the non-perpendicular angle of incidence that requires the consideration of all electro-optical tensor elements for a proper analysis are some of the typical error sources [Park, 2008, Wang *et al.*, 1995].

Transmission measurements

When detecting changes in the light polarization using a transmission geometry, the data evaluation is simplified compared to the Teng-Man method, in particular when using perpendicular angle of incidence [Abel *et al.*, 2013(a)]. Already early electro-optical measurements of bulk crystals were performed in transmission with static applied fields [Johnston and Weingart, 1965]. The drastically reduced interaction length between the light and the electro-optical material requires lock-in amplification techniques in order to study thin films [Abel *et al.*, 2013a, Adachi *et al.*, 1983, Hoerman *et al.*, 1999, Reitze *et al.*, 1993]. Although the method is typically used to obtain effective Pockels values rather than single tensor elements, it can be extended to a full tensor analysis. An example of obtaining single tensor elements is discussed in Sec. 4.3.

Standard transmission and reflection measurements have also been used for investigating the optical changes of oxides with SMT [Jerominek *et al.*, 1993, Verleur *et al.*, 1968]. In that case, the total reflected and transmitted power of a laser beam is detected, without considering changes of the polarizations. The technique is not used as an electro-optic characterization method since the phase transitions are thermally induced. By using electrodes and electrically triggering the SMT [Ruzmetov *et al.*, 2009] a full electro-optic investigation might however be possible.

Prism coupling

A very sensitive method to measure the refractive index of thin films with a resolution of $\Delta n - 5 \times 10^{-4}$ is prism coupling [Hunsperger, 2009]. With this method, a guided planar waveguide mode in the film is excited by using a prism in contact to the film and a laser beam at a specific angle of incidence (Fig. 12(b)). When the angle of incidence is varied, the reflected signal shows sharp minima when film modes are excited. The refractive index can be deduced from multiple of these specific angular positions. Options for an electro-optical characterization of the film by applying an electric field between the prism and the sample [Herminghaus *et al.*, 1991, Liu *et al.*, 2009, Rabiei and Gunter, 2004] are commercially available. However, the method cannot be easily applied to oxide thin films on Si samples: no guided modes exist in such material stack due to the higher refractive index of silicon compared to oxide materials. In addition, film thicknesses of several 100 nm are required to support multiple modes, which is more than the film thicknesses typically needed for integrated optics.

Integrated devices

The fabrication of test devices such as Mach-Zehnder interferometers (Fig. 12(c)) [Chmielak *et al.*, 2011, Palmer *et al.*, 2013, Petraru *et al.*, 2003, Xiong *et al.*, 2014] or resonant structures like ring resonators [Lee *et al.*, 2011, Xiong *et al.*, 2012] can be used to determine the electro-optical properties of a material embedded in this structure. While devices can reveal even small Pockels effects (< 2 pm/V [Lee *et al.*, 2011]), the design and processing is an important overhead compared to a pure material characterization technique and can possibly induce modification in the EO material itself. This is in particular true for structures in oxides on silicon where non-typical design and process steps are involved in the device fabrication [Abel *et al.*, 2013c, Rabiei *et al.*, 2013, Xiong *et al.*, 2012, Xiong *et al.*, 2014].

Alternative techniques

Beside these most common methods, a variety of other techniques for characterizing the electro-optical properties of thin films exist, such as the utilization of diffraction patterns caused by a local variation of the refractive index with interdigitated electrodes [Tayebati *et al.*, 1996], the inclusion of electro-optically active materials into cavities [Hou *et al.*, 2007, Uchiki and Kobayashi, 1988], or by detecting the interference pattern of a double slit [Kalluri *et al.*, 1996]. These methods are more specialized, and do not provide any general advantage for the analysis of nonlinear oxide thin films on silicon compared to previously described measurements.

4.2. *EO-characterization with silicon substrates*

In order to properly characterize and utilize electro-optic field effects in thin films, such as the Pockels effect, it is necessary to generate a significant electric field within the active layer. Many studies of electro-optic thin films have been performed with insulating oxide substrates [Petraru *et al.*, 2002, Teng and Man, 1990] or with thick insulating buffer layers [Enami *et al.*, 2007, Xiong *et al.*, 2012], both of which hardly influence the electric field lines between the electrodes and thus allow a simple approximation of the field in the active region.

However, if the film is close to or in direct contact with a semiconducting silicon substrate, the field is partially screened by free electrons and holes. The strength of the remaining electric field in the active region depends on the doping of the silicon, the geometry of the cross section, and the frequency of the applied signal (Figs. 13(a) and (b)). To generate in-plane electric field lines in the electro-optic film, low-doped silicon substrates or insulating buffer layers with a thickness larger than the gap between the electrodes are required.

In order to obtain an out-of-plane electric field, the requirements are reversed: Highly doped silicon without any insulating buffer layer can be used as bottom electrode (Fig. 13(c)). Alternatively, conductive oxides between silicon and the electro-optic oxide such as $LaNiO_3$ [Qiao and Bi, 2010] and $SrRuO_3$ [Dekkers *et al.*, 2009] may be implemented. Due to their crystal structure, conductive oxides are often compatible with the

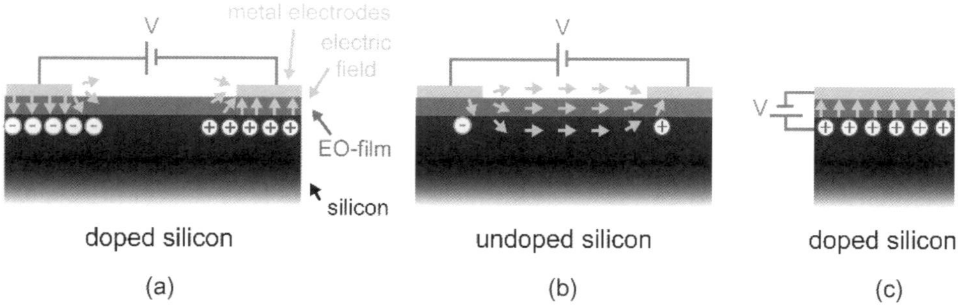

Fig. 13. (a) Screening of the electric field by charges in the semiconducting, doped silicon substrate. (b) Non-vanishing in-plane electric field in the active region for undoped silicon, where only few free carriers are available to screen the electric field. (c) Out-of-plane electric field utilizing doped silicon as a bottom electrode.

epitaxial deposition processes of high-quality electro-optic active films. While conductive electrodes close to the active region are acceptable for thin film characterization, they can typically not be used in integrated photonic devices because of their high optical absorption. The high absorption is in particular critical when light is travelling along the interface, for example in integrated waveguides. A compromise between conductivity, switching speed, optical losses, and voltage requirements has to be made in the device design process.

4.3. *Example of EO-characterization: Epitaxial $BaTiO_3$ on silicon*

The following section describes the experimental determination of the Pockels effect in more detail than the previous section with the example of a ~130 nm thin barium titanate film integrated on a silicon substrate. Such layer consists of multiple 90° domains to relax strain due to the symmetry mismatch between cubic silicon and tetragonal $BaTiO_3$ as indicated in Fig. 5(a) (Sec. 3.1). Despite the long history of analyzing the example ferroelectric barium titanate over the past century, the electro-optic response for thin films on silicon has only recently been investigated [Abel *et al.*, 2013a].

Measurement and analysis procedure

One way to characterize nonlinear optical properties of thin layers is to determine changes in the polarization of light transmitted through the film upon applying an electric field. Figure 14 shows a sensitive experimental setup using lock-in detection scheme, which is capable of detecting small rotations of transmitted light of only 10^{-5} degrees. Such sensitivity allows to probe extremely small changes of the refractive index and enables the electro-optic analysis of even very thin layers (~10 nm).

By systematically changing measurement parameters such as the applied voltage, the orientation of the analyzer, and the incident polarization, different electro-optic contributions such as the Pockels effect, the Kerr effect, and electro-absorption effects can be separated. The orientation of the electric field relative to the crystalline axes can easily be changed in the experiment by depositing differently oriented pairs of electrodes on the film surface. The investigation of the induced phase shift as a function of the angle between the applied electric field, incident polarization, and crystallographic orientation is crucial to interpret the electro-optic effect properly. In particular, it allows to probe single elements of the electro-optic tensor. Figure 15 illustrates the importance to consider the tensorial

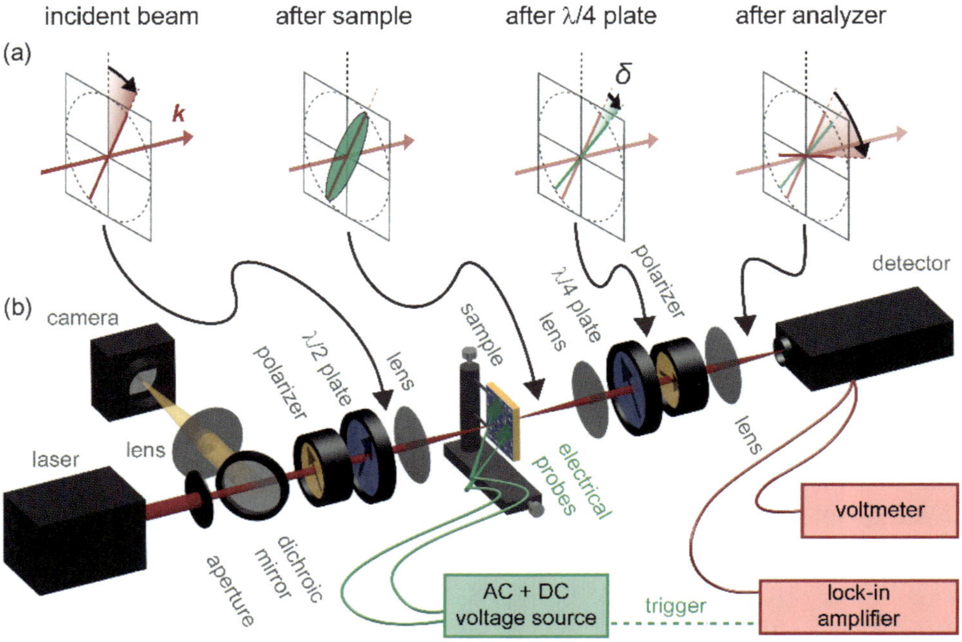

Fig. 14. Characterization of electro-optical properties via transmission measurements. (a) Polarization states at various positions of an initially linear-polarized laser beam in an electro-optic characterization setup with transmission geometry. The electro-optic response of the sample causes a rotation δ of the linearly polarized beam. (b) Schematics of the optical and electrical elements. Experimental details and operation principle are described in Ref. [Abel *et al.*, 2013a].

nature of the Pockels effect when investigating the film: the induced phase shift due to the electro-optical effect changes by more than one order of magnitude when the electric field is rotated by 45°. This behavior can only be consistently understood when taking into account the crystalline symmetry, the ferroelectric domain structure (see below), and the Pockels tensor of the layer [Abel *et al.*, 2013a].

Electro-optical visualization of ferroelectricity

Thin ferroelectric films often consist of multiple crystallographic domains, which establish during the deposition process due to lattice mismatches and different thermal expansion coefficients (Sec. 3.1) and multiple ferroelectric domains, as a consequence of energy minimization by reducing external electric fields [Rabe *et al.*, 2007]. The macroscopic electro-optic behavior strongly depends on the ferroelectric configuration of a film (Sec. 2.4): the net electro-optical response of two anti-parallel domains with the same size will cancel out. By applying an electric field above the coercive field of

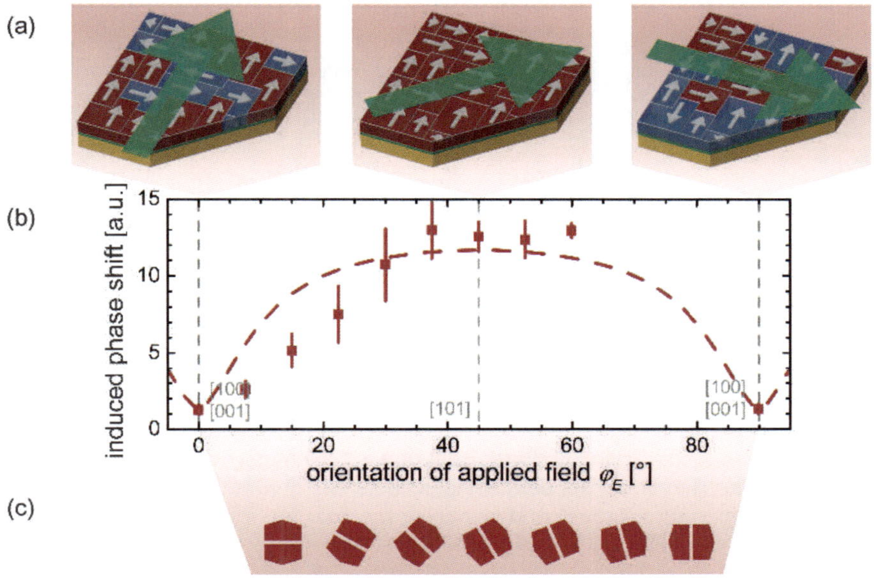

Fig. 15. (a) Schematics indicating the different domain configurations in ~130 nm $BaTiO_3$ thin film on silicon for various orientations of the electric field (green), with ferroelectric domains oriented towards the field (red) and randomly oriented ones (blue). (b) Electro-optically induced phase shift for different orientations of the electric field. The dashed curve shows the simulated response based on the experimentally determined Pockels tensor. (c) Illustration of various electrode orientations used to experimentally realize different orientations of the electric field. Data from [Abel *et al.*, 2013a].

the material, the ferroelectric polarization can be flipped and anti-parallel domain pairs can be eliminated. Hence, a ferroelectric hysteresis loop is observed in the optical measurements upon varying the applied electric field (Fig. 16). Besides extracting optical nonlinear properties, electro-optic transmission measurements provide an unambiguous way to characterize ferroelectric properties such as the coercive field, the remnant polarization and the temporal stability of the domain configuration.

5. Applications for EO-Thin Oxide Films on Silicon

5.1. *Data communication*

Due to their strong optical nonlinear properties, electro-optic oxides have been used for decades in integrated optics for switching, routing, and modulation. In particular lithium niobate has evolved to a mature material platform which has remained state-of-the-art for high-speed, low-loss modulation up to date [Janner *et al.*, 2009, Wooten *et al.*, 2000]. Active devices have also been demonstrated for other electro-optical oxides such as $BaTiO_3$

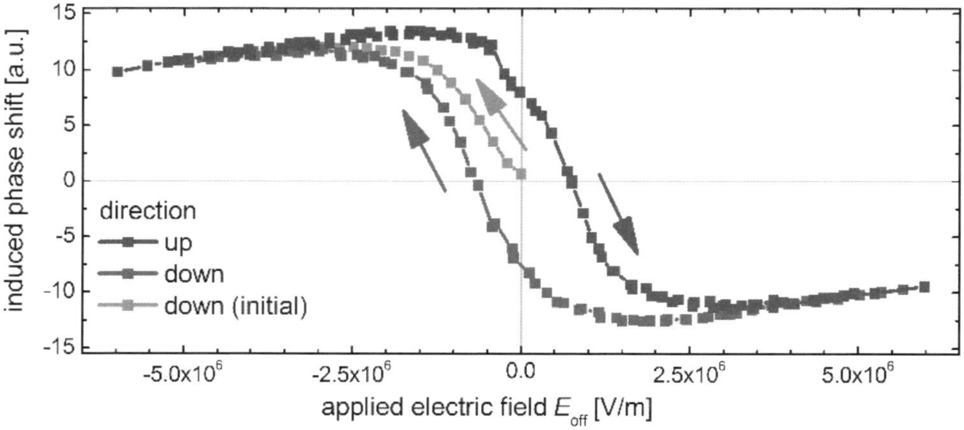

Fig. 16. Electro-optically induced phase shift of a ~130 nm $BaTiO_3$ thin film on silicon as a function of the applied field. The data set is collected at 25°C after heating the film above its bulk Curie temperature (green) and sweeping the field up (red) and down (blue) again. Measurement details in Ref. [Abel *et al.*, 2013a].

and PLZT thin films on oxide substrates [Adachi and Wasa, 1991, Gill *et al.*, 1997, Petraru *et al.*, 2002, Wessels, 2007].

In recent years silicon photonics evolved quickly as a new platform for integrated photonics. Triggered by the transparency of silicon at the relevant telecommunication wavelengths and the possibility for low-cost manufacturing in current CMOS (complementary metal-oxide-semiconductor) fabrication, a technology for intra-chip data connection has rapidly advanced [Beausoleil *et al.*, 2008, Haurylau *et al.*, 2006, Miller, 2009]. Manifold novel devices were designed to guide, couple, filter, or modulate light [Horst *et al.*, 2013, Reed *et al.*, 2010, Taillaert *et al.*, 2006] — all based on silicon. Despite the great progress in the field of silicon photonics, many challenges such as the integration of light sources and efficient detectors still need to be solved [Liu *et al.*, 2010, Michel, 2010]. Modulators operating at high speeds of more than 40 Gb/s have been shown [Reed *et al.*, 2010, Xiao *et al.*, 2013], but the plasma-dispersion effect utilized in these devices typically causes high insertion losses. Furthermore, active tuning of resonant structures in order to compensate temperature drifts and variations during the fabrication remains a challenge. Local heaters and athermal claddings are potential solutions [Atabaki *et al.*, 2010, Teng *et al.*, 2009], but they suffer from an increased power consumption, and cannot be used to account for process variabilities, respectively.

The integration of optical nonlinear materials into the silicon photonic platform could thus enable novel high-speed modulators and zero-power tuning elements, and add additional degrees of freedom when engineering integrated photonic circuits [Abel, 2014]. Since bulk silicon does not show any Pockels effect, a hybrid platform of nonlinear materials and silicon photonics is required. The first realization of this concept has recently been shown with nonlinear polymers [Leuthold *et al.*, 2009, Palmer *et al.*, 2013]. The limited thermal stability of organics during the device fabrication remains however a challenge.

Epitaxial barium titanate thin films on silicon

More stable, nonlinear oxides have only very lately been integrated into silicon photonic circuits: The first passive structures were demonstrated with a slot-waveguide cross-section where the active $BaTiO_3$ layer is sandwiched between two silicon layers in order to increase the optical power in the active region (Fig. 17) [Abel *et al.*, 2013c]. Xiong *et al.* [2014] developed a similar fabrication process with almost identical waveguide cross-section. Both groups could demonstrate passive devices such as grating couplers, ring resonators, and Mach-Zehnder interferometers (Fig. 17(c)). Although the $BaTiO_3$-based devices in principle show the expected performance, they

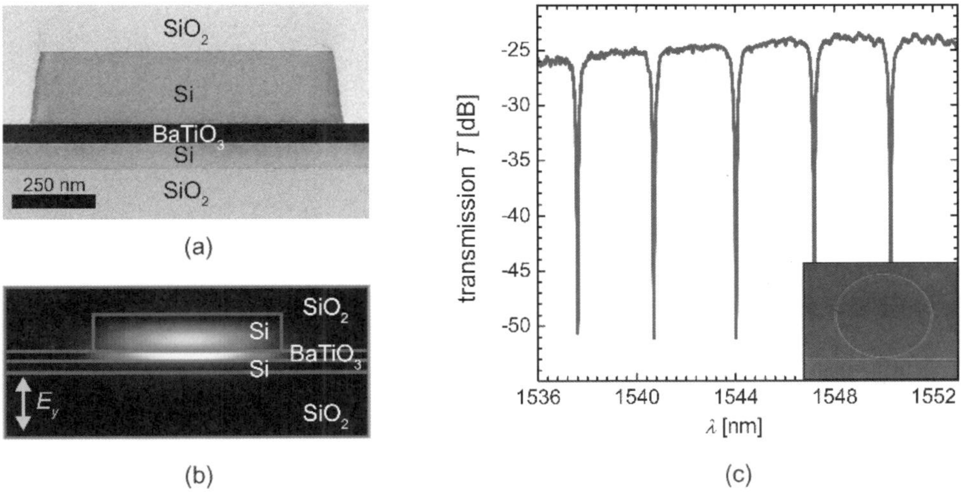

Fig. 17. Passive $BaTiO_3$/Si waveguides: (a) bright-field scanning transmission electron microscopy (STEM) cross section of a partially etched slot waveguide structure. (b) Distribution of the optical power for a TM-like mode with the electric field mainly oriented along the y-direction. A major part of the optical power is confined in the $BaTiO_3$ layer. (c) Transmission spectrum of a ring resonator with a cross section as shown in (a). The inset shows a scanning electron microscopy (SEM) image of the device (radius of ring: $30\mu m$). Data from [Abel, 2014, Abel *et al.*, 2013b].

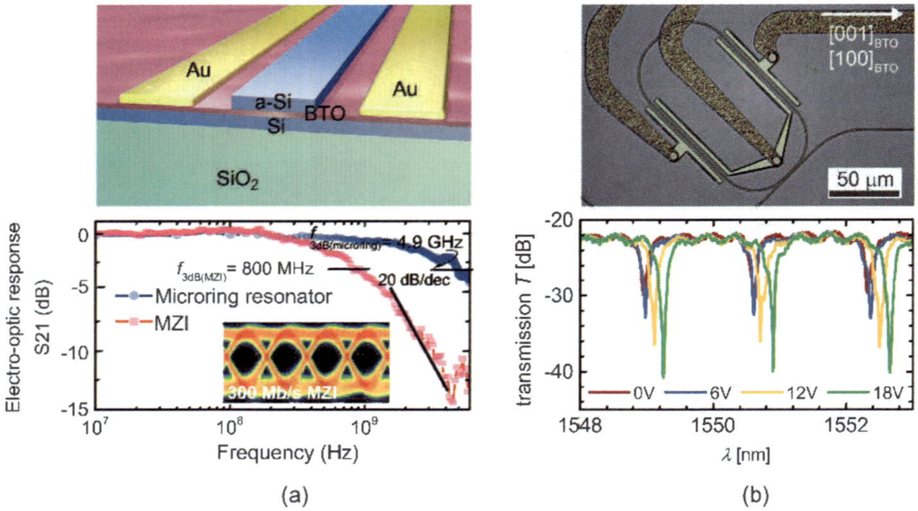

Fig. 18. (a) Cross-section (top) and high-speed measurement results (bottom) of a $BaTiO_3$-based ring resonator and Mach-Zehnder modulator. Adapted from [Xiong *et al.*, 2014] with permission of OSA, The Optical Society. (b) SEM image (top) and observed spectral shift of the resonances (bottom) in active race-track resonator with similar cross section as shown in (a) [Abel, 2014, Abel 2016].

suffer from rather high propagation losses in the order of 50 to 100 dB/cm [Abel, 2014, Xiong *et al.*, 2014] whose origin is not understood up to now.

In addition to passive structures, also switching and modulation in active devices was observed [Abel, 2014, Abel, 2016, Xiong *et al.*, 2014]. Ring- and Mach-Zehnder modulators with a bandwidth of 5 GHz and 0.8 GHz, respectively, were demonstrated (Fig. 18a) [Xiong *et al.*, 2014]. The corresponding effective Pockels coefficient in the $BaTiO_3$ layer is 213 pm/V, which is almost a magnitude of order higher than in $LiNbO_3$ modulators. Besides applications for high-speed switches, Pockels-effect based devices might also be used as low-speed tuning elements with extremely low power consumption, which are necessary in highly scaled, resonant silicon photonic circuits. Figure 18b shows the first results of a $BaTiO_3$-based race track resonator whose resonance can be tuned by 18 pm/V after applying an electric field (Fig. 18b) [Abel, 2014]. However, a clear experimental proof to exclude the contribution of other physical effects and to identify the Pockels effect as the origin for the switching behavior in integrated devices is currently still missing.

The results on $BaTiO_3$/Si hybrid devices are a promising first step towards the integration of highly efficient electro-optic oxides into silicon photonic structures. Various challenges such as the reduction of propagation losses and switching speeds above 40 Gb/s have however still to be solved before silicon photonics can clearly benefit from the integration of $BaTiO_3$ thin films.

Lithium niobate thin films on silicon

For the past decades lithium niobate ($LiNbO_3$) has been the standard oxide material for electro-optic applications. Even though the Pockels coefficients of $LiNbO_3$ are by more than an order of magnitude lower than in $BaTiO_3$, they are quite large compared to many alternative materials. The advantage of $LiNbO_3$ over $BaTiO_3$ is the availability of large wafers and the low mismatch between the electrical and optical propagation constants [Wooten *et al.*, 2000]. Due to the low index contrast between $LiNbO_3$ waveguides and the cladding, the bending radii have to be kept in the mm-range which results in a large footprint of the devices.

State-of-the art $LiNbO_3$ devices are based on bulk single crystals. The recent integration of $LiNbO_3$ on silicon substrates tries to combine both the high integration density of silicon photonics with the mature optical nonlinear technology of lithium niobate. Since epitaxial growth for obtaining high-quality crystalline layers directly on silicon is not possible, processes to bond thin layers of $LiNbO_3$ on silicon have recently been developed (Sec. 3.2). Using this technology, active silicon photonic devices with $LiNbO_3$ cladding have been fabricated (Fig. 19) [Chen *et al.*, 2013, Lee *et al.*, 2011], and a tunability of up to 12.5 pm/V in 30 μm ring resonators has been achieved. These results are in particular promising as they

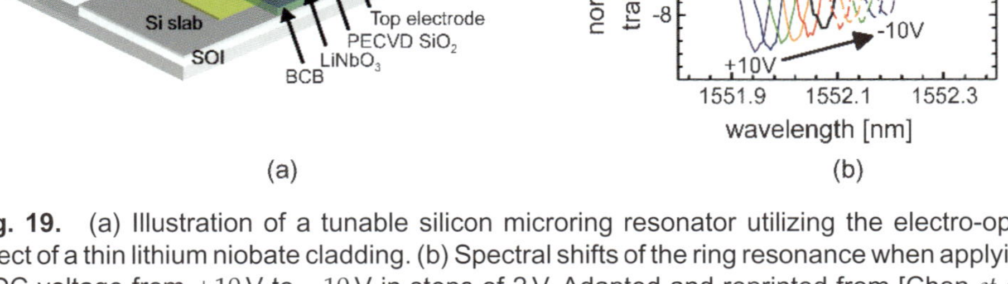

(a) (b)

Fig. 19. (a) Illustration of a tunable silicon microring resonator utilizing the electro-optic effect of a thin lithium niobate cladding. (b) Spectral shifts of the ring resonance when applying a DC voltage from $+10$ V to -10 V in steps of 2 V. Adapted and reprinted from [Chen *et al.*, 2013] with permission of OSA, The Optical Society.

provide a similar tunability as state-of-the-art silicon microring resonators [Dong *et al.*, 2010, Li *et al.*, 2011] at zero power consumption. Zero power consumption is a clear advantage compared to standard silicon photonic tuners that are based on heaters or the plasma-dispersion effect [Dong *et al.*, 2010, Li *et al.*, 2011], and shows the added value of complex oxides to the silicon photonics platform. The recent demonstration of a $LiNbO_3$/Si modulator operating at 9 Gb/s also demonstrates the applicability of this hybrid platform for high-speed switching in addition to low-speed tuning [Chen *et al.*, 2014].

In a different approach, lithium niobate thin films are bonded on thermally oxidized silicon wafers [Rabiei *et al.*, 2013]. Successively, $LiNbO_3/Ta_2O_5$ rib-waveguides and electro-optic modulators were fabricated from this layer stack. Although the waveguides are not in direct contact to any silicon photonic structures, the demonstrated technique allows to overcome thermal problems of $LiNbO_3$-wafer based devices, and could enable the co-integration of electrical and photonic circuits in the silicon and $LiNbO_3$ layer, respectively, on the same wafer.

Vanadium dioxide for optical switching

The semiconductor-to-metal transition in vanadium dioxide has a large impact on the optical properties and motivated the integration of VO_2 into silicon photonic structures [Briggs *et al.*, 2010, Ryckman *et al.*, 2012, Ryckman *et al.*, 2013]. The phase transition causes changes of the real and complex part of the refractive index by $\Delta n > -1.06 + 2.62i$ between the monoclinic crystal structure at room temperature and the rutile crystal structure at 100°C (Fig. 2) [Briggs *et al.*, 2010]. These changes are very large compared to the variation of the optical properties in optical nonlinear materials such as lithium niobate and barium titanate. They can thus be utilized for ultra-compact optical switches such as ring resonators with a diameter of only 3μm [Ryckman *et al.*, 2012, Ryckman *et al.*, 2013]. Since the spectral shift of the resonances induced by the phase transition is a couple of nm and thus relatively large, the resonators are applicable as efficient broadband switches, which are less prone temperature variations and process instabilities. Besides phase-shifting devices, also absorption switches of only $0.5\,\mu$m length relying on the variation of the complex part of the refractive index in VO_2 have been demonstrated [Ryckman *et al.*, 2013].

Current VO_2/Si hybrid photonic devices are typically designed as standard silicon photonic structures that are partially covered with ~70 nm thin VO_2 layers (Fig. 20) [Briggs *et al.*, 2010, Ryckman *et al.*, 2012, Ryckman

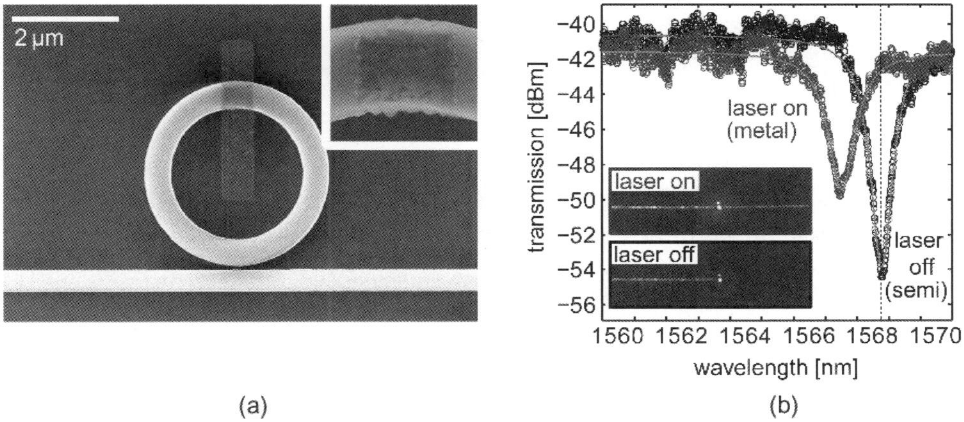

Fig. 20. (a) SEM image of a hybrid VO_2/silicon photonic ring resonator with a diameter of 3μm. The structural phase transition in the VO_2 is triggered by a laser beam focused on the active region. The inset shows the magnification of the active VO_2 (shaded in red) film on the waveguide. (b) Transmission spectra of the ring with VO_2 in the semiconducting/insulating and conductive state. The vertical, infrared radiation of the device is shown in the inset. Adapted from [Ryckman *et al.*, 2012] with permission of OSA, The Optical Society.

et al., 2013]. Light evanescently couples into the VO_2 cladding, enabling active switching. Although larger VO_2 pads might increase the switching performance of the devices, the interaction length is often restricted to few μm due the relatively large optical losses even in the semiconducting state of VO_2 of $\sim 5 \times 10^4$ dB/cm [Briggs *et al.*, 2010]. The short interaction length thus keeps the insertion losses of the devices low, e.g. <2 dB for small absorption modulators [Ryckman *et al.*, 2013].

Increasing the switching speed is one of the major challenges for VO_2-based photonic devices: The switching time of devices relying on a thermally activated semiconductor-to-metal phase transition (SMT) [Briggs *et al.*, 2010, Ryckman *et al.*, 2012] was lately reduced by utilizing a photo-induced SMT [Ryckman *et al.*, 2013]. Limited by the length of the pump beam triggering the transition in the experiment, switching speeds of $\sim$30 ns were reached. Faster transitions at time scales of $\sim$100 fs in blanket films have already been shown [Cavalleri *et al.*, 2004, Wall *et al.*, 2012], which in principle allows operation in the THz regime. In contrast to this athermal, fast SMT, the reverse transition from the metallic rutile to the semiconducting monoclinic phase is a thermally mediated relaxation process at larger time scales [Ryckman *et al.*, 2013]. The switching time for this process observed in current photonic structures is $\sim$70 ns. By optimizing the thermal design and the device geometry, for example by reducing the VO_2 film thickness or using cladding layers, the time constant might be reduced to a few ps

[Ryckman *et al.*, 2013], as observed in blanket VO_2 films [Lysenko *et al.*, 2007, Zhang and Ramanathan, 2011]. The relaxation time could in principle even further be reduced to sub-picosecond time scales when not completing the SMT [Pashkin *et al.*, 2011, Ryckman *et al.*, 2013]. The utilization of such techniques might ultimately enable high-speed VO_2 photonic modulators.

Besides increasing the speed of the devices, finding a way to electrically drive the optical switches remains a challenging task. The application of current temperature-controlled [Briggs *et al.*, 2010, Ryckman *et al.*, 2012] or all-optical switches [Ryckman *et al.*, 2013] in dense, integrated photonic networks is limited due to slow switching speeds and high-optical powers required to drive the transition. However, since electric control of the SMT in VO_2-based transistors has already been shown [Ruzmetov *et al.*, 2009, Yang *et al.*, 2011] and since there is no fundamental limitation to use a similar mechanism in photonic devices, the outlook for future highs-speed electric controlled VO_2-based switches is very promising. Indeed, major progress has very recently been achieved, resulting into first demonstrations of electro-optical switching in VO_2 based silicon photonic devices [Markov, 2015, Joushaghani, 2015].

5.2. *Sensing*

The detection of electric fields at high-frequency has become an important task for designing and characterizing modern electric circuits and devices, detecting radio-frequency circuits, and receiving high-power microwaves [Chen and Reano, 2012, Hsu *et al.*, 2007]. Since metallic probes suffer from a limited bandwidth and tend to perturbate the electric fields, the electro-optic effect in dielectric materials has been exploited to detect electric signals [Frankel, 1996, Kuo and Tang, 2005, Sun *et al.*, 2007]. The principle of this method is to optically detect the variation of the refractive index in the electro-optic material caused by the electric field that should be detected. Since the Pockels effect is present even in the THz regime the technique can be used even at very high frequencies.

Electro-optic field detectors have been demonstrated with bulk electro-optic crystals using free-space laser beams [Frankel, 1996, Kuo and Tang, 2005], with microresonators made from optical nonlinear polymers attached to optical fibers [Sun *et al.*, 2007], and with integrated photonic circuits on lithium niobate wafers [Fuwen *et al.*, 2004]. Recently, an electric field sensor based on $LiNbO_3$/Si hybrid waveguides on silicon substrates has been shown [Chen and Reano, 2012]: the optical field in the silicon ring resonator is evanescently coupled to the $LiNbO_3$ layer (Fig. 21), which causes a

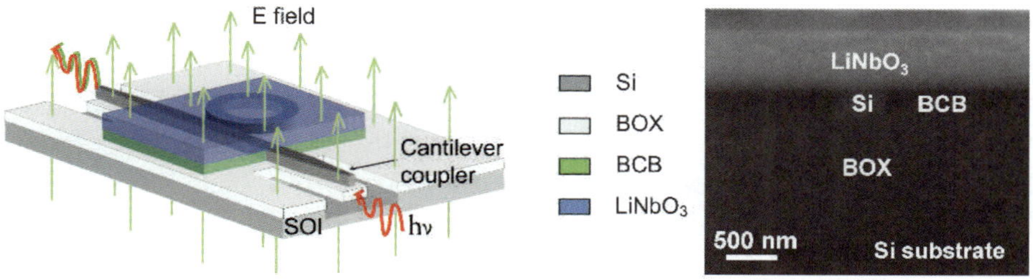

(a) (b)

Fig. 21. (a) Silicon ring resonator covered with a thin layer of nonlinear $LiNbO_3$ to be used as an electric field sensor. (b) SEM cross section of the waveguide in the field sensor. Reprinted from [Chen and Reano, 2012] with permission of OSA, The Optical Society.

shift of the resonance wavelength when applying an electric field with a sensitivity of 4.5 $Vm^{-1}Hz^{-0.5}$ due to the Pockels effect in $LiNbO_3$. Although the operation frequency of 1.86 GHz is still lower than bulk solutions, the work shows a first step of utilizing electro-optic materials in silicon photonic circuits for electric field sensing. This sensing application might enable novel devices such as antennas for radio-frequency detection and electrical circuits that are invulnerable to electromagnetic pulses [Hsu *et al.*, 2007] directly integrated on CMOS circuits.

6. Conclusion and Outlook

Despite the long history of research on optical nonlinear oxides, strong efforts to integrate those materials as thin films on silicon substrates have only recently been made. The development was strongly motivated by the success of silicon photonics, which allows the design and fabrication of dense, integrated photonic structures with standard CMOS fabrication processes. By combining this technology with the class of optical nonlinear oxides, novel devices such as high-speed modulators and zero-power tuning elements will provide additional functionalities to a silicon photonics baseline technology. Initial work on photonic devices made from hybrid $BaTiO_3$/Si and $LiNbO_3$/Si material systems, respectively, have demonstrated the feasibility to utilize complex oxides in integrated photonics. Much attention has recently also been brought to VO_2/Si photonic structures that successfully exploit the phase transition in VO_2 between two crystalline phases with very different optical properties for switching applications.

These results represent first concepts where electro-optical active oxides on silicon could enhance the current silicon photonics technology, but many more are expected to come. Some of the future devices might be inspired by bulk electro-optic structures, such as advanced electric-field sensors and acousto-optic modulators, which could enable frequency shifting in silicon photonics. Other devices may rely on fundamentally different physical phenomena, for example in plasmonics modulators based on surface-plasmons at the interface between metals and optical nonlinear oxides [Dicken *et al.*, 2008, Krasavin and Zayats, 2010a, Krasavin and Zayats, 2010b]. Also oxides showing a SMT could be inserted into plasmonics structures for active switching [Ooi *et al.*, 2013]. At the moment, those new types of devices have only been simulated and experimental realizations are still missing.

The design and the fabrication of novel photonic applications in silicon photonics using electro-optically active oxides requires a continuous, parallel effort in the field of materials: various aspects of the technology such as the electronic to optical design including device and circuit integration have to be encompassed. Lots of advances have already been made in epitaxially integrating high-quality oxide materials directly on silicon wafers, which have resulted into crystalline perfection close to bulk oxides. Based on this work, electro-optic active films such as $BaTiO_3$ have been integrated on silicon, and a strong electro-optic response has recently been demonstrated. However, compared to bulk crystals, the Pockels coefficients of such layers remain significantly lower and their optical propagation losses are higher than expected. When using a different integration route, for example the transfer of thin $LiNbO_3$ layers via bonding, the optical nonlinear response and losses seem to be fairly close to the bulk properties. However, due to the low bonding yield no large-scale integration for dense integrated optical circuits is currently possible.

For the success of hybrid oxide/silicon photonic devices it is therefore key to understand, control, and enhance current integration routes of optical nonlinear oxides. Selecting a well suited integration route, while improving the electro-optic response of thin films compared to those of bulk materials using strain engineering or superlattices [Lu and Knize, 2004]: this is an exciting task for materials scientists to look into a research area that will impact silicon technology in the next decades.

References

Abarkan, M., Aillerie, M., Salvestrini, J., Fontana, M. and Kokanyan, E. (2008), *App. Phy. B* **92**(4), 603–608.

Abel, S. (2014), Electro-optic photonic devices based on epitaxial barium titanate thin films on silicon, PhD thesis, Universite de Grenoble.

Abel, S., Stoferle, T., Marchiori, C., Caimi, D., Czornomaz, L., Stuckelberger, M., Sousa, M. Offrein, B. J. and Fompeyrine, J. (2015), *J. Lightwave Technol.* Vol. 34, No. 8, pp. 1688–1693.

Abel, S., Caimi, D., Sousa, M., Stöferle, T., Rossel, C., Marchiori, C., Chelnokov, A. and Fompeyrine, J. (2012), Electro-optical properties of barium titanate films epitaxially grown on silicon, *in* F. H. Teherani, D. C. Look and D. J. Rogers, eds, 'Proc. SPIE', Vol. 8263, SPIE, p. 82630Y.

Abel, S., Stöferle, T., Marchiori, C., Rossel, C., Rossell, M. D., Erni, R., Caimi, D., Sousa, M., Chelnokov, A., Offrein, B. J. and Fompeyrine, J. (2013a), *Nat. Commun.* **4**, 1671.

Abel, S., Sousa, M., Rossel, C., Caimi, D., Rossell, M. D., Erni, R., Fompeyrine, J. and Marchiori, C. (2013b), *Nanotechnology* **24**(28), 285701.

Abel, S., Stöferle, T., Marchiori, C., Caimi, D., Czornomaz, L., Rossel, C., Rossell, M., Erni, R., Sousa, M., Siegwart, H., Hofrichter, J., Stuckelberger, M., Chelnokov, A., Offrein, B. J. and Fompeyrine, J. (2013c), Electro-optical active barium titanate thin films in silicon photonics devices, *in* 'Advanced Photonics 2013', Optical Society of America, p. IW4A.5.

Abel, S., Stoferle, T., Marchiori, M., Daniele, C., Czornomaz, L., Stuckelberger, M., Sousa, M., Offrein, B. J. and Fompeyrine, J. (2016) *J. Lightwave Tech.* **34**(8) 1688–1693.

Adachi, H., Kawaguchi, T., Setsune, K., Ohji, K. and Wasa, K. (1983), *Appl. Phys. Lett.* **42**(10), 867–868.

Adachi, H. and Wasa, K. (1991), *Ultrasonics, Ferroelectrics, and Frequency Control, IEEE Trans.* **38**(6), 645–655.

Almeida, V. R., Xu, Q., Barrios, C. A. and Lipson, M. (2004), *Opt. Lett.* **29**(11), 1209–1211.

Ashman, C. R., Först, C. J., Schwarz, K. and Blöchl, P. E. (2004), *Phys. Rev. B* 69, 075309.

Atabaki, A. H., Shah Hosseini, E., Eftekhar, A. A., Yegnanarayanan, S. and Adibi, A. (2010), *Opt. Express* **18**(17), 18312–18323.

Bachelet, R., de Coux, P., Warot-Fonrose, B., Skumryev, V., Fontcuberta, J. and Sánchez, F. (2011), *J. App. Phy.* **110**(8), 086102.

Baek, S. H., Park, J., Kim, D. M., Aksyuk, V. A., Das, R. R., Bu, S. D., Felker, D. A., Lettieri, J., Vaithyanathan, V., Bharadwaja, S. S. N., Bassiri-Gharb, N., Chen, Y. B., Sun, H. P., Folkman, C. M., Jang, H. W., Kreft, D. J., Streiffer, S. K., Ramesh, R., Pan, X. Q., Trolier-McKinstry, S., Schlom, D. G., Rzchowski, M. S., Blick, R. H. and Eom, C. B. (2011), *Science* **334**(6058), 958–961.

Banaszek, K., U'Ren, A. B. and Walmsley, I. A. (2001), *Opt. Lett.* **26**(17), 1367–1369.

Beausoleil, R., Kuekes, P., Snider, G. S., Wang, S.-Y. and Williams, R. S. (2008), *Proc. IEEE* **96**(2), 230–247.

Bernasconi, P., Zgonik, M. and Gunter, P. (1995), *J. Appl. Phys.* **78**(4), 2651–2658.

Boyd, R. W. (2003), *Nonlinear Optics*, 2nd edn, Academic Press.

Braun, W. and Ploog, K. H. (2003), *J. Cryst. Growth* **251**(1–4), 68–72. Proceedings of the 12th International Conference on Molecular Beam Epitaxy.
Briggs, R. M., Pryce, I. M. and Atwater, H. A. (2010), *Opt. Express* **18**(11), 11192–11201.
Cavalleri, A., Dekorsy, T., Chong, H. H. W., Kieffer, J. C. and Schoenlein, R. W. (2004), *Phys. Rev. B* **70**, 161102.
Chen, L. and Reano, R. M. (2012), *Opt. Express* **20**(4), 4032–4038.
Chen, L., Wood, M. G. and Reano, R. M. (2013), *Opt. Express* **21**(22), 27003–27010.
Chen, L., Xu, Q., Wood, M. G. and Reano, R. M. (2014), *Optica* **1**(2), 112–118.
Chmielak, B., Waldow, M., Matheisen, C., Ripperda, C., Bolten, J., Wahlbrink, T., Nagel, M., Merget, F. and Kurz, H. (2011), *Opt. Express* **19**(18), 17212–17219.
Choi, K. J., Biegalski, M., Li, Y. L., Sharan, A., Schubert, J., Uecker, R., Reiche, P., Chen, Y. B., Pan, X. Q., Gopalan, V., Chen, L.-Q., Schlom, D. G. and Eom, C. B. (2004), *Science* **306**(5698), 1005–1009.
Cook Jr, W. R. (n.d.*a*), 5.3.2.14 tetragonal, 4mm (c4v), *in* 'Landolt-Börnstein — Group III Condensed Matter', Vol. 30A, Springer, pp. –.
Cook Jr, W. R. (n.d.*b*), 5.3.2.9 trigonal (rhombohedral), 32 (d3), *in* 'Landolt-Börnstein — Group III Condensed Matter', Vol. 30A, pp. –.
Cuniot-Ponsard, M. (2011), *Ferroelectrics - Material Aspects*, InTech, chapter Strontium Barium Niobate Thin Films for Dielectric and Electro-Optic Applications, pp. 498–518.
Dalton, L. R., Sullivan, P. A. and Bale, D. H. (2010), *Chem. Rev.* **110**(1), 25–55. PMID: 19848381.
Damjanovic, D., Brem, F. and Setter, N. (2002), *Appl. Phys. Lett.* **80**(4), 652–654.
Dekkers, M., Nguyen, M. D., Steenwelle, R., te Riele, P. M., Blank, D. H. A. and Rijnders, G. (2009), *App. Phy. Lett.* **95**(1), 012902.
Demkov, A. A. and Posadas, A. B. (2014), *Integration of Functional Oxides with Semiconductors*, Springer New York.
Dicken, M. J., Sweatlock, L. A., Pacifici, D., Lezec, H. J., Bhattacharya, K. and Atwater, H. A. (2008), *Nano Lett.* **8**(11), 4048–4052.
Diest, K., Dionne, J. A., Spain, M. and Atwater, H. A. (2009), *Nano Lett.* **9**(7), 2579–2583.
Doll, J. C. and Pruitt, B. L. (2013), *Piezoresistor Design and Applications*, Springer New York.
Dong, P., Shafiiha, R., Liao, S., Liang, H., Feng, N.-N., Feng, D., Li, G., Zheng, X., Krishnamoorthy, A. V. and Asghari, M. (2010), *Opt. Express* **18**(11), 10941–10946.
Dubourdieu, C., Bruley, J., Arruda, T. M., Posadas, A., Jordan-Sweet, J., Frank, M. M., Cartier, E., Frank, D. J., Kalinin, S. V., Demkov, A. A. and Narayanan, V. (2013), *Nat. Nanotechnol.* **8**(10), 748–754.
Enami, Y., Derose, C. T., Mathine, D., Loychik, C., Greenlee, C., Norwood, R. A., Kim, T. D., J. Luo, Y. T., Jen, A. K.-Y. and Peyghambarian, N. (2007), *Nat Photon* **1**(3), 180–185.
Frankel, M. (1996), *Optical and Quantum Electronics* **28**(7), 783–800.
Fukumoto, H., Imura, T. and Osaka, Y. (1988), *Jap. J. App. Phys.* **27**(8A), L1404.
Fuwen, Z., Fushen, C. and Kun, Q. (2004), *Microwave and Optical Technology Letters* **40**(4), 302–305.

Gill, D. M., Conrad, C. W., Ford, G., Wessels, B. W. and Ho, S. T. (1997), *Appl. Phys. Lett.* **71**(13), 1783–1785.
Gösele, U. and Tong, Q.-Y. (1998), *Ann. Rev. Mater. Sci.* **28**(1), 215–241.
Gu, X., Lubyshev, D., Batzel, J., Fastenau, J. M., Liu, W. K., Pelzel, R., Magana, J. F., Ma, Q., Wang, L. P., Zhang, P. and Rao, V. R. (2009), *J. Vac. Sci. Technol., B* **27**(3), 1195.
Guiller, A., Moussavou, A.-G., Guilloux-Viry, M., Perrin, A., Fompeyrine, J., Sauleau, R. and Mahdjoubi, K. (2008), *Ferroelectrics* **362**(1), 95–104.
Günter, P. (1987), *Ferroelectrics* **75**(1), 5–23.
Günter, P. and Huignard, J.-P., eds (2006), *Photorefractive Materials and Their Applications 1 — Basic Effects*, Springer Science+Business Media, Inc.
Günter, P. and Huignard, J.-P., eds (2007a), *Photorefractive Materials and Their Applications 2 — Materials*, Springer.
Günter, P. and Huignard, J.-P., eds (2007b), *Photorefractive Materials and Their Applications 3 — Applications*, Springer Science + Business Media, Inc.
Günter, P. and Zgonik, M. (1991), *Opt. Lett.* **16**(23), 1826–1828.
Haeni, J., Theis, C. and Schlom, D. (2000), *J. Electroceram.* **4**(2), 385–391.
Haertling, G. H. and Land, C. E. (1971), *J. Ameri. Ceram. Soc.* **54**(1), 1–11.
Haurylau, M., Chen, G., Chen, H., Zhang, J., Nelson, N. A., Albonesi, D. H., Friedman, E. G. and Fauchet, P. M. (2006), *IEEE J. Sel. Topics Quantum Electron.* **12**(6), 1699–1705.
Hellberg, C. S., Andersen, K. E., Li, H., Ryan, P. J. and Woicik, J. C. (2012), *Phys. Rev. Lett.* 108, 166101.
Herman, M. A., Richter, W. and Sitter, H. (2004), *Epitaxy*, Springer-Verlag.
Hermet, P., Veithen, M. and Ghosez, P. (2009), *J. Phys.: Condens. Matter* **21**(21), 215901.
Herminghaus, S., Smith, B. A. and Swalen, J. D. (1991), *J. Opt. Soc. Am. B* **8**(11), 2311–2317.
Hoerman, B. H., Nichols, B. M., Nystrom, M. J. and Wessels, B. W. (1999), *Appl. Phys. Lett.* **75**(18), 2707–2709.
Hoerman, B., Nichols, B. and Wessels, B. (2003), *Opt. Commun.* **219**(1–6), 377–382.
Horst, F., Green, W. M., Assefa, S., Shank, S. M., Vlasov, Y. A. and Offrein, B. J. (2013), *Opt. Express* **21**(10), 11652–11658.
Hou, A., Liu, H., Sun, J., Zhang, D. and Yi, M. (2007), *Optics & Laser Technology* **39**(2), 411–414.
Hsu, R. C. J., Ayazi, A., Houshmand, B. and Jalali, B. (2007), *Nat Photon* **1**(9), 535–538.
Huang, C. H.-J. (1995), *Integrated Ferroelectrics* **6**(1–4), 355–362.
Huang, H.-C., Dadap, J. I., Malladi, G., Kymissis, I., Bakhru, H. and Osgood, R. M. (2014), *Opt. Express* **22**(16), 19653–19661.
Hunsperger, R. G. (2009), *Integrated Optics — Theory and Technology*, 6th edn, Springer Science + Business Media.
Irie, M. (2000), *Chem. Rev.* **100**(5), 1683–1684. PMID: 11777415.
Jalan, B., Engel-Herbert, R., Wright, N. J. and Stemmer, S. (2009), *J. Vac. Sci. Technol., A* **27**(3), 461–464.
Jang, J. W., Chung, S. J., Cho, W. J., Hahn, T. S. and Choi, S. S. (1997), *J. Appl. Phys.* **81**(9), 6322–6327.

Janner, D., Tulli, D., Garcia-Ruperez, J.-Granda, M., Belmonte, M. and Pruneri, V. (2009), *Laser Photonics Rev.* **3**(3), 301–313.

Jerominek, H., Picard, F. and Vincent, D. (1993), *Opti. Eng.* **32**(9), 2092–2099.

Johnston, A. R. and Weingart, J. M. (1965), *J. Opt. Soc. Am.* **55**(7), 828–833.

Joushaghani, A., Jeong, J., Paradis, S. Alain, D., J. Stewart Aitchison, and Joyce K. S. Poon, (2015), *Opt. Express* 23, 3657–3668.

Kalluri, S., Garner, S., Ziari, M., Steier, W. H., Shi, Y. and Dalton, L. R. (1996), *Appl. Phys. Lett.* **69**(2), 275–277.

Kana, J. K., Ndjaka, J., Vignaud, G., Gibaud, A. and Maaza, M. (2011), *Opti. Communi.* **284**(3), 807–812.

Kang, T. D., Xiao, B., Avrutin, V., Ozgur, U., Morkoc, H., Park, J. W., Lee, H. S., Lee, H., Wang, X. and Smith, D. J. (2008), *J. Appl. Phys.* **104**(9), 093103.

Kay, H. F. and Vousden, F. (1949), *Philos. Mag.* **40**(7), 1019 – 1040.

Khatib, D., Jannot, B. and Jullien, P. (1993b), *Ferroelectrics Lett.* **15**(1), 1–7.

Khatib, D., Jullien, P. and Jannot, B. (1993a), *Ferroelectrics* **145**(1), 181–193.

Kim, D.-Y., Moon, S. E., Kim, E.-K., Lee, S.-J., Choi, J.-J. and Kim, H.-E. (2003), *Appl. Phys. Lett.* **82**(9), 1455–1457.

Kim, S.-I., Kim, D.-H., Kim, Y., Moon, S. Y., Kang, M.-G., Choi, J. K., Jang, H. W., Kim, S. K., Choi, J.-W., Yoon, S.-J., Chang, H. J., Kang, C.-Y., Lee, S., Hong, S.-H., Kim, J.-S. and Baek, S.-H. (2013), *Adv. Materi.* **25**(33), 4612–4617.

Kinase, W., Yoshikawa, M. and Ohnishi, N. (1986), *Ferroelectrics* **67**(1), 159–168.

Kityk, I. V., Ebothe, J., Hichou, A. E., Addou, M., Bougrine, A. and Sahraoui, B. (2002), *J. Phys. Condens. Matter* **14**(21), 5407.

Krasavin, A. V. and Zayats, A. V. (2010a), *Appl. Phys. Lett.* **97**(4), 041107. Journal article.

Krasavin, A. V. and Zayats, A. V. (2010b), *Opt. Express* **18**(11), 11791–11799.

Kuhlow, B. (n.d.) (2006), *Springer Materials — The Landolt-Börnstein Database*, Vol. 1A2: Laser Fundament. Part 2, Chapter 7.1 Modulators.

Kuo, W.-K. and Tang, D.-T. (2005), *Rev. Sci. Instrum.* **76**(5), 055111.

Lee, Y. S., Kim, G.-D., Kim, W.-J., Lee, S.-S., Lee, W.-G. and Steier, W. H. (2011), *Opt. Lett.* **36**(7), 1119–1121.

Leuthold, J., Freude, W., Brosi, J.-M., Baets, R., Dumon, P., Biaggio, I., Scimeca, M., Diederich, F., Frank, B. and Koos, C. (2009), *Proceedings of the IEEE* **97**(7), 1304–1316.

Li, G., Zheng, X., Yao, J., Thacker, H., Shubin, I., Luo, Y., Raj, K., Cunningham, J. E. and Krishnamoorthy, A. V. (2011), *Opt. Express* **19**(21), 20435–20443.

Liu, J., Sun, X., Camacho-Aguilera, R., Kimerling, L. C. and Michel, J. (2010), *Opt. Lett.* **35**(5), 679–681.

Liu, W. C., Mak, C. L. and Wong, K. H. (2009), *J. Phys. D: Appl. Phys.* **42**(10), 105114.

Lu, Y. and Knize, R. J. (2004), *J. Phys. D: Appl. Phys.* **37**(17), 2432.

Lysenko, S., Rúa, A., Vikhnin, V., Fernández, F. and Liu, H. (2007), *Phys. Rev. B* **76**, 035104.

Ma, H., Levy, J., Biegalski, M. D., Trolier-McKinstry, S. and Schlom, D. G. (2009), *J. Appl. Phys.* **105**(1), 014102.

Markov, P., Robert E. Marvel, Hiram J. Conley, Kevin J. Miller, Richard F. HaglundJr., and Sharon M. Weiss, (2015), *ACS Photonics*, 2(8), pp. 1175–1182 DOI: 10.1021/acsphotonics.5b00244.

McDaniel, M. D., Ngo, T. Q., Posadas, A., Hu, C., Lu, S., Smith, D. J., Yu, E. T., Demkov, A. A. and Ekerdt, J. G. (2014), *Adv. Mater. Interf.* pp. n/a–n/a.

McKee, R. A., Walker, F. J. and Chisholm, M. F. (1998), *Phys. Rev. Lett.* **81**(14), 3014–3017.

McKee, R. A., Walker, F. J., Conner, J. R., Specht, E. D. and Zelmon, D. E. (1991), *Appl. Phys. Lett.* **59**(7), 782–784.

Meng, X. Y., Wang, Z. Z., Zhu, Y. and Chen, C. T. (2007), *J. Appl. Phys.* **101**(10), 103506.

Michel, J. (2010), Active Ge based devices for silicon photonics, Winterschool Manterndorf, Manterndorf, Austria.

Mikheev, E., Kajdos, A. P., Hauser, A. J. and Stemmer, S. (2012), *Appl. Phys. Lett.* **101**(25), 252906.

Miller, D. A. B. (2009), *Proc. IEEE* **87**(7), 1166–1185.

Mortimer, R. J. (2011), *Ann. Rev. Mater. Res.* **41**(1), 241–268.

Niu, F., Teren, A. R., Hoerman, B. H. and Wessels, B. W. (2000), *Mat. Res. Soc. Symp. Proc.* **637**, E1.9.

Niu, G., Yin, S., Saint-Girons, G., Gautier, B., Lecoeur, P., Pillard, V., Hollinger, G. and Vilquin, B. (2011), *Microelectron. Eng.* **88**(7), 1232–1235.

Nolte, D., ed. (1995), *The Photorefractive Effect and Materials*, Electronic Materials: Science and Technology, Springer.

Norga, G. J., Marchiori, C., Guiller, A., Locquet, J. P., Rossel, C., Siegwart, H., Caimi, D., Fompeyrine, J. and Conard, T. (2005), *Appl. Phys. Lett.* **87**(26), 262905.

Norga, G. J., Marchiori, C., Rossel, C., Guiller, A., Locquet, J. P., Siegwart, H., Caimi, D., Fompeyrine, J., Seo, J. W. and Dieker, C. (2006), *J. Appl. Phys.* **99**(8), 084102–7.

Nye, J. F. (1960), *Physical Properties of Crystals*, Oxford University Press.

Ooi, K. J., Bai, P., Chu, H. S. and Ang, L. K. (2013) *Nanophotonics* **2**(1), 13–19.

Palmer, R., Alloatti, L., Korn, D., Schindler, P., Baier, M., Bolten, J., Wahlbrink, T., Waldow, M., Dinu, R., Freude, W., Koos, C. and Leuthold, J. (2013), *IEEE Photon. Technol. Lett.* **25**(13), 1226–1229.

Park, D. H. (2008), Characterization of Linear Electro-Optic Effect of Poled Organic Thin Films, PhD thesis, University of Maryland.

Park, J., Bogorin, D., Cen, C., Felker, D., Zhang, Y., Nelson, C., Bark, C., Folkman, C., Pan, X., Rzchowski, M., Levy, J. and Eom, C. (2010), *Nat. Commun.* **1**, 94.

Park, J. W., Baek, S. H., Bark, C. W., Biegalski, M. D. and Eom, C. B. (2009), *Appl. Phys. Lett.* **95**(6), 061902.

Pashkin, A., Kübler, C., Ehrke, H., Lopez, R., Halabica, A., Haglund, R. F., Huber, R. and Leitenstorfer, A. (2011), *Phys. Rev. B* **83**, 195120.

Petraru, A., Schubert, J., Schmid, M. and Buchal, C. (2002), *Appl. Phys. Lett.* **81**(8), 1375–1377.

Petraru, A., Schubert, J., Schmid, M., Trithaveesak, O. and Buchal, C. (2003), *Opt. Lett.* **28**(24), 2527–2529.

Posadas, A., Berg, M., Seo, H., de Lozanne, A., Demkov, A. A., Smith, D. J., Kirk, A. P., Zhernokletov, D. and Wallace, R. M. (2011), *App. Phys. Lett.* **98**(5), 053104.

Qiao, L. and Bi, X. (2010), *Phys. Status Solidi a* **207**(11), 2511–2516.

Rabe, K. M., Ahn, C. H. and Triscone, J.-M. (2007), *Physics of Ferroelectrics*, Topics in Applied Physics, Springer-Verlag, Berlin Heidelberg.

Rabiei, P. and Gunter, P. (2004), *Appl. Phys. Lett.* **85**(20), 4603–4605.
Rabiei, P., Ma, J., Khan, S., Chiles, J. and Fathpour, S. (2013), *Opt. Express* **21**(21), 25573–25581.
Reed, G. T., Mashanovich, G., Gardes, F. Y. and Thomson, D. J. (2010), *Nature Photon.* **4**(8), 518–526.
Reiner, J. W., Kolpak, A. M., Segal, Y., Garrity, K. F., Ismail-Beigi, S., Ahn, C. H. and Walker, F. J. (2010), *Adv. Mater.* **22**(26–27), 2919–2938.
Reitze, D. H., Haton, E., Ramesh, R., Etemad, S., Leaird, D. E., Sands, T., Karim, Z. and Tanguay, Jr., A. R. (1993), *Appl. Phys. Lett.* **63**(5), 596–598.
Ruglovsky, J. L., Li, J., Bhattacharya, K. and Atwater, H. A. (2006), *Acta Mater.* **54**(14), 3657–3663.
Ruzmetov, D., Gopalakrishnan, G., Deng, J., Narayanamurti, V. and Ramanathan, S. (2009), *J. Appl. Phys.* **106**(8), 083702.
Ryckman, J. D., Diez-Blanco, V., Nag, J., Marvel, R. E., Choi, B. K., Haglund, R. F. and Weiss, S. M. (2012), *Opt. Express* **20**(12), 13215–13225.
Ryckman, J. D., Hallman, K. A., Marvel, R. E., Haglund, R. F. and Weiss, S. M. (2013), *Opt. Express* **21**(9), 10753–10763.
Sakayori, K.-I., Matsui, Y., Abe, H., Nakamura, E., Kenmoku, M., Hara, T., Ishikawa, D., Kokubu, A., ichi Hirota, K. and Ikeda, T. (1995), *Jpn. J. Appl. Phys.* **34**, 5443–5445.
Salama, C. A. T. and Siciunas, E. (1972), *J. Vac. Sci. Technol. A* **9**(1), 91–96.
Saleh, B. E. A. and Teich, M. C. (2007), *Fundamentals of Photonics*, Wiely series in pure and applied optics, second edn, John Wiley & Sons, Inc., Hoboken, New Jersey.
Sando, D., Hermet, P., Allibe, J., Bourderionnet, J., Fusil, S., Carrétéro, C., Jacquet, E., Mage, J.-C., Dolfi, D., Barthélémy, A., Ghosez, P. and Bibes, M. (2014), *Phys. Rev. B* **89**, 195106.
Schildkraut, J. S. (1990), "Determination of the electrooptic coefficient of a poled polymer film". *Appl. Opt.* 29, 2839–2841
SITOGA (2014), 'European project funded under 7th fwp, silicon cmos compatible transition metal oxide technology for boosting highly integrated photonic devices with disruptive performance'.
Soref, R. A. (1992), *J. Appl. Phys.* **72**(2), 626–630.
Spirin, V. V., Lee, C. and No, K. (1998), *J. Opt. Soc. Am. B* **15**(7), 1940–1946.
Sun, H., Pyajt, A., Luo, J., Shi, Z., Hau, S., Jen, A. K.-Y., Dalton, L. R. and Chen, A. (2007), *Sensors Journal, IEEE* **7**(4), 515–524.
Taillaert, D., Laere, F. V., Ayre, M., Bogaerts, W., Thourhout, D. V., Bienstman, P. and Baets, R. (2006), *Jpn. J. Appl. Phys.* **45**(8A), 6071–6077.
Takeda, K., Muraishi, T., Hoshina, T., Kakemoto, H. and Tsurumi, T. (2009), *Mat. Sci. Eng. B* **161**(1–3), 61–65.
Tang, P., Meier, A. L., Towner, D. J. and Wessels, B. W. (2005), *Opt. Lett.* **30**(3), 254–256.
Tayebati, P., Trivedi, D. and Tabat, M. (1996), *Appl. Phys. Lett.* **69**(8), 1023–1025.
Teng, C. C. and Man, H. T. (1990), *Appl. Phys. Lett.* **56**(18), 1734–1736.
Teng, J., Dumon, P., Bogaerts, W., Zhang, H., Jian, X., Han, X., Zhao, M., Morthier, G. and Baets, R. (2009), *Opt. Express* **17**(17), 14627–14633.

Tompkins, H. G. and Irene, E. A., eds. (2005), *Handbook of Ellipsometry*, William Andrew Inc.

Uchiki, H. and Kobayashi, T. (1988), *J. Appl. Phys.* **64**(5), 2625–2629.

Vaithyanathan, V., Lettieri, J., Tian, W., Sharan, A., Vasudevarao, A., Li, Y. L., Kochhar, A., Ma, H., Levy, J., Zschack, P., Woicik, J. C., Chen, L. Q., Gopalan, V. and Schlom, D. G. (2006), *J. Appl. Phys.* **100**(2), 024108–9.

Valdmanis, J. A. (1990), *Measurement of High-Speed Signals in Solid State Devices (Chapter 4)*, Academic Press, INC., San Diego.

Veithen, M. and Ghosez, P. (2002), *Phys. Rev. B* **65**, 214302.

Veithen, M., Gonze, X. and Ghosez, P. (2004), *Phys. Rev. Lett.* **93**(18), 187401.

Veithen, M., Gonze, X. and Ghosez, P. (2005), *Phys. Rev. B* **71**, 125107.

Verleur, H. W., Barker, A. S. and Berglund, C. N. (1968), *Phys. Rev.* **172**, 788–798.

Wall, S., Wegkamp, D., Foglia, L., Appavoo, K., Nag, J., Haglund, R., Stähler, J. and Wolf, M. (2012), *Nat Commun* **3**, 721.

Wang, D., Li, S., Chan, H. and Choy, C. (2011), *Curr. Appl. Phys.* **11**(3, Supplement), S52–S55.

Wang, F., Furman, E. and Haertling, G. H. (1995), *J. Appl. Phys.* **78**(1), 9–15.

Wei, X., Li, Y., Zhu, J., Liang, Z., Zhang, Y., Huang, W. and Jiang, S. (2005), *Appl. Surf. Sci.* **252**(5), 1442–1448.

Weis, R. and Gaylord, T. (1985), *Applied Physics A* **37**(4), 191–203.

Wessels, B. W. (2007), *Annu. Rev. Mater. Res.* **37**, 659–679.

Wooten, E., Kissa, K., Yi-Yan, A., Murphy, E., Lafaw, D., Hallemeier, P., Maack, D., Attanasio, D., Fritz, D., McBrien, G. and Bossi, D. (2000), *J. Sel. Topics Quantum Electron.* **6**(1), 69–82.

Xiao, X., Xu, H., Li, X., Li, Z., Chu, T., Yu, Y. and Yu, J. (2013), *Opt. Express* **21**(4), 4116–4125.

Xiong, C., Pernice, W. H. P. and Tang, H. X. (2012), *Nano Lett.* **12**(7), 3562–3568.

Xiong, C., Pernice, W., Ngai, J., Reiner, J., Kumarh, D., Walker, F., Ahn, C. and Tang, H. (2014), *Nano Lett.* **14**(3), 1419.

Yang, Z., Ko, C. and Ramanathan, S. (2011), *Ann. Rev. Mater. Res.* **41**(1), 337–367.

Yariv, A. and Yeh, P. (1984), *Optical Waves in Crystals*, John Wiley & Sons, New York.

Yoon, J.-G. and Kim, K. (1996), *Appl. Phys. Lett.* **68**(18), 2523–2525.

You, P., Lu, C., Ye, W., Hao, L., Zhu, J. and Zhou, Y. (2013), *App. Phys. Lett.* **102**(5), –.

Zgonik, M., Bernasconi, P., Duelli, M., Schlesser, R., Günter, P., Garrett, M. H., Rytz, D., Zhu, Y. and Wu, X. (1994), *Phys. Rev. B* **50**(9), 5941–5949.

Zhang, J., Yao, K. L., Liu, Z. L., Gao, G. Y., Sun, Z. Y. and Fan, S. W. (2010), *Phys. Chem. Chem. Phys.* **12**, 9197–9204.

Zhang, Y. and Ramanathan, S. (2011), *Solid-State Electronics* **62**(1), 161–164.

Metals on Silicon

Chapter 11

Silicide Contacts

Ahmet S. Özcan
IBM Semiconductor Research and Development Center, Hopewell Junction, NY 12533, USA

1. Scope

In this chapter, we will review the fundamental materials science of silicide thin films which are used as low resistance contacts in microelectronic CMOS devices. The historic evolution of silicide materials used by the industry will provide the basis to compare the properties of currently used Ni silicide to the prior Co and Ti silicide contacts. After a brief section on formation kinetics of metal silicide films, we will focus on defects with examples from state-of-the-art CMOS technology development. Discussion on defects will illustrate the necessity to understand and control the fundamental materials properties that are in play during the formation of silicide contacts, as well as during the degradation. We also present several material improvements that have been reported and adapted by the semiconductor industry. The next section will switch gears to device performance and the effects of scaling on contact resistance. Here, the emphasis will be on parasitic resistance in CMOS devices and optimization of silicide contacts to improve performance. The chapter will conclude with discussions on the future of contacts in the next generation of devices including the issues for devices beyond the conventional transistor.

2. Introduction

Silicides have been used in state of the art complementary metal-oxide semiconductor (CMOS) devices for more than two decades as contacts to

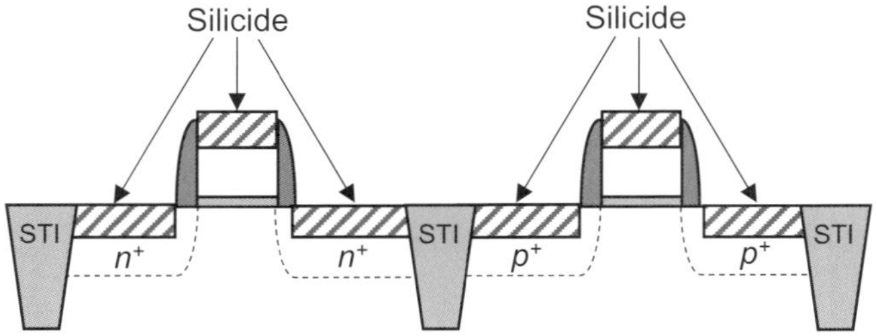

Fig. 1. Schematic illustration of silicide contacts on *n* and *p*-type MOSFETs with STI (shallow trench isolation).

source, drain and gates in the transistor (Figure 1), mainly due to their low resistivity and low contact resistance [Maex, 1993; Gambino and Colgan, 1998; d'Heurle and Miglio, 2000; Zhang and Östling, 2003; Lavoie *et al.*, 2007]. In addition to their use as contacts between the metallization levels and the semiconductor junctions, silicides have been employed as local interconnects [Mann *et al.*, 1995] and in passive devices such as in electronic fuses (eFUSE) [Kothandaraman *et al.* 2002]. This chapter will focus on the silicide applications for contacts and particularly discuss the metal silicides which were widely utilized by the semiconductor industry.

Silicide films form via the reaction between the metal (e.g. Ni) and the semiconductor (e.g. Si) through a rapid thermal annealing (RTA) process. Typically, metal thin films are sputter deposited (Physical Vapor Deposition — PVD) on semiconductor surfaces, rather than the co-deposition of metal and semiconductor simultaneously. The process for silicide formation, which includes surface preparation, metal film deposition and annealing, is critical for controlling the final properties of the silicide films. Integration of the silicide process in the CMOS device formation flow is also non-trivial and will be discussed further in detail.

2.1. *Historical evolution of the transistor and silicide contacts*

Success of CMOS technology has been the result of ingenious miniaturization and the continuous innovation in materials for many decades. The scaling ideas described by Moore [Moore, 1965] and Dennard [Dennard *et al.*, 1974] have set the industry on a course of developing advanced process technologies and new materials on a regular pace to improve device performance, power and density, generation after generation. In order to the place the evolution of silicides in context, historical evolution of technology nodes and devices must be briefly reviewed.

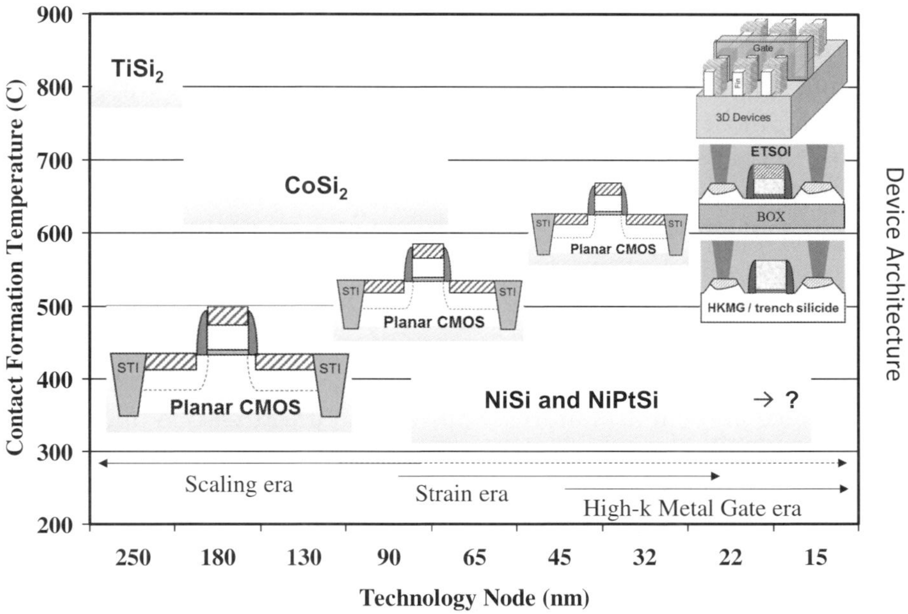

Fig. 2. Evolution of silicide contacts through the technology nodes and schematic illustration of changing device architectures.

As illustrated in Figure 2, the timeline of the technological advances in devices can be loosely divided into three periods: 1-Scaling era, 2-Strain era, 3-High-k metal-gate (HKMG) era. It should be emphasized that "scaling" has not stopped with the emergence of the "strain era", but it has ceased to be the dominant knob for performance enhancements, which drove innovations in channel strain engineering as well as new materials in the gate stack. As illustrated in the cartoon schematics in Figure 2, CMOS architecture is changing drastically with more variations across the industry. This variety in the device offerings is partly due to the emergence of mobile devices and the growing need for low-power products in contrast to the products that demand high performance. Starting with the 28 nm node and beyond, we start to see fully depleted devices and 3-dimensional (3D) devices (e.g. tri-gate/finFET) in manufacturing as the industry is going through a major transformation. Despite the desire of the industry, the rate of changes in the post scaling era has been very rapid and complex, which hiked up the cost for development and manufacturing. On the other hand, these challenges have created many interesting problems for the materials scientists that require a deep understanding for the root cause in order to overcome those challenges. As we will discuss with examples in the later sections, these challenges have been different than the limitations which were faced as a result of scaling or miniaturization of the devices.

During the classical scaling era, which we roughly end with the 65 nm node, most of the silicide problems were associated with "narrow linewidth" effects [Cabral *et al.*, 1997; Mann *et al.*, 1995; Lavoie *et al.*, 2003]. During that period, the gate has been the smallest structures that have "tested" the feasibility and robustness of the silicide films. For example in the case of Ti silicide, the difficult nucleation of the desired low resistance phase ($TiSi_2$- C54) has been the limiting issue in narrow line-widths [Mann *et al.*, 1995]. The scarcity of the C54 phase nucleation sites caused narrow lines to remain partially in the high resistance C49 phase, and therefore caused device and circuit failures. Solutions to improve C54 nucleation in narrow lines included alloying Ti with other transition metals such as Ta [Cabral *et al.*, 1997; Ozcan *et al.*, 2002]. Scaling driven issues continued for Co silicide which exhibited void defects in narrow poly silicon gates [Lavoie *et al.*, 2003].

After the switch to Ni based silicides, which coincides with the strain era, feature size and scaling driven challenges mostly left their place to new issues created by the strain performance elements, such as embedded SiGe junctions and nitride stress liners. Incompatibility of Co silicide with SiGe was also a factor in the decision to switch to Ni silicide, which is mainly due to the immiscibility of Ge in the low resistance $CoSi_2$ phase [Lavoie *et al.*, 2003]. In addition to continuous focus on channel strain engineering, introduction of high-k metal gates mostly imposed new restrictions for the silicide process and the integration such as new thermal budget considerations [Chudzik *et al.*, 2007]. These new restrictions have been mostly addressed by integration solutions rather than fundamental materials modifications within the silicide module [Shang *et al.*, 2012].

As CMOS technologies are scaled further, the source-drain (S/D) extrinsic resistance (Rext) has started to strongly limit MOSFET drive current and performance [Kim *et al.*, 2005; Ohuchi *et al.*, 2007]. Even though the gate length scaling has significantly slowed after the "scaling era", the pitch (i.e. gate to gate distance) has become smaller and therefore the contact length has continued to shrink aggressively. As a result, the contact resistivity requirements to satisfy device performance have become increasingly stringent. According to earlier predictions by International Technology Roadmap for Semiconductors (ITRS), Ni based silicides would have to be replaced with "band-edge" silicide materials in a dual silicide scheme (i.e. different silicide material for N and P FET devices). These predictions have fueled the research on alternative silicide materials, such as Er, Yb and Pt silicides due to their low barrier heights [Nishi *et al.*, 2007; Alptekin *et al.*, 2009; Besser *et al.*, 2008]. In addition, there has been alternative efforts to engineer Ni silicide to improve interface resistance and meet device requirements [Luo *et al.*, 2010; Zhang *et al.*, 2007]. Today, down to 20 nm technology node, Ni based silicides are still predominantly used by the industry, which is evidence that continues improvement of the Ni silicide contacts and junctions have been able to meet the device performance requirements.

2.2. *SALICIDE process*

Formation of silicide contacts in CMOS devices rely on a self-aligned process which has tremendous advantages. From a simplistic point of view, the semiconductor device fabrication consists of film depositions, patterning and removals, which are repeated hundreds of times. The total cost of manufacturing a "chip" depends on the number steps and the complexity of processes. Therefore, processes that are self-aligned, which do not require photolithography, significantly reduce the cost of manufacturing. Figure 3(a) illustrates the self-aligned silicide formation process steps which has been unchanged for nearly three decades. Silicide process can be "self-aligned" if two important criteria are met:

(1) Selective reaction with the substrate (e.g. Si, SiGe, Si:C) that can form a low resistance silicide phase, and no reaction with dielectric materials (oxides/nitrides).
(2) Selective etch process which can remove the unreacted metal from undesired locations (e.g. spacers) without etching silicide.

These criteria were met for Ti, Co and Ni through many generations of devices. The selectivity of the wet etch process depends on the silicide phase and could pose challenges for optimizing the formation temperature. An example will be discussed in the next section to illustrate this further.

With the introduction of HKMG, there has been different integration approaches for the silicide contacts. In the "gate-last" integration scheme [Shang *et al.*, 2012; Auth *et al.*, 2008; Cabral *et al.*, 2004], where the "dummy" gate is replaced after the junction formation, silicide contacts are formed later in the process flow. This is due to the high thermal budget required during the HKMG module which is more than what NiSi can withstand. Also, in this process flow, gate material is a metal (e.g. Al or W), therefore, silicidation is only needed for making contacts to active areas and

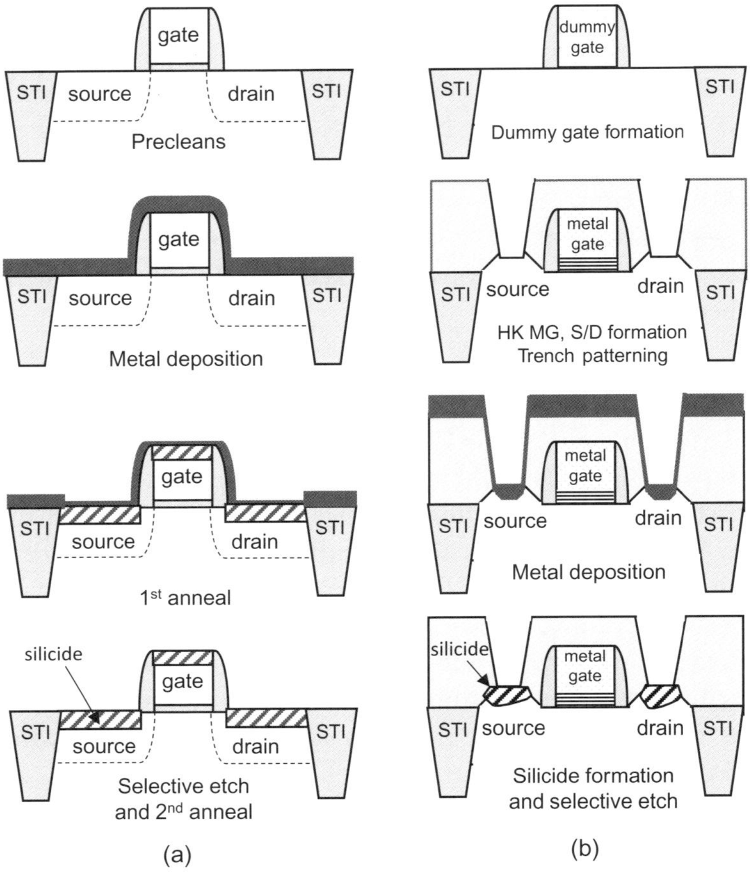

Fig. 3. Self-aligned silicide formation process steps. (a) conventional SALICIDE process, (b) trench silicide formation in high-k replacement metal gate flow.

there is no need for poly-silicon silicidation. Figure 3(b) illustrates a typical silicide process flow for "gate-last/silicide-last" integration. This process is also called "trench silicide" formation because silicide metal is deposited into contact trenches which were formed using patterning. After silicide formation, the trenches are filled with tungsten followed by a planarization scheme.

Trench silicide process has major implications for engineering contacts:

(1) No silicidation of the gate (therefore, problems associated with narrow line poly silicidation disappear and requirements for sheet resistance changes).
(2) Selective etch is not mandatory (Metal on the top surface can be removed during planarization).

(3) Contact (interface) resistance becomes the critical parameter (silicide sheet resistance does not play a major role any more due to the absence of gate silicide and presence of tungsten metal in the trenches).

Three points listed above have fundamentally changed the criteria to select the silicide material and the tools to optimize the process. Formation of silicide in narrow high aspect ratio trenches also drove changes and innovations in process technology especially in the thin film deposition and pre-cleaning. Silicide formation on non-planar devices, (e.g. FinFET) will add further complexity and constraints which will discussed in the last section.

3. Challenges and Solutions

Once the front-end process has been completed and devices are formed, wafers can be tested electrically. Post silicide formation is typically the first place in the process flow to check device functionality and defectivity since test probes can make contact to silicided pads. More comprehensive testing is done after the first level of metallization (typically called Metal 1 or M1 test) and after building more metal levels in the back-end-of-the-line, SRAM functionality can be tested, depending on the wiring configuration. During the development of a product on a new technology node, yield and performance are the key deliverables. Yield ramp-up depends on identifying and reducing defects at a fast rate. Since the industry is working on very aggressive timelines to develop every new technology generation, early detection of defects is key. In addition to electrical tests at different levels, there are many other inline inspections at sector level to evaluate defects. These include automated inline optical defect inspections as well as top-down SEM reviews. It is desirable to detect defects early at sector rather than at later stages (e.g. M1 or SRAM test) not only because it is easier to isolate the defect but also for reducing the time to get feedback from actions to solve it (i.e. reduced cycle of learning). For example it may take several months to build wafers up to SRAM test, hence, any solution that has been tried on experiments will be tested and evaluated after months. In contrast, if the detection point is in sector, progress can be made in days, including understanding the root cause and solving the issue rapidly.

In this section we will discuss examples of silicide defects which are detectable at sector, at M1 test and after SRAM test.

3.1. *Aqua Regia attack — Grainy silicide defect*

As mentioned earlier, strain engineering has become an important knob to increase transistor performance and embedded SiGe technology has been widely adopted by the industry to boost PFET channel mobility in MOSFETs [Lee *et al.*, 2005; Ohta *et al.*, 2005; Zhang, D. 2005]. In this scheme, S/D areas are recessed and epitaxial $Si_{1-x}Ge_x$ film is grown to exert a compressive force on the channel and enhance hole mobility. As a result of this process, silicide contacts need to form on the $Si_{1-x}Ge_x$ substrate, unless a layer of Si cap is grown on top. The presence of Ge raises the concerns for compatibility of silicide material, formation kinetics, selective etch process and thermal stability. Ni-based silicides are "compatible" with $Si_{1-x}Ge_x$ substrate in the sense that low resistance Ni germanide phases exist which can concurrently form with Ni silicide phases [Zhang, S.-L. 2003]. However, presence of Ge poses challenges especially for morphological stability and the selective etch [Seger 2005].

As discussed earlier, NiPt alloys with typically 5 to 10% atomic concentration of Pt have been implemented by most of the industry since 65 nm node in order to improve the phase and morphological stability of NiSi. The addition of Pt; being a noble metal; limits the options for selective etch chemistries. The typical NiPt process involves wet selective etching to remove unreacted NiPt by a hydrochloric acid (HCl) based aqueous solution such as Aqua Regia (mixture of HCl and HNO_3). The HCl-based process has two possible platinum dissolution reaction paths [Cotton and Simon, 1997].

The selectivity of the wet chemistry is based on the difference in etch rates between unreacted metal (i.e. Ni or NiPt alloy) and Ni-silicide phases. For a single step RTA process, the goal is to form the final phase (i.e. NiSi) and the selectivity needs to be with respect to the monosilicide phase. However if the film is not fully converted into NiSi and residual metal-rich phases (e.g. Ni_2Si) remains, the wet etch chemistry may "attack" the silicide (i.e. etch metal-rich grains locally). Figure 4 shows examples of the selective etch attack. Analysis of these defects show that the wet chemistry etches out metal and converts regions into porous oxide [Imbert *et al.*, 2008]. These defects are easily detected due to the elemental contrast between metal and insulator under the electron beam inspection. The electrical failure mechanisms for such defects can be multiple, including contact opens if a tungsten contact lands on an oxide defect and high resistive fails, since the remaining silicide film is very thin or even a grain is completely converted into oxide.

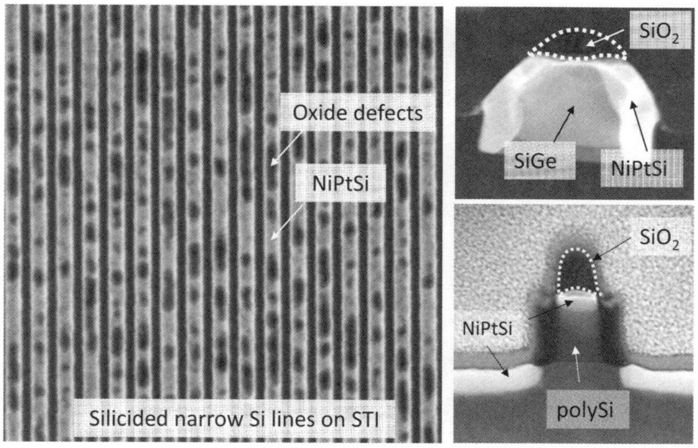

Fig. 4. Examples of Aqua regia attack on silicided active Si lines and poly-Si gate.

Another indirect consequence of such defects is the saturation of defect inspection signals at sector which masks other defects and delay the process development. Therefore, even if a defect has very little impact on device performance or yield (i.e. "cosmetic" defect), it deserves focus to be understood or eliminated, otherwise it can prevent the detection of a more important defect.

An important observation for the grainy defect was the fact that it predominantly affected PFETs with embedded SiGe source and drain terminals. Hence, one must consider the impact of Ge on the silicide formation to understand the root cause. *In-situ* studies of NiPt silicide formation on SiGe shows that, the nucleation of the mono-silicide (i.e. NiSi) phase is delayed and residual metal-rich silicide phase remains until higher temperatures compared to formation on Si (Figure 5). Therefore, selective etch chemistry may attack the silicide grains which are rich in metal, resulting in a grainy or spotty defect morphology. The obvious solution is to re-optimize the RTA process in order to ensure full NiSi formation on NFETs and PFETs. However, increasing the temperature or time (or overall thermal budget) may not be desirable due to the higher risk of encroachment defects which cause leakage or shorts that are detrimental for devices. In fact, precisely because of these risks there has been alternative solutions developed to overcome the selective etch attack problem. The considered solutions have been twofold: (1) Alternative selective chemistries, (2) Solutions to address the difficult formation of the monosilicide phase on SiGe. One of the alternative chemistries for NiPt strip is high temperature sulfuric peroxide which also shows significantly better selectivity to metal-rich silicide phases [Imbert *et al.*, 2008; Chen 2009]. On the other hand, the solutions to improve

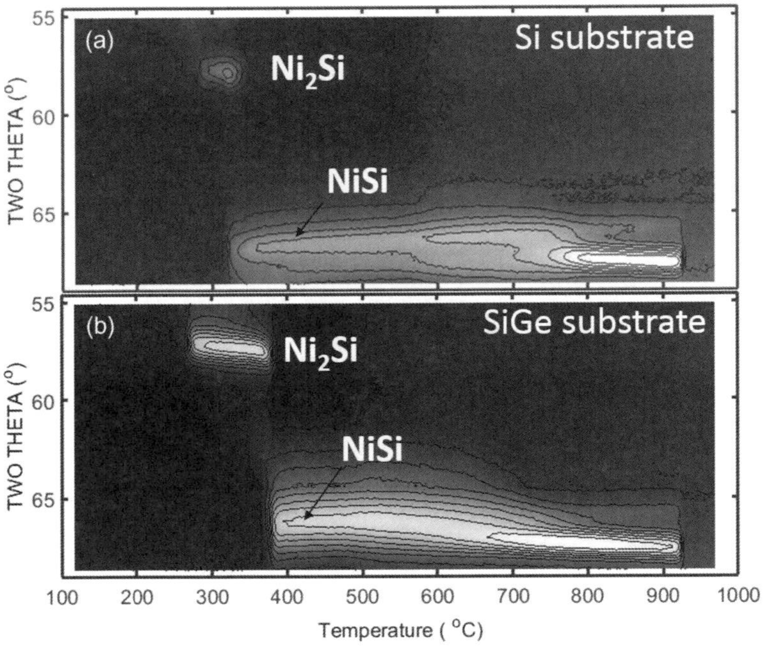

Fig. 5. *In-situ* XRD data for 10 nm NiPt10% on (a) Si and on (b) SiGe30%, annealed at 3 C/s.

NiSi formation on SiGe included pre-amorphization implants (PAI) prior to NiPt deposition which lowers the formation temperature of NiSi [Ozcan *et al.*, 2013; Deshpande *et al.*, 2010]. Alternatively, engineering the source-drain epitaxy to include a Si capping layer is an obvious choice which eliminates the problem, since the silicide formation behaves similar to the NFET with Si substrate. However, selective Si cap development is also not simple and may not be desirable from device performance perspective.

3.2. *Nitride stress liner and the spotty NiSi defect*

Another defect which resembles the selective etch attack in terms of morphology, is the spotty silicide defects that may form during the middle-of-the-line process with the presence of a nitride stress liner (Figure 6). Nitride stress liners have been employed for device performance boost at least since 90nm node. The SiN film can be engineered from very compressive to tensile mainly by modulating the amount of H bonding in the film, which has allowed the use of dual stress liners to boost PFET and NFET devices simultaneously [Yang *et al.*, 2005]. The processing temperature is in the range of 400 to 500°C, which is assumed to be low enough for NiSi to withstand

(a) (b)

Fig. 6. Spotty silicide defect examples: (a) TEM cross section of NiPt silicide on SiGe after stress nitride liner deposition (3.5 GPa compressive), (b) Top-down SEM inspection that shows many defects (dark spots) after the nitride liner deposition.

without agglomeration, but in fact it is sufficient for Ni and Si diffusion to take place (note that NiSi formation occurs in this temperature range).

The thermal degradation of silicide films depends on many parameters including film thickness, substrate and alloying (e.g. Pt concentration). For example the presence of Ge in PFET active areas (embedded SiGe), increases the risk for early degradation. On the other hand, the effect of capping layers on morphological degradation of silicides have not been investigated in detail. It is also challenging to detect defects that form post processing during downstream processing. For example, inline optical scattering and top-down SEM inspection will be blind to silicide defects that form once a few hundred nanometer thick nitride film covers them. Typically after electrical testing (M1 or functional test), these type of defects can be localized and isolated. Detailed failure analysis gives clues about the problem and potential root causes.

In order to understand the impact of stress liners on silicide films, blanket wafers can be utilized to simplify the characterization. In order to simulate the effects seen on fully processed device wafers, the substrate, dopants, film properties and thermal budgets should be mimicked as much as possible. For this study, epitaxial SiGe has been grown on blanket Si and 10 nm thick NiPt film has been deposited afterwards. After silicide formation anneal and selective etch, SiN films with different stress levels were deposited on different wafers. As a reference, one control wafer had no nitride film on top of the silicide. The morphological degradation of NiSi has been monitored by *in situ* optical scattering during annealing. Samples cut out from wafers were annealed in purified N_2 up to 900°C at 3 C/s rate and scattered light intensity was collected simultaneously. Sharp

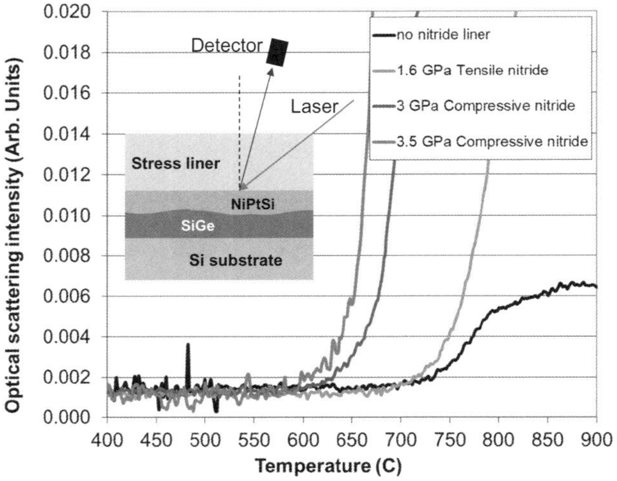

Fig. 7. *In situ* optical scattering results of 20 nm NiPtSi films with and without stress nitride liner deposited at the surface, annealed at 3 C/s. Inset illustrates the experimental setup and the film stack.

increase in the scattered intensity may indicate roughening such as film agglomeration. This type of *in-situ* monitoring is very quick and useful for relative comparisons between different conditions. Results of these *in situ* studies in Figure 7 show that the presence of a stress liner strongly impacts the degradation behavior of NiSi. Compared to the "no liner" control, the 3.5 GPa compressive film causes NiSi to degrade more than 100°C earlier during the ramp anneal. The results also show that the amount of stress also plays a role in the severity of the degradation. Compressive nitride liner with 3 GPa film is slightly more stable compared to the 3.5 GPa nitride liner. On the other hand, the sign of the stress (i.e. compressive vs. tensile) is not the main factor for the degradation mechanism since the tensile liner sample also shows worse thermal stability compared to the no liner control which is confirmed by other studies [Li *et al.*, 2011].

These results agree with the observations on fully integrated wafers. Higher the stress in SiN, the more defectivity in NiSi were generally reported. However, the value of the stress directly impacts the channel mobility and it is critical for device performance. Therefore, replacing the stress liner with a neutral or reduced stress film is not desirable. The *in situ* optical scattering technique described above is invaluable for screening different solutions in a very short time. It is significantly cheaper and faster to prepare blanket wafers with few layers compared to fully integrated device wafers, and, the optical scattering setup provides quick relative feedback between different experimental conditions. For example, impact of film thickness, deposition conditions, dopants, substrate on silicide degradation can be investigated rapidly.

Since the stress nitride liner is critical for device performance, solutions related to the modification of the liner process pose a tradeoff. Alternatively, the silicide film itself could be engineered to withstand the adverse effects of the stress liner. Root-cause analysis of the problem indicate that the defect formation mechanism is related to enhanced atomic diffusion (Ni and Si/Ge) at the interface between silicide and stress nitride. Normally, without the stress liner, Ni silicide degradation on SiGe starts with grain boundary grooving at the silicide — SiGe substrate interface. However, in the current problem, the silicide — SiGe interface remains relatively unchanged but grain boundary grooving or voids form at the silicide surface (Figure 6). Therefore the solution to improve the robustness of the silicide must address the silicide — nitride liner interface. One of the ways to reduce or block diffusion through grain boundaries is to introduce impurities. Small atoms like C or N can be very effective in stuffing the grain boundaries. In order to test these hypotheses, silicide films were implanted with C or N at very low energy to keep impurities close to the nitride interface and prevent any damage in the junction. Optical scattering results as well as topdown SEM inspection results confirmed that, indeed, introduction of impurities such as C or N can reduce or prevent the formation of spotty silicide defects during the deposition high stress nitride films (Figure 8).

3.3. *Silicide stringers and electrical shorts*

Silicide formation in small dimensions such as narrow lines has been a challenge for many generations of CMOS devices. For example, Ti silicide was replaced by Co silicide because of the difficult nucleation of the desired C54-TiSi2 phase in narrow lines. Co silicide formation was also problematic in narrow poly-Si lines due to voids that caused high resistance or opens in the circuits. Even though Ni silicide formation does not have the same problems as Ti and Co silicides in narrow lines, the aggressive scaling in CMOS devices continues to pose new challenges for Ni silicide as well. Some of these challenges emerge because of the large volume changes during silicide formation and the presence of mechanical constraints exerted by sidewall spacers or STI around the narrow Si lines.

The fast diffusion of Ni may cause non-uniform silicide thickness in different geometries. Narrow lines separated by oxide or poly-Si gates with sidewall spacers have abundant source of extra metal on the edges, such as

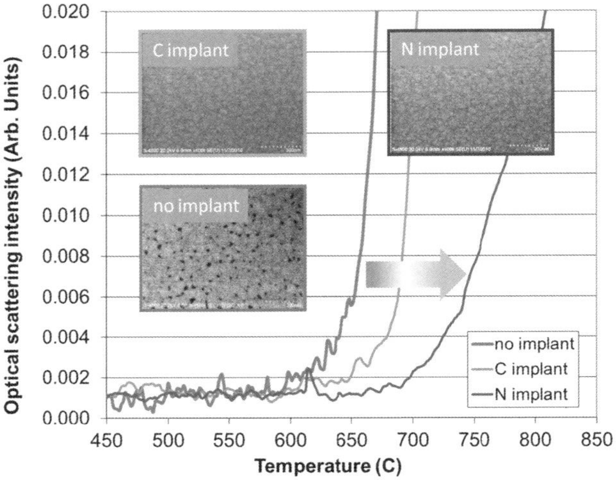

Fig. 8. *In-situ* optical scattering results of implanted silicide films, all with 3.5 GPa compressive nitride liner. Inset images are top-down SEM inspection right after stress nitride deposition.

the sidewall terraces or in divots where oxide and Si lines meet. Therefore, extra amount of Ni, which is in addition to the Ni deposited on the top surface, is supplied during the formation anneal. This effect is referred as the "reverse line-width effect" which suggests thicker silicide formation in narrow lines [Lavoie *et al.*, 2003].

One of the consequences of deep silicide formation in narrow lines, below the sidewall spacers, is extrusion of silicide stringers onto sidewall terraces. Figure 9 shows electron microscope cross sections and top-down views of narrow silicided Si and poly-Si lines and the silicide stringer defects on the spacers and on STI. The length of the stringers ranges from few nanometers to roughly 10 nm. Analysis of the stringers by TEM and Energy Dispersive X-ray (EDX) line scans confirms the presence of Ni, Pt and Si. The abundant presence of silicon, nitride and oxide around the stringers makes the TEM-EDX or electron energy loss spectroscopy (EELS) analysis in-conclusive about the exact composition or phase of the stringers. However, the resistance of these defects to wet etch such as Aqua Regia (HNO_3-HCl mixture) which can easily remove Ni and NiPt alloys, indicates that they are in silicide phase or the composition of the stringers should be sufficiently Si rich since the Aqua Regia is ineffective to remove them.

Extrusion of silicide stringers at low temperatures, especially below 400C is unexpected. Experimental results indicate that the presence of the stringer defects is related to the silicide depth with respect to the

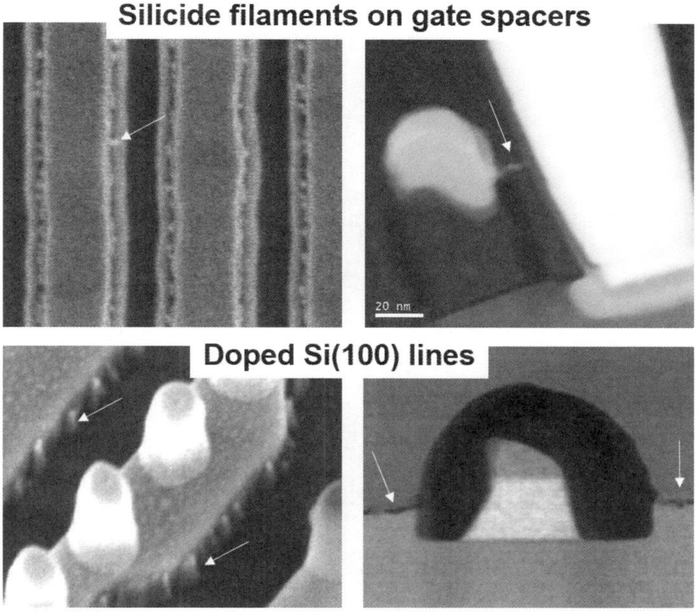

Fig. 9. Top images show silicide stringer defects on poly lines. Left: top-down SEM view, right: TEM cross section of a defect which shorts the gate to the W contact. Bottom images are examples of silicide stringers on narrow Si lines. Left: top-down view of Si lines on STI, right: cross section of a line.

sidewall spacers. At the same formation temperature and Ni deposition thickness, defects do not form if there is large enough spacer pull-down without significant silicide growth below the spacers. Given these results, one must consider the volume changes during silicide formation. Since Ni is the dominant diffusing species during silicide formation, the growth front proceeds downwards into the Si substrate. Figure 10 schematically illustrates the growth of the metal-rich Ni_2Si and the mono-silicide (NiSi) phases in a narrow gate with spacers. If the volume of the silicide phases are compared to the volume of the Si consumed, one finds that Ni_2Si and NiSi phases occupy approximately 61% and 21% larger volume consecutively. It should be stressed that the volume change comparison must be made with respect to the volume of Si consumed. Without any constraints, the expansion of the growing phase can be accommodated by the free top surface. However, if one considers the same scenario for narrow lines with sidewall spacers, the height of the spacers and the depth of the silicide will have a different interaction. Considering the required volume changes during silicide growth, the mechanical constraint set by the spacers would make the growth of the silicide extremely difficult below the spacer

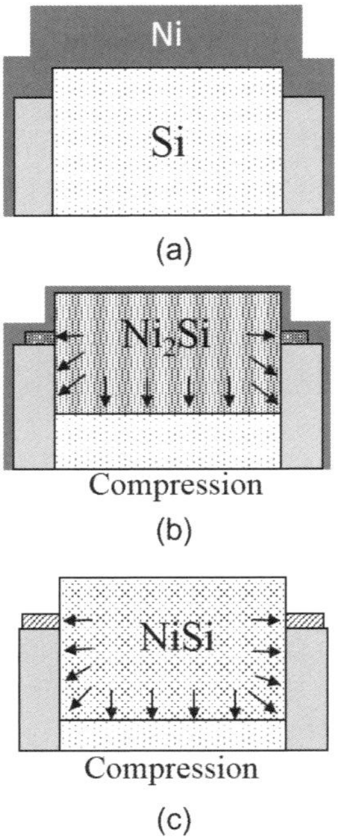

Fig. 10. Schematic illustration of the Ni silicide reaction in the narrow poly gate and the volume changes. (a) After Ni deposition, (b) Ni_2Si formation (~60% volume change), (c) NiSi formation (~20% volume change).

pull-down level. Therefore, the extrusion of silicide stringers onto sidewall terraces could be a mechanism to reduce the stress during expansion which has enhanced Si diffusion onto the spacers.

In order to further test the relation between silicide stringers and difficult expansion in a constrained volume, silicide reaction temperature can be used to modulate the thickness and depth of the silicide in narrow lines. Formation anneal can strongly affect the amount of metal diffusion. Figure 11 shows cross sectional and topdown views of two samples which had silicide formation anneals at 300°C and 400°C. The 300°C anneal results show thinner NiSi formation on the narrow gates which looks like an inverse "U". The inverse "U" shape is inherited by the metal deposition and reaction around the poly-Si gate. These results do not show any stringer formation compared to the 400°C anneal results which show thicker silicidation on the poly gates and stringers randomly observable in cross section and topdown SEM analysis.

Implementation of low RTA process in order to prevent the formation of stringers require a NiPt strip which is selective to the phases that form at low temperature (i.e. metal-rich phases). This challenge has been discussed in

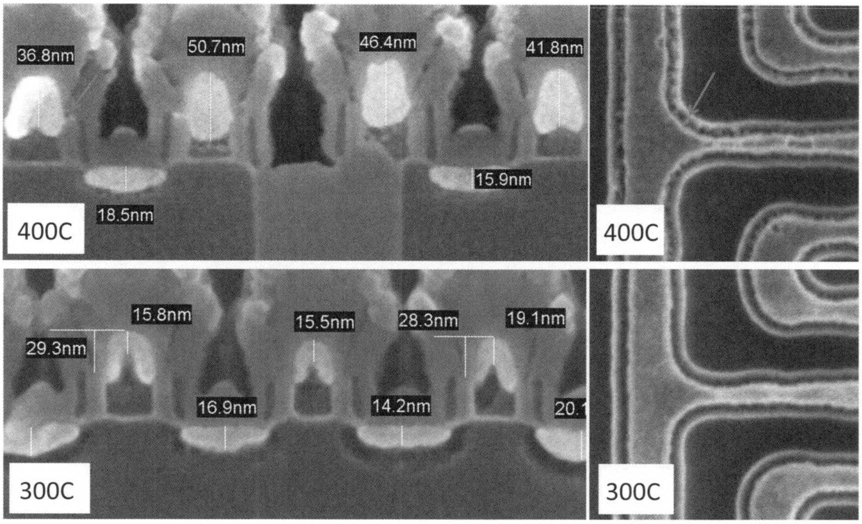

Fig. 11. Cross-sectional (left) and topdown (right) SEM analysis of test structures after silicide formation and selective etch. 400C anneal shows thicker silicide in poly-Si gates and stringer defects (indicated by red arrows).

the first defect example (i.e. grainy silicide defect) in detail. Hence, these two defects highlight the important concept of "the process window". In the current example we see two defect mechanisms that limits the optimization of silicide formation anneal: more "stringer" defects at higher temperature and more "silicide attack" by wet etch for lower temperature. If there is no RTA process window that allows a defect free solution, one must change another process knob (such as selective etch chemistry) to widen the window. Having a wide process window is also important for manufacturability to allow tool or process variations without compromising yield.

3.4. *Silicide encroachment defects*

Even though Ni silicide has significant advantages over Ti and Co silicides in the advanced technology nodes, two fundamental material properties have significantly challenged the microelectronics industry. To sum up: 1- Ni diffusion in Si is extremely fast, 2- Low resistance NiSi phase is meta-stable.

The first one, fast diffusion of Ni, is especially important during the initial formation of Ni silicide. As discussed earlier, for narrow linewidth devices or in the presence of a divot at the edge of active areas, there is plenty of excess Ni ready to react. Therefore controlling the thermal budget during the first RTA is critical to limit Ni diffusion and achieve the desired silicide profile in the junction. This is quite challenging as the junction gets shallower at every new generation.

The second one, meta-stable nature of NiSi, needs to be considered particularly for the post silicide thermal budget. The stable $NiSi_2$ phase is not only a problem because of its higher resistivity, but the fact that its volume is approximately 40% larger than NiSi implies a morphological defect that can cause leakage or shorts in devices.

When Ni silicide was first introduced, there were many reports of "NiSi pipes" as a severe yield limiter [Strane *et al.*, 2007; Johnson *et al.*, 2007; Lai *et al.*, 2011]. The term "NiSi pipe" was coined to describe deep lateral encroachment defects which can short source or drain of the device to the channel. In extreme cases source to drain could be shorted via a NiSi pipe defect. Figure 12 shows an example of a pipe defect which was isolated after tedious failure analysis and characterization.

Before going into the details of the formation mechanism it is important to discuss the electrical failure mode and the detection methodology. As opposed to the prior defect examples, such as grainy silicide or stringers, NiSi pipe defects typically occur at a very low density (e.g. 1 in 100.000 transistors or even 1 parts-per-million, ppm) and standard inline defect monitoring tools fail to detect these rare occurrences within the sector. As a result, typically, these defects have been caught only after SRAM test (after fourth metal layer or later in the BEOL), which makes the learning cycles very long. Because of the very low density, the electrical detection at M1 level is very difficult even on specifically designed test structures. If the test

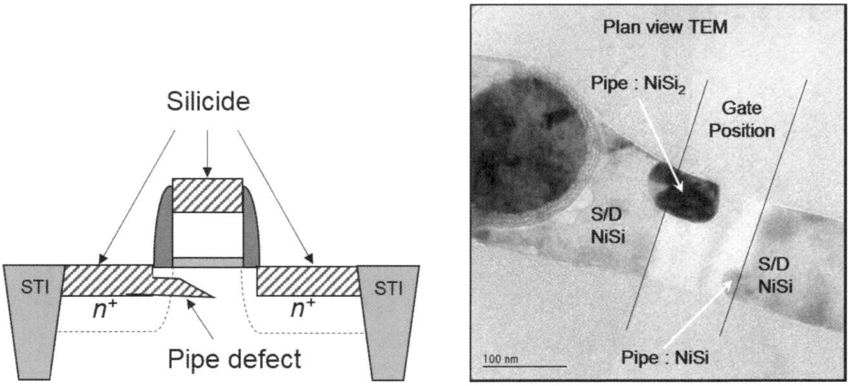

Fig. 12. Illustration of silicide pipe defects (left) and failure analysis of an actual defect. Sample was de-layered (all metal and insulator levels down to silicide including the gate were removed) in order to characterize the defect.

structure (e.g. an array of transistors) is designed too large, the total leakage of the array will be high and the leakage contribution of one pipe defect will be too small to detect. On the other hand, if the array size is small, it may not catch a pipe defect at all.

One of the most successful methods for early pipe detection is inline voltage contrast (VC) inspection. Inline VC inspection, which is non-destructive, has been proven to be very useful in detecting other yield limiting defects such as missing spacer and contact opens [Patterson *et al.*, 2007 and 2008]. This technique relies on scanning a defined area with an electron beam and comparison of the charging contrast between different zones. As a result of the electron beam interaction with the material, certain areas will appear lighter or darker depending on if it is insulated (retaining charge) or grounded (not charged). This contrast difference is visible to the SEM and location of the defect is flagged with high resolution. Hence, defects are not only quantified but also localized precisely. The inspection can be done right after silicide formation on bulk Si technologies. For SOI technologies, the inspection is more challenging since transistors as isolated (i.e. floating) from the substrate. However, specific VC test structures can be designed to overcome these problems.

The rare occurrence and the orientation of the pipe defects along certain Si crystal directions provide hints about the root cause. It is no surprise that presence of crystalline defects, such as dislocations, is one of the most common root causes for pipe defects. Diffusion of Ni through a dislocation can be orders of magnitude higher than in bulk Si, which is already high. This is evidenced by the atomic probe tomography observations which detected Ni in dislocation loops more than 100 nm deeper from the as-deposited surface [Hoummada *et al.*, 2011]. Not only dislocations but also end-of-range damage induced by ion implantation could trigger silicide encroachment. Therefore, a lot of efforts have been placed in reducing or eliminating crystalline defects either through implant adjustments (e.g. low temperature cryogenic implantation) or through anneals to heal defects prior to silicidation.

Stabilization of NiSi through Pt alloying has been very effective in preventing pipes [Strane *et al.*, 2007]. Detailed TEM and diffraction characterization of pipe defects have indicated that defects are in $NiSi_2$ phase. Therefore delaying the formation of $NiSi_2$ using Pt (note that no $PtSi_2$ phase exists) would reduce the chance for pipe formation. Since silicide pipe defects may form during downstream processing due to the high thermal budget in the MOL or BEOL, increasing the thermal stability of NiSi is critical for prevention of morphological defects including encroachment.

4. Silicide Contact Resistance

The primary focus of silicide development over the years has shifted from yield improvements to device performance optimization due the shrinking contact size and the increasing contribution of silicide contact resistance to the total external resistance. The sheet resistance of silicide is no longer the critical parameter to optimize in the advanced nodes with metal gates. These changes have triggered significant research and development efforts to combat contact resistance both in academia and industry. Before discussing the approaches that address the contact resistance problem, we need to describe the resistance components in devices and illustrate how scaling affects the contribution of silicide to source/drain interface resistance.

Figure 13 shows the resistance elements in a CMOS device. The total R_{ext} is divided into five main regions: (1) R_{ov}: extension to gate overlap, (2) R_{spread}: spreading resistance (3) R_{ex}: extension resistance under the spacers, (4) R_{co}: silicide to silicon contact resistance (5) R_{lumped}: metal contact (W), contact to silicide interface (e.g Ti/TiN liners) and silicide resistance lumped together. Junction doping profile and activation not only impacts overlap and spreading resistance but also strongly affects the silicide contact

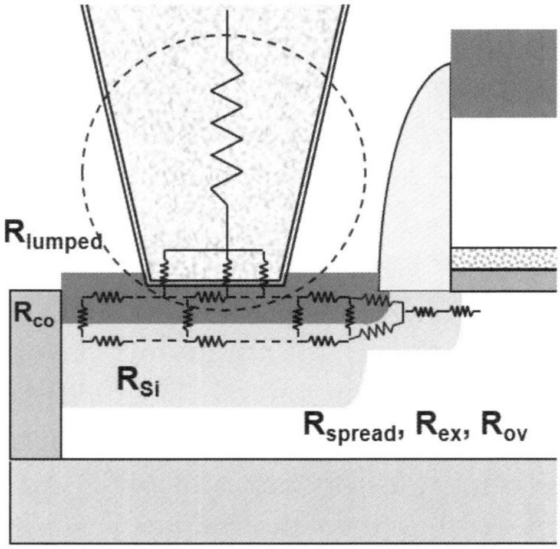

Fig. 13. Resistive elements of a planar CMOS device.

resistance. The specific contact resistivity is described by:

$$\rho_C \propto e^{\frac{2\Phi_B}{q\hbar}\sqrt{\frac{m^*\varepsilon_r}{N}}}$$

where ϕ is Schottky barrier height, N is active doping concentration, m^* is effective mass of the carrier, q is elementary charge, h is Planck's constant, ε is the relative permittivity of the semiconductor. In order to achieve low contact resistivity, the Schottky barrier height needs to be low and active doping concentration in the semiconductor needs to be high.

In order to appreciate the impact of scaling on contact resistance, one can calculate the silicide-silicon interface resistance for a given area and specific contact resistivity value. For example for a 1×1 micron square area and $\rho_C = 1 \times 10^{-7}\,\Omega\,\text{cm}^2$, assuming short contact limit (i.e. contact length is smaller than the transfer length [Ting and Chen, 1971]), the resistance is only $10\,\Omega$. If we shrink one side of the device to 100 nm (1000 nm by 100 nm), for the same ρ_C, the resistance increases to $100\,\Omega$. If the contact area further shrink to 10 nm on one side (1000 nm by 10 nm area), the resistance cost is $1000\,\Omega$. If we systematically calculate and plot the contact resistance as a function of contact area for different specific resistivity values, the trend becomes clear and one can see the effect of both scaling and the resistivity (Figure 14). Considering that the total external resistance of a state of the art device is few hundred ohms, the sharp increase in the contact resistance as the contact length gets smaller is a serious a device performance limiter.

Measurement of the specific contact resistivity is very challenging especially for the sub $1 \times 10^{-8}\,\Omega\,\text{cm}^2$ values because of the potential sources of errors. One can refer to other excellent reviews for detailed descriptions of the measurement structures and the methodologies [Zhang and Zhang, 2014]. Even though these measurement techniques are invaluable to study and improve silicide contact resistance, they should be viewed in relative terms. The absolute value of the silicide specific contact resistivity on a transistor could be different because of the differences in structure and the impact of integration. The silicide growth could be significantly different between the test structure and the S/D of a transistor, such as potentially the case for the Cross Bridge Kelvin structure [Proctor *et al.*, 1983]. If S/D junctions are formed by *in situ* doped epitaxy, it is another challenge to replicate the dopant concentration and profile on the measurement structure. Because of these challenges there are still efforts to improve or re-design new measurement structures to accurately measure specific contact resistivity.

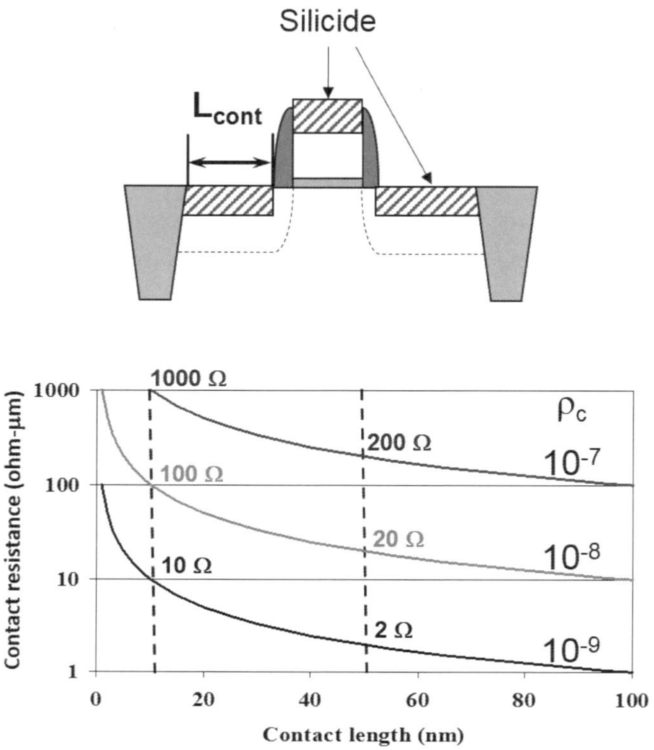

Fig. 14. Contact resistance calculation as a function of contact length for various specific contact resistivity values (Ω-cm^2).

The equation that describes the specific contact resistivity and the scaling trends plotted above highlight the key parameters to optimize in order to reduce contact resistance:

(1) Schottky barrier height
(2) Active dopant concentration
(3) Contact area

In order to achieve low resistivity contacts for *n* and *p* regions, one can use two silicide materials with low Schottky barrier heights for electrons and holes. However this scheme (i.e. dual silicide integration) is very expensive to integrate and require the optimization of two different materials. Processing of rare earth materials (e.g. Yb, Dy, Er) is also very challenging as these materials oxidize readily. Therefore, mid-gap silicides (such as Ti, Co, Ni silicides) are preferred as a good compromise.

Active dopant concentration is also important for lowering the source-drain resistance and reduces the other parasitic resistance components of the transistor. In addition to the efforts to increase dopant concentration in the junction through implants or during epitaxial growth, there has been great progress in the activation anneal itself, utilizing ultrafast annealing techniques. In the current technology nodes that are being manufactured, millisecond range laser or flash lamp annealing is indispensable.

Finally, another important knob for the reduction of contact resistance is the silicide contact area optimization. Especially in the sub 32 nm technology nodes with "silicide last" or "trench silicide" integration, the area impact has become more pronounced. As we reviewed in the early section on silicide formation, the trench silicide integration requires the formation of contacts through lithography defined trenches or holes. Instead of the whole active source-drain regions, silicide forms only partially on the active region. Therefore, the silicided area in a trench silicide scheme is smaller compared to the one formed by conventional SALICIDE. In order to compensate the decrease in the lateral dimension, the silicide shape can be engineered to increase the area or the Schottky barrier height must be reduced further.

Even though Ni based silicides have been considered "midgap" and earlier ITRS predictions had assumed that contact resistivity requirements cannot be met beyond 32 nm node, NiPt based silicides were successfully implemented in 22/20 nm technology nodes. The unusually low contact resistivity values obtained for NiPt based silicides both for *n* and *p* regions is due to the optimization of the silicide and substrate interface. When silicidation occurs on a heavily doped substrate only a small amount is incorporated within the silicide film and dopants segregate to the interfaces (grain boundaries, substrate interface etc.). The relatively low temperatures during Ni silicide formation (<500°C) does not induce significant dopant diffusion in Si. However, dopant diffusion in Ni silicide (both Ni_2Si and NiSi) can be orders of magnitude faster according to the diffusion coefficient measurements [Blum *et al.*, 2008 and 2010], which allows efficient transport of dopants through the silicide material. Based on these observations, dopant segregation schemes were developed to exploit the positive effects and improve contact resistivity.

Mainly there are two schemes to engineer silicide interface using dopant segregation:

(1) Pre-silicide implantation and dopant segregation (also called SIDS for 'silicide-induced-dopant-segregation')
(2) Implantation into silicide and anneal for dopant segregation (also called SADS for 'silicide-as-dopant-source')

The first scheme is easier to implement since it is utilized many times during S/D implantation for NMOS and PMOS devices. However, as the junction becomes more and more shallow, controlling the energy distribution and keeping the dopants only within the volume that is consumed by the silicide is a challenge. The second scheme requires implantation into a metallic film (i.e. silicide) which is more problematic for manufacturing since Ni contamination is not allowed in implanters dedicated for FEOL. This issue could be overcome by the use of a screen oxide on silicide. The other difficulty with the SADS scheme is the thermal budget required to segregate dopants without degrading the silicide morphology. Studies have shown that 500 to 600°C for about 30 sec. could be sufficient to segregate dopants. These temperatures may not be acceptable especially for Ni silicide and less so for NiPt alloy silicide. Another issue could be the integration with high-k metal gate integration that further limits the thermal budget requirements. These thermal budget issues could be overcome by the use of ultrafast annealing techniques such as the millisecond laser anneal processes that already have been implemented in the industry. Figure 15 shows that As diffusion within NiPt silicide can be realized with laser anneal without any significant impact on the silicide morphology.

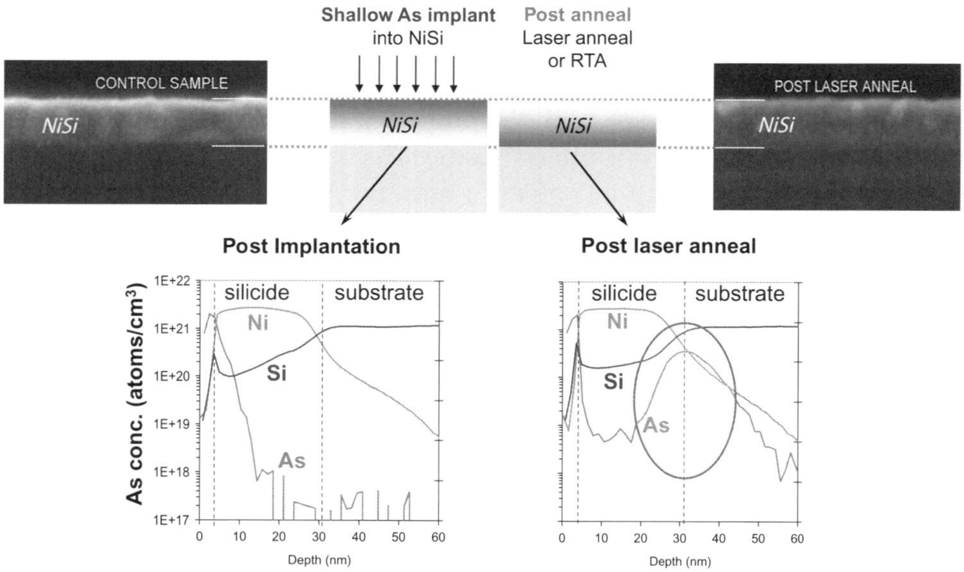

Fig. 15. Demonstration of dopant segregation using millisecond laser anneal. SEM cross-sections (top) do not show morphological change, however, SIMS results (bottom) clearly shows the segregation of As to the silicide-substrate interface after laser anneal.

Low Schottky barrier height values were reported for both SIDS and SADS applications by various groups for both polarities [Zhang *et al.*, 2007; Qiu *et al.*, 2008; Luo *et al.*, 2009].

The reported barrier heights were 100 mV or less with respect to either band edge which shows that interface engineering could be a more attractive choice compared to dual silicide integration with the use of rare earth metals.

5. Conclusions and Future Trends

According to the ITRS 2007 update, the aggressive scaling in CMOS devices will continue and gate dimensions down to 5 nm needs to be achieved. Despite many debates on the future trends for 7 nm node and beyond, making low resistance contacts to devices will become even more critical. Whether it is carbon based electronics or Si nanowire based devices, there needs to be more efforts in contact formation in very small dimensions. Most studies for these advanced devices focused on mobility and other transistor characteristics while utilizing very large contacts in order to demonstrate feasibility. In order to bridge these promising results to the industry one must also demonstrate low resistance contact formation in the dimensions that matter.

In addition to the use of silicides, alternative schemes have been proposed to achieve low SBH contacts using a thin layer of dielectric between the metal and the semiconductor [Ang *et al.*, 2012]. Metal-insulator-semiconductor (MIS) contacts can avoid the Fermi level pinning close to the mid-gap of the semiconductor (e.g. Si or Ge) and as long as the dielectric layer is thin enough current can tunnel through with low resistance. Furthermore, by using appropriate metal oxides (such as La or Al oxides), an interface dipole may form and decrease the SBH even more. Even though the concept for MIS contacts is demonstrated, reported contact resistivity values are not better than optimized silicide contacts. The manufacturability for MIS contacts remain as the ultimate challenge since the tunneling current depends exponentially on the thickness of the dielectric. Even though atomic layer deposition tools exist to precisely control the thickness of these layers, given all the variability in the process flow, it is quite risky to implement a scheme with exponential dependence on a given process. In fact, in the coming generations with extremely small contact dimensions, one needs to look for more forgiving solutions in terms of variability and process window.

Despite these challenges the MIS type contacts are attractive to replace conventional silicides in finFET devices and even more so if the silicide formation is performed through trenches or contact openings. One of the challenges in finFET and ETSOI devices is the limited supply of available Si which increases the risk for lateral silicide encroachment toward the channel. The risk can alleviated by raised epitaxial source drain and utilizing other improvements used in prior art such as Pt alloying, low temperature process and millisecond anneal. Since the encroachment risks mainly arise due to the fast Ni diffusion, alternative silicide materials with high melting point and where Si is the dominant diffusing species, could also be considered. Then the remaining question is the contact resistance optimization and whether the new material can be engineered to meet the device requirements.

As we are at the cusp of major changes in the device architecture and devices beyond the conventional transistor are on the horizon, it is natural to expect more exciting times coming for the silicide contacts for scientists and engineers.

References

Alptekin, E., Mehmet, C. O. and Veena, M. (2009), *Electron Device Letters, IEEE*, **30**(9), p. 949.

Ang, K.-W., Majumdar, K., Matthews, K., Young, C. D., Kenney, C., Hobbs, C., Kirsch, P. D. *et al.* (2012), IEDM Electron Devices Meeting, *IEEE International*, p. 18.

Auth, C., Cappellani, A., Chun, J. S., Dalis, A., Davis, A., Ghani, T *et al.* (2008), *IEEE VLSI Technology*, 2008 Symposium p. 128.

Besser, Paul R., *et al.* (2008), *ECS Transactions*, **13**(1), p. 377.

Blum, I., Portavoce, A., Mangelinck, D., Daineche, R., Hoummada, K., Lábár, J. L., Carron, V. and Perrin, C. (2008), *J. Appl. Phys.*, **104**(11), p. 114312.

Blum, I., Portavoce, A., Chow, L., Mangelinck, D., Hoummada, K., Tellouche, G. and Carron, V. (2010), *Appl. Phys. Lett.*, **96**(5), p. 054102.

Cabral, Jr C., Clevenger, L. A., Harper, J. M. E., d'Heurle, F. M., Roy, R. A., Lavoie, C., Saenger, K. L., Miles, G. L., Mann, R. W. and Nakos, J. S. (1997), *Appl Phys Lett*, **71**, p. 3531.

Cabral, C., Lavoie, C., Ozcan, A. S., Amos, R. S., Narayanan, V., Gusev, E. P., Jordan-Sweet, J. L. and Harper, J. M. E. (2004), *J. Electrochem. Soc.*, **151**(12), F283–F287.

Chen, Yi Wei, Nien Ting Ho, Jerander Lai, Tsai, T. C., Huang, C. C., Tzou, S. F. and James MM Chu. (2009), *Solid State Phenomena* **145**, p. 211.

Chudzik, M. *et al.* (2007), *VLSI Technology, IEEE* p. 194.

Cotton, Simon A., (1997), *Chemistry of Precious Metals*, (Springer-Verlag).

Dennard, R. *et al.* (1974), *IEEE J. Solid State Cir.*, **9**(5), p. 256.

Deshpande, S. V., Ozcan, A., Wall, D., Kim, E. and Gluschenkov, O. (2010), *Junction Technology (IWJT)*, p. 1.
d'Heurle, F. M. and Miglio, L. (eds) (2000), *Silicides: Fundamentals and Applications*, (World Scientific, Singapore).
Gambino, J. P. and Colgan, E. G. (1998), *Mater Chem Phys*, **52**, p. 99.
Hoummada, K., Mangelinck, D., Gault, B. and Cabié, M. (2011), *Scripta Materialia*, **64**(5), p. 378.
Imbert, B., Zoll, S., Garnier, P., Pernet, B., Galpin, D., Beneyton, R., Juhel, M., Mur, P., Carron, V. and Thomas, O. (2008), *Mater. Sci. Eng B*, **154**, p. 155.
Johnson, K. D., Lin, X., Pan, L. and Miner, B. (2007), *Microscopy and Microanalysis* **13**(S02), p. 786.
Kim, S.-D., Narasimha, S. and Rim, K. (2005), IEDM Technical Digest. *IEEE International*, p. 149.
Kothandaraman, C. *et al.* (2002), *IEEE Elec. Dev. Lett.*, **23**(9), p. 523.
Lai, J., Chen, Y. W., Ho, N. T., Lin, J. F., Huang, C. C. and Wu, J. Y. (2011), Interconnect Technology Conference and 2011 Materials for Advanced Metallization (IITC/MAM), 2011 *IEEE International*, p. 1.
Lavoie, C., d'Heurle, F. M., Detavernier, C. and Cabral, Jr C. (2003), *Microelec. Eng.* **70**(2), p. 144.
Lavoie, C., d'Heurle, F. M. and Zhang, S.-L. (2007), 'Silicides', in Doering, R. and Nishi, Y. *Handbook of Semiconductor Manufacturing Technology*, 2nd edn, CRC Press, Boca Raton, FL.
Lee, W.-H., Waite, A., Nii, H., Nayfeh, H. M., McGahay, V., Nakayama, H., Fried, D. *et al.* (2005), *IEDM Technical Digest*, p. 61
Li, M. Y., Chen, J. M., Liu, C. C. and Lin, J. F. (2011), *Electron Device Letters, IEEE*, **32**(12), p. 1725.
Luo, Jun, *et al.*, (2010), *Journal of Vacuum Science & Technology B*, **28**(1), C1I1–C1I11.
Maex, K. and van Rossum, M. (eds) (1995), Properties of Metal Silicides, EMIS Datareviews Series No. 14, INSPEC, London.
Mann, R. W. *et al.*, (1995), *IBM J. Res. Develop.*, **39**(4), p. 403.
Nishi, Y., Tsuchiya, Y., Kinoshita, A., Yamauchi, T. and Koga, J. (2007), IEDM. *IEEE International*, p. 135.
Ohta, H., Kim, Y., Shimamune, Y., Sakuma, T., Hatada, A., Katakami, A., Soeda, T., Kawamura, K., Kokura, H., Morioka, H., Watanabe, T., Hayami, J. O. Y., Ogura, J., Tajima, M., Mori, T., Tamura, N., Kojima, M. and Hashimoto, K. (2005), *IEEE International Electron Devices Meeting*, IEDM Technical Digest, p. 247.
Ohuchi, K., Lavoie, C., Murray, C., D'Emic, C., Lauer, I., Chu, J. O., Yang, B., Besser, P., Gignac, L., Bruley, J., Singco, G. U., Pagette, F., Topol, A. W., Rooks, M. J., Bucchignano, J. J., Narayanan, V., Khare, M., Takayanagi, M., Ishimaru, K., Park, D.-G., Shahidi, G. and Solomon, P. (2007), *Tech Dig Intl Electron Devices Meet*, p. 1029.
Özcan, A. S., Ludwig Jr, K. F., Cabral C., Jr, Lavoie, C. and Harper, J. M. E. (2002), *J. Appl. Phys.*, **92**(12), p. 7210.
Ozcan, A. S., Wall, D., Jordan-Sweet, J. and Lavoie, C. (2013), *Appl. Phys. Lett.*, **102**(17), p. 172107.

Patterson, O. D. *et al.* (2008), *IEEE Transactions on Semiconductor Manufacturing*, **21**, p. 322.

Patterson, O. D., Kang, H. H., Strane, J., Lovoie, C., Barth, K., Ouyang, X. and Wu, K. (2007), *International Symposium for Testing and Failure Analysis*, **33**, p. 270.

Proctor, S. J., Linholm, L. W. and Mazer, J. A. (1983), *IEEE Trans on Electron Devices*, **30**, p. 1535.

Seger, Johan. (2005), *Interaction of Ni with SiGe for electrical contacts in CMOS technology* (Ph.D. Thesis).

Shang, H. *et al.* (2012), *VLSI Technology (VLSIT), Symposium on.* IEEE, p. 129.

Strane, Jay, *et al.* (2007), VLSI Technology, Systems and Applications, VLSI TSA, p. 1.

Ting, C.-Y. and Chen, C. Y. (1971), *Solid-State Electronics*, **14**, p. 433.

Yang, H. S., Malik, R., Narasimha, S., Li, Y., Divakaruni, R., Agnello, P. *et al.* (2004), In Electron Devices Meeting, IEDM Technical Digest. *IEEE International* p. 1075.

Zhang, S.-L. and Zhang, Z. (2014), *Metallic Films for Electronic, Optical and Magnetic Applications: Structure, Processing and Properties.* Woodhead Publishing p. 244.

Zhang, D., Nguyen, B. Y., White, T., Goolsby, B., Nguyen, T., Dhandapani, V., Hildreth, *et al.* (2005), Symposium on VLSI Technology, Digest of Technical Papers. 14-16 p. 26.

Zhang, S.-L. and Östling, M. (2003), *Metal silicides in CMOS technology: past, present and future trends*, Critical Reviews in Solid-State and Materials Sciences, **28**, p. 1.

Zhang, S. L. (2003), *Microelectronic Engineering*, **70**(2), p. 174.

Zhang, Z., Qiu, Z., Liu, R., Ostling, M. and Zhang, S. L. (2007), *Electron Device Letters, IEEE*, **28**(7), p. 565.

Index